HEALTHCARE TECHNOLOGIES SERIES 50

Explainable Artificial Intelligence in Medical Decision Support Systems

IET Book Series on e-Health Technologies

Book Series Editor: Professor Joel J.P.C. Rodrigues, College of Computer Science and Technology, China University of Petroleum (East China), Qingdao, China; Senac Faculty of Ceará, Fortaleza-CE, Brazil and Instituto de Telecomunicações, Portugal

Book Series Advisor: Professor Pranjal Chandra, School of Biochemical Engineering, Indian Institute of Technology (BHU), Varanasi, India

While the demographic shifts in populations display significant socio-economic challenges, they trigger opportunities for innovations in e-Health, m-Health, precision and personalized medicine, robotics, sensing, the Internet of things, cloud computing, big data, software defined networks, and network function virtualization. Their integration is however associated with many technological, ethical, legal, social, and security issues. This book series aims to disseminate recent advances for e-health technologies to improve healthcare and people's wellbeing.

Could you be our next author?

Topics considered include intelligent e-Health systems, electronic health records, ICT-enabled personal health systems, mobile and cloud computing for e-Health, health monitoring, precision and personalized health, robotics for e-Health, security and privacy in e-Health, ambient assisted living, telemedicine, big data and IoT for e-Health, and more.

Proposals for coherently integrated international multi-authored edited or co-authored handbooks and research monographs will be considered for this book series. Each proposal will be reviewed by the book Series Editor with additional external reviews from independent reviewers.

To download our proposal form or find out more information about publishing with us, please visit https://www.theiet.org/publishing/publishing-with-iet-books/.

Please email your completed book proposal for the IET Book Series on e-Health Technologies to: Amber Thomas at athomas@theiet.org or author_support@theiet.org.

The Institution of
Engineering and Technology

Explainable Artificial Intelligence in Medical Decision Support Systems

Edited by
Agbotiname Lucky Imoize, Jude Hemanth,
Dinh-Thuan Do and Samarendra Nath Sur

The Institution of Engineering and Technology

Published by The Institution of Engineering and Technology, London, United Kingdom

The Institution of Engineering and Technology is registered as a Charity in England & Wales (no. 211014) and Scotland (no. SC038698).

First published 2022

The Institution of Engineering and Technology
Futures Place
Kings Way, Stevenage
Hertfordshire SG1 2UA, United Kingdom

www.theiet.org

British Library Cataloguing in Publication Data
A catalogue record for this product is available from the British Library

ISBN 978-1-83953-620-5 (hardback)
ISBN 978-1-83953-621-2 (PDF)

Typeset in India by MPS Limited

Cover Image: Jorg Greuel / Photodisc via Getty Images

Contents

About the editors

Agbotiname Lucky Imoize (Senior Member, IEEE) received the B.Eng. degree (Hons.) in Electrical and Electronics Engineering from Ambrose Alli University, Nigeria, in 2008, and the M.Sc. degree in Electrical and Electronics Engineering from the University of Lagos, Nigeria, in 2012. He is a lecturer with the Department of Electrical and Electronics Engineering, University of Lagos. Before joining the University of Lagos, he was a lecturer at the Bells University of Technology, Nigeria. He worked as the Core Networks Products Manager at ZTE, Nigeria, from 2011 to 2012 and as a Network Switching Subsystem Engineer at Globacom, Nigeria, from 2012 to 2017. He was awarded the Fulbright Fellowship as a visiting research scholar at the Wireless@VT Laboratory, Bradley Department of Electrical and Computer Engineering, Virginia Tech., USA, where he worked under the supervision of Prof. R. Michael Buehrer from 2017 to 2018. He is currently a research scholar with Ruhr University Bochum, Germany, under the Nigerian Petroleum Technology Development Fund (PTDF) and the German Academic Exchange Service (DAAD) through the Nigerian-German Postgraduate Program. He has co-edited two books and coauthored over 100 papers in peer-reviewed journals and conferences. His research interests include beyond 5G and 6G wireless communications, chaotic communications, and wireless security systems. He is the Vice Chair of the IEEE Communication Society, Nigeria Chapter. He is a registered engineer with the Council for the Regulation of Engineering in Nigeria (COREN) and a member of the Nigerian Society of Engineers.

D. Jude Hemanth received his B.E. degree in ECE from Bharathiar University in 2002, M.E. degree in communication systems from Anna University in 2006, and Ph.D. from Karunya University in 2013. His research areas include computational intelligence and image processing. He has authored over 100 research papers in reputed SCIE indexed International Journal and Scopus indexed International Conferences. His Cumulative Impact Factor is more than 130. He has published 27 edited books with reputed publishers such as Elsevier, Springer, and IET. He has served as an associate editor of SCIE Indexed International Journals such as the Journal of Intelligent and fuzzy systems. He serves as an editorial board member/guest editor of many journals with leading publishers such as Elsevier (Soft Computing Letters), Springer (Multidimensional Systems and Signal Processing, SN Computer Science, Sensing, and Imaging), and Inderscience (IJAIP, IJICT, IJCVR, IJBET). He is the series editor of "Biomedical Engineering" book series in Elsevier and "Robotics & Healthcare" book series with CRC Press. He has received a project grant of 35,000 UK Pound from the Government of the UK (GCRF scheme) with collaborators from the University of Westminster, UK. He has also completed one funded research project from CSIR, Govt. of India, and one ongoing funded project from DST, Govt. of India. He also serves as the "Research Scientist" of Computational Intelligence and Information Systems (CI2S) Lab, Argentina; LAPISCO research lab, Brazil; RIADI Lab; Tunisia and Research Centre for Applied Intelligence, University of Craiova, Romania. He has also been the organizing committee member of several international conferences globally, such as Portugal, Romania, the UK, Egypt, and China. He has delivered more than 100 Keynote talks/ Invited Lectures in International Conferences/workshops. He holds professional membership with IEEE Technical Committee on Neural Networks (IEEE Computational Intelligence Society) and IEEE Technical Committee on Soft Computing (IEEE Systems, Man and Cybernetics Society). Currently, he is working as an associate professor in the Department of ECE, Karunya University, Coimbatore, India.

Dinh-Thuan Do (Senior Member, IEEE) received the B.S., M.Eng., and Ph.D. degrees from Vietnam National University (VNU-HCMC) in 2003, 2007, and 2013, respectively, all in communications engineering. Prior to joining the University of Colorado Denver (USA), he also worked with The University of Texas at Austin (USA), Asia University (Taiwan) and Ton Duc Thang University (Vietnam). His research interests include signal processing in wireless communications networks, non-orthogonal multiple access, full-duplex transmission, machine learning for wireless networks, and reconfigurable intelligent surfaces (RIS).

Dr Thuan received the Golden Globe Award from the Vietnam Ministry of Science and Technology in 2015 (top 10 excellent scientists nationwide). He is currently an editor of IEEE Transactions on Vehicular Technology, Computer Communications (Elsevier), EURASIP Journal on Wireless Communications and Networking (Springer), ICT Express and KSII Transactions On Internet And Information Systems. He was a lead guest editor of The Special Issue "Recent Advances For 5G: Emerging Scheme of Noma in Cognitive Radio And Satellite Communications" in electronics in 2019; Guest Editor of a Special Issue on "Power Domain-Based Multiple Access Techniques In Sensor Networks," International Journal of Distributed Sensor Networks (IJDSN) in 2020 and a guest editor of a special issue on "UAV-Enabled B5G/6G Networks: Emerging Trends and Challenges" in physical communication (Elsevier) in 2020; guest editor: Special Issue On "Advanced Machine Learning For Future Internet of Things of 5G Networks," International Journal of Distributed Sensor Networks (IJDSN), 2021; lead guest editor of a special issue on "Enabling Reconfigurable Intelligent Surfaces (RIS) for 6G Cellular Networks," Electronics in 2021. His publications include over 100 SCIE/SCI-indexed journal articles and over 50 international conference papers. He is the sole author of one textbook, one edited book and eight book chapters.

Samarendra Nath Sur (Senior Member IEEE) (M,2016, SM'2020) was born in Hooghly, West Bengal, India, in 1984. He received B.Sc. degree in Physics (Hons.) from the University of Burdwan in 2007. He received M.Sc. degree in electronics science from Jadavpur University in 2007 and M.Tech. degree in digital electronics and advanced communication from Sikkim Manipal University in 2012 and Ph.D. degree in MIMO signal processing from National Institute of Technology (NIT), Durgapur. Since 2008, he has been associated with the Sikkim Manipal Institute of Technology, India, where he is currently an assistant professor in the Department of Electronics & Communication Engineering. His current research interests include broadband wireless communication (MIMO and spread spectrum technology), advanced digital signal processing, and remote sensing, radar image/signal processing (soft computing). He is a senior member of the Institute of Electrical and Electronics Engineers (IEEE), IEEE-IoT, IEEE Signal Processing Society, Institution of Engineers (India) (IEI), and International Association of Engineers (IAENG). He has published more than 77 SCI/Scopus-indexed international journal and conference papers. He is the recipient of the University Medal & Dr S.C. Mukherjee Memorial Gold Centered Silver Medal from Jadavpur University in 2007. He is also a regular reviewer of repute journals, namely IEEE, Springer, Elsevier, Taylor and Francis, IET, Wiley, etc. He is currently editing several books with Springer Nature, Elsevier, and Routledge, & CRC Press. He is also serving as a guest editor for topical collection/special issues of the journal like Springer Nature, MDPI, and Hindawi.

Preface

The book presents Explainable Artificial Intelligence (XAI) in Medical Decision Support Systems (MDSSs). The book presents well-structured chapters from industry experts and researchers across the globe, resulting in diverse and high-quality work for the readers. The book provides researchers and academicians with new insights into the real-world scenarios of the deployment, application, management, and associated benefits of XAI in MDSS. The book critically examines the limitations of the existing MDSS and proffers solutions to revamp the conventional systems architecture. Specifically, the book examines the application of XAI-driven solutions toward addressing critical issues in traditional MDSS systems. These solutions have been critically analyzed and compared in terms of the required computational resources, design complexity, system performance, and overall efficiency.

Also, the book presents the design of efficient and explainable learning models for MDSS applications and discusses XAI-based analytics for patient-specific MDSS. The book discusses the critical MDSS security and privacy issues affecting all parties in the healthcare ecosystem and provides practical XAI-based solutions to address these problems. The book is written in basic and easy-to-understand English with several colored illustrations and tables for efficient reading and understanding. Finally, the book comprises 18 chapters of experimental findings, reviews, and case studies.

Chapter 1 introduces explainable artificial intelligence (XAI) in medical decision systems (MDSS), focusing on healthcare systems. Chapter 2 considered explainable artificial intelligence in medical decision support systems in the context of its applicability, prospects, legal implications, and challenges. Chapter 3 torchlights explainable artificial intelligence-based frameworks for medical decision support systems. Chapter 4 presents a prototype interface for detecting mental fatigue with EEG and XAI frameworks in Industry 4.0. In Chapter 5, the idea of applying XAI for medical image segmentation in medical decision support systems was discussed comprehensively.

In Chapter 6, XAI robot-assisted surgeries for future MDSS are discussed extensively. Chapter 7 presents the prediction of erythemato squamous disease using an ensemble learning framework. Chapter 8 examines security-based explainable artificial intelligence in healthcare systems. The concept of explainable dimensionality reduction modeling with deep learning for diagnosing hypertensive retinopathy was covered in Chapter 9. Chapter 10 dissects how to understand cancer patients with diagnostically influential factors using high-dimensional data embedding. In

Chapter 11, explainable neural networks in diabetes mellitus prediction have been presented.

Chapter 12 presents a KNN and ANN model for predicting heart diseases. Chapter 13 X-rayed how artificial intelligence-enabled Internet of Medical Things can be harvested for COVID-19 pandemic data management. Chapter 14 examines deep neural networks for the identification of lead molecules in antibiotics discovery. Chapter 15 conducted statistical tests with differential privacy for medical decision support systems. Chapter 16 gives an automated decision support system for diagnosing sleep diseases using machine intelligence techniques. Chapter 17 gives XAI methods for precision medicine in medical decision support. Finally, Chapter 18 provides an overview of the psychology of explanation in medical decision support systems. The psychological perspectives on explanation in healthcare systems with a binocular focus on MDSS are highlighted.

Bochum, North Rhine-Westphalia, Germany
Agbotiname Lucky Imoize

Acknowledgments

Agbotiname Lucky Imoize: First, I express my profound gratitude to God for His faithfulness and wisdom to edit this book. This book would not have been possible without the support of the Department of Electrical Engineering and Information Technology, Institute of Digital Communication, Ruhr University Bochum, Germany, and the University of Lagos, Nigeria. I acknowledge the sponsorship from the Nigerian Petroleum Technology Development Fund (PTDF) and the German Academic Exchange Service (DAAD) through the Nigerian-German Postgraduate Program. Special thanks to my beloved wife Kelly, and our sons Lucius, Luke, and Lucas. Also, I am indebted to the Deeper Life Bible Church, Essen Region, North Rhine-Westphalia, Germany, for their unwavering support. Last, my sincere gratitude to the Institution of Engineering and Technology (IET) for their editorial support.

Chapter 1

Explainable artificial intelligence (XAI) in medical decision systems (MDSSs): healthcare systems perspective

Oluwafisayo Babatope Ayoade[1], Tinuke Omolewa Oladele[2], Agbotiname Lucky Imoize[3,4], Joseph Bamidele Awotunde[2], Adetoye Jerome Adeloye[2], Segun Omotayo Olorunyomi[5] and Ayorinde Oladele Idowu[1]

Abstract

The healthcare sector is very interested in machine learning (ML) and artificial intelligence (AI). Nevertheless, applying AI applications in scientific contexts is difficult due to explainability issues. Explainable AI (XAI) has been studied as a potential remedy for the problems with current AI methods. The usage of ML with XAI may be capable of both explaining models and making judgments, in contrast to AI techniques like deep learning. Computer applications called medical decision support systems (MDSS) affect the decisions doctors make regarding certain patients at a specific moment. MDSS has played a crucial role in systems' attempts to improve patient safety and the standard of care, particularly for non-communicable illnesses. They have moreover been a crucial prerequisite for effectively utilizing electronic healthcare (EHRs) data. This chapter offers a broad overview of the application of XAI in MDSS toward various infectious diseases, summarizes recent research on the use and effects of MDSS in healthcare with regard to non-communicable diseases, and offers suggestions for users to keep in mind as these systems are incorporated into healthcare systems and utilized outside of contexts for research and development.

[1]Department of Computing and Information Science, School of Pure and Applied Sciences, College of Science, Bamidele Olumilua University of Education, Science and Technology, Nigeria
[2]Department of Computer Science, Faculty of Information and Communication Sciences, University of Ilorin, Nigeria
[3]Department of Electrical and Electronics Engineering, Faculty of Engineering, University of Lagos, Nigeria
[4]Department of Electrical Engineering and Information Technology, Institute of Digital Communication, Ruhr University, Germany
[5]Ekiti State Data Center, Old Governor's Office, Nigeria

Keywords: Explainable artificial intelligence; Healthcare medical decision support systems; Non-communicable diseases; Sickle cell disease; Diabetes mellitus; COVID-19 pandemic

1.1 Introduction

The healthcare industry is increasingly utilizing big data, cloud techniques, Internet of Things (IoT), and artificial intelligence (AI)-based technology for a range of applications, including clinical automatons, automated medical diagnostics, precision medicine, and individualized healthcare [1]. The application and usability of these systems in the medical field are thought to be of greater interest to nurses, hematologists, and other healthcare professionals than patients, despite the fact that performance and accuracy metrics of these technologies and AI-based systems may be of interest to recipient of the health facilities. Nevertheless, medical doctors and other healthcare workers might not have the requisite training to fully understand some of these technologies and other AI-based systems [2,3].

As a result, there is a situation that prevents the use of cutting-edge technology that may automate labor-intensive repetitive analysis tasks, on the first hand, and while, on the other hand, precludes the benefits of linked productivity. It also restricts the quantity of information that may be utilized to assist the development, validation, and iteration of AI-based systems for the delivery of healthcare solutions in conjunction with particular healthcare activities. Medical decision support systems (MDSSs) and the models that underpin them are the subject of a research area known as explainable AI (XAI), which focuses on techniques and approaches that make it simpler to understand and communicate how these technologies operate [4].

Researchers have contributed to the study in a variety of fields as a result of the advancement and widespread usage of technology, with the majority of studies focusing on how it is used in focused areas such as business, healthcare, and education. In line with this, researchers must examine the link between the works articulating these benefits and the reality of technology implementation in real-world settings in order to assess the benefits of integrating new technologies in healthcare [2–5].

The increasing use of various AI-based technologies and the resulting legal constraints have increased interest in XAI research in the healthcare industry. The shift from an information technology-based economy to a digital economy is a subject that is actively being debated when looking at contemporary structural management prototypes [6,7]. In a high-tech, astonishingly swift information environment, a corporation is being formed to handle the issue of the digital economy. This ecosystem enables the development of goods that may be personalized for customers and provide value for manufacturers. Information technology addresses the information-based strategy for the present and upcoming operations of the firm [8].

The development of novel resources like big data, the IoT, AI, and machine learning (ML), as well as the possibility of their analysis and construction based on interpretation technologies and endwise business processes, are what is paramount in this situation [8,9]. AI and big data examples include their capacity to classify unrelated things utilizing a range of sign, classification, and explanatory systems, change and exist in real time, as they come from several sources, etc. It is possible to forecast seasonal diseases and establish prescription supply levels using data on the patterns of explicit healthcare demand in online apothecaries. In order to replace the conventional method of developing research-based medical information systems, a new business model has been developed. This approach makes good use of IT services. As a result, scientific research and medical decision-making are linked to the usage of AI, ML, and big data in the healthcare industry [2,10,11].

Healthcare MDSS (HMDSS), often referred to as evidence-adaptive decision support system (DSS), is one of the numerous examples of descriptive technologies for decision making. HMDSS assist medical personnel with understanding medical research and data so that they can draw conclusions from it [12,13]. These initiatives seek to enhance the way medical information is presented so that it can be applied more effectively in a range of situations. They achieve this by focusing on actual medical decisions and fusing the expertise of medical professionals with computerized medical data. The center of interaction of hematologists, geneticists, biomedical science, healthcare organizations, physicians, etc. is where a body of knowledge concerning effective MDSSs (HMDSS) for healthcare is continually growing [14].

Given that each individual has a different genetic makeup and set of physical characteristics, it is evident that care must be given while selecting the right treatments for both communicable and non-communicable diseases. The knowledge base needs to be updated frequently because medical conditions, diagnosis, recommendations, and treatment algorithms are always evolving. Furthermore, it is undeniable that personalized medicine is on the increase and will improve as a result of sophisticated information technology [15]. Instead of being a set of rules in this case, the knowledge base with HMDSS is a particular software solution that uses cloud and classified-explanatory technologies. The disparity between HMDSS research, its relationship with other fields like XAI, its application by medical personnel, and its beneficial contribution to the effectiveness of the healthcare sector are all areas that have received little attention in healthcare settings notwithstanding the depth of study on multi-criteria decision-making techniques, healthcare procedures, and related domains [14–16].

Complete electronic medical record (EMR) systems with features and resources that facilitate outcome and sharing of information are crucial elements of healthcare. When used properly, HMDSS can support patient-specific decisions that are made in accordance with professional medical guidance [17]. After these benefits and in spite of spending vast sums of money on technical deployment, most of the African countries are still having problems embracing EMR, in contrast to other developed nations. A key barrier to the introduction and adoption of the HMDSS as a component of EMR in various African countries is the poor adoption of EMR and personalized medicine as sources of medical data. How can the branch succeed if the HMDSS source fails in its utilization? [18–20]

Consequently, to support healthcare decision-makers in their future deliberations over non-communicable disorders and delivering archived ready-made transcriptions of HMDSS in the healthcare sector to academic academics, this chapter presents an overview of healthcare MDSSs (HMDSS), XAI, and the usage of single or combined HMDSS with the support of XAI.

1.2 Overview of HMDSSs

The majority of human acts are decided upon throughout the day. Optimal decision-making is seen to be an art. According to studies, the majority of individuals act weaker than they should. One may argue that all human events and activities, regardless of the area of life, are the outcome of decision-making processes [19]. Today, decision-making is acknowledged as a problem-solving activity since it is a process that is linked to problem-solving. In other words, a person experiences mental difficulties when their desired situation departs from their current reality. In such a case, the person first tries to alter the position or state they are in, and then they are willing to modify the surroundings around them in order to attain their goals [21,22].

Given the importance of making an informed decision at the right time, the availability of a system to aid people in making judgments is highly significant. DSSs are systems that take part in even the most fundamental corporate decision-making processes in addition to providing information. DSS is a computer-based information processing system that was primarily developed to address managerial and commercial needs. Several authors claimed that DSSs may be used to any system that supports decision-making [22,23]. DDSs may also be thought of as information systems that assist with administrative, institutional, likewise corporate processes that are somehow related to decision-making. In situations when things are changing fast and it is challenging to forecast or foretell what will happen in the future, DSSs are essential [13].

One of the main issues in public health is medical mistakes, which are seen as dangers to the safety of patients. In healthcare, patient security plays a significant role. Information technology advancements have been proposed by researchers as a viable technique to enhance the caliber of healthcare services and patients' health. Medical decision support systems (HMDSSs) are among the most significant and practical information systems. Medical decision-making is, in reality, a wonderful area of DSS deployment [22–24].

Currently, the field of health encompasses a vast body of knowledge that genuinely calls for expert advice and assistance, particularly in light of the ongoing expansion of medical knowledge across various facets of the healthcare system. All three phases of primary, secondary, and tertiary prevention are covered by these features, which also encompass diagnosis, medication, treatment, and follow-up [16]. Medical decision support system (HMDSS) is interactive software built on the foundation of expert systems to help and support the decision-making of doctors, healthcare professionals, and other staff members working in more generalized areas of health-care systems. It should be mentioned that HMDSS links health observations with health information to enhance healthcare practitioners' decision-making. The use of AI in both public and private health-care systems is demonstrated by the HMDSS [12,25].

HMDSSs are regarded as active systems of knowledge that produce patient-specific medical recommendations utilizing two or more categorization orders. This proves that HMDSS is a DSS that focuses on knowledge management in health-care matters to arrive at a medical recommendation based on a limited number of difficulties. The primary objective of developing modern HMDSSs is to support medical professionals, including doctors, nurses, geneticist, at various points throughout professional care systems. As a result, the healthcare professionals and personnel must actively interact with HMDSS to make the best possible diagnosis and analysis based on patient data [25].

Earlier guidelines and convictions of HMDSS were based on using it to diagnose patients with a medical doctor. In the past, medical doctors would only act in line with HMDSS results after giving it information and waiting for it to reach the correct decision. The modern methodology for using HMDSS forces medical doctors and healthcare professionals to interact with HMDSS and simultaneously uses both knowledge to better analyze patient data and arrive at a more accurate diagnosis and more accurate healthcare services, in contrast to earlier methods that only used these two. Usually, HMDSS categorizes and provides doctors and healthcare personnel with suggestions as well as a list of desired outputs. In this vein, doctors and other healthcare professionals explicitly choose pertinent data and reject unhelpful system recommendations [13].

Healthcare workers are only seen as a support for medical sciences experts, healthcare services, healthcare personnel, diagnosis, and treatment in HMDSSs, which are not intended to replace medical doctor. These technologies make it easier to provide particular diagnoses, prescriptions, and treatment. They also eliminate the need for expert consultation, which considerably lowers healthcare system costs and improves the quality of healthcare services [26].

In light of this, the use of information technology, such as the HMDSS, will surely help and support healthcare administrators, policy makers, staff members, and medical professionals. Several components of the healthcare system are using HMDSS to improve their services and cut down on medical error rates. This chapter research's goal is to spread awareness of HMDSS, its theoretical foundations, and its health-related benefits.

1.2.1 MDSSs in healthcare system

Computerized tools called MDSSs were developed to help nurses, hematologists, geneticists, doctors make decisions about their patients, such as those involving diagnosis and treatment. HMDSSs, which are computer applications, are developed to assist healthcare professionals in making scientific decisions regarding certain patients [22,26,27]. In other words, MDSS are active knowledge systems that provide suggestions for a given case based on at least two patient data points. These software programs help doctors make better decisions about things like preventative, acute, and chronic care, diagnostics, ordering specialist tests, and prescribing procedures. They accomplish this by using pertinent medical and scientific data, rules from a knowledge base, and applicable information [25].

A MDSS provides a clinician with advice, assessments, and prescriptions by fusing information regarding baseline demographics with a reliable knowledge base, or help that is tailored to the patient. Patient data can be manually entered into computer systems by patients, healthcare providers, or clinic employees. Alternately, you can query EMRs to get patient characteristics (EMR). Physicians can discover and choose the optimal course of action with the aid of these decision-support technologies [12]. The decision-support services offered are built on sophisticated algorithms and outcomes evaluation methods that scour information repositories for the most recent improvements in best practice. Whatever the definition of MDSS, it is critical to recognize that the field is unregulated and expanding swiftly. Particularly in terms of the efficiency of medical personnel, the standard of treatment, and patient outcomes, MDSS has a lot of potential to be helpful. If systems are not adequately designed and tested, they run the danger of being hazardous [14].

Different types of professionals must make medical decisions on patient data with a hazy understanding of the patients' state of health. Computer technologies have been created to support both veterinary and human healthcare practitioners in this decision-making process in order to help manage this ambiguity. These technologies were created for a number of reasons, which including enhanced information retrieval, patient record analysis, and intelligent systems that directly support decision-making via ML. Some computational techniques have been developed since the 1950s. A 1979 evaluation that looked at the benefits and drawbacks of the early clinical algorithms, databanks, and mathematical models that allowed computer-based clinical decision support systems also contained documentation of these early tests [17].

Since the historical book release titled "To Err Is Human" in 2000, MDSS and computer-based physician order entry systems have been crucial in evaluating and enhancing patient care. With MDSS, it has been discovered that patient outcomes and healthcare expenses are both improved [18]. They have shown to decrease analytical errors and their diagnostic procedures have shown to enable more accurate diagnoses by warning the clinician of potentially dangerous prescription combinations. MDSS may be used in a variety of ways in clinical settings [24]. Among the principal applications are the following:

(i) supporting the patient's decision-making;
(ii) deciding on the best treatment plans for certain individuals;
(iii) assisting overall health strategies by calculating the clinical and financial results of various therapy approaches;
(iv) calculating therapeutic effects in situations when unmethodical studies are either not feasible or not possible.

For instance, according to [23] as cited by [24], a study of 100 patient studies was undertaken, and the results showed that MDSS improved diagnosis in 64% and patient outcomes in 13% of the studies examined. In the same year, decision support systems considerably enhanced medical practice in 68% of all trials, according to a thorough analysis of 70 separate instances undertaken by Duke University. The success of the two analyses was due to the MDSS properties listed below:

(i) seamless incorporation into the clinical process;
(ii) nature of electronic medical reports;
(iii) not before or after the patient participation, but at the time and place of therapy, decision assistance should be offered;
(iv) substitution of indicated care for care evaluation.

In two areas of healthcare, the pharmacy and diagnostic (out-patient departments – OPD) sectors, MDSS have had a particularly significant influence. Batch-based order checking systems are being used more often by pharmacies to check orders for potentially hazardous drug interactions and notify the patient's ordering physician about them. OPDs have used MDSS to evaluate both likely treatment trajectories and common medicare conditions in order to develop treatment plans that achieve the perfect balance among patient care, medication prescriptions, and financial expenses.

1.2.2 Basis of HMDSS

Healthcare professionals, staff, patients, and other people can benefit from expertise and skillfully selected facts about a single person and given at the right times to enhance medical care and wellness. The Ministries of Health of various countries has long expressed concerns about the quality of healthcare in African nations and has long promoted the use of health information technology (IT), such as electronic MDSS (e-health), to raise medical standards [28]. The governments of several African nations have also supported the use of EMRs, and adoption of electronic health records has been modest but growing (e-health). However, it is crucial to keep in mind that these health IT systems are merely a tool, not a goal in and of itself, to enhance the quality of treatment [28–30].

An alternative viewpoint on MDSS, commonly referred to as evidence-adaptive DSSs, assists users in making selections informed on medical evidence and research. These systems emphasize on medical judgments in real-world settings by merging electronic medical data with physicians' discretion, enhancing how medical knowledge is presented so that it can be applied more effectively. Even though decision-making technologies, its multi-criteria (MC) techniques, in such domains, and medical approaches, the discrepancy between MDSS research and the practical application of end users and its contribution to organizational performance has received little attention [30].

In this instance, a number of MDSS strategies may have avoided the pharmaceutical interaction. These instruments could include pop-up warnings about possible drug interactions when a new prescription is provided, risk assessment standards for the patient's painkiller, clinical recommendations for the treatment of sickle cell disease (SCD), or requests for timely follow-up. This hypothetical example demonstrates how EMRs are the foundation for raising patient safety and the caliber of healthcare, but MDSS is necessary for completely achieving these objectives [31,32].

1.2.3 Characterizing and categorizing HMDSS

MDSSs come in a very wide variety of styles across the healthcare industry. Throughout the last 10 years, there has also been a substantial change in the fundamental principles of architecture and strategy. A variety of healthcare MDSS qualities are linked to or directly impacted by the scientific effectiveness, functionality, error avoidance, possibility of acceptability in the medical community, system portability, cost-efficiency, and other characteristics of health sector. As a result, it is essential to explain HMDSSs in a way that makes it easy for us to understand the range of decision support systems. Understanding categorization and the earlier mentioned common healthcare decision-making processes offer a powerful collection of foundational assumptions that are beneficial to the creators and reviewers of HMDSSs. As a result, having defined HMDSSs in varieties of phenomenon, this chapter will attempt to categorize HMDSS by fusing a number of resources to produce a thorough categorization that captures important aspects of HMDSS strategy and purpose with regard to non-communicable illness in healthcare.

According to Figure 1.1, Refs. [33,34] asserted that the five axes of DSS – context axes, knowledge axes, decision support axes, information delivery axes,

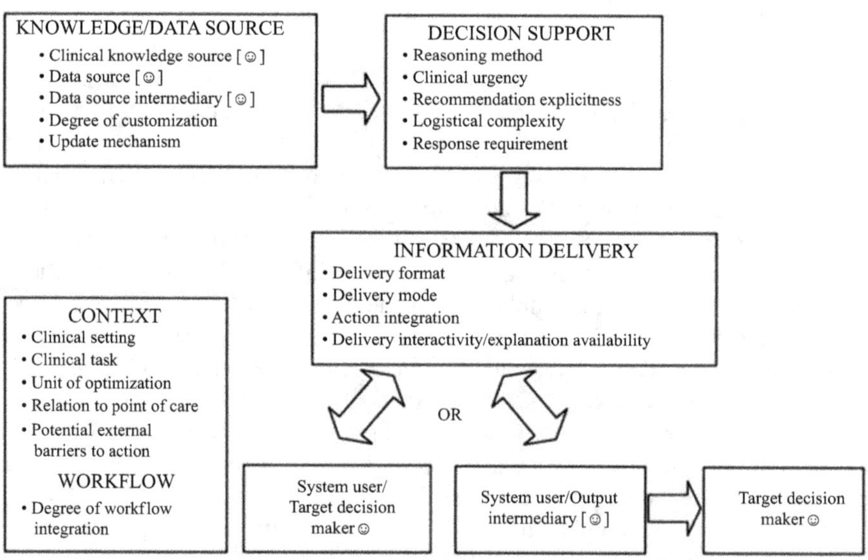

Figure 1.1 Outline of MDSS taxonomy axes

and workflow – may be used to classify the HMDSS. Each of them has a unique subpart that contributes to the system's development. From the figure, the smileys notations ☺ and |☺| symbols represent humans and possible human roles within the systems, respectively. The following are thorough explanations of them.

1.2.3.1 Context axes

The generalized and relevance of a HMDSS are influenced by the environment in which it is used and evaluated. These are their main areas of influence on the healthcare industry:

(a) *Clinical setting*: The healthcare facilities' inpatient and outpatient settings are included in this. Here, the HMDSSs may be utilized in an inpatient or outpatient context or independently of any healthcare organization (e.g., a website for smoking termination). Medical students may participate in educational settings of such.

(b) *Clinical task*: The target task of MDSSs, such as prevention or screening, diagnosis, therapy, drug dosage or prescription, and test ordering, has conventionally been the foundation for evaluation. This include chronic illness management and health-related activities (like exercises) to these tasks since MDSSs for chronic diseases may be more prone to failure. Under this category, we have the following procedures:

  (i) *Diagnostic assistance*: The HMDSS offers likely diagnoses depending on the patient's information and the system's database of knowledge. It is possible to combine complicated data retrieval systems, such as electrocardiogram (ECG) with diagnostic aid. It aims to pinpoint "what is true" with relation to a specific patient.

  (ii) *Therapy consultation and criticism*: The order-entry process used by medical doctors may contain this capability as an example. It evaluates the course of treatment, checks for faults and inconsistencies, compares possible drug interactions, and stops the prescription of allergic medications. It has been demonstrated that the therapeutic significance of HMDSSs is greatly increased by the need that a medical doctor explains a valid justification for any deviation from the guidelines. The HMDSS can offer an ideal treatment plan and support adhering to it employing patient information acquired from the EMR together with protocols and evidence-based recommendations. These HMDSSs answer the issue "what to do" with a patient and often make recommendations for more analytical analysis (i.e. which X-rays, CT scans, to tests-order, and so on) [33]. With the help of such software, you may ask more questions and give even more detailed recommendations for subsequent therapy (and diagnosis).

  (iii) *Medication dosage or prescription*: To guarantee that a prescription conforms with guidelines and recommendations, HMDSS can lower the number of dangerous pharmaceuticals in a prescription and shorten the time it takes to maintain therapeutic control. If the system is linked to an EMR, it can prevent the administration of drugs that have negative side effects. These systems are well accepted because they incorporate automated order input forms, electronic transmission to pharmacies, and effortless incorporation into the clinician's routine practice. One of a doctor's most common behaviors is to prescribe medication, which is also one of the clinical duties where HMDSSs are most frequently used.

(iv) *Alerts and reminders*: Real-time auditory, visual, or tactile warnings can be sent by an expert system (e.g. e-mail, SMS, pager) that is incorporated with a monitoring equipment or healthcare information system (such as a test center information system or EMR). Reminder systems are made to prepare the medical practitioner to important chores that must be completed before a certain occurrence (like not taking drug therapy before to major surgery or fasting prior to actually endoscopy).

(v) *Information gathering*: Locating pertinent information in large knowledge systems or on the Internet

(vi) *Understanding and recognizing images*: Today, a portion of the interpretation of scientific pictures from CT scans, magnetic resource imaging (MRI), angiograms, etc. may be done automatically. More crucially, HMDSS can serve as a tool for mass screening in which software identifies photos that need the clinicians' specific attention.

(vii) Others are choosing testing procedures, preventing, detecting, using a skilled laboratory system and treating lingering illnesses.

1.2.3.2 Knowledge axes

The sources, quality, and personalization of the information and data provided by the HMDSS are the topical issues of this axes. They use knowledge base techniques in relation to the following:

(a) *Scientific sources of information*: It may come from reputable sources (including directives from national or professional societies, observational studies, and randomized supervised investigations, or by involving medical professionals who will ultimately use the software).

(b) *Data source*: An EMR, a medical device (such as blood pressure monitor), or another data source can provide the patient-specific information. It is possible that the information will be given by a person or a paper chart. The data must then be fed through a data input intermediate into the system. This feature has a considerable impact on the likelihood that HMDSS will be used in real-world applications. It has been demonstrated that automatically sending data to the system via computers is preferable (e.g., via an EMR). A doctor who feeds data into a data source is referred to as a data source intermediate (see above). Patients themselves might serve as intermediaries.

(c) *Data coding*: It is preferable to utilize a widely used coding scheme, such as version 10 of the International Classification of Diseases (ICD-10) or Systematized Nomenclature of Medicine (SNOMED), for a number of reasons (including financing and epidemiology). Obviously, plain text might also be used for the data.

(d) *Personalized data*: HMDSS has a higher likelihood of being clinically relevant and beneficial the more patient-specific targeted suggestions it generates, according to age, gender, co-occurring diseases, etc.
(e) *System updating*: As already mentioned, the knowledge base needs to be instantly created and contemporary. Both knowledge-based and knowledge-free systems are included in HMDSS. A knowledge-based system's knowledge base, inference engine, and communication mechanism make up the majority of its components. They can reason using the information they get from the many sources mentioned above because they have professional clinical knowledge of very particular facts and actions. These systems commonly employ probabilistic links between acquired knowledge and data.

The inference engine, which mixes and it links patient data to knowledge-based standards, is the brain of AI in knowledge-based systems. In essence, the inference engine makes inferences and comes to new inferences using the data. In contrast, the non-knowledge-based systems rely on the concepts of ML, such as neural networks or genetic algorithms, where computers acquire new abilities via previous experience and/or search for patterns in a patient's scientific data.

1.2.3.3 Decision support axes

The most crucial aspect of HMDSS is undoubtedly addressing an appropriate decision-making process.

(a) *Reasoning approach*: HMDSS reasoning engines include, among others:
   (i) *Rule-based systems*: Different expert knowledge bases are utilized by rule-based systems in the form of expressions and it can be broken down into IF–THEN rules (production rules). Such a system is an illustration of an empirical technique, whereby distinct logical claims in the form of production rules are acquired through professionals' observation, interview, and debriefing, and then merged in an effort to mimic knowledgeable thinking. This method was originally employed in the MYCIN (an AI technology designed for DSS) to choose the best antibiotic treatment for a patient.
   (ii) *Neural networks*: A non-knowledge-based adaptive HMDSS called an artificial neural network employs ML-based models to learn from understandings and spot outlines in clinical data.
   (iii) *Bayesian network*: The Bayesian network is a common example of a knowledge-based decision-making system, referred to as a causal probabilistic network or belief network at times. Using the Bayes theorem's conditional probability, it shows probabilistic relationships between groups of data, like disease and symptom data. In this network, the causality of the linkages is needed explicitly. The fact that clinical knowledge oftentimes struggles to express itself clearly A major barrier is "what is the impact" and "what the cause" to this network's capacity to accurately forecast how an illness will progress over time and how several diseases will interact.

(b) *Model-based systems*: Patient-specific modeling is the most recent accomplishment.
(c) *Logical condition*: Decisions are made using logic based on the value of a specific variable. If the value is inside or outside of the predetermined parameters, the decision-making process will provide various results.
(d) *ML and data mining procedures*: The database of the system indicates that these strategies are based on probabilistic decision-making. Large and well-designed databases are desirable because they enable accurate retrieval of patients who are comparable to the current patient. The optimum therapy for the present patient is chosen based on an analysis of how those patients responded to various therapies.
(e) *Genetic algorithm*: Iterative techniques are used by this non-knowledge-based approach to reorganize itself and offer the best possible outcome depending on patient data.
(f) *Medical emergencies*: The provision of decision assistance for hastily needed choices. According to the dictum "cure first what kills first," HMDSS should devote its resources initially to problems with a clear clinical importance. Better patient outcomes and medical doctor performance are the result of this trait.
(g) *Emphasis on explicitness*: In-depth instructions that clearly lay out a particular course of action are more likely to be followed by end users.
(h) *Adequate prompt responses*: It is possible to inquire about the utilizing physician's response to the HMDSS's suggestions. Numerous strategies can be used to do this, such as acknowledging the advice, outlining any replacement actions performed, and providing a justification for failure.

1.2.3.4 Information delivery axes

The goal of these axes is to give the user access to recently produced information. They contain the following subsections:

(a) *Delivery format*: Using extra technical tools such as phone, pager, or e-mail, or using paper-based, electronic (online or incorporated into an EMR), or electronic methods.
(b) *Delivery mode*: Whether it is a notification, a prompt, or a request for optimization, the ideas may be provided upon the decision-request maker's request or alternatively, they may be provided without the decision-agreement maker's request. The software is initially inactive since it requires the doctor to make extra effort to obtain a diagnostic or therapeutic evaluation, recognize when the advice would be beneficial, and "go to the program" to input data. The so-called "push systems," which provide solutions autonomously, might be more advantageous, and extensively used. Because of their work in data management, they actively assist decision-making (e.g., monitoring, EMR supervision). The findings of the system's decision analysis are provided without the need for further work on the part of the physician since system decision logic is in a sense incorporated inside the patient database that has already been compiled from a variety of sources. How to minimize "alarm

fatigue," which occurs when a physician is too alerted to tiny anomalies that are ordinarily observed and understood, is an important issue to think about in this situation.

(c) *Integration of actions*: The HMDSS must give the decision-maker the tools necessary to quickly implement the advised actions. For instance, the program is able to offer direct connections to order entry forms and therapy planning section of the EMR, as well as suggestions for therapeutic review. Similar to verifying a mark, the therapy adjustment activity should be completed in a few clicks. The utility and acceptance of HMDSS are clearly increased by action integration.

(d) *The availability of justification*: In order to clarify its suggestions and give evidence, the system offers links to books, articles, or knowledge base entries.

1.2.3.5 Workflow axes

Despite the fact that HMDSS is sometimes referred to as a procedure, it actually functions more as a technology intervention at the point of care and may result in a disruption. Systems that function in unison with the institution's workflow are more commonly used and more successful in improving practitioners' performance.

1.3 Case study of XAI enabled with MDSSs in various infectious diseases

1.3.1 SCD

Even though the "era of data" has access to improved computer and storage capabilities, the challenges associated with understanding these enormous data sets have risen significantly. Many industries, including those in education, health, business, and organizations, are developing intelligent systems to address these issues by utilizing relevant concepts and methods including data mining, data science, ML, and even AI. One of the most active areas in computer science is AI. As a result, it imparts insight and importance to the data [35,36].

Ref. [36] asserts that the daily growth in data velocity and the development of digital technology result in the continuous generation of new data across a range of data collecting tactics. In the area of AI, computers help people understand data in a way that is comparable to a person. In this field of AI, computational learning and pattern recognition are included. ML, a subfield of AI, is impacted by advances in human knowledge technology. This has to do with changes made to systems that do various AI-related tasks, such as robotics, planning, prediction, analysis, and recognition [36,37].

The quantity of data and knowledge that our civilization is able to produce and retain is growing exponentially every day, but our capacity to absorb it is not keeping up [4]. It is crucial to employ current technology to balance the seemingly incompatible aims of scalability and usability in data and information interpretation in order to overcome these difficulties. According to [38], several data and information analysis techniques have been developed and used all over the world with the intention of making them more complicated. Particularly in a number of organizational sectors, the most prominent of which is the healthcare sector, which encompasses several subsections such as genetics, human genetics, and medical genetics, humans must be involved in the early stages of data analysis and information generation. A primary objective and top priority for many governmental and commercial sector companies is to improve data analysis processes.

The need to provide new and appropriate services, as well as the need to improve productivity and adhere to rules, all have an impact on the illumination of medical data analysis. In the current digital era, healthcare systems are swamped with data, yet having access to this data and expertise is insufficient to fully capitalize on improving patient health. The ability to combine, adjust, or anonymize data from many sources, including EHRs, health surveys, administrative data, physician notes, and consultant reports, to mention a few, is therefore required by both private and public healthcare organizations. The patient would receive a complete medical assessment of their condition as a result [39]. Various patient data are combined, examined, and reported on using healthcare data analytic technologies. Additionally, healthcare organizations make wiser clinical, administrative, and financial choices that improve patient care and engagement. The importance of utilizing information and communication technology (ICT) to change patient medical data with regard to their health cannot be overstated [40].

In order to improve healthcare diagnosis, management as well as appropriate decision-making, the optimal framework for patient medical data analysis and decision support systems must be effectively managed as IT in data analysis in the health sector increases. When the healthcare industry transitions to a value-based paradigm, medical consultants, hematologists, geneticists, and others may find themselves in danger of not respecting patient decisions. As a result, reviewing medical data to guide judgments is no longer a good idea but rather a need [41–43].

SCD is a red blood cell (RBC) disorder brought on by the deficiency of oxygen. The end consequence is a group of diseases known as hemoglobinopathies, the most prevalent of which are thalassemia and sickle cell anemia (SCA). There are no additional treatment options for adults with SCD, a congenital abnormality of the hemoglobin structure, which can only be treated in infants by bone marrow or cord blood transplantation [44]. Hemoglobinopathies (abnormal Hb or anemia), according to [44], are brought on by genes that affect around 5% of the world's population. The prevalence of sickle cell gene carriers is higher compared to the prevalence of affected newborns because healthier individuals have thalassemia than SCA (who acquire just one mutant gene from their parents). In Ref. [45] it is said that among the diseases for which ML algorithms have been used to model, forecast, and diagnose include SCD, cancer, and malaria. One of the most serious medical illnesses in the world, SCD, has a variety of symptoms that can vary from moderate to catastrophic.

SCD, for instance, is one of the most prevalent genetic RBC illnesses in humans in Nigeria and affects people of all ages, resulting in hemolysis and vaso-occlusive crises. Studies show that Nigeria, where over 150,000 births occur annually, has the largest number of people with SC illness. The RBCs of patients with SCD exhibit aberrant hemoglobin, such as hemoglobin S or sickle hemoglobin [46,47]. Hemoglobin, a protein that transports oxygen throughout the body, is found in human RBCs. SCD is a genetically based non-communicable illness that is passed on from parent to child [48]. Like Ebola, Zika, tuberculosis, or the typical cold and flu, it is not contagious. Patients with SCD have long-term acute pain [vaso-occlusive episodes (VOEs)], long-term suffering, multiple organ damage (including kidney failure, heart failure, and others), a short lifespan, bone infections, and stroke, to name a few symptoms and problems [49,50].

Apparently, a single-point mutation (glutamic acid substitution) of valine at position 6 in the hemoglobin component known as beta globin (-globin) causes sickle hemoglobin (HbS) to deform or transform, according to [45,48]. According to [48], those who have two copies of the HbS mutation are homozygous (HbSS) and have the sickle cell phenotypic form of the disease, as opposed to those who have one copy of the gene, who are heterozygous carriers (HbAS), who do not. In order to lower the mortality rate and other issues that SCD patients face, it is crucial to give medical officers, hematologists, and healthcare managers a platform to strategize on how to make life meaningful and fulfilling for these patients by integrating ML algorithms into the analysis of SCD medical datasets [51].

In order to address practical issues like significant variation, low accuracies, the presence of feature noises and biases, as well as medical doctors' decisions based on the data gathered, the involvement of healthcare MDSSs with the support of explanatory AI should be developed [52]. As a result, the availability of enormous databases containing pertinent genomic data would enable researchers to focus on developing new approaches for diagnosing and treating human genetic diseases like SCD. Due to the size and diversity of the data, it may be challenging to effectively employ genomic databases for SCD in this situation without the proper development of advanced data analysis tools [53].

In order to provide proper management and follow-up on SCDs, this section provides a review of SCD along with its correlation and the participation of explicable AI in healthcare decision support systems. The effectiveness of these explanatory AI technologies will reassure medical experts, other healthcare workers, researchers, and patients who are suffering from this condition.

1.3.1.1 Historical perspective with SCD

Doctors have found it difficult and time-consuming to correctly diagnose SCD and forecast the patient's chances of survival and life expectancy over time. According to studies, conventional or clinical tests take a long time to discover the presence of SCD or its variants. As a result, explanatory AI technologies are being employed more often to provide SCD patients with a non-clinical diagnosis. In 30–50% of cases, SCD is regarded to be avoidable [36,44]. In order to provide accurate, objective, and systematic forecasts of human blood cells, advanced computer approaches such as XAI are needed [2].

One cannot overestimate the importance of medical data analysis and medical decision support systems to any nation's economy [39]. One of the challenges faced the healthcare sector is decision-making and medical data analysis issues, particularly in the context of non-communicable disorders like SCD. When compared to the inadequate local medical infrastructure that is appropriate for this human genetic concern in the examination of medical data, the birth rate in Africa continent is quite high and at a critical stage [54]. When learning from huge medical histories and their datasets like SCD, difficulties including missing parameter values (incompleteness), systematic or data random noise, and inappropriate parameter selection have all arisen. It will be possible to handle the difficulties that medical datasets like SCD bring using a HMDSS with support of XAI technologies [55].

1.3.1.2 Synopsis of SCD

A patient with SCD inherits mutant hemoglobin (Hb) genes from both parents and has hematological abnormalities in RBCs. SCD is a non-communicable illness. SC problems are brought on by the RBCs' lack of oxygen [44]. In Figure 1.2, some of

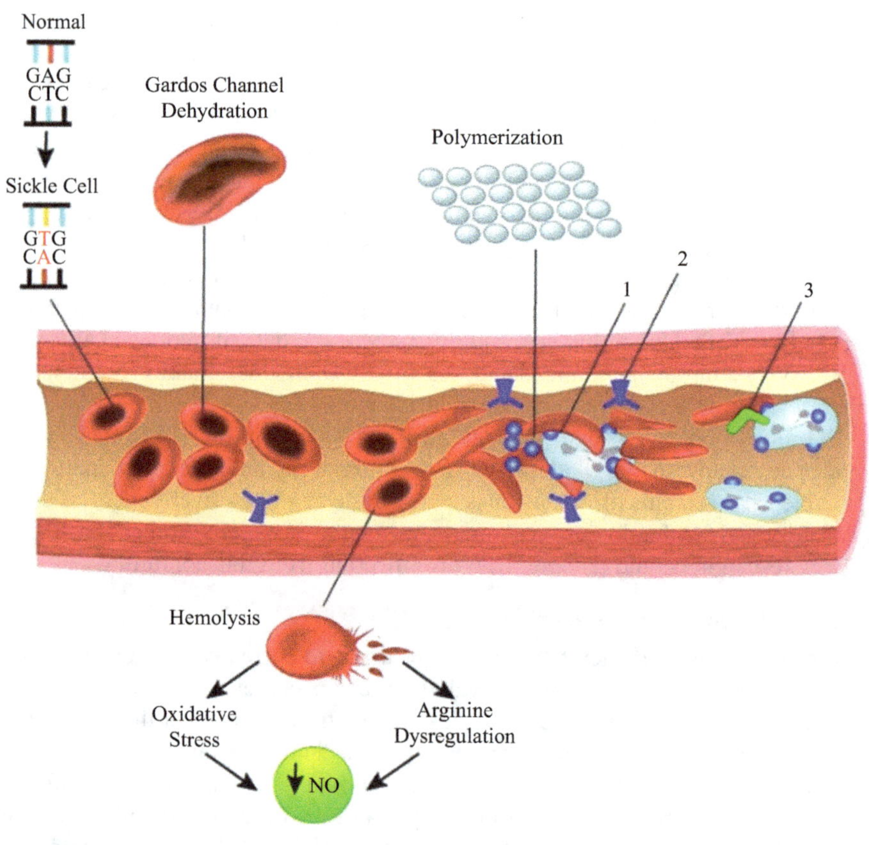

Figure 1.2 Sickle erythrocytes from single gene mutation (GAG–GTG and CTC–CAC) resulting in a defective hemoglobin due to deoxygenation exposure (1: mononuclear cells and platelets {light blue and dark blue}; 2: ligands; and 3: receptors {dark green}; NO: nitric oxide) [56]

the pathophysiologic elements of the disease are shown in a streamlined manner. The pathophysiology of the illness is described in depth in several researches. The basic assumption that sickle cells are only responsible for creating vascular blockage or vaso-occlusion is no longer correct after red cells assume the pathognomonic sickle cell form after being subjected to deoxygenation. SCD is brought on by a single gene mutation that results in complex physiologic abnormalities, whereas vaso-occlusion is essential to understanding the illness and can bring about localized hypoxia and inflammation. The illness manifests itself in many ways as a result of these alterations. In addition to vaso-occlusion, anemia, and hemolysis, it is now known that SCD is a disorder characterized by issues with arginine metabolism, increased inflammation, hypercoagulability, and oxidative stress.

According to [45], SCD is one of the most hazardous diseases in the world, with a wide spectrum of symptoms that can be anything from mild to fatal. SCD affects people of all ages, is one of the most prevalent genetic RBC illnesses in Nigeria, and causes hemolysis and vaso-occlusive crises. The sickled erythrocytes and irregularities that develop as a result of low oxygen levels in RBCs are seen in Figure 1.3. With approximately 150,000 parturitions (births) per year, African continent especially Nigeria has the highest number of individuals living with SC disease, according to studies. The survey found that hemoglobin S or sickled hemoglobin was present in the RBCs of SCD patients, along with other hemoglobin abnormalities [46,47]. Table 1.1 shows the fundamental medical signs of SCD in both children and adults, along with the many possible symptoms.

1.3.1.3 Importance of XAI support in HMDSS for SCD

In areas including patient deterioration, readmissions, mortality, improved documentation, sickness diagnosis, patient relocation, and chronic care management, predictive modeling in healthcare has been driven by digital transformation. Thanks to advancements in big data, AI, cloud computing, and the IoTs, consequently, mixing data from various sources, offer massive computing, and store shared resources are made possible by the modern technologies. Hence, connect data to one another through devices, sensors, software, and other technologies, and exchange data with them [2]. AI's success in a range of applications results in a large number of autonomous systems and accurate projections for decision support. In healthcare, predictive modeling encompasses forecasting illness states and their trajectories, hospital readmission, adverse medication responses, and drug–drug interactions, using the interrelatedness of organ dysfunctions to predict the survival of sepsis patients, and finding off-label usage of pharmaceuticals.

Healthcare predictive modeling may be challenging, counterintuitive, and frequently difficult to explain. Due to their opaque nature in the healthcare field, these traits prevent predictive solutions from being adopted widely or having much value in the clinical situation. In order to improve their capacity to be understood, predictive solutions in healthcare require a successful strategy. A greater understanding of the model, enhanced value of its output, and improved patient care outcomes may result from improving model explainability [5].

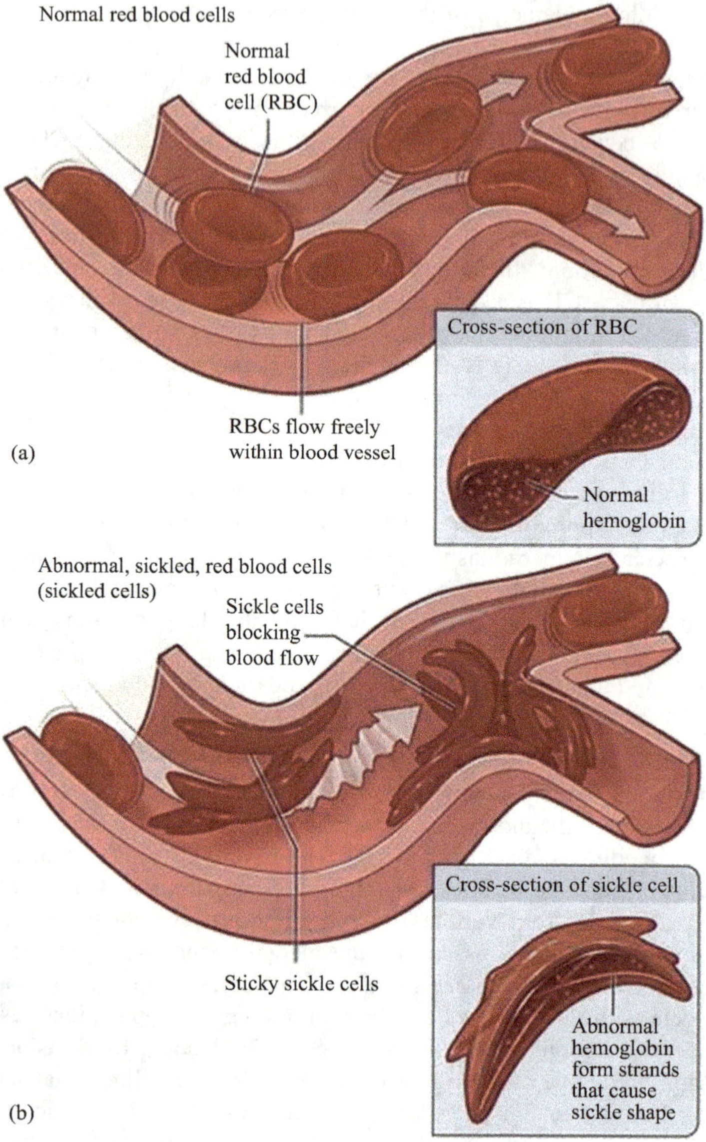

Figure 1.3 Normal and sickled erythrocytes showing abnormality forming irregular stiff rods [39]

In some functional areas of healthcare, like personalized medicine for SCD, AI can help doctors make better clinical judgments or even take the role of human judgment in some situations. With the use of pertinent clinical questions developed in collaboration with healthcare experts, cutting-edge AI algorithms like XAI may

Table 1.1 Medical indicators of children and adult SCD (adapted from [38,49])

Children with SCD **Indications/ symptoms**	**Infants**	**Children**	**The program is able to offer direct connections to order entry forms**
	Chest, stomach, and limbs/joints pain anemia, dactylitis, mild cyanosis, fever, enlarged spleen, upper respiratory infections that are common	Pain (acute or chronic), severe anemia, infections jaundice, inadequate nutrition and metabolism, failure in school, and premature puberty	significant joint pain, persistent leg ulcers, retinal disease, thromboembolic side effects, neurological problems, drug tolerance/dependence
		Complications	
• CNS • Eye • Lung • Heart • Spleen • Liver • Kidney • Gall bladder • Genitals • Bones/ joints • Skin	• Stroke, retinopathy caused by retinal artery blockage, ACS, and asthma • hypertrophy of the left ventricle, Cardiomyopathy, acute sequestration of the spleen, reduced immunity (such is sepsis or a viral infection) • Hyposthenia Priapism, perivascular necrosis, proteinuria-renal neurocognitive, cholelithiasis, persistent sores, usually on the ankles		• Hemorrhagic stroke and recurrent ischemic stroke Progressive retinal disease, persistent ACS, chronic pulmonary illness, pulmonary hypertension, an early form of coronary artery disease, heart attack, Auto-infarction, Asperenia with function • liver sequestration liver failure brought on by excessive transfusions of iron Nephropathy, Urinary tract infections frequently • Chronic leg ulcers, cholelithiasis, priapism, avascular necrosis, and early loss of bone density

Keys: CNS, central nervous system; ACS, acute chest syndrome.

uncover clinically pertinent information concealed in the vast amount of healthcare data. Most clinical trials for medical research include well-defined hypotheses or particular questions. Given a collection of characteristics, the AI algorithms are taught to anticipate specific events, and via the prediction, insights may be gained [3]. Recent developments in deep learning have attracted a lot of interest. Deep learning is a neural network with numerous hidden layers that allow for the exploration of more intricate nonlinear patterns to enhance prediction accuracy. Understanding forecasting outcomes and how AI systems function to produce predictions are challenging tasks.

This portion of the chapter discusses the numerous data sources used to support SCD predictive modeling in healthcare, along with its benefits and drawbacks. Additionally, it would include some examples of predictive modeling in healthcare. The necessity of explainability in AI for HMDSS is next discussed in order to win the confidence of medical experts in predictive modeling for SCD. To explain internal decisions, behaviors, and actions to the interacting humans, a smart AI system must include an explanation model. Additionally, the information-based and instance-based clarifications in XAI using HMDSS for SCD healthcare prediction modeling will be covered.

1.3.2 Diabetes mellitus (DM)

Every year, emergency departments in hospitals serve millions of patients. Although a sizable number of the patients are not emergencies, forcing hospitals to assign medical staff where it is not absolutely necessary, making inefficient use of personnel and handling genuine patient emergency situations [57,58]. In addition, there is an increasing shortage of doctors in rural areas, which results in underserved patients, in especially because of the changing demographics and an increase in the number of elderly patients [59,60]. Future application of AI-based models in healthcare as a proponent of systems medicine could be one way to solve these issues [61].

One of the most prevalent non-communicable diseases and a leading cause of morbidity and mortality worldwide is DM [62]. By 2030, diabetes is expected to overtake heart disease as the seventh biggest cause of death worldwide, affecting an estimated 422 million people [63,64]. In every nation, there are more persons with type 2 DM (T2DM), the most prevalent kind of the disease [62]. Between 1990 and 2010, the prevalence of diabetes nearly tripled in the United States [65], moreover, 1.7 million new adult cases were detected in 2012 [66]. There are 26% more Americans who have the condition than those who are 65 or older (4). For diabetic patients, hypoglycemia (HG) is known as the primary limiting factor in effective glycemic control [67,68]. Significantly detrimental consequences have been found on cardiovascular safety and quality of life [69,70]. Additionally, it raises the economic costs associated with managing HG and its effects through the use of healthcare resources, and additionally from patient disengagement and lost output [71].

Many diabetic patients are uninformed when HG begins, specifically those with recurrent bouts, notwithstanding the possibility of fatal and major adverse consequences. Finding patients who are particularly at risk for HG may offer the chance to take action and lower the frequency of occurrences [72]. It is difficult to identify HG using computer-based techniques and EMRs, due to the inconsistent use of HG diagnosis codes and the possibility of undercounting. Only an unstructured, narrative (text-based) style may be used to document the signs or symptoms of a given episode of HG, especially if they are not severe.

DM is diagnosed, treated, and monitored by physicians with relatively low rates of adherence to guidelines [73]. Alternative approaches to real-time patient monitoring are necessary due to time restraints, patient overcrowding, and complex requirements. MDSSs and monitoring systems, which are rapidly developing, offer an efficient answer to these issues. The goal of chronic illness means of combating is no longer to cure the patient, but to improve the patient's adherence to the treatment regimen, and to enhance the quality of life through collaboration. Because of this, the value of safeguarding, preserving, and enhancing health has been prioritized over the treatment of symptoms, fostering the idea of "self-care" [74].

Diabetes is a persistent medical condition. Patients with diabetes frequently experience difficulties during treatment, such as psychological issues and trouble adjusting to lifestyle modifications. In addition to the normal discomfort brought on by symptoms, consequences, and therapies. Future concerns for the patient have an impact on his social, cognitive, and emotional health [75]. In people with health issues like diabetes that require substantial treatment and support, self-efficacy beliefs are crucial, adjusting one's lifestyle and picking up new abilities to deal with the disease progression. Diabetics are required to possess considerable self to manage sophisticated diabetes treatment and management. By raising their levels of self-efficacy, diabetics can enhance their self-care habits [76].

Substantial proof studies carried out by professionals at the national and worldwide level serve as the foundation for the creation of DM standards and guidelines. The criteria for diagnosis are laid out in DM management guidelines, and the standards for diagnostic process are standardized [77]. The disease is then categorized based on the patient's medical history, physical assessment, and test results. The recommendations make recommendations for which tests should be run and which patient risk factors should be examined. Following diagnosis and differential diagnosis, the doctor is given various treatments based on the prescriptions provided by the evidence-based studies. These recommendations include a methodical approach to therapy and follow-up in accordance with the diagnosis and outlook [78].

Primary care physicians are crucial in the identification, management, and control of DM illness. According to the World Health Organization, primary healthcare is crucial in lowering chronic illness deaths and morbidity. All nations are working to improve primary care to ensure the management of chronic conditions. Therefore, adjusting the doctor's strategy to the most recent DM management recommendations and creating a consistent strategy will result in notable improvements in blood sugar control. The patient's life will be longer and of higher quality if their diabetes is under management. Patient applications to primary healthcare providers will decrease as a result of this. As a result, doctors will spend more time on each patient.

For the management of chronic diseases, a number of computerized MDSSs have been created [79]. The MDSS is used in primary care for a number of purposes, including depression, hypertension, and drug reviews [80–82]. Additionally, a number of AI-based and rule-based DSSs have been created for the diagnosis of diabetes [83,84].

According to research, measures started initially in pregnancy can lower the rate of DM in expectant mothers who are overweight or obese [85–87]. Applying remedies in every situation, meanwhile, can be expensive and time-consuming. A ML-based MDSS can be useful in supplying a strong and impartial computerized tool to support physicians in identifying women at risk of DM. By enabling focused intervention, it would significantly cut down on time and expense. Clinical settings offer significant promise for the MDSS, especially in light of the fact that many doctors have used telemedicine to keep a distance from patients during the COVID-19 pandemic [88].

Despite the literature on their effective deployment, MDSSs offer a potential to significantly improve healthcare delivery. There is a lack of MDSS data, particularly MDSSs that based on ML-based algorithms. In addition to system accuracy, authors in [89] noted that for a MDSS to be adopted and implemented into healthcare setting, efficiency and attractiveness are crucial. In order to easily get system outputs while juggling a heavy clinical workload, a MDSS should be time-saving, intuitive, and simple to operate. Additionally, they emphasized that black-boxes are unacceptable for MDSSs. This is consistent with the findings of authors in [90], who said that explainability is a crucial element for a MDSS to be successfully incorporated in practical application. A ML-based system can mirror the pattern in the training data, as demonstrated by a well-known example by authors in [91] but not transcend to clinical practice since it is at odds with medical understanding.

According to their computer model, people with a history of asthma were less likely than the general population to pass away from bronchitis. This is due to the vigorous treatment that patients who had asthma and had pneumonia typically received, which reduces their risk. Even if the system accurately recorded the training data, using it in clinical settings without knowing why the model performed this way would indeed be troublesome. AI-based model that is comprehensible and explainable can solve these issues and this is called XAI. The use of XAI in MDSS has been shown to offer numerous advantages, including boosting acceptability and credibility of the system, as well as raising the causality hypothesis and decisions assurance. However, the published literature as a whole clearly lacks the applicability and implementation of XAI in MDSSs [90].

1.3.3 Hypertensive retinopathy (HR)

Vision loss is a result of the retinal condition known as HR, which is brought on by persistently high blood pressure (hypertension). Millions of people around the world are afflicted with HR disease as a result of high blood pressure [92]. The irregularities, such as tortuous retinal arteries, abscess formation, cotton wool patches, hemorrhages, and enlargement of the optic disc (OD), are brought on by hypertension. Ophthalmologists can investigate these indicators of retinopathy brought on by HR by taking digital pictures with a fundus camera [93]. There are no early warning indications for this condition, and frequently, When the condition results in blindness or vision loss, HR is discovered at a later stage. As a result, hypertension people should often get their eyes examined. Figure 1.4 displays

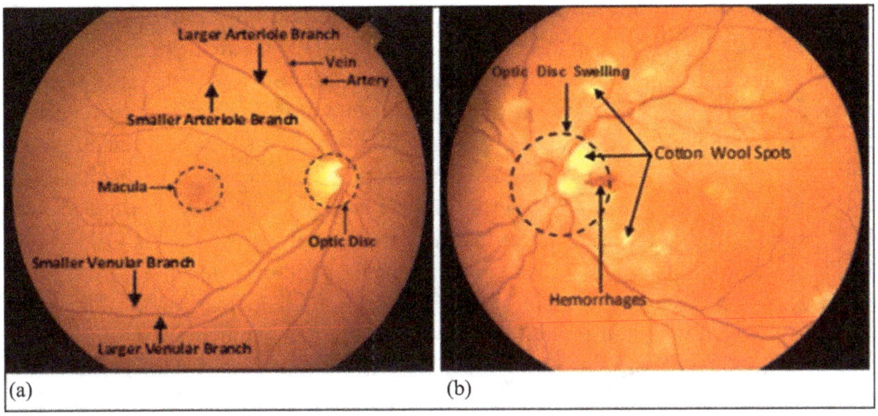

Figure 1.4 (a) A basic fundus retinal image, including the arteries, veins, OD, and macula. Additionally, smaller and larger venular branches and arteriole branches are visible; (b) fundus retinal picture was damaged by HR, and there were some hemorrhages, cotton wool spots, and OD edema

normal and HR pictures along with the fundamental fundus constituents and alterations, which are visible on the fundus because HR is present.

Systems for computer-aided diagnosis are frequently employed in the healthcare sector. Various retinal disorders are detected with computerized diagnostic techniques and are beneficial to both patients and ophthalmologists. These automated devices allow the ophthalmologists to adhere to the disease's treatment strategy. A form of computer-aided diagnostic system that can automatically detect and grade HR using retinal fundus images must therefore be developed. HR is also used as a marker for injury to certain target organs. Through HR indications, doctors can foresee the likelihood of heart illness, stroke, and even mortality [94]. The development of the HR signs and symptoms typically occurs in more established phases of the illness. These indicators support clinical therapy and rules for patients with hypertension [95,96]. There are various levels and grades in HR. Arteriovenous ratio (AVR) is the foundation for HR phases [97], which is regarded as a useful metric for the identification and evaluation of HR.

The evaluation of medical images is an essential tool for the computer-aided diagnosis of many diseases. Due to their dependability and adaptability, DL-based AL-based models are more prevalent in disease analysis than traditional image processing methods [98]. Computer vision has a promising possibility to evaluate these retinal disorders through image segmentation for early identification, diagnosis, and treatment of some retinal diseases that are linked to blindness or visual loss [99]. For quicker screening and computer-aided analysis of numerous retinal illnesses connected to retinal anatomy, DL-based technologies are widely acknowledged [100]. Due to their intricate structure, retinal blood vessels serve as crucial biomarkers for identifying and assessing various retinal dysregulation and associated disorders [101]. Medical image categorization, one of many deep learning techniques, aids ophthalmologists and other medical professionals in making many challenging diagnoses. Segmentation techniques lessen the need for manual illness image recognition and DL-based semantic segmentation is a cutting-edge method for classifying medical images pixel-by-pixel, whether for diagnosis or symptom screening [102]. Due to the aberrant proliferation or deterioration of these capillaries, diabetes and hypertension cause changes in the retinal vessels, which can be found by precisely segmenting the retinal blood vessels [103].

There is a diagnostic load placed on medical specialists because to the laborious and time-consuming manual investigation and identification of this retinal vasculature; obviously, automated technologies could enable quicker diagnostics [104]. The thickness, surface roughness, formation, and removal of these retinal vessels can all be used as diagnostic cues [105]. A retinal condition known as HR is linked to hypertension and gets worse with elevated blood pressure. A retinal condition known as HR is linked to hypertension and gets worse with elevated blood pressure, specifically, localized and arteriolar shortening [106]. Vascular changes due to diabetes and hypertension can be subtle, necessitating a sharp eye for detail in retrospective analysis to detect the important changes brought on by the disease [107]. Deep learning has the ability to diagnose many illnesses and assist medical professionals in various medical applications [108,109].

Similar to this, resilient architecture and DL-based classification techniques can identify minute variations in the retinal blood vessels and can assist doctors in identifying the associated changes to make quicker and more accurate diagnosis. There are a number of deep feature-based techniques for detecting retinal blood vessels [110]. The creation of low-cost, reliable techniques that can precisely detect vasculature with slight alterations is nevertheless necessary. Additionally, current segmentation-based approaches [111] solely concentrate on segmentation rather than offering a comprehensive solution for the identification of HR.

1.3.4 Carcinoma

Non-melanoma skin cancers (NMSCs), which can afflict both sexes, are the fifth most prevalent type of cancer globally. According to estimates, there are more than a million new cases of NMSC each year, with squamous cell carcinoma (SCC) accounting for about 20% of all skin cancers [112]. Each year, more than 1.8 million new cases of NMSCs are recorded in the United States, and skin cancers are common, with cutaneous SCC being the most common type [113,114]. SCC is more common in Asian Indians and African Americans, and additionally, it is the second most common among Asians of Chinese/Japanese descent and Hispanic descent [115]. SCC has been identified as a keratinocyte-related cancer type. The skin condition carcinoma-in-situ, commonly known as actinic keratosis (AK), has been linked in multiple studies to the development of SCC. About 5–10% of all high-risk SCC cases are extremely challenging to identify and treat, with radiation or surgery being the preferred treatments in the majority of these situations.

Therapies for such high-risk metastatic skin cancer are less likely to be effective, especially in an elderly population [116], demonstrating the urgent need for an effective but organized diagnosis and treatment plan for SCC [117]. The amount of microarray data is increasing, and the knowledge it provides about the genes that control for a point mutation is being employed for variation classification and investigation, in addition to other possibilities. Microarrays are a relatively new approach that analyzes gene expression in samples by placing hundreds of DNA probes on a small chip that are matched to target genes. Comparing cancer and healthy tissues was one of the approach's main applications, among other factors, diverse cancer subtypes, people with different prognostications [118]. In terms of recognizing microarray samples, support vector machines (SVMs) [119], ANNs [120], logistic regression, Naive Bayes [121], and other commonly used ML techniques performed wonderfully [122].

Many research use metabolomic data to understand the metabolites that characterize each organism state, and how those metabolites behave in different environments. Systems biology is incredibly dependent on the "omics" field. It has recently been widely adopted in a range of sectors due to its concentration on small molecules and relationships, such as the detection and acquisition of biomarkers, drug development, and personalized healthcare, among others [123]. Pioneering omics data investigations have produced tools for normalization like NOREVA [124–126], and ANPELA, a combined approach for data label-free quantification (LFQ) [126]. Various areas of scientific research have benefited greatly from these techniques.

Using ML-based techniques, previous studies have identified vital biomarkers in the search for genes with greater SCC risk prediction value, and this has helped scientists find compounds that have better predictive value [127]. Scientists are increasingly using AI-based and ML techniques to investigate the genetic diversity of cancer. It can be used to increase the accuracy of diagnoses, generate efficient biomarkers, as well as the success of cancer treatments [128]. AI refers to a robot's capacity to replicate human behavior, which is especially beneficial when working with enormous amounts of data. ML-based is one of the most significant applications of AI models, which enables computers to acquire knowledge through observation without explicit programming [129,130]. ML models can be viewed as a modeling technique that relies on gaining expertise together with performance improvement. These models are meant to assist in locating advantageous components and their relationships [131].

Over the past several years, AI has improved, transitioning from a mainly intellectual to a practical implementation condition. There are now strong prospects for the use of AI in many different industries, ML has already been used to analyze survival in cancer studies, in particular and predict models for advanced nasopharyngeal cancer, breast cancer, pancreatic cancer, and a number of other malignancies [132]. While ML algorithms in particular seem to be effective in producing results and predictions, and they suffer from opacity, which makes it difficult to understand their fundamental workings. It exacerbates the issue since it offers serious risks to entrusting crucial decisions to a system that is unable to adequately defend itself.

To solve this problem, XAI proposes a paradigm change toward more transparent and understandable AI. Its goal is to create a set of tactics that produce more comprehensible models while maintaining high accuracy and achievement. The requirement to justify the model's judgments or forecasts to users and experts led to the creation of XA and it is attracting increasing amounts of interest in AI. Numerous XAI techniques have been established with diverse methodologies, and several classification schemes for XAI models have been presented by authors in [133], based on the size, length of the information extraction process, or model AI.

1.3.5 COVID-19 pandemic

Governments all over the world have adopted policies for social isolation, quarantine, and ultimately lockdowns due to COVID-19's extremely contagious existence [134]. Nigeria was not exempt from the significant health and economic problems brought on by these occurrences [135]. To aid in decision-making on the part of the federal and local governments, it is necessary to test the huge majority of the population. COVID-19 tests are only available to healthcare experts and patients with serious diseases due to a lack of resources. Consequently, it does not reach the vast majority of people. Since the pandemic started, the digital health landscape has captured everyone's interest to offer potential health solutions in this period of unprecedented medical crisis in order to lessen the effects of this epidemic [136].

Although AI techniques have previously been used to support clinical decisions, "Emergency AI" is currently in demand. There are chances for Automation decisions based on gathered vitals, test results, medication orders, and diagnoses throughout the patient care route [137]. There are still crucial factors to take into account while creating and verifying AI models in light of the continuously expanding datasets. AI techniques can be used to comprehend patient categories, direct clinical decision-making, and enhance both patient- and operation-centered results. This viewpoint emphasizes the advantages of these tools as seen in many therapeutic contexts and explains the importance of ML and AI techniques. When thoughtfully constructed, they might be enhanced during the COVID-19 pandemic.

ML models have been shown to predict the presence of clinical factors from medical imaging with remarkable accuracy. However, these complex models can be difficult to interpret and are often criticized as "black boxes". Prediction models that provide no insight into how their predictions are obtained are difficult to trust for making important clinical decisions, such as medical diagnoses or treatment. Explainable ML (XML) methods, such as Shapley values, have made it possible to explain the behavior of ML algorithms and to identify which predictors contribute most to a prediction. Incorporating XML methods into medical software tools has the potential to increase trust in ML-powered predictions and aid physicians in making medical decisions. Specifically, in the field of medical imaging analysis, the most used methods for explaining deep learning-based model predictions are saliency maps that highlight important areas of an image. However, they do not provide a straightforward interpretation of which qualities of an image area are important.

It has been demonstrated that ML-based models can accurately predict the existence of clinical variables from medical imaging [138]. ML models can deliver better predictions thanks to technologies that can capture intricate correlations between features, which may lead to "black box" models that are challenging to understand [138]. These techniques take patient MR images as inputs and output the patient's projected likelihood of a particular result without revealing how the prognosis was made. As a result, even if these techniques can surpass conventional predictive models, a possible deterrent to using these predictions in clinical decision-making is the lack of openness in how they are made. Understanding what variables affect a model's prediction and gaining insight into these ML "black box" models is important. The application of XML techniques is receiving a lot of attention.

Numerous research focus on the diagnosis of COVID-19 by chest imaging, including radiography and computer tomography (CT). While precise in recognizing occurrences that are positive, most of these instruments are unable to distinguish COVID-19 from other lung illnesses [139]. Additionally, studies demonstrate implausible ideal outcomes, which are probably the result of issues with data leakage and a lack of transparency in the experimentation process, thus, making their work neither marketable nor replicable. By examining [140,141], it is obvious that the majority of methods rely on advanced DL and computer vision methods as well as image data. To assist doctors and field experts, model interpretability and prediction explanations are, however, given little to no importance.

A straightforward and incredibly effective ML method for identifying the COVID-19 utilizing basic features based on patient questionnaires was described by authors in [142]. Age, symptoms including fever and sore throat, and proven contact with an infected person were among the factors used. The test sets' results reveal an area under curve (AUC) measure of 0.9, which is regarded as a high standard result in the medical industry. The most important variables, such as cough, fever, and contact with an infected person, were discovered through Shapley additive explanations. This method has two important drawbacks: the first one cannot recognize patients without symptoms; additionally, the second is that it might be biased toward negative cases. Since the authors note that patients with negative test results may underestimate their symptoms, this tool can undoubtedly aid field agents on early trials in better managing COVID-19 tests in limited supply and guiding patient ward placement.

AI-based approaches should be encouraged and implemented as standard practice in the implementation of DL techniques for multimedia categorization, notably for issues with medical diagnosis. In actuality, when increasingly complex neural network topologies emerge, their inability to be debugged and inability to provide human-centered justification for their decisions continue to limit their utility. One of the key instruments in overcoming what continues to be a significant barrier for future AI is XAI, the "black-box technique," which involves creating easily interpretable models to deal with practical problems, to increase human comprehension of AI-based models [143].

In terms of clinical trials and medication repositioning, AI should be an addition to human procedures, to ensure that ML/DL tactics are adapted to particular contexts, it is carried out in multidisciplinary teams. Moreover, it is essential to research and intensify the challenges of building innovative approaches for molecular regeneration, recommended practices for data exchange are also mentioned [144]. Clinical trials and medication repurposing are two topics. Although there are commonalities between clinical research and computationally predicted drug repurposing, notwithstanding the promised applications of AI in this area, we were unable to locate any specific proof that computational guidance was used to conduct clinical trials. As far as we are aware, the bulk of the research analyzed did not follow DL predictions with clinical proof. The collections of possibilities produced by computational approaches for possible pandemics should be encouraging, this can later be put to the test in studies or inspections. However, there is additional work to be done in this area of inquiry.

Regarding text modeling approaches, the majority of them integrated NLP approaches with a specially honed strategy for COVID-19-related data and after that displayed the conclusions drawn from such data. In our perspective, studies on string patterns did not provide anything fresh to test analysis or clinical testing. In terms of time series analysis, taking into account studies employing COVID-19 forecasts, the best model to utilize is not generally agreed upon. The anticipated mortality toll from COVID-19 is fluctuating, and a wide range of forecasting variances have been shown by all prediction models. Sadly, the paucity of information on COVID-19, particularly at the start of the pandemic, and additionally, policy changes have increased the significance of research that uses limited datasets or enhances the precision of mathematical model (such as the SIR model) [145], that anticipate smaller data. There is not an approach that consistently outperforms others, and it depends on preparation and fine-tuning. Utilizing data fusion approaches, as in [146,147], to combine time series from several data sources is a correct approach that we recommend. During the COVID-19 pandemic, Figure 1.5 depicted various areas where AI-based models could be applied.

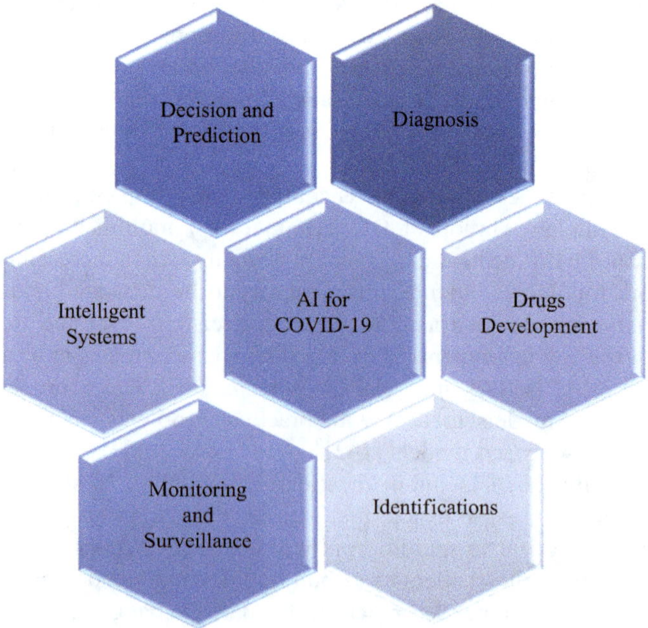

Figure 1.5 Some fundamental functions of AI in combating COVID-19 pandemic

Additionally, pattern identification, forecasting, describing, and classifying medical data are all areas where AI is particularly helpful. Infections with COVID-19 are projected, diagnosed, treated, be improved so that decision-making by government officials, medical professionals, and other policymakers can produce useful results. AI can assist in COVID-19 patient discovery, prediction, prevention, and monitoring of outbreaks, but it cannot replace scientific knowledge. The AI-based model can be used for cleaning up the environment, helping to make monoclonal antibodies and other treatments, managing health care, conducting business and trade, increasing transparency, and formulating policy, among other things.

1.4 XAI research trends and open issues

XAI intends to assist people in comprehending the reasoning behind a machine's choice and determining whether or not it is reliable. XAI is therefore unavoidably a paradigm for bridging AI and human intelligence, with the intention of facilitating and increasing human participants' acceptability of AI systems [148]. XAI can be thought of as "AI for citizens" in this context.

Although intelligent systems have a lot of potential, the concerns about providing such intelligent systems too much power without even being able to adequately explain the decision-making process underlying such complex systems to domain specialists are raised by the XAI research program (doctors, attorneys, financial specialists, etc.) a language and a format that they can understand. This not only clarifies the reasoning behind particular conclusions reached by such technologies but also inspires the study and urges academics to develop more humanistic (human-like) alternatives, and improved knowledge of the brain as natural phenomena for cognitive processing. Furthermore, user rights must be preserved because machines are increasingly replacing humans in making decisions in many aspects of daily life. Machine intelligence still struggle to process complex and deeper knowledge or abstract information unless it is first transformed into an algorithmic format. (attributes, results, and labels).

In several AI application fields, the aforementioned crucial issue has taken on a significant importance. For instance, a patient's intervention may be influenced by a decision made by a computerized diagnosis process; therefore, doctors must comprehend the rationale behind the decision and assess the associated risks. If we take autonomous drones for medical use, the reasons, circumstances, and locations of automated treatment, management, or monitoring by drones must be known to the doctors. Thus, a reliable XAI system becomes a crucial requirement before AI can be used to solve almost any real-world issue. Currently, a lot of study is being done on how to solve these kinds of issues.

Modern AI methods like DL have their origins in the simulation of the human brain. Finding a way to match human intellect is the fundamental objective in order to make DNNs understandable, to figure out a means to create a human-made "brain" that can comprehend, at least at a higher degree of functionality, the neuronal activity in the human brain, connect the multilayered brain knowledge stream processors to the deep structures.

The properties of contemporary orthodox DL and the human brain differ significantly in two ways. First, the human brain lacks the ability to maintain very precise characteristics, more resembling an analog circuit. Second, unlike the meticulously "handcrafted" topologies of the contemporary standard DL, synapses in the human brain are densely linked. Therefore, it is odd that the mainstream DL literature has harsh criticism for supposedly "handcrafted" aspects [149], nevertheless is reluctant to acknowledge that the designs it is advancing are "handcrafted," very problem-specific as well as a variety of contextual, including stride, kernel sizes, the number of layers, and others.

Given the aforementioned issues, XAI can contribute to a mutually advantageous bridge between DL and neurology. On the one hand, neurology and cognitive science can assist in developing XAI models that are logical and easier for humans to comprehend [150,151]. The mechanisms of intelligence in the human brain can also be better understood using XAI models derived from DNNs, on the other hand [152,153]. The quest of fully comprehending how human intellect derives from neurons could be recast as the ultimate objective of XAI.

1.4.1 XAI perspective in healthcare

Putting more of an emphasis on prototype-based models is one possible avenue for future study [154,155] rather than on deeply engrained, abstract frameworks. Models based on prototypes are not new like the work of authors in [156] putting the simplest first (and very effective kNN example), by using ANNs of the RBF variety and IF...THEN criteria. Tibshiran recognized the potency of prototype-based models in the study of authors in [156], but, however, these have not yet been produced in the context of DL, where they can combine a more complex infrastructure with a manner of representation that is easily comprehensible. Despite being effective, the kNN method is not a learning method in the strictest sense because it needs all the data to be present and stored. A certain amount of sparsity is required, which might come from straightforward unsupervised learning methods like clustering or more intricate end-to-end auto-encoders.

There is a widespread misperception that the only type of learning is parameterized learning achieved through cost (or loss) function reduction. In reality, humans pick up prototypes from data samples utilizing resemblance to learn. According to this reasoning, the position becomes the central focus of the learning in prototype-based models. as opposed to the parameters/weights-centered approach that predominates in the mainstream, and the characteristics of the prototypes in the feature/data space. Additionally, there is a fundamental distinction between statistical learning and similarity (i.e., two different methods for analyzing the divergence and difference between two data items). While statistical measures require a large number of data points or samples, similarity can be defined over two (conceivably limitless) number of separate observations for the data.

The construction of Turing's type-B random machines is another viable path (or chaotic machinery) [157,158], additionally random Boltzmann machines, which might result in a generic AI. New cognitive neuroscience results will be included into XAI models, which will legitimize XAI investigation, likewise, such cross-disciplinary application will make XAI valuable for others outside of the field of AI. but potentially assist in resolving century-old problems on how to comprehend human intelligence. The following are open questions in this field: (a) how should the network/model architecture be determined?; (b) how should features be extracted and represented?; (c) what are the most accurate distance measurements, and what are the consequences?; (d) which way of optimization works the best?; and (e) how to select the group of prototypes that best capture the data (if an approach based on prototypes is employed)?

1.5 Conclusion and future directions

XAI is a recent research area that emphasizes ML interpretability and seeks to develop a more open AI. The main objective is to provide a set of interpretable methods and frameworks that produce representations that are easier to comprehend while retaining superior prediction accuracy. Unfortunately, there is not a standardized approach of what is explainable should accomplish. Although some scholars discriminate between the terms interpretability and explainability, others use them identically. The basic objective of XAI is to create a collection of approaches that offer models that are easier to understand while still having highly predicted accuracy. MDSS are computer programs created to influence physician decisions on a range of patients as soon as necessary and, most often, in real-time. The atmosphere for using XAI and MDSS innovation is rapidly changing, and most researchers have tried to see how these innovations is applicable in healthcare systems so as to make their adoptions in healthcare environments acceptable. Therefore, this chapter covers the current requirements and development issues for XAI and MDSS. The summary of potential areas for open issues, and further study comes towards the end of the chapter. The most recent statistics on the application of and effects of MDSSs and XAI in practice were given, and offered recommendations for users to take into account as these systems start to be integrated into commercialized products, and put into practice outside of research and development environments. Therefore, the chapter will provide wider knowledge for both academic researchers and healthcare policy makers and give access to a historical ready-made transcription of HMDSS in the healthcare sector, this can assist them in their future decision-making processes with respect to non-communicable diseases.

Acknowledgment

The work of Agbotiname Lucky Imoize is supported in part by the Nigerian Petroleum Technology Development Fund (PTDF) and in part by the German Academic Exchange Service (DAAD) through the Nigerian-German Postgraduate Program under grant 57473408.

References

[1] M. Ghassemi, L. Oakden-Rayner, and A. L. Beam, "The false hope of current approaches to explainable artificial intelligence in health care," *Lancet Digit. Heal.*, vol. 3, no. 11, pp. e745–e750, 2021, doi: 10.1016/S2589-7500(21)00208-9.

[2] C. C. Yang, "Explainable artificial intelligence for predictive modeling in healthcare," *J. Healthc. Informatics Res.*, vol. 6, no. 2, pp. 228–239, 2022, doi: 10.1007/s41666-022-00114-1.

[3] J. Amann, A. Blasimme, E. Vayena, D. Frey, and V. I. Madai, "Explainability for artificial intelligence in healthcare: a multidisciplinary perspective," *BMC Med. Inform. Decis. Mak.*, vol. 20, no. 1, pp. 1–9, 2020, doi: 10.1186/s12911-020-01332-6.

[4] U. Pawar, D. O'Shea, S. Rea, and R. O'Reilly, "Explainable AI in healthcare," In *2020 International Conference on Cyber Situational Awareness, Data Analytics and Assessment, Cyber SA* 2020, July, 2020, doi: 10.1109/CyberSA49311.2020.9139655.

[5] O. Biran and C. Cotton, "Explanation and justification in machine learning: a survey," In *IJCAI-17 Work. Explain. AI*, pp. 8–13, 2017.

[6] D. Wang, Q. Yang, A. Abdul, B. Y. Lim, and U. States, "Designing theory-driven user-centric explainable AI," In *CHI*, pp. 1–15, 2019.

[7] J. A. Akrimi, A. R. Ahmad, L. E. George, and S. Aziz, "Review of artificial intelligence," *Int. J. Sci. Res.*, India Online, vol. 2, no. 2, pp. 487–505, 2013, doi: 10.32628/ijsrset1207625.

[8] U. Pawar, D. O'Shea, S. Rea, and R. O'Reilly, "Incorporating explainable artificial intelligence (XAI) to aid the understanding of machine learning in the healthcare domain," *CEUR Workshop Proc.*, vol. 2771, pp. 169–180, 2020.

[9] N. H. Alharbi, R. O. Bameer, S. S. Geddan, and Hajar M. Alharbi, "Recent advances and machine learning techniques on sickle cell disease," *Fut. Comput. Informat. J.*, vol. 5, no. 1, pp. 46–59, 2020, doi: 10.54623/fue.fcij.5.1.4.

[10] M. J. Panaggio, D. M. Abrams, F. Yang, T. Banerjee, and N. R. Shah, "Can subjective pain be inferred from objective physiological data? Evidence from patients with sickle cell disease," *PLoS Comput. Biol.*, vol. 17, no. 3, pp. 1–12, 2021, doi: 10.1371/journal.pcbi.1008542.

[11] Y. Zhang, Y. Weng, and J. Lund, "Applications of explainable artificial intelligence in diagnosis and surgery," *Diagnostics*, vol. 12, no. 237, pp. 1–18, 2022, doi: 10.3390/diagnostics12020237.

[12] I. I. Kukhtevich, V. V. Goryunova, T. I. Goryunova, and P. S. Zhilyaev, "Medical decision support systems and semantic technologies in healthcare," In *Advances in Economics, Business and Management Research*, 2020, vol. 148, pp. 370–375, doi: 10.2991/aebmr.k.200730.068.

[13] R. T. Sutton, D. Pincock, D. C. Baumgart, D. C. Sadowski, R. N. Fedorak, and K. I. Kroeker, "An overview of clinical decision support systems: benefits, risks, and strategies for success," *Digit. Med.*, vol. 3, no. 1, pp. 1–10, 2020, doi: 10.1038/s41746-020-0221-y.

[14] S. Aljarboa and S. J. Miah, "Acceptance of clinical decision support systems in Saudi healthcare organisations," *Inf. Dev.*, pp. 1–21, 2021, doi: 10.1177/02666669211025076.

[15] Vi. Bellotti and K. Edwards, "Intelligibility and accountability – human considerations in context-aware systems," *Hum. Comput. Interact. – Taylor Fr.*, vol. 10, pp. 193–212, 2009, doi: 10.1207/S15327051HCI16234.

[16] V. Petrauskas, G. Damuleviciene, A. Dobrovolskis, J. Dovydaitis, and A. Janaviciute, "XAI-based medical decision support system model," *Int. J. Sci. Res. Publ.*, vol. 10, no. 12, pp. 598–607, 2020, doi: 10.29322/IJSRP.10.12.2020.p10869.

[17] T. Jodie, S. Martins, M. Michel, *et al.*, "Evaluation of the acceptability and usability of a decision support system to encourage safe and effective use of opioid therapy for chronic, noncancer pain by primary care providers," *Pain Med.*, vol. 11, no. 4, pp. 575–585, 2010, http://0-ovidsp.ovid.com.lib.exeter.ac.uk/ovidweb.cgi?
T=JS&PAGE=reference&D=emed12&NEWS=N&AN=358905180

[18] J. Kim, Y. M. Chae, S. Kim, S. H. Ho, H. H. Kim, and C. B. Park, "A study on user satisfaction regarding the clinical decision support system (MDSS) for medication," *Healthc. Inform. Res.*, vol. 18, no. 1, pp. 35–43, 2012, doi: 10.4258/hir.2012.18.1.35.

[19] J. Razmak, C. H. Bélanger, and W. Farhan, "Managing patients' data with clinical decision support systems: a factual assessment," *J. Decis. Syst. – Taylor Fr.*, vol. 27, no. 3, pp. 123–145, 2018, doi: 10.1080/12460125.2018.1533159.

[20] D. Zikos and N. Delellis, "MDSS-RM: a clinical decision support system reference model," *BMC Med. Res. Methodol.*, vol. 18, no. 1, pp. 1–14, 2018, doi: 10.1186/s12874-018-0587-6.

[21] Y. Li, C. Bai, and C. K. Reddy, "A distributed ensemble approach for mining healthcare data under privacy constraints," *Inf. Sci. (Ny).*, vol. 330, pp. 245–259, 2016, doi: 10.1016/j.ins.2015.10.011.

[22] T.-Y. Leong, "Decision support systems in healthcare: emerging trends and success factors," In *Appl. Decis. Support with Soft Comput.*, New York, NY: Springer, pp. 151–179, 2003, doi: 10.1007/978-3-540-37008-6_6.

[23] A. X. Garg, N. K. J. Adhikari, H. McDonald, *et al.*, "Effects of computerized clinical decision support systems on practitioner performance and patient outcomes: a systematic review," *J. Am. Med. Assoc.*, vol. 293, no. 10, pp. 1223–1238, 2005, doi: 10.1001/jama.293.10.1223.

[24] M. Alther and C. K. Reddy, "Clinical decision support systems," *OR Insight*, vol. 10, no. 2, pp. 18–32, 2007, doi: 10.1057/ori.1997.9.

[25] D. Dinevski, U. Bele, and T. Sarenac, "Clinical decision support systems," *Telemed. Tech. Appl.*, vol. 37, no. 6, pp. 491–498, 2021, doi: 10.1159/000519420.

[26] M. A. Musen, Y. Shahar, and E. H. Shortliffe, "Clinical decision support systems," In *Health Information System*, second ed., pp. 221–235, 2002, doi: 10.32628/cseit1952264.

[27] K. Sadegh-Zadeh, “Clinical decision support systems: a virtual survey,” *Philos. Med.*, vol. 119, pp. 705–722, 2015, doi: 10.1007/978-94-017-9579-1_20.

[28] P. Status, U. Acceptance, and I. Diseases, “Health evaluation through logical programming (HELP),” *Clinfowiki*, pp. 1–4, 2015, http://clinfowiki.org/wiki/index.php/Health_Evaluation_through_Logical_Programming_(HELP)

[29] J. M. Kirigia, A. Seddoh, D. Gatwiri, L. H. K. Muthuri, and J. Seddoh, “E-health: determinants, opportunities, challenges and the way forward for countries in the WHO African Region,” *BMC Public Health*, vol. 5, no. 137, pp. 1–11, 2005, doi: 10.1186/1471-2458-5-137.

[30] C. Division, “Implementing e-health in developing countries: guidance and principles,” In *International Telecommunication Union*, September, pp. 1–53, 2008.

[31] WHO, “An assessment of e-health projects and initiatives in Africa,” In *E-Health Telemed. Wolrd Heal. Organ.*, pp. 1–35, 2010.

[32] WHO, “The Promise of eHealth in the African Region,” Press Release, September, 2013, Available: http://www.afro.who.int/en/media-centre/pressreleases/item/5816-the-promise-of-ehealth-in-the-african-region.html

[33] I. Sim and A. Berling, “A framework for classifying decision support systems,” In *AMIIA 2003 Symposium*, 2003, pp. 599–603.

[34] E. H. Shortliffe and J. J. Cimino, *Biomedical Informatics – Computer Application in Healthcare and Biomedicine*, vol. 3. 2006.

[35] C. Grosan and A. Abraham, *Machine Learning*, vol. 17, 2011, doi:10.1007/978-3-642-21004-4_10.

[36] R. L. Kumar, In Wang Y, Poongodi T, Imoize AL, editors (eds.) , *Internet of Things, Artificial Intelligence and Blockchain Technology*, Cham: Springer; 2021.

[37] G. Roth, “Machine learning with Python: an introduction,” JavaWorld, pp. 1–5, 2019, https://www.javaworld.com/article/3322898/application-development/machine-learning-with-python-an-introduction.html

[38] O. B. Ayoade, *Comparative Analysis of Selected Machine Learning Algorithms for Predicting Sickle Cell Disease*, Department Comput. Sci. Fac. Commun. Inf. Sci. Univ. Ilorin, Kwara State, Niger, December, pp. 1–270, 2021.

[39] N. I. of H. NIH, *Health Information for the Public – Sickle Cell Disease (SCD),* National Heart Lung and Blood Institute, 2016.

[40] N. I. of H. NIH, *The Management of Sickle Cell Disease*, Natl. Hear. Lung Blood Inst., no. 02–2117, pp. 1–206, 2015, Available: http://www.nhlbi.nih.gov

[41] L. M. Gunder and S. A. Martin, *Essentials of Medical Genetics for Health Professionals*, USA: Jones & Bartlett Learning, LLC, 2011.

[42] S. E. Roger and H. R. Rodney, “Some medical and social aspects of the treatment for genetic-metabolic diseases,” *Ann. Am. Acad. Polit. Soc. Sci.*, vol. 399, pp. 30–37, 2017.

[43] M. Saad and Z. Salem, “Basic concepts of medical genetics, formal genetics,” *Egypt. J. Med. Hum. Genet.*, vol. 15, no. 1, pp. 99–101, 2014, doi: 10.1016/j.ejmhg.2013.10.001.

[44] World-Health-Organization, "Sickle-cell anemia," *World Heal. Organ.*, vol. 11, no. April, pp. 1–5, 2020.

[45] P. L. Stephenson, M. V. Taylor, and C. Anglin, "Sickle cell disease," *J. Consum. Health Internet*, vol. 19, no. 2, pp. 122–131, 2015, doi: 10.1080/15398285.2015.1026706.

[46] M. W. Darlison and B. Modell, "Sickle-cell disorders: limits of descriptive epidemiology," *Lancet (London, England)*, vol. 381, no. 9861, pp. 98–9, 2013, doi: 10.1016/S0140-6736(12)61817-0.

[47] J. R. Frost, R. K. Cherry, S. O. Oyeku, *et al.*, "Improving sickle cell transitions of care through health information technology," *Am. J. Prev. Med.*, vol. 51, pp. 17–23, 2016, doi: 10.1016/j.amepre.2016.02.004.

[48] C. P. Rivera, A. Veneziani, R. E. Ware, and M. O. Platt, "Sickle cell anemia and pediatric strokes: computational fluid dynamics analysis in the middle cerebral artery," *Exp. Biol. Med.*, vol. 241, pp. 755–765, 2016, doi: 10.1177/1535370216636722.

[49] J. Kanter and R. Kruse-Jarres, "Management of sickle cell disease from childhood through adulthood," *Blood Rev.*, vol. 27, no. 6, pp. 279–87, 2013, doi: 10.1016/j.blre.2013.09.001.

[50] S. D. Grosse, I. Odame, H. K. Atrash, D. D. Amendah, F. B. Piel, and T. N. Williams, "Sickle cell disease in Africa: a neglected cause of early childhood mortality," *Am. J. Prev. Med.*, vol. 41, no. 6 SUPPL.4, pp. S398–S405, 2011, doi: 10.1016/j.amepre.2011.09.013.

[51] O. S. Platt, D J. Brambella, W. F. Rose, *et al.*, "Mortality in sickle cell disease-life expectancy & risk factors," *N. Engl. J. Med.*, vol. 330, no. 23, pp. 1639–1644, 2012.

[52] D. Divya, K. N. Rao, Si. G. Ratnam, and D. Sowjanya, "Supervised machine learning algorithms for analysis on sickle cell anemia," *High Technol. Lett.*, vol. 26, no. 11, pp. 994–1004, 2020.

[53] T. M. Sabu, "Bioinformatics," In *Fundam. Concepts Bioinforma.*, pp. 1–155, 2003.

[54] A. D. Hardie, L. Ramos-Duran, and J. U. Schoepf, "Cardiac MR assessment of myocardial iron deposition in sickle cell disease: risk factors and association with cardiac function," *J. Cardiovasc. Magn. Reson.*, vol. 1, pp. 48–48, 2010, doi: 10.1186/1532-429X-12-S1-P274.

[55] G. D. Magoulas and A. Prentza, *Machine Learning in Medical Applications*, New York, NY: Springer, vol. 204, no. 9, pp. 300–307, 2015, doi: 10.1007/3-540-44673-7.

[56] R. V. Gardner, "Sickle cell disease: advances in treatment," *Ochsner J.*, vol. 18, no. 4, pp. 377–389, 2018, doi: 10.31486/toj.18.0076.

[57] C. O'Keeffe, S. Mason, R. Jacques, and J. Nicholl, "Characterising non-urgent users of the emergency department (ED): a retrospective analysis of routine ED data," *PLoS One*, vol. 13, no. 2, p. e0192855, 2018.

[58] K. Eastwood, K. Smith, A. Morgans, and J. Stoelwinder, "Appropriateness of cases presenting in the emergency department following ambulance

service secondary telephone triage: a retrospective cohort study," *BMJ Open*, vol. 7, no. 10, p. e016845, 2017.

[59] K. E. Deligiannidis, "Primary care issues in rural populations," *Phys. Assist. Clin.* vol. 4, no. 1, pp. 11–19, 2019.

[60] W. S. Eidson-Ton, J. Rainwater, D. Hilty, *et al.*, "Training medical students for rural, underserved areas: a rural medical education program in California," *J. Health Care Poor Underserved*, vol. 27, no. 4, pp. 1674–1688, 2016.

[61] J. Baumbach and H. H. Schmidt, "The end of medicine as we know it: introduction to the new journal, systems medicine," *Syst. Med.* vol. 1, no. 1, pp. 1–2, 2018.

[62] I. D. Federation and I. D. Atlas, *International Diabetes Federation. IDF Diabetes Atlas*, 6th ed, Brussels: International Diabetes Federation, 2013.

[63] I. D. Oladipo, A. O. Babatunde, J. B. Awotunde, and M. Abdulraheem, "An improved hybridization in the diagnosis of diabetes mellitus using selected computational intelligence," In *International Conference on Information and Communication Technology and Applications* 2020 Nov 24, Cham: Springer, pp. 272–285.

[64] World Health Organization. World Health Organization Diabetes Fact Sheet. 2018 [Updated 2018 October 30].

[65] Centers for Disease Control and Prevention. National Diabetes Fact Sheet: National Estimates and General Information on Diabetes and Prediabetes in the United States, 2011. Atlanta, GA: US Department of Health and Human Services, Centers for Disease Control and Prevention, 2011; vol. 201, no. 1, pp. 2568–2569.

[66] Centers for Disease Control and Prevention (CDC). National Diabetes Statistics Report, 2014. Centers for Disease Control and Prevention. US Department of Health and Human Services, Atlanta, GA, 2014.

[67] V. J. Briscoe and S. N. Davis, "Hypoglycemia in type 1 and type 2 diabetes: physiology, pathophysiology, and management," *Clin. Diabetes,* vol. 24, no. 3, pp. 115–121, 2006.

[68] N. N. Zammitt and B. M. Frier, "Hypoglycemia in type 2 diabetes: pathophysiology, frequency, and effects of different treatment modalities," *Diabetes Care*, vol. 28, no.12, pp. 2948–2961, 2005.

[69] P. F. Hsu, S. H. Sung, H. M. Cheng, *et al.*, "Association of clinical symptomatic hypoglycemia with cardiovascular events and total mortality in type 2 diabetes: a nationwide population-based study," *Diabetes Care*, vol. 36, no. 4, pp. 894–900, 2013.

[70] P. Vexiau, P. Mavros, G. Krishnarajah, R. Lyu, and D. Yin, "Hypoglycaemia in patients with type 2 diabetes treated with a combination of metformin and sulphonylurea therapy in France," *Diabetes Obesity Metabol.* vol. 10, pp. 16–24, 2008.

[71] L. Jönsson, B. Bolinder, and J. Lundkvist, "Cost of hypoglycemia in patients with Type 2 diabetes in Sweden," *Value Health*, vol. 9, no. 3, pp. 193–198, 2006.

[72] X. Li, S. Yu, Z. Zhang, *et al.*, "Predictive modeling of hypoglycemia for clinical decision support in evaluating outpatients with diabetes mellitus," *Curr. Med. Res. Opin.*, vol. 35, no. 11, pp. 1885–1891, 2019.

[73] Ö. Kart, V. Mevsim, A. Kut, İ. Yürek, A. Ö. Altın, and O. Yılmaz, "A mobile and web-based clinical decision support and monitoring system for diabetes mellitus patients in primary care: a study protocol for a randomized controlled trial," *BMC Med. Inform. Dec. Mak.* vol. 17, no. 1, pp. 1-0, 2017.

[74] P. Aggleton and H. Chalmers, "Models and theories. Five. Orem's self-care model," *Nursing Times*, vol. 81, no. 1, pp. 36–39, 1985.

[75] M. J. Pearce, K. Pereira, and E. Davis, "The psychological impact of diabetes: a practical guide for the nurse practitioner," *J. Am. Assoc. Nurse Practition.* vol. 25, no. 11, pp. 578–583, 2013.

[76] K. A. Weinger, J. O. Yi, F. R. Pouwer, H. E. Ad, and F. J. Snoek, "The confidence in diabetes selfcare scale," *Diabetes Care*, vol. 26, no. 3, pp. 713–718, 2003.

[77] İ. Satman, Ş. İmamoğlu, C. Yılmaz, S. Akalın, S. Salman, and N. Dinççağ, "Diabetes Mellitus ve komplikasyonlarının tanı, tedavi ve izlem kılavuzu," Miki Matbacılık: Türkiye Endokrinoloji ve Metabolizma Derneği, Ankara, Mayıs, 2014, pp. 15–25.

[78] American Diabetes Association, "Standards of medical care in diabetes—2015 abridged for primary care providers," *Clin. Diabetes: Publ. Am. Diabetes Assoc.*, vol. 33, no. 2, p. 97, 2015.

[79] P. S. Roshanov, S. Misra, H. C. Gerstein, *et al.*, "Computerized clinical decision support systems for chronic disease management: a decision-maker-researcher partnership systematic review," *Implement. Sci.*, vol. 6, no. 1, pp. 1–6, 2011.

[80] R. Anchala, E. Di Angelantonio, D. Prabhakaran, and O. H. Franco, "Development and validation of a clinical and computerised decision support system for management of hypertension (DSS-HTN) at a primary health care (PHC) setting," *PLoS One,* vol. 8, no. 11, p. e79638, 2013.

[81] B. T. Kurian, M. H. Trivedi, B. D. Grannemann, C. A. Claassen, E. J. Daly, and P. Sunderajan, "A computerized decision support system for depression in primary care," *Primary Care Companion J. Clin. Psychiatry,* vol. 11, no. 4, 140, 2009.

[82] M. C. Meulendijk, M. R. Spruit, P. A. Jansen, M. E. Numans, and S. Brinkkemper, "STRIPA: A rule-based decision support system for medication reviews in primary care," *ECIS 2015 Research-in-Progress Papers*, 2015, pp. 1–13.

[83] S. Rahaman, "Diabetes diagnosis decision support system based on symptoms, signs and risk factor using special computational algorithm by rule base," In *2012 15th International Conference on Computer and Information Technology* (*ICCIT*), 2012 Dec 22. New York, NY: IEEE, pp. 65–71.

[84] S. F. Jaafar and D. M. Ali, "Diabetes mellitus forecast using artificial neural network (ANN)," In *2005 Asian Conference on Sensors and the*

International Conference on New Techniques in Pharmaceutical and Biomedical Research, 2005 Sep 5. New York, NY: IEEE, pp. 135–139.

[85] J. A. Quinlivan, L. T. Lam, and J. Fisher, "A randomised trial of a four-step multidisciplinary approach to the antenatal care of obese pregnant women," *Austral. New Zealand J. Obst. Gynaecol.* vol. 51, no. 2, pp. 141–146, 2011.

[86] Y. Sun and H. Zhao, "The effectiveness of lifestyle intervention in early pregnancy to prevent gestational diabetes mellitus in Chinese overweight and obese women: a quasi-experimental study," *Appl. Nurs. Res.*, vol. 30, pp. 125–130, 2016.

[87] C. Wang, Y. Wei, X. Zhang, *et al.*, "A randomized clinical trial of exercise during pregnancy to prevent gestational diabetes mellitus and improve pregnancy outcome in overweight and obese pregnant women," *Am. J. Obst. Gynecol.*, vol. 216, no. 4, pp. 340–351, 2017.

[88] D. M. , S. Nikpay, and R. S. Huckman, "The business of medicine in the era of COVID-19," *Jama*, vol. 323, no. 20, pp. 2003–2004, 2020.

[89] E. H. Shortliffe and M. J. Sepúlveda, "Clinical decision support in the era of artificial intelligence," *Jama*, vol. 320, no. 21, pp. 2199–2200, 2018.

[90] A. M. Antoniadi, Y. Du, Y. Guendouz, *et al.*, "Current challenges and future opportunities for XAI in machine learning-based clinical decision support systems: a systematic review," *Appl. Sci.*, vol. 11, no. 11, p. 5088, 2021.

[91] R. Caruana, Y. Lou, J. Gehrke, P. Koch, M. Sturm, and N. Elhadad, "Intelligible models for healthcare: predicting pneumonia risk and hospital 30-day readmission," In *Proceedings of the 21th ACM SIGKDD International Conference on Knowledge Discovery and Data Mining*, 2015, pp. 1721–1730.

[92] World Health Organization, *A Global Brief on Hypertension: Silent Killer, Global Public Health Crisis: World Health Day 2013*, Geneva: World Health Organization, 2013..

[93] S. Akbar, M. U. Akram, M. Sharif, A. Tariq, and U. ullah Yasin, "Arteriovenous ratio and papilledema based hybrid decision support system for detection and grading of hypertensive retinopathy," *Comput. Methods Prog. Biomed.*, vol. 154, pp. 123–141, 2018.

[94] T. Y. Wong and P. Mitchell, "Hypertensive retinopathy," *N. Eng. J. Med.* vol. 351, no. 22, pp. 2310–2317, 2004.

[95] G. Mancia, R. Fagard, K. Narkiewicz, *et al.*, "Practice guidelines for the management of arterial hypertension of the European Society of Hypertension (ESH) and the European Society of Cardiology (ESC): ESH/ESC Task Force for the Management of Arterial Hypertension," *J. Hypertens*, vol. 31, no. 10, pp. 1925–1938, 2013.

[96] C. Guerrero-García and A. F. Rubio-Guerra, "Combination therapy in the treatment of hypertension," *Drugs Context*, vol. 7, pp. 212531, 2018.

[97] J. C. Parr and G. F. Spears, "Mathematic relationships between the width of a retinal artery and the widths of its branches," *Am. J. Ophthalmol.*, vol. 77, no. 4, pp. 478–483, 1974.

[98] S. Kazeminia, C. Baur, A. Kuijper, *et al.*, "GANs for medical image analysis," *Artif. Intell. Med.*, vol. 109, pp. 101938, 2020.

[99] K. Mittal and V. Rajam, "Computerized retinal image analysis-a survey," *Multimed. Tools Appl.*, vol. 79, no. 31, pp. 22389–22421, 2020.

[100] M. Badar, M. Haris, and A. Fatima, "Application of deep learning for retinal image analysis: a review," *Comput. Sci. Rev.*, vol. 35, pp. 100203, 2020.

[101] D. L. Castro, D. Tegolo, and C. Valenti, "A visual framework to create photorealistic retinal vessels for diagnosis purposes," *J. Biomed. Inform.*, vol. 108, pp. 103490, 2020.

[102] N. Tajbakhsh, L. Jeyaseelan, Q. Li, J. N. Chiang, Z. Wu, and X. Ding, "Embracing imperfect datasets: a review of deep learning solutions for medical image segmentation," *Med. Image Anal.*, vol. 63, pp. 101693, 2020.

[103] Q. Fu, S. Li, and X. Wang, "Mscnn-am: a multi-scale convolutional neural network with attention mechanisms for retinal vessel segmentation," *IEEE Access*, vol. 8, pp. 163926–163936.

[104] Y. Guo and Y. Peng, "BSCN: bidirectional symmetric cascade network for retinal vessel segmentation," *BMC Med. Imaging*, vol. 20, no. 1, pp. 1–22, 2020.

[105] M. Miri, Z. Amini, H. Rabbani, and R. Kafieh, "A comprehensive study of retinal vessel classification methods in fundus images," *J. Med. Signals Sens.*, vol. 7, no. 2, pp. 59, 2017.

[106] W. Dai, N. Dong, Z. Wang, X. Liang, H. Zhang, and E. P. Xing, "Scan: structure correcting adversarial network for organ segmentation in chest x-rays," In *Deep Learning in Medical Image Analysis and Multimodal Learning for Clinical Decision Support,* Cham: Springer, pp. 263–273, 2018.

[107] Y. D. Kim, K. J. Noh, S. L. Byun, *et al.*, "Effects of hypertension, diabetes, and smoking on age and sex prediction from retinal fundus images," *Sci. Rep.*, vol. 10, no. 1, pp. 1–4, 2020.

[108] M. Arsalan, M. Owais, T. Mahmood, J. Choi, and K. R. Park, "Artificial intelligence-based diagnosis of cardiac and related diseases," *J. Clin. Med.*, vol. 9, no. 3, pp. 871, 2020.

[109] M. Owais, M. Arsalan, T. Mahmood, J. K. Kang, and K. R. Park, "Automated diagnosis of various gastrointestinal lesions using a deep learning-based classification and retrieval framework with a large endoscopic database: model development and validation," *J. Med. Internet Res.*, vol. 22, no. 11, pp. e18563, 2020.

[110] N. Tsiknakis, D. Theodoropoulos, G. Manikis, *et al.*, "Deep learning for diabetic retinopathy detection and classification based on fundus images: a review," *Comput. Biol. Med.*, vol. 135, pp. 104599, 2021.

[111] W. L. Alyoubi, W. M. Shalash, and M. F. Abulkhair, "Diabetic retinopathy detection through deep learning techniques: a review," *Inform. Med. Unlocked*, vol. 20, pp. 100377, 2020.

[112] M. Cives, F. Mannavola, L. Lospalluti, *et al.*, "Non-melanoma skin cancers: biological and clinical features," *Int. J. Mol. Sci.,* vol. 21, no. 15, pp. 5394, 2020.

[113] U. Leiter and C. Garbe, "Epidemiology of melanoma and nonmelanoma skin cancer—the role of sunlight," In *Sunlight, Vitamin D and Skin Cancer*, New York, NY: Springer, 2008, pp. 89–103.

[114] A. Esteva, B. Kuprel, R. A. Novoa, *et al.*, "Dermatologist-level classification of skin cancer with deep neural networks," *Nature*, vol. 542, no. 7639, pp. 115–118, 2017.

[115] H. M. Gloster Jr and K. Neal, "Skin cancer in skin of color," *J. Am. Acad. Dermatol.*, vol. 55, no. 5, pp. 741–760, 2006.

[116] D. Melzer, L.C. Pilling, and L. Ferrucci, "The genetics of human ageing," *Nat. Rev. Genet.*, vol. 21, no. 2, pp. 88–101, 2020.

[117] V. R. Yanofsky, S. E. Mercer, and R. G. Phelps, "Histopathological variants of cutaneous squamous cell carcinoma: a review," *J. Skin Cancer*, vol. 2011, 210813, 2011.

[118] A. Gluba, J. Rysz, and T. Pietrucha, "Microarray technology in the study of genetic determinants of cardiovascular diseases," *Central Eur. J. Med.*, vol. 4, no. 1, pp. 1–10, 2009.

[119] Q. X. Yang, Y. X. Wang, F. C. Li, *et al.*, "Identification of the gene signature reflecting schizophrenia's etiology by constructing artificial intelligence-based method of enhanced reproducibility," *CNS Neurosci. Ther.*, vol. 25, no. 9, pp. 1054–1063, 2019.

[120] S. Cui, Q. Wu, J. West, and J. Bai J, "Machine learning-based microarray analyses indicate low-expression genes might collectively influence PAH disease," *PLoS Comp. Biol.*, vol. 15, no. 8, pp. e1007264, 2019.

[121] S. O. Folorunso, J. B. Awotunde, N. O. Adeboye, and O. E. Matiluko, "Data classification model for COVID-19 pandemic," In *Advances in Data Science and Intelligent Data Communication Technologies for COVID-19 2022*, Cham: Springer, pp. 93–118, 2022.

[122] S. O. Folorunso, J. B. Awotunde, F. E. Ayo, and K. K. Abdullah, "RADIoT: the unifying framework for iot, radiomics and deep learning modeling," In *Hybrid Artificial Intelligence and IoT in Healthcare,* Singapore: Springer, pp. 109–128, 2021.

[123] A. Cambiaghi, M. Ferrario, and M. Masseroli, "Analysis of metabolomic data: tools, current strategies and future challenges for omics data integration," *Brief. Bioinform.*, vol. 18, no. 3, pp. 498–510, 2017.

[124] J. Fu, Y. Zhang, Y. Wang, *et al.*, "Optimization of metabolomic data processing using NOREVA," *Nat. Protocols*, vol. 17, no. 1, pp. 129–151, 2022.

[125] B. Li, J. Tang, Q. Yang, *et al.*, NOREVA: normalization and evaluation of MS-based metabolomics data," *Nucleic Acids Res.*, vol. 45, no. W1, pp. W162–W170, 2017.

[126] J. Tang, J. Fu, Y. Wang, *et al.*, "ANPELA: analysis and performance assessment of the label-free quantification workflow for metaproteomic studies," *Brief. Bioinform.*, vol. 21, no. 2, pp. 621–636, 2020.

[127] M. X. Li, X. M. Sun, W. G. Cheng, *et al.*, "Using a machine learning approach to identify key prognostic molecules for esophageal squamous cell carcinoma," *BMC Cancer*, vol. 21, no. 1, pp. 1–1, 2021.

[128] Z. Dlamini, F. Z. Francies, R. Hull, and R. Marima, "Artificial intelligence (AI) and big data in cancer and precision oncology," *Comput. Struct. Biotechnol. J.*, vol. 18, pp. 2300–2311, 2020.

[129] T. Davenport, A. Guha, D. Grewal, and T. Bressgott, "How artificial intelligence will change the future of marketing," *J. Acad. Market. Sci.*, vol. 48, no. 1, pp. 24–42, 2020.

[130] Y. Dong, J. Hou, N. Zhang, and M. Zhang, "Research on how human intelligence, consciousness, and cognitive computing affect the development of artificial intelligence," *Complexity*, vol. 2020, pp. 1–10, 2020.

[131] A. Bohr and K. Memarzadeh, "The rise of artificial intelligence in healthcare applications," In *Artificial Intelligence in Healthcare*, Academic Press, pp. 25–60, 2020.

[132] I. H. Sarker, "Machine learning: algorithms, real-world applications and research directions," *SN Comput. Sci.*, vol. 2, no. 3, pp. 1–21, 2021.

[133] A. Adadi and M. Berrada, "Peeking inside the black-box: a survey on explainable artificial intelligence (XAI)," *IEEE Access*, vol. 6, pp. 52138–52160, 2018.

[134] J. B. Awotunde, S. Oluwabukonla, C. Chakraborty, A. K. Bhoi, and G. J. Ajamu, "Application of artificial intelligence and big data for fighting COVID-19 pandemic," *Decis. Sci. COVID-19*, pp. 3–26, 2022.

[135] S. O. Folorunso, J. B. Awotunde, O. O. Banjo, E. A. Ogundepo, and N. O. Adeboye, "Comparison of active COVID-19 cases per population using time-series models," *Int. J. E-Health Med. Commun. (IJEHMC).*, vol. 13, no. 2, pp. 1–21, 2021.

[136] J. B. Awotunde, R. G. Jimoh, M. AbdulRaheem, I. D. Oladipo, S. O. Folorunso, and G. J. Ajamu, IoT-based wearable body sensor network for COVID-19 pandemic. In *Advances in Data Science and Intelligent Data Communication Technologies for COVID-19*, 2022, pp. 253–275.

[137] S. Debnath, D. P. Barnaby, K. Coppa, *et al.*, "Machine learning to assist clinical decision-making during the COVID-19 pandemic," *Bioelectron. Med.*, vol. 6, no. 1, pp. 1–8, 2020.

[138] M. M. Banoei, R. Dinparastisaleh, A. V. Zadeh, and M. Mirsaeidi, "Machine-learning-based COVID-19 mortality prediction model and identification of patients at low and high risk of dying," *Critical Care*, vol. 25, no. 1, pp. 1–4, 2021.

[139] M. Ghaderzadeh and F. Asadi, "Deep learning in the detection and diagnosis of COVID-19 using radiology modalities: a systematic review," *J. Healthc. Eng.*, vol. 2021, pp. 1–10, 2021.

[140] H. Mohammad-Rahimi, M. Nadimi, A. Ghalyanchi-Langeroudi, M. Taheri, and S. Ghafouri-Fard, "Application of machine learning in diagnosis of COVID-19 through X-ray and CT images: a scoping review," *Front. Cardiovasc. Med.*, vol. 8, pp. 638011, 2021.

[141] M. Roberts, D. Driggs, M. Thorpe, *et al.*, "Common pitfalls and recommendations for using machine learning to detect and prognosticate for COVID-19 using chest radiographs and CT scans," *Nat. Mach. Intell.*, vol. 3, no. 3, pp. 199–217, 2021.

[142] Y. Zoabi, S. Deri-Rozov, and N. Shomron, "Machine learning-based prediction of COVID-19 diagnosis based on symptoms," *NPJ Digital Med.*, vol. 4, no. 1, pp. 1–5, 2021.

[143] L. Longo, R. Goebel, F. Lecue, P. Kieseberg, and A. Holzinger, "Explainable artificial intelligence: concepts, applications, research challenges and visions," In *International Cross-Domain Conference for Machine Learning and Knowledge Extraction 2020 Aug 25*, Cham: Springer, pp. 1–16.

[144] J. B. Awotunde, S. A. Ajagbe, M. A. Oladipupo, *et al.*, "An improved machine learnings diagnosis technique for COVID-19 pandemic using chest X-ray images," In *International Conference on Applied Informatics 2021 Oct 28*, Cham: Springer, pp. 319–330, 2021.

[145] J. B. Awotunde, R. O. Ogundokun, A. E. Adeniyi, K. M. Abiodun, and G. J. Ajamu, "Application of mathematical modelling approach in COVID-19 transmission and interventions strategies," In *Modeling, Control and Drug Development for COVID-19 Outbreak Prevention*, Cham: Springer, pp. 283–314, 2022.

[146] K. M. Abiodun, J. B. Awotunde, D. R. Aremu, and E. A. Adeniyi, "Explainable AI for fighting COVID-19 pandemic: opportunities, challenges, and future prospects," In *Computational Intelligence for COVID-19 and Future Pandemics*, 2022, pp. 315–32.

[147] Y. Kim, P. C. Kyriakidis, and N. W. Park, "A cross-resolution, spatiotemporal geostatistical fusion model for combining satellite image time-series of different spatial and temporal resolutions," *Remote Sensing*, vol. 12, no. 10, pp. 1553, 2020.

[148] P. P. Angelov, E. A. Soares, R. Jiang, N. I. Arnold, and P.M. Atkinson, "Explainable artificial intelligence: an analytical review," *Wiley Interdiscip Rev. Data Mining Knowl. Discov.*, vol. 11, no. 5, pp. e1424, 2021.

[149] I. J. Goodfellow, J. Pouget-Abadie, M. Mirza, *et al*, *Generative Adversarial Nets (Advances in Neural Information Processing Systems)*, Red Hook, NY: Curran, pp. 2672–2680, 2014.

[150] R. M. Byrne, "Counterfactuals in Explainable Artificial Intelligence (XAI): evidence from human reasoning." In *IJCAI* 2019 Aug 10, pp. 6276–6282.

[151] J. E. Taylor and G. W. Taylor, "Artificial cognition: how experimental psychology can help generate explainable artificial intelligence," *Psychono. Bull. Rev.*, vol. 28, no. 2, pp. 454–475, 2021.

[152] M. A. Vu, T. Adalı, D. Ba, *et al.*, "A shared vision for machine learning in neuroscience," *J. Neurosci.*, vol. 38, no. 7, pp. 1601–1607, 2018.

[153] J. M. Fellous, G. Sapiro, A. Rossi, H. Mayberg, and M. Ferrante, "Explainable artificial intelligence for neuroscience: behavioral neurostimulation," *Front. Neurosci.*, vol. 13, pp. 1346, 2019.

[154] P. Angelov and E. Soares, "Towards explainable deep neural networks (xDNN)," *Neural Netw.*, vol. 130, pp. 185–194, 2020.

[155] P.P. Angelov and X. Gu, "Toward anthropomorphic machine learning," *Computer*, vol. 51, no. 9, pp. 18–27, 2018.

[156] J. Bien and R. Tibshirani, "Prototype selection for interpretable classification," *Ann. Appl. Stat.*, vol. 5, no. 4, pp. 2403–2424, 2011.

[157] R. Jiang and D. Crookes, "Shallow unorganized neural networks using smart neuron model for visual perception," *IEEE Access*, vol. 7, pp. 152701–152714, 2019.

[158] C. S. Webster, "Alan Turing's unorganized machines and artificial neural networks: his remarkable early work and future possibilities," *Evol. Intell.*, vol. 5, no. 1, pp. 35–43, 2012.

Chapter 2

Explainable artificial intelligence (XAI) in medical decision support systems (MDSS): applicability, prospects, legal implications, and challenges

Joseph Bamidele Awotunde[1], Emmanuel Abidemi Adeniyi[2], Sunday Adeola Ajagbe[3], Agbotiname Lucky Imoize[4,5], Olukayode Ayodele Oki[6] and Sanjay Misra[7]

Abstract

The healthcare sector is very interested in machine learning (ML) and artificial intelligence (AI). Nevertheless, applying AI applications in scientific contexts is difficult because of the issues with explainability. Explainable AI (XAI) has been studied as a possible remedy for the issues with current AI methods. The usage of machine learning (ML) with XAI may be capable of both explaining models and making judgments, in contrast to AI techniques like deep learning. Computer applications called medical decision support systems (MDSS) affect the decisions doctors make regarding certain patients at a specific moment. MDSS have played a crucial role in systems' attempts to advance patient wellbeing and the standard of care, particularly for non-communicable illnesses. Moreover, they have been a crucial prerequisite for the effective utilization of electronic healthcare (EHRs) data. This chapter bargains a comprehensive impression of the application of AI and XAI in MDSSs, summarizes recent research on the use and effects of MDSS in healthcare, and offers suggestions for users to keep in mind as these systems are

[1]Department of Computer Science, Faculty of Information and Communication Sciences, University of Ilorin, Nigeria
[2]Department of Computer Science, Precious Cornerstone University, Nigeria
[3]Department of Computer Engineering, Ladoke Akintola University of Technology LAUTECH, Nigeria
[4]Department of Electrical & Electronics Engineering, Faculty of Engineering, University of Lagos, Nigeria
[5]Department of Electrical Engineering and Information Technology, Institute of Digital Communication, Ruhr University, Germany
[6]Information Technology Department, Walter Sisulu University, South Africa
[7]Department of Computer Science and Communication, Ostfold University, College, Norway

integrated into healthcare systems and utilized outside of contexts for research and development.

Keywords: Machine learning; Explainable artificial intelligence; Healthcare medical decision support systems; Artificial intelligence; Healthcare diagnostics

2.1 Introduction

In traditional diagnostics, a practitioner examines potential lesions manually in a medical environment [1]. This manual inspection takes time and requires the doctor's full concentration, who is required to look over countless patient data, X-rays, and images from a single medical treatment [2]. However, in the past few years, in areas including therapeutic diagnostics, business, forensics, scientific exploration, and education, there is an increased interest in deep learning and artificial intelligence (AI)-based information withdrawal from patient data and images [3]. In these areas, it is frequently required to comprehend the rationale behind the model's choices so that a human can verify the consequence of the choice [4]. To avoid unanticipated negative consequences on decision-making, rules and regulations are pushing toward transparency standards for information sources [5]. Users have the right to information on machine-generated choices thanks in specific to the General Data Protection Regulations (GDPR) of the European Union [6]. Therefore, people who are impacted by resolutions made by an AI-based system could try to comprehend the factors that led to the decision result.

The past few years have seen a significant increased interest in healthcare diagnostics due to AI-based imaging information extraction. In order to help doctors, make decisions that are transparent, intelligible, and explicable, the authors in [7] stressed the significance of adopting explainable AI (XAI) in the medical industry. They projected that the acceptance of AI-based systems in the medical industry would be supported by the ability to explain AI-based decisions. The authors in [7] speak to the significance of using XAI to make judgments simple and understandable. In their investigation of the use of XAI in medical settings, the authors in [8] reached the conclusion that in order to remove obstacles to ethical standards, explainability must be included as a condition.

This can guarantee that the patient remains the focus of the treatment, and with the assistance of medical experts, patients may make informed decisions about their health that are their own. To reliably and understandably explain the classifier's predictions, the authors in [9] created the local interpretable model-agnostic explanations (LIME) procedure. The inclusion of text and graphic explanations for various models demonstrated the model's adaptability. When choosing between models, it assisted users of all skill levels and assessed their level of confidence and upgraded unreliable models by giving information about their forecasts. Additionally, authors in [4] address the impact of explainability on confidence in AI and machine learning (ML) techniques through the enhanced understandability and certainty of AI-based judgments based on deep learning on diagnostic medical datasets. Additionally, they investigate the usage of XAI to relate the recognition methods of two DL prototypes, convolutional neural network (CNN) and multilayer perceptron.

The development of MDSSs in healthcare systems is to reduce medical errors, increase real-time decision assistance, and lessen life-threatening events brought on by delayed or incorrect medical judgments in order to improve the outcomes for critically sick patients. Computer-assisted "dynamic information schemes" MDSSs use two or more patient data points to produce case-specific recommendations [10]. There is compelling evidence that MDSSs can enhance a doctor's decision-making abilities [11]. The MDSS must be data-driven, quick, and knowledgeable for best clinical decision-making. A MDSS is evidence based if it derives its information from and continuously updates, the most recent data from mindfulness sources and scholarly publications [12]. The IF–THEN rule is a general type of evidence in an evidence-based MDSS. According to the rule, IF an antecedent (i.e., a collection of circumstances) exists, THEN a result is anticipated or a course of action is required.

A MDSSs effectiveness and efficiency depend on how well the knowledge base is developed, which makes sure that current information is stored and that advice may be retrieved [13,14]. Using a knowledge base with two parts should allow for the gradual expansion of domain knowledge and clinical records [15]. A database with patient-related static and dynamic data, such as old electronic medical records (EMRs), is one of the components. The other is a rule base with rules that specify the relationship between input and output variables using an IF–THEN structure.

Red blood cells that have sickle shapes are referred to as having sickle cell disease (SCD). The most prevalent and frequently the most severe form of SCD is called sickle cell anemia (SCA), corresponds to a monogenic, recessive genetic disorder that affects different organs and disrupts the shape of red blood cells. Each year, this illness affects about 300,000 children worldwide, among the group of currently known genetic diseases, and is regarded as one of the most common conditions [16–18]. Due to a significant advancement in SCD care during the past few decades, a median survival of more than 60 years has been reported, indicating a considerable increase in life expectancy [19,20]. More research on the experience of looking in their respiratory function is required as there is an increasing number of individuals living with SCA. The establishment of innovative medicines and clinical care may be informed by knowledge of the processes behind lung injury.

Therefore, this chapter examines how XAI-enabled MDSSs may be used to identify and forecast a variety of diseases, the prospects, challenges, and ethical implications of XAI were also presented. The specific contributions of the chapter are:

(i) a broad overview of the application of AI-based and XAI in medical decision support systems (MDSSs);
(ii) summarizes recent research on the use and effects of MDSS in healthcare systems, and offers suggestions for users to keep in mind as these systems are incorporated into healthcare systems;

(iii) the challenges and prospects of explainable technologies for supporting medical decisions using AI were also discussed; and
(iv) presents the ethical and implications of XAI in MDSS in healthcare systems generally.

2.1.1 Chapter organization

Section 2.2 presents overview of MDSSs in healthcare systems. Section 2.3 discusses the AI in MDSS. Section 2.4 presents XAI. Section 2.5 discusses ethical effects and implications of XAI for MDSSs in healthcare systems, while Section 2.6 concludes the chapter with future direction and research gaps.

2.2 MDSS overview in healthcare systems

MDSS represents a paradigm transformation in modern healthcare. MDSS are utilized by doctors to support their intricate decision-making strategies. Since they were originally used in the 1980s, MDSS have advanced quickly [21]. They are now frequently delivered via automated medical workflows such as EMRs, which have been made possible by the rising global deployment of modern EMRs. Despite these improvements, it is still unclear how MDSS will affect the doctors who utilize them, outcomes for patients, and expenditures [22,23]. The successful experiences of MDSS have been widely publicized during the past decade or more; however, significant failures have also taught us that MDSS is not without dangers.

MDSS strives to improve medical judgments by giving them access to patient data, targeted clinical expertise, and other health-related resources. A typical MDSS is composed of program created to directly support healthcare decision-making, where a computerized clinical knowledge base is used to compare the characteristics of each specific patient, and the clinician is then given patient-specific evaluations or suggestions for a determination [24]. In today's world, point-of-care MDSSs are mostly employed by the health professional to integrate their expertise with that of others or recommendations made by the MDSS. However, there are more and more MDSS being created with the capacity to use data, and data that is normally unavailable to or incomprehensible to people.

The earliest computer-based MDSSs date back to the 1970s. They took a lot of time, had poor system interoperability, and, at the same time, were typically limited to intellectual activities [25,26]. Concerns about physician autonomy, the use of computers in medicine, and other matters of ethics and law were also raised, and individuals would be in error if they followed the advice of a flawed system's "explainability," etc. [27]. Currently, MDSS frequently utilizes online applications or computerized provider order entry (CPOE) systems' connection with EMRs [28]. You may administer them either through a computer, tablet, or smartphone, but also other technologies, like smart wearable equipment and biometric surveillance, monitoring, and tracking devices technologies. These devices might or might not be connected to EHR databases or create outcomes immediately on the device [29].

MDSSs have been categorized and grouped into a number of groups and types, including the delivery method active or passive, and the timing of the interaction

[30,31]. Expert systems that are knowledge-based or not are frequently used to describe MDSS. The creation of rules (IF–THEN statements) originates in knowledge-based systems, as the computer retrieves information to assess the rule, creating a result or a result [31]; MDSSs rules can be created using evidence that is based on the literature, practice, or patient-directed data. Even without knowledge, MDSS still needs a data source; however, the choice uses statistical pattern recognition, AI, or ML, instead of being designed to adhere to professional medical expertise [30]. Ignoring the fact that the use of AI in healthcare is increasingly rising, non-knowledge-based MDSSs are challenging to implement, such as having trouble understanding the justifications used by AI to generate recommendations (black boxes) [32], and the difficulties of accessing data [33]. They have not yet been widely implemented. Both versions of MDSS share characteristics with slight variations, as shown in Figure 2.1.

They are made up of: (i) base: the set of guidelines integrated into the framework (knowledge-based), the process used to make decisions (non-knowledge based),

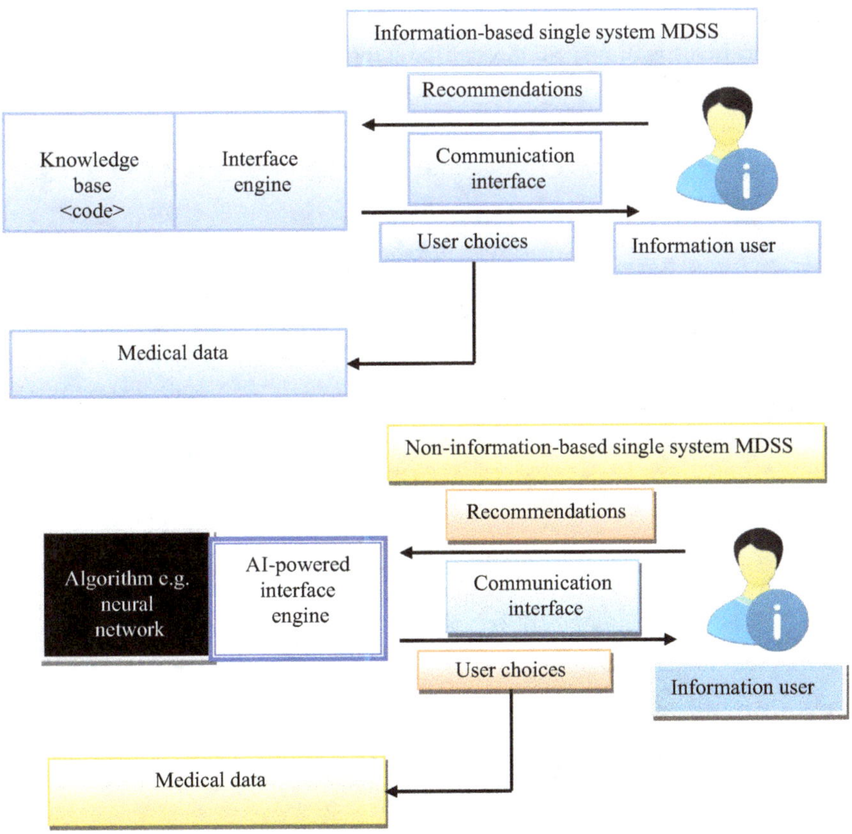

Figure 2.1 Essential connections between knowledge-based and non-knowledge-based MDSS

in addition to the facts accessible; (ii) inference engine: consider the rules set by AI or software, with data structures that employ patient medical data as an input to generate output or take measures and present it to the user (such as a doctor); and (iii) via the transmission medium: the entrance interface of an EHR, an application, or a website that the end user uses to engage with the systems.

2.2.1 Importance and prospects of MDSSs

MDSSs are instruments that combine current patient data with known clinical knowledge to improve patient treatment; they cover a wide range of tactics in favor of numerous subjects [34]. MDSSs are intended to support the doctor–patient relationship at many stages, including the first consultations, diagnosis, and follow-up. It is anticipated that a properly outfitted MDSS will have a considerable significant influence on patient care at all stages. The ever-increasing time pressures on clinicians will be lessened with MDSS. Figure 2.2 displays the importance of MDSSs in healthcare systems.

The followings are the major important and advantages of using MDSS.

2.2.1.1 Clinical management

According to investigations, MDSS can be integrated and consistent with clinical recommendations [35]. Given that it has been established that traditional medical

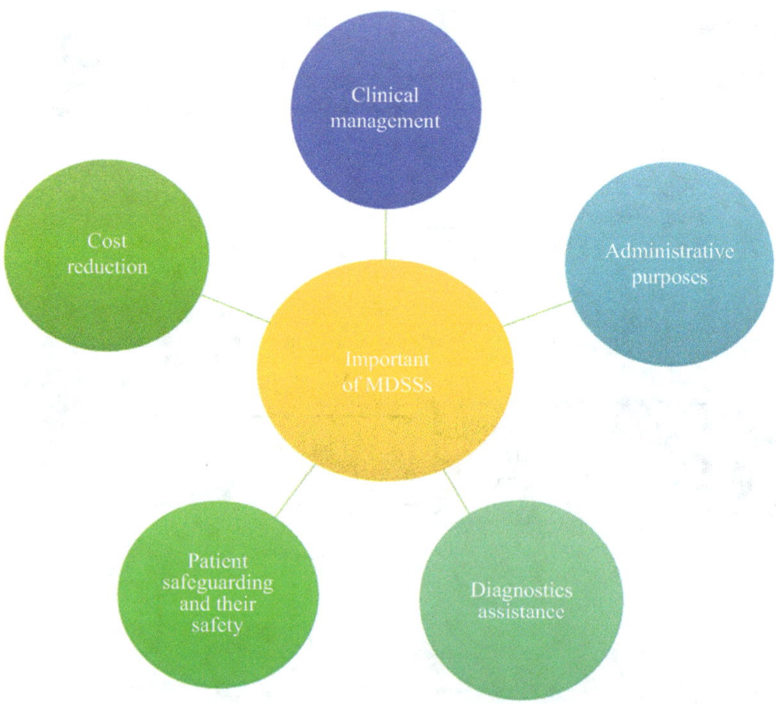

Figure 2.2 The important of MDSSs in healthcare industry

practices and these treatment pathways have low clinician compliance and are difficult to implement in practice [36,37]. It has not proven to be accurate for doctors to learn, take in, and apply new guidelines [38]. However, MDSS may actually have guidelines that mistakenly contain rules. Such MDSS could take the form of alerts for specific processes, notifications for diagnostics, standard order sets for a given case, among others. Additionally, MDSS can help with patient management for research and therapy processes [39], monitoring and arranging orders, following up on referrals, and making sure precautionary care is provided [40].

Additionally, MDSS can notify doctors when a patient is due for follow-up care or has not adhered to their treatment plan, and help support the identification of patients who meet the requirements for research eligibility [41,42]. If a patient's medical record satisfies the criteria for a clinical investigation, doctors are informed at the time of treatment by a Cleveland Clinic-developed and used MDSS [43]. The message requires that the user complete a form so that their identity can be verified and their consent to be contacted, sends the research coordinator the patient's medical record, and produces a patient records sheet for a clinical trial.

2.2.1.2 Cost reduction

For health systems, MDSS can be cost effective through intervention strategies [44], reducing the length of stay for patients, CPOE-integrated systems recommending less expensive drug substitutes [45], or cutting down on test duplication. A CPOE-rule that restricted the scheduling of blood counts was put into place in a pediatric cardiovascular critical care unit (ICU), panels for chemistry and coagulation to a 24-h gap [46]. Without affecting length of stay (LOS) or death, this lowered laboratory resource consumption at anticipated cost reductions of $717,538 annually.

The MDSS may advise the user about cheaper pharmaceutical alternatives or conditions that are covered by insurance. In Germany, majority of hospital patients are switched to hospital-prescribed pharmaceuticals. A drug-switch algorithm was developed by Heidelberg Hospital and added to their CPOE system after discovering that one in five replacements were wrong [47]. By easily and successfully transferring 91.6% of 202 medication appointments, the MDSS could improve safety, lighten provider burden, and lower costs.

2.2.1.3 Administrative purposes

MDSS offers assistance with patient triage, ordering of treatments and testing, and therapeutic and diagnostic classification. To assist doctors in selecting the most relevant diagnosis code, simulation based can produce a streamlined list of clinical features (s). A MDSS was created to remedy incorrect ICD-9 emergency department (ED) entrance code [the International Classification of Diseases (ICD) is a system of standardized codes for describing diseases and diagnoses] [48]. The instrument has an anatomical user interface (visual, interactive human body representation), ED doctors can more quickly and accurately locate diagnostic admission codes when they are connected to ICD codes.

Medical record quality can be directly improved by MDSS. An improved alerting system was included in an obstetric MDSS, much better documentation of labor induction indications compared to the control hospital and gestational weight [49]. Due to its ability to substantially support diagnostic testing, accurate documentation is essential. For instance, to ensure that patients acquired the proper immunizations following major surgery, an MDSS was created to reduce the threat of infections (such are meningococcal, *Haemophilus influenzae*, and pneumococcal), which follow spleen ectomy. Conversely, the study's findings revealed that 71% of patients with the keyword "splenectomy" in their EHR did not include it in their medical records complaint list (it was the event that caused the MDSS alert) [50]. Then, to improve the splenectomy problem list paperwork, an additional MDSS was created [51] and increase the usefulness of the initial MDSS for immunization.

2.2.1.4 Patient safeguarding and their safety

MDSS is frequently used in strategies to decrease drug mistakes. Drug–drug interaction (DDI) errors are described as frequent and avoidable. In a hospital setting, up to 65% of patients are susceptible to one or more risky behaviors [52]. CPOE systems are commonly created with pharmacovigilance software that includes dosage protections, repetition of treatments, and DDI verification [53]. One of the most widely used types of decision support is the warnings that these systems produce [54]. However, studies have discovered a significant amount of variation in the way DDI notifications are presented (for instance, disruptive or active) [55,56]. Whichever methods are employed to identify DDIs are recommended by the authors in [57]. There is no established protocol for carrying out which notifications to providers in systems, and they frequently provide various degrees of irrelevant alerts. The US Office of the National Coordinator for Health Information Technology has compiled a list of "elevated" DDIs for CDS. It has been deployed to various degrees in MDSSs from several nations, such as the United Kingdom, Belgium, and Korea [58].

Additionally, MDSS enhances patient safety by providing systems for various medical events reminders, and not simply problems brought on by taking medicine. There are a lot of examples, the number of hypoglycemic incidents could be reduced in the ICU by using a MDSS for blood glucose testing [59]. In accordance with a local glucose monitoring procedure, this MDSS continuously reminded nurses to take a glucose reading, which outlined the frequency of readings based on the patient's particular demographics and historical glucose thresholds [59]. In general, MDSS uses CPOE to target patient safety. Although such autonomous alarms, drug–event monitoring, and other technologies have generally been successful in reducing medication and dosage errors, incompatibilities, and other methods [60]. Patient safety may be regarded as a secondary goal, (or prerequisite) of practically all MDSS variants, regardless of the main objective driving their installation.

Electronic drug dispensing systems (EDDS) are among the other systems aimed at improving patient safety and point-of-care (BPOC) bar-coded systems for administering medication [61]. These are frequently combined to form a "closed loop," in which each stage of the process is located (prescription, transcription, distribution, and administration), is robotic, and takes place within an interacted system. Radio-frequency identification (RFID) is used to automatically identify the drug at administration or cross-checked with patient data and medicines using barcodes. The tangible advantage is the decrease of medication carriage errors at the "bedside," which represents additional target for MDSS (contrasting to additional upstream). Due in part to expensive and complex technological constraints, utilization is rather minimal [62]. Studies indicate that these strategies are effective at reducing errors, though [63]. According to authors in [61], several of these techniques can be combined with CPOE and MDSS at the same time. Thus, significantly reduced rates of ordering errors, overdose, and medication allergy diagnosis in prescribing [61]. As with most MDSS, omission by providers can still result in mistakes, or intentionally avoid the innovation [64].

2.2.1.5 Diagnostics assistance

MDSSs utilise diagnosing decision-support systems (DDSS). These techniques have typically included a computerized "consultation" or assessment phase, when they can be provided with user input or data and produce a list of plausible or likely diagnoses [65]. Due to a number of variable needs, DSS has sadly not (yet) had as much of an influence as other forms of MDSS, such as biased and unfavorable clinician perceptions (frequently because of data availability gaps), with inadequate system implementation that necessitates human data entry [66,67]. The latter is getting a lot better thanks to improved EHR integration and uniform language like Snomed Medical Concepts.

The MDSS developed by authors in [68] is a nice illustration of an efficient DDSS for applying fuzzy approach in the diagnostics of peripheral neuropathy. Through 24 input areas, including symptoms and results of diagnostic tests, when it came to recognizing motor, sensory, mixed neuropathies, or normal instances, they outperformed professionals by 93%. While this is quite helpful, especially in locations where accessibility to seasoned clinical professionals is limited, and systems that can support specialized diagnostics are also wanted. A probable diagnosis is given by the electronic reference-based DDSS called DXplain based on clinical symptoms [50]. In a randomised well-ordered investigation with 87 populaces in emergency medicine, a standardized diagnosis clinical trial comprising the patients who were randomly allocated to utilize the method displayed much better accuracy for 30 clinical symptoms (84% versus 74%) [69].

Given the frequency of diagnostic mistakes, especially in patient healthcare [70], there is a great deal of hope that MDSS and IT technologies would enhance diagnosis [71]. Now, diagnostic systems are being created using methods like ML-based models that are not knowledge-based, which can open the door to a diagnosis that is more precise. A notable illustration of the possibilities is the Triage and Diagnostic System powered by AI in the Babylonia United Kingdom, but also of the tasks that must be completed before these technologies are prepared for widespread use [72].

2.2.2 The challenges and pitfalls of MDSS

Despite encouraging preliminary findings of the aforementioned important and advantages of MDSSs in healthcare sectors, the beyond standard alerts, recalls, overview displays, and automatic knowledge discovery systems, the bulk of MDSS have not offered any more capabilities [73]. The significant proportion of hospitals in the United States have not yet implemented MDSS in any way. EHR platforms that contain decision support tools are referred to be "encompassing" (therapeutic advice, clinical recollections, drug–allergy warnings, drug–drug interaction warnings, pharmacological and toxicological warnings, and drug–dosing help); just 1.5% of the 2,952 hospitals investigated were classified as "complete" [74]. There has not been much evidence of outcomes improvement in MDSS research, and any such consequences have only been statistically significantly at low levels [73]. Only 20% of 148 clinical studies on MDSS adoption had an impact on clinical results, according to a meta-analysis [75]. Increases in morbidity consequences, such as the frequency of admissions, deep vein thrombosis, cardiovascular events, and surgical site infections, nonetheless, mortality or pharmacologic adverse events were not significantly affected [75]. That is to say, MDSS will need to be improved before they can consistently deliver clinically useful information.

The vast majority of scientific and clinical data sets are stored in various databases around the world in incredibly huge files ("data storage tower"). To create forms that doctors and researchers may readily analyze from countless data points and pertinent information, silos of information need to be interconnected. To determine which sources of evidence might be pertinent to a certain query, sophisticated machines must use established languages/phenotypes. If the massive quantity of healthcare data presently housed in institutional EHR systems is viewed as an additional evidence silo, afterward, how would interactions be made between patient records and major works for use in both science and medicine? A genuinely practical MDSS would integrate data opinions and knowledge derived from countless years of investigation, clinical assays, blood tests, and follow-up information are used to reach the scientific endpoint [76]. The concept of collaborations in healthcare is embodied by the bridging of information between various fields.

The process of entering data, or how they really enter the system, is the first problem. Some systems demand that the user manually enters patient data or query the system. This "double data entry" not only delays the patient precaution procedure but it also takes time, and time is precious in the outpatient context. If the system is not accessible and/or requires a protracted logon, it takes considerably longer. By linking the MDSS and integrating the EMR and healthcare information service, much of this interruption can be reduced [77]. As was already indicated, a number of commercial solutions provide embedded decision support features. This means that the decision support system can act on the data if they have previously been placed into the medical record, and in fact, various systems may be able to use several supplementary systems simultaneously. Despite the fact that not all MDSSs are integrated, this is positive, and such interconnections would be challenging without industry standards for incorporation of supplementary systems [78,79].

Additionally, there are a few standalone systems, such as some diagnosis and drug interaction systems. As a result, patient information must be input twice: once into the health records system and once more, into the system that supports decisions. The double data entering can limit the effectiveness of such solutions for numerous doctors [80,81]. Who is responsible for entering data into medical centers that are connected or even independent programs are a related query. Although they frequently engage with hospital systems, doctors are not necessarily the primary selection [82]. The physician can get warnings and notices from decision support systems (DSSs) considerably more effectively, which is one of the reasons MDSS and physician order input are linked.

Not simply order entry but also notification procedures are at issue. In the case study that was previously noted, the alert was disregarded by the medical provider [83,84]. These systems may be beneficial, but without cooperation between the computer science specialists and the physicians, their full potential cannot be realized. Vocabularies may not appear to be such a challenging subject, but frequently, clinicians would not necessarily use a system until they give it a shot. A system having a regulated vocabulary, such as a DSS, computerized patient record, they are conscious of the fact that either the system does not comprehend what they are attempting to communicate or, much worse, that it utilizes various words for the same thought or the same words for completely different meanings. The issue is that there are no standards for clinical terminology that are widely accepted. Additionally, errors can have a significant effect because the majority of DSSs have a controlled vocabulary.

The followings are some of the notable challenges of MDSSs in healthcare systems.

2.2.2.1 Overlapping processes

Workflow for clinicians may be hampered by MDSS, particularly with standalone systems. Several early MDSS were created as platforms that demanded the provider to get information from sources other than their regular workstation or to preserve knowledge. If MDSS are developed without considering how people interpret information and behave, they might also interrupt workflow. As a result, MDSS were created utilizing the "think-aloud" technique to simulate clinicians' workflows, and design a more user-friendly system [85].

Workflow disruptions might result in longer task completion times and higher cognitive strain, and spending less time in person with patients. Even when existing information systems are fully connected with MDSS, there may be a gap between engagements with a computer workstations and in-person encounters. According to studies, clinicians who have more practical experience are less willing to employ MDSS and more prone to overturn it [86].

2.2.2.2 Effect on user ability

Healthcare professionals, pharmacists, and nurses were the only ones used to double-check orders prior to CPOE and MDSS. MDSS may provide the appearance that checking an order's accuracy is seamless or not essential [85]. Dispelling this notion is crucial. A MDSS's prospective long-term impact on users must also be taken into account. A MDSS can have a training effect over time, so that it might not be necessary to use the MDSS anymore. The "carry-over effect" was first used to describe MDSS that have an educational focus [87]. On the other hand, providers could become overly dependent or trusting on a MDSS for a certain duty [88]. This may be comparable to repeatedly utilizing a calculator for mathematical computations, and after which one's mental math abilities decline. Due to the user's decreased individuality, it could be challenging, and if they move to a setting alone without MDSS, they will be less prepared to perform that task.

2.2.2.3 Impact on operations of inaccurate content and poor data quality

Due to their reliance on data from outside, dynamic systems, EHRs and MDSSs may have previously undiscovered flaws [89]. Some MDSS modules, for instance, might promote procurement even when the hospital is short on supplies. Several specialists stated in a research by authors [90] that at their hospital, inventories of pneumococcal vaccines or hemoccult testing soon deplete, however, the MDSS is not informed of this. If they are not maintained or used properly, prescription and issue lists might be troublesome. The prescription list at one location can be a list of derogations, which indicates that they either be or not be used by patients (and hence needs to be questioned directly) [90]. Only CPOE orders are used to establish other drug lists, consequently, manual verification that patients are taking the medication is still necessary. Technologies that make it simple to tell these apart are desirable. It is a significant area where PHRs could provide a remedy, by getting information about patient adherence to medication straight from them.

Users may create alternatives in improperly developed systems that endanger data, such as inputting inaccurate or generic data [90]. A centralized, sizable clinical data repository serves as the foundation for the MDSS existing knowledge. Data quality can influence the effectiveness of decision assistance. If the system's data collection or input is not regulated, the data can actually be corrupted. A system for usage at the area of concern could be created by you, however, when used with data and surroundings from the actual world will not be used effectively. It cannot be overstated how important it is to use informational standards like ICD, SNOMED, and others.

2.2.2.4 Financial difficulties and challenges

Roughly 74% of those with an MDSS said that it is still challenging to preserve economic stability [91]. Initial setup and integration costs for new systems might be high. As new employees must be educated to utilize the system, ongoing expenditures may always be a problem, and system changes are necessary to stay informed with current knowledge. The findings of cost evaluations of MDSS deployments are ambiguous, conflicting, and scant [92,93]. Numerous variables affect how cost-effective an intervention is, incorporating environmental-specific issues are both institutional and socioeconomic [92]. Cost–benefit analysis alone has its limitations, with difficulties such a lack of uniform measures [93]. To increase our knowledge of the financial impacts of MDSS, additional work has to be done in the developing topic of research and investigation.

2.2.2.5 Limitations of interconnectivity and portability

Despite more than three decades of continual research, MDSS, as well as EHRs in general, have interoperability problems. Numerous MDSS are clumsy standalone systems, might be a part of a system that is unable to efficiently communicate with other systems. Why is it so difficult to develop transportability? In addition to programming challenges that may make interoperability challenging, another issue is the variety of clinical data sources [94,95]. Transferring private patient information is frowned upon or seen as posing a risk. In a good way, interoperability guidelines are continuously developed and improved, including Health Level 7 (HL7) and Fast Healthcare Interoperability Resources (FHIR). These have already been used by for-profit EHR manufacturers [96]. Numerous government agencies, healthcare institutions, and organizations that support informatics are actively supporting and some countries even require that these interoperability guidelines be used in healthcare systems [97,98].

Additionally, the cloud provides a potential remedy for interoperability (as well as other EHR problems like data synchronization, software updates, etc.) [99]. Open architecture, more recent standards, and more adaptable system communication are all features of cloud EHRs [100]. Another widespread misunderstanding is that data stored in the cloud seems to be more susceptible, but this is not always the case. Web-based EHRs are utilized to keep data in sophisticated warehouses with other security measures. National data security regulations must be followed, for instance the Health Insurance Portability and Accountability Act (HIPAA) in the United States and the General Data Privacy Regulation (GDPR) in Europe, and the Personal Information Protection and Electronic Documents Act (PIPEDA) in Canada, to mention just some [101]. They have the same potential for safety (or vulnerability) as a conventional server-based design [101]. In contrast to server-based records, there are frequently fewer users who have accessibility to unprotected data and information in the cloud warehouses [101].

2.3 AI in MDSS

An expansive theoretical grouping of computer programs that aim to duplicate, replicate, or enhance human decision-making is known as AI [102]. Expert systems based on rules (ES) and ML technologies are the two main areas of AI. When combined with one or more human operators, ES and ML systems can enhance the total system's quality, effectiveness, or safety. The ML system is frequently trained by human operators as well, and the procedure is similar to training a pet. The operator offers instances and encouragement for wise decisions, as well as instances and criticism for errors in judgment. A sufficiently big collection of instances that fit (correct) or fail (incorrectly) can be used to teach or infer trends by the machine. These examples may have already been graded or contributed by one or more specialists. Software for mammograms or an automated hematology laboratory system could be built using ML [103,104]. Other times, if and when the machine is unable to, the human operator can take control successfully interpret the data (for instance, in the event of confusion, a personality automobile may hand over control to the driver, overloaded, or stipulated safety variables are outperformed, or a bank's credit card administrator may individually survey a potential customer if the software is unable to make a decision about whether to approve credit).

The use of ML in diagnostic healthcare goods, including as tumor diagnosis, medicine recommendations, and safety devices, was perhaps first broadly accepted in those areas [105,106]. Substantial ML-based tools first appeared in pathology and the lab, where skilled pathologists could oversee training and outcome confirmation. This reserved the "scholarly intermediary" (i.e., the certified technician or licensed pathologist) "up to date" on the diagnostic items and services provided by the labs. The Clinical Laboratory Improvement Amendments (CLIA) programs of the 1988 CMS were created with the goal of ensuring that skilled human operators adhered to specified quality assurance procedures so that computerized laboratory equipment delivered reliable, accurate findings [107,108]. The goal of CLIA procedures is to maintain the status of the domain brain as the authorized expert in charge of the diagnostic tools [102].

For accurate and sensitive diagnosis, many newly developed and in-production ML-based imagery analytic tools rely on expert human operator management and judgment. Despite the fact that ML-based growth identification techniques may be "close to faultless," there is such a huge variety of physiology of the human body and the morphology of disease that algorithms cannot always produce conclusive choices [109]. However, the human expert may get tired. The ML technology can help in verifying the work of the human operator while they are busy or preoccupied. Therapeutic applications have evolved as ML-based diagnostic systems gain popularity. Smart infusion pump systems are now the go-to solution for assuring the security of IV medications [110]. Although the majority of those systems rely more on professional AI than ML, their dependability and resilience have paved the way for more sophisticated ML-based solutions like inserted insulin pumps and newly developed synthetic pancreas with restricted gadgets. It is impossible to overestimate the importance of sealed ML technologies for vital medicinal treatment. A restricted embedded defibrillator, pacemaker, or pancreas gradually loses its "trained intermediary" or expert operator. To enable the release of such products on the market, decades of policy and program alterations were required. One might fairly anticipate that for the life-critical personal and business applications like self-driving cars, ML-based "smart" is being developed for vehicles that will ultimately strengthen society's willingness to enable additional ML-based, automated robots, barring any significant or catastrophic defeats healthcare diagnostic and therapeutic devices. Every robotics and stability improvement that can be made will be required to meet the favorable potential for increasing safety, versatility, accessibility, and economy for continuously expanding, continuously aging, and widely scattered populations of citizens and patients. There have been many good options in healthcare described above, and as a result of apps, ubiquitous storage, and cloud computing infrastructure, technology in healthcare is growing quickly. The broad categories of healthcare AI (HAI)

(i) pattern recognition and perception (malignancies, cracks, foreign particles, abnormalities in the gait);
(ii) sound and language processing preparedness and prevention (vocal style robotics, neurological diagnostic booths, self-service deficiency detection);
(iii) guidance from knowledgeable clinicians to help other careers or patients (guidelines for medications, improve medical standards and procedures, and testing and treatment of stroke);
(iv) self-operating surgical tools or improvements (automaton surgery, cardiac surgery, and sensory enhancement);
(v) sealed medical technology (artificial pancreas, AED).

For a range of purposes, AI has been incorporated into the majority of electronic health records [111]. The majority of them are supplementary devices to speed up medical choices, lessen or completely eradicate inaccuracies, and/or enhance the standard of care, pricing, or patient pleasure. For instance, the majority of smart intravenously pump systems now include modules and auxiliary equipment scanners for RFID technology, barcode scanners, and support the enforcement of the "Five Rights": the patient's identity, the correct drug, the proper amounts, and the right place, at the right moment [112].

2.3.1 The basis of AI in healthcare systems

Nowadays, AI technologies have had a significant impact on the medical industry, sparking an ongoing debate on whether AI practitioners would someday substitute human surgeons [113,114]. Although AI can help doctors create good medical judgment or perhaps require people judgment in some operational areas of medicine, we do not in the near future anticipate that technology will replace human physicians (e.g., radiology). The swift growth of big data analytical approaches and the expanding availability of medical data have made recent practical deployments of AI in healthcare possible [115]. Powerful AI approaches can uncover important clinical facts concealed in the huge amount of data when directed by pertinent clinical inquiries, which can help medical decision making.

2.3.1.1 Healthcare data

AI systems need to be "programmed" utilizing data before they can be employed in telemedicine, collected from patient evaluation, including health checks, treatment plan, treatment allocation, and so on. This will allow them to identify set of features that are equivalent to one another and identify relationships between subject features and desired outcomes. These clinical data frequently exist as demographic information; medical records include, but are not limited to, digital records from medical supplies, medical testing, diagnostic tests data, and photos. Particularly, a sizable component of the AI literature analyzes data from electrodiagnosis, genetic testing, and diagnostic imaging at the diagnosis stage.

2.3.1.2 AI devices

The two main categories of intelligent machines are the following. The first category includes ML methods that examine structured data, such as genetic, imaging, and EP data. The ML techniques used in medical applications attempting to categorize individual traits or forecast the possibility that an illness might appear. The second category consists of approaches for natural language processing (NLP) that extracts data from unstructured origins, such as medical notes and medical journals in order to supplement and enhance structured medical data. The NLP techniques aim to turn texts into structured information that is machine readable and can be analyzed by classification methods. The diagram shown in Figure 2.3 illustrates the path from the collection of clinical data, through NLP data enhancement and ML data processing, to medical decision making for improved communication. Despite how potent AI approaches may be, their application must ultimately support medical practice and be driven by medical symptoms.

2.3.2 The role of AI in MDSS

AI is the area of research and development that contributes more efficiently and positively to the medical field. Many different automated diagnostic systems have been created with AI's help [116]. Nowadays, medical centers use these technologies on a large scale. Both patients and medical professionals have found them to be quite effective in creating decisions. These systems are developed using a variety of approaches. In different approaches, the methods for acquiring input data and presenting result information are diverse. MDSS are any computer programs

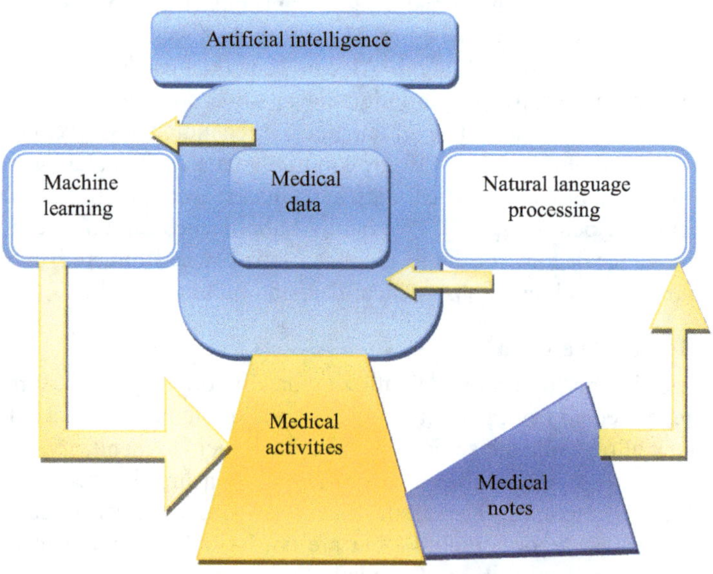

Figure 2.3 The progression of healthcare records from its creation to NLP, data enrichment, ML, and medical decision-making

that aid professionals in medical decision-making. Being able to facilitate the generation and use of medical knowledge is a key feature of AI. We can create systems with the ability to learn and generate new medical experience using AI.

The use of AI is crucial in decision support systems. AI-based decision support systems have the capacity to adapt to novel environments and learn over time [117]. The information needed for decision-making in cognitive computing and machine assistance programs is gathered using a variety of techniques. Neural networks, knowledge-based processes, fuzzy rule-based strategies, genetic programming, and statistical techniques, and others are some of these methodologies. The choice of a specific methodology is influenced by a variety of elements, such as the following:

(i) What is the area of concern?
(ii) What would the remedy be?
(iii) Data volume at hand.
(iv) Selection and goal of the scientist.

Medical technology requires computer-aided programs that can gather individual health-related information and translate them into pain intensity for the evaluation of pain [118]. Patients' quality of life is impacted by pain, and because there are not adequate evaluation techniques, some patients cease requesting for more medication when their pain worsens [119]. The vital surveillance of the individual following surgery also requires proper assessment of the medication dosage, as overdosing can occasionally pose a threat to life. It is significantly more effective, efficient, and cost-effective to employ a MDSS to assess the degree of discomfort and make a diagnosis.

Medical decision systems are frequently used during operations. Nowadays, minimally invasive surgery is the favored approach. The ability to bring a laser beam trans endoscopically inside bodily cavities will be made possible by the creation of a dependable flexible fiber or wave guide. It creates a potent surgical tool for operations by fusing the endoscopic approach with the beneficial laser interaction with tissue. It has reduced costs, faster healing, and less postoperative discomfort [120]. Different methodological branches of the MDSSs are represented by the image in Figure 2.4. Systems that support medical decisions can be roughly divided into two categories. MDSSs that are not knowledge-based and those that are knowledge based.

2.3.2.1 Knowledge-based MDSS

Most of the principles in the knowledge-based MDSS are expressed as IF–Then clauses. Generally, the data is connected to these rules. For instance, if the level of pain intensity reaches a specific point, create a warning, etc. There are typically three main components to knowledge-based systems: knowledge foundation, inference guidelines, and a communication method. The rules are stored in the knowledge base, the inference engine applies the rules to the patient data, and the communication mechanism is used to show the results to the users and accept input from them. When it comes to treating chest pain, for example, adaptive suggestions

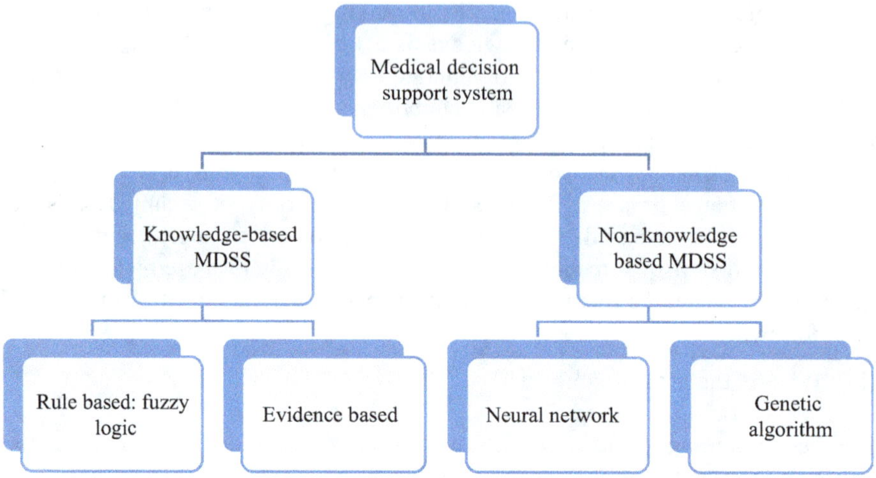

Figure 2.4 Methodological branches of MDSS

from a knowledge-based server are demonstrably much more successful than other options [121].

They are the kind of MDSS that hospitals and clinics most frequently use. They may even be able to use case-based reasoning if they have clinical expertise on a specifically defined task. Expert systems typically represent their knowledge as a collection of rules. Knowledge-based approaches are occasionally combined with variation planning to effectively deliver high-quality medical services and carry out patient care processes. Utilizing object-oriented analysis, UML methodologies, and the creation of generalized fuzzy ECA (GFECA) rules, this knowledge-based administration system handles variation.

2.3.2.2 Types of knowledge based

2.3.2.2.1 Fuzzy logic rule based

It is a type of knowledge base that has produced a number of crucial strategies and procedures for diagnosing the illness and treating the patient's discomfort. For instance, the right ventricular mass (RVM) learning technique is used to reduce pain in patients who are unable to verbally express themselves. The vector machine approach, an evolution of object recognition, can help medical professionals measure the discomfort [122]. The classifier based on fuzzy logic rules is particularly effective in terms of high levels of accurate diagnosis and positive predictive value. For instance, a fuzzy logic rule-based classifier has an average accuracy rate of 95% when predicting outcomes for conditions like appendicitis [123].

2.3.2.2.2 Rule-based systems & evidence-based systems

They frequently translate the expertise of subject–matter specialists into phrases that may be assessed as rules. The working knowledge will be compared to the rule base by integrating rules until a consensus is drawn once a significant number of rules have been assembled into a rule basis. For keeping a lot of data and knowledge, it is useful. Experts find it challenging to translate their expertise into clear standards, though. Scientific proof practice seemed to be the ideal method for bridging the gap between doctors and MDSSs. It is demonstrated to be an extremely effective tool for enhancing patient services and health experience. It might decrease costs and raise standards of quality and safety [124].

2.3.2.3 Non-knowledge-based MDSS

Non-knowledge-based MDSS are MDSS without a knowledge base. Rather, these systems made the use of a type of AI known as ML. The two primary groups of non-knowledge-based MDSSs are then further separated.

2.3.2.3.1 Neural network

The nodes and weighted linkages in neural networks are used to determine the association between the symptoms and the diagnosis. This satisfies the requirement that input rules not be written. However, the system is unable to specify why a particular use of the data is being made of the data. Therefore, its dependability and transparency may be the cause. It has been found that the self-organizing procedure of training the neural network, in which it is not provided with any prior knowledge about the subgroups, it is expected to recognize, is capable of obtaining pertinent information from the input data in order to create subsets that correlate to class. In addition, just a tiny fraction of the accessible data is needed to train the model [125].

2.3.2.3.2 Genetic algorithms

They are found on the process of evolution. A selection algorithm assesses the elements of a problem-solving strategy. When a proper solution is not seen, the process is repeated using the top-performing solutions. The generic system employs an iterative process to generate the best possible outcome for a task [126]. None of the cases examined in this study included genetic algorithms, which indicates that the researcher missed the chance to benefit from genetic algorithms. Additionally, it describes the potential for applying genetic algorithms to construct MDSSs.

2.3.3 Related work of AI in MDSS

The authors in [127] a MDSS's best usage of AI approaches was examined in the study. These technologies are designed to aid doctors in their diagnostic processes by enhancing decision-making, minimizing clinical mistakes, enhancing patient comfort, and lowering costs. However, the efficiency and precision of these systems greatly rely on the core AI approach that was employed; the same clinically relevant issue can be fixed using a variety of AI techniques, each of which may result in a different set of results. In order to determine the fundamental requirements for the appropriate application of smart approaches inside such tools, the paper evaluates a number of studies that used AI methods in clinical decision support systems. A yes/no inquiry strategy was used in this study and is based on evidence from earlier research projects. This investigation's goal is to make it easier to choose the most advantageous and practical AI method to integrate into the MDSS in order to produce the best results.

In [33], the authors in the Age of AI deliver medical decision support. Medical experts have long dreamed of the day when technology might help with challenging therapeutic choices. About six decades back, the first article on this topic emerged in the scholarly literature, and since then, the idea of computer-based clinical decision support has dominated informatics study. The potential of deep learning in healthcare was underlined in two recent JAMA perspectives. Such innovative data analysis techniques have a lot to offer in terms of deciphering huge and intricate data sets. Regardless of the specific analytic approach that they employ, this perspective is concentrated on the subset of decision support systems that are intended to be utilized dynamically by physicians as they seek to make decisions.

The authors in [102] worked on MDSSs and AI in hospital strategy. An extensive theoretical group of software applications that aim to imitate, simulate, or enhance human decision-making is known as AI. Expert systems based on rules (ES) and ML systems are the two main areas of AI. ML AI is built on a diverse foundation and plan, where the computer program is either trained to detect or deduce preferred patterns or subjected to trial-and-error investigation to get intended outcome. ES and ML tools can also be merged into a distinct invention or a collection of products, similar to the features found in the majority of modern antivirus program.

The authors in [107] give a thorough analysis of computational and AI-based T1D management decision support systems. The repositories IEEE Xplore, ScienceDirect, and PubMed were used to find the documents. There were no time limitations placed on the search. About 562 articles remained to be reviewed after repetitions and off-topic materials were removed. Based on algorithm assessment using actual human data, in silico experiments, or clinical investigations, we selected 61 of those articles for thorough review. The paper discusses the effectiveness and possible uses of each group of decision support system, as well as the AI techniques utilized in these systems.

In [103], authors worked on a decision assistance system using AI for type 1 diabetes care. Insulin deficiency and dysfunctional pancreatic beta cells are hallmarks of type 1 diabetes (T1D). In several medical professions, advanced analytics is being included using ML AI techniques. Here, we present an approach that recommends weekly insulin dosages for MDI-treated persons with T1D. In order to use a set of 12 cases to train a k-nearest neighbours decision model (KNN-DSS) to assess the causes of hyperglycemia or hypoglycemia and the necessary insulin adjustments possible suggestions, we use a special virtual platform3 to produce over 50,000 glucose samples. When tested on actual human data, the KNN-DSS algorithm obtains an overall agreement with board-certified endocrinologists of 67.9% and generates safe suggestions, according to endocrinologist assessment. According to the study, persons with T1D may use the KNN-DSS to enhance glycaemic outcomes and avoid potentially fatal consequences by early identifying risky insulin regiments.

Similarly, in [128], the authors' goal was to determine if computerized decision support systems based on AI are helpful in health and social care settings. To find pertinent randomized control trials carried out between 2013 and 2018, a comprehensive literature survey was carried out. Health Business Full-text Elite, ProQuest Public Health, PsycINFO, Sciencedirect, Cochrane Library, ASSIA, MEDLINE, EMBASE, CINAHL, and Emerald, and publications sites were among the databases that were explored. Three categories of search phrases were conceptualized: terms associated with AI, computerized decision support, and health and social care. Since there is little evidence of the usefulness of data-driven AI in supporting decision-making in healthcare setting, our work offers crucial insights into how a substantial scientific basis in this developing subject needs to be established moving forward.

In order to support intelligent decision making in the current unpredictable environment by the authors in [129], the study identifies problems and future research possibilities in forecasting and training, decision making, and optimization. The results of this review demonstrate how AI has improved operations research decision-making. Synergies, contrasts, and overlaps between AI, DSSs, and OR are presented in this review. Along with the underlying theories, an explanation of the literature based on the methods used to construct the DSS is also provided. Theory-building and application-based techniques, as well as taxonomies based on the AI, DSS, and OR fields, make up the bulk of the classification. In this evaluation, previous studies were adjusted for prognostic potential, use of big data sets, number of parameters taken into consideration, growth of training functionality, and confirmation in the decision-making paradigm.

2.3.4 AI weakness in healthcare system

The worlds of medicine and biological research are gradually altering as a result of AI [130,131]. One of the most exciting fields for AI applications has long been medicine [132]. Numerous clinical decision support systems have been proposed and created by researchers since the middle of the twentieth century. The 1970s saw a lot of success for rule-based techniques, which have been demonstrated to interpret ECGs, diagnose diseases, select effective medications, provide interpretations of clinical reasoning, and help doctors come up with diagnostic hypotheses in challenging patient cases. However, because they demand explicit representations like any literature, rule-based methods require the development of decision trees and human-authored changes, are costly to implement, and can be brittle [133]. The effectiveness of the methods is also constrained by the depth of existing clinical experience, and elevated connections among diverse bits of knowledge written by various specialists are challenging to encode. Furthermore, it was challenging to put in place a system that combines deterministic and stochastic reasoning to focus on the pertinent clinical context, rank diagnostic hypotheses, and suggest treatment.

Although AI has the potential to change healthcare, there are still many technological obstacles to overcome. Care should be made to gather data that is typical of the intended patient group because ML-based algorithms rely substantially on the accessibility of copious quantities of high training examples. For instance, various healthcare settings' data may have different kinds of skew and distortion, which could prevent a model trained on one hospital's data from generalizing to another [134,135]. It has been shown that when the diagnostics activity has a low inter-expert unanimity, agreement diagnosis might greatly increase the effectiveness of the ML models trained on the data [136]. For processing heterogeneous data, adequate data curation is required. Additionally, in order to acquire the global standard of patients' clinical status, professionals must personally evaluate each patient's scientific notes, which is unaffordable for the general public. Subsequently, a gold benchmark that imputed the patients' actual conditions using NLP methods and diagnostic codes has been presented. The reliability of the prediction models will be improved, increasing the safety of using them in life-and-death decisions by using sophisticated algorithms that can address the quirks and peculiarities of different datasets.

Numerous effective ML techniques produce outcomes that are challenging for unaided humans to understand [137]. Even while these models are capable of performing better than humans, it is difficult to communicate intuitive ideas that underlie the models' results, to spot model flaws, or to extrapolate new biological insights from these computational "black boxes." Use saliency maps to illustrate the significance of each image region or show convolution filters that are recent methods for explaining image classifier model. Nevertheless, for deep neural network models trained on data other than images, model comprehension remains significantly more difficult, and this is the subject of ongoing research efforts.

Modern AI solutions would not be able to fully realize their promise unless they are included into clinical operations. But studies have shown that using AI to healthcare is not a simple task [138]. It is well known that clinical information systems have a number of unforeseen consequences, such as notify fatigue, increased physician burdens, disturbance of interpersonal interactions (incorporating doctor–patient interaction), and the creation of particular risks that call for heightened monitoring to identify. For instance, doctors are more likely to forgo the diagnosis when a patient has a severe problem since the CAD tool used for mammograms results in a false negative. It is difficult to pinpoint the ideal medical workflow that will enhance the benefits of AI-assisted diagnostics, even though many CAD models can be changed to stabilise the understanding and accuracy required for each scientific use case. The perspective of healthcare professionals and patients demonstrates the need for careful formulation and construction, which are frequently absent when integrating digital technologies into clinical environments.

2.4 XAI

The conventional doctor–patient connection may be harmed by the implementation of AI-based technologies for MDSSs in healthcare situations, which is found on trust and openness in medical recommendations and clinical choices. When a diagnosis or treatment decision is no longer decided purely by a doctor, yet judgments made in a major way by a machine-utilizing algorithms to lose their transparency [2]. The most typical use of ML-based techniques in healthcare decision-making is skill learning. These algorithms belong to a class that is very broad (artificial neural networks, classifiers, etc.), which are adjusted using examples to improve how well they classify new, undiscovered cases. Asking for justification for a decision is useless [139,140]. Experts in statistics or data science may be able to comprehend an AI algorithm's mathematical elements in great depth. However, this "developer's interpretation" falls short of expectations of the fate of naive users or human beings [2].

The issue that AI-based medical decision-making offers to the information rights of impacted people has been acknowledged by the European Union (EU) and has written a seminal work emphasizing the necessity for automated decision explanations so that they can be clearly explained to the impacted patients [141]. The paradigm of XAI, which is piqueing scientific curiosity, holds the key to the problem [142]. This is in line with efforts made by the US military to create models that can be explained, making judgments made by autonomously machines comprehensible [143]. Alchemists from the middle ages have been linked to ML-based models since they rely on trial and error and lack a thorough grasp [144]. According to the same logic, this section concentrates on the need for XAI to be able to fully explain to the domain expert the decisions taken by an AI in an MDSS environment. A doctor, for instance, in the instance of AI-based medical choices concerning a disease's diagnostic, management, or prediction.

2.4.1 The basis of XAI

AI is based on the idea that it can mimic human intelligence by learning from the collected data from diverse sources and performing tasks that humans can complete, recognizing patterns, or predicting outcomes [145]. NLP, financial technology, autonomous driving, recommender systems in social media and e-commerce, and question-answering software all make extensive use of AI and ML methods. Additionally, AI is gradually altering the landscape of medical research.

The field of XAI is not new and has been in existence for a while [146], and 1980–1990 saw a lot of research into AI systems. Predicate logic graphs and other formal representations of human knowledge served as the foundation for these systems. For instance, directed acyclic graphs (DAG) with an approximation method like Bayes [147], Dempster-Shafer theory, and fuzzy reasoning [148]. The GeneOntology knowledge base, for instance, is one of the most effective knowledge-based platforms in the world [149]. The fact that the AI must first have a knowledge representation developed manually is a key constraint of these approaches. This issue appears to be resolved by algorithms that can train themselves to behave in an apparently intelligent manner. However, the majority of ML-based algorithms in use today neither produce knowledge nor are they knowledge-based. When viewed through the lens of knowledge-based AI, these systems compromise performance for clarity and comprehensibility. Therefore, artificial skills-based systems (AS) would be a better moniker for the majority of ML-based systems and several "AI" systems.

A transparency AI-based decision may ideally be reached when a reliable scientific theory explaining how the fundamental ML system operates is provided. The trustworthy system can then get its results by making logical deductions based on this notion. In order to accurately predict the placements of planets in astronomy, this is comparable to knowing Kepler's laws. The range and precision of a forecast can be determined in systems based on a reliable scientific concept. A justification (explain) for the outcome may also be provided [150]. For instance, Newton's law of gravity can be used to deduce Kepler's three laws of astronomy. These rules can be used to determine why a certain planet is in a given place.

For situations when a theory of evolution is not provided or even known, ML-based techniques are frequently used. For instance, systems based on ML-based models are created to diagnose patients using several measures of gene expression, even when the precise biological mechanisms causing the disease are only vaguely comprehended. In these circumstances, the ML research is happy to be a diagnostic system's "efficiency" that can be determined by examining the accuracy of its forecasts on a little amount of data that was not utilized throughout the formation (tuning, instruction, learning, and adaption) of the so-called "test data," of the system [151,152]. That is, the model is trained to do a certain task using a collection of carefully chosen training and test data, making a medical diagnosis, for instance. The capacity to extrapolate to the novel, uncharted scenarios is assessed in this way. A measure of performance on unobserved data is used in this methodology to estimate trustworthiness. But generally speaking, and when the model was created, these "unseen" data were already available.

It is clear what skill-based ML systems cannot do for data with a pattern that is substantially close to the training data, the system will operate effectively. The skill-based ML system will fall short of data with a different structure and could not even be aware that such facts are outside the algorithm's area of expertise. This is comparable to the epicycle model of planetary motion in astronomy. One could describe it as a planet move empirical Fourier series, with a succession of alternately higher and lesser circles [153]. Under "normal" conditions, and for brief intervals, the epicycle model can reasonably estimate a planet's position [154]. It is unknown, nevertheless, when the predictions are accurate or false.

Asking for an explanation of a decision is meaningless for skill-based ML systems. For instance, "associative memory" systems keep a recollection of all instances and their diagnosis (database). The majority of the diagnoses from the most similar cases are assigned after looking for the cases that are the most similar to the current case in question. The k-nearest neighbor classification technique is an illustration of this kind of model [155]. The only thing that can be learned from attempts to thoroughly evaluate skill-based models is the mathematical model that underlies them. For instance, patient A's analysis is D since A is most like patient X, who previously had the same finding. Fairness and antidiscrimination against minorities are also important, along with other moral obligations. In skill-based systems, some behaviors, such not hurting people, cannot be guaranteed or enforced.

2.4.2 The role of XAI in MDSSs

Whether explainability is a necessary property for MDSSs is a hotly contested intellectual concern. This argument first appears to be dominated by two opposing viewpoints, if one side contends that focusing on performance with strong validation is sufficient [156]. While the opposing side emphasizes the value of explainability and even makes a case for philosophical grounds for adopting explainability [157,158]. However, a third viewpoint that is becoming more prevalent does not contest the significance of explainability, but instead emphasizes that only by using interpretable techniques will the advantages of explainability be realized. This viewpoint's proponents contend that black-box explanations are found on questionable theoretical premises, as well as the fact that such post-hoc assumptions of the fundamental models would not result in trustworthy explainability [159–161].

Developed from earlier works [161] and we contend that if properly confirmed by both the current analyses' findings and outcomes and with methodologically solid explanations for a black-box system that are intuitive and simple for consumers to understand, then there would not be any defense for excluding them. It is clear that when faced with a black box, establishing trust in a system that is intended to aid emergency responders in their decision-making is disproportionately challenging, more so if, as in the case of the current use case, they are legally and likely responsible for the choice made [162]. This could lead to disregarding the black box system, as was the case in the first validating study of the system, or even worse, by upsetting the users and degrading performance. These factors, in our opinion, are relevant to MDSSs generally.

However, it is possible that those who support the idea of explainability will ignore technological barriers while developing justifications for black-box algorithms. Finding proven and technologically solid explanations for black-box models are not simple, specifically suited for the widely used applications of artificial neural networks. In this situation, it is crucial to carefully assess the benefits and drawbacks of various explanation-generating strategies, and the ability of these strategies to be comprehended by end users should also be considered [163]. Regulators must also be familiar with the benefits and restrictions of comprehensible and interpretable processes. However, regardless of the type of explainability used for a MDSS, the authors of [160] contend that it is essential to involve end users in the design process and customize the system to meet their demands, as well as to educate them about the system's functionality, the data and methods employed, and pointing out any biases and restrictions in the training data that could impact how well the machine performs when applied to a larger population. These actions are advised for reliable AI [164] and, additionally, they are crucial in our perspective for promoting MDSS user confidence. Moreover, adding justifications should not be used as an explanation for skipping crucial clinical confirmation and validation. Explainability implementation is a fresh stage of the creative process, where it is guaranteed that all explanations and the clients can use and comprehend their interpretations. It also aids in pointing out situations in which they might not accurately represent the classification procedure used by the system.

In the past 50 years, the rule-based strategy that depended on the curation of healthcare information and the creation of strong decision rules has attracted considerable interest in the diagnosis of diseases and clinical decision support. Medical applications of AI techniques like ML and DL, which take into account complicated interactions between features, have recently been proved to be promising [165]. Recent advancements in AI have revolutionized several aspects of healthcare, including diagnosis and surgery. These methods have proven successful in these sectors. Some diagnosis tasks based on deep learning are even more accurate than those performed by human doctors. The black-box aspect of the DL model, however, restricts the models' ability to be explained and prevents their widespread application in medicine. In the transdisciplinary fields of AI and medicine, there are many researchers that have come to the conclusion that the explainability of the AI model, not its accuracy, is the key to its deployment in the clinical setting. Before being accepted and implemented into healthcare profession, medical AI applications should be described. Therefore, XAI is necessary for the acceptability of medical AI applications, and research into medical XAI is motivated [166].

Basically, XAI plays a key role in medical diagnosis. Table 2.1 is the review of the literature on the use of XAI in medicine for diagnosis. We surveyed 20 studies

Table 2.1 Examining the literature on the application of XAI in medicine for diagnosis

Reference	Goal	XAI approach/types	AI techniques	Results
[167]	Diagnosis of allergies	Condition prediction (IF–THEN). Rules based	kNN, SVM, AdaBag MLP and RF	Accuracy: 86.39% Sensitivity: 75%
[168]	Treatments for breast cancer	Adaptive dimension reduction	Cluster analysis	N/A
[169]	Treatment of spine	LIME (explanation by simplification)	One-class SVM, binary RF	F1: 80 ± 12% MCC: 57 ± 23% BSS: 33 ± 28%
[170]	Alzheimer's disease	SHAP, Fuzzy using feature relevance, rule-based	Two-layer model with RF	1st layer: accuracy: 93.95% F1-score: 93.94% 2nd layer: 87.08% F1-score: 87.09%
[171]	Hepatitis	SHAP, LIME, partial dependence plots (PDP)	LR, DT, kNN, RF, SVM,	Accuracy of 91.9%
[172]	Chronic injury	LIME	CNN-based technique: pre-trained VGG-16	F1-score: 94%, Precision: 95% Recall: 94%
[173]	Fenestral otosclerosis	CNN-based technique: proposed otosclerosis logical neural network (LNN) technique	Displaying DL representations	AUC: 99.5%, Sensitivity: 96.4% Specificity: 98.9%

(Continues)

Table 2.1 (Continued)

Reference	Goal	XAI approach/types	AI techniques	Results
[174]	Lymphedema (Chinese EMR)	Counterfactual multi-granularity graph supporting facts extraction technique	Graph neural network, counterfactual reasoning	F1-score: 99.02%, Precision: 99.04% Recall: 99.00%
[175]	Clinical diagnosis	Entity-aware CNN	Bayesian network ensembles	Top-3 sensitivity: 88.8%
[176]	Glioblastoma multiforme (GBM) diagnosis	LIME	VGG-16	Accuracy: 97%
[177]	Diagnosis of pulmonary nodule	Visually interpretable network (VINet), CAM, LRP and VBP	CNN	Accuracy: 82.15%
[178]	Diagnosis of Alzheimer disease	Context-free grammar	Naïve Bayes (NB), grammatical evolution	Accuracy: 81.5%; Brier: 0.178; F1-score: 85.9%; ROC: 0.913
[179]	Diagnosis of lung cancer	LIME and natural language explanation	Neural network and RF	N/A
[180]	Diagnosis of COVID-19 using chest X-ray	GSInquire	CNN-based model: proposed COVID-Net	Accuracy: 93.3% Sensitivity: 91.0%
[181]	Analysis of colorectal cancer	Explainable collective fuzzy class membership measure	CNN	Accuracy: 91.08%; F1-score: 91.26%; Precision: 91.44% Recall: 91.04%
[182]	Analysis of phenotyping psychiatric disorders	Explainable deep neural network (EDNN) technique	DNN	White matter accuracy: 90.22% Sensitivity: 89.21%Specificity: 91.23%
[183]	Post-stroke hospital discharge disposition	LR, LIME	LR, RF, RF with AdaBoost and MLP	Accuracy: 71.0% F1-score: 59.0%; Precision: 64%; Recall: 26.0%
[184]	Choosing a diagnosis and treatment for breast cancer	Case-based reasoning (CBR) approach	kNN, distance-weighted kNN, rainbow boxes-inspired algorithm	Accuracy: 80.3%
[185]	Diagnosis of Alzheimer's	An interpretable ML ideal: light high-order communication archetypal with dismissal choice	DT, RF and SVM	AUC: 0.81;Sensitivity: 84.0% Specificity: 67.0%
[186]	Automatic recognition of instruments in laparoscopy videos	Activation maps	CNN	F1-score: 97.0%; Precision: 96.0%; Sensitivity: 86.0%

on XAI in diagnostic. This section of the chapter included research that was examined from the perspectives of study objectives, XAI approaches, XIA procedures, and performances, as can be shown in Table 2.1. We discovered that CNNs are the most widely used deep learning method, with 25% (5/20) of articles using VGG-16 and other CNN-based models; additionally, the most prevalent XAI method in these publications was LIME, with 30% (6/20) of the research that was done in the published articles included in this study involved LIME. Most studies using XAI approaches followed the pipeline and used post-hoc methods. They developed the medical applications and helped the doctors make decisions by first applying DL techniques like CNN-based techniques or complicated ML random forest (RF) model, and then by explaining the AI model using post-hoc techniques including XAI assessments employed in the studies covering 45% (9/20) of the papers survey.

2.5 Ethical effects and implications

AI-powered technologies are becoming more prevalent in healthcare, the ethical concerns related to this impending paradigm shift must be investigated. The "Principles of Biomedical Ethics" by Beauchamp and Childress are a frequently used and well-suited ethical foundation when evaluating biomedical ethical dilemmas [187,188]. Introducing equity as well as independence, humanitarianism, non-maleficence, and four fundamental concepts [187]. Although not all bioethical approaches are principlist guideline obtainable, it is a very practical, fundamental structure that is well-liked in both academic and clinical environments [189]. Therefore, in the sections that follow, we evaluate explainability in light of the previously discussed principles.

Explainability has significance for both patients and doctors with regard to autonomy [190]. Informed consent is one of the main safeguards for patients' autonomy, which is an independent, typically written consent by which the patient gives the doctor permission to carry out a certain medical act [191]. A thorough understanding of the nature and hazards of a medical process is the foundation for proper informed consent, and not unreasonably interfering with the patient's resolution to have the surgery. Currently, there is no ethically accepted position on whether informed consent should entail disclosure of the employment of a mysterious medical AI algorithm. Currently, there is no ethically accepted position on whether informed consent should entail disclosure of the employment of a mysterious medical AI algorithm, which could contravene the adherence to clinical advice. If a patient afterwards learned that a clinician's prescription came from a mysterious AI system. This can prompt the patient to question the advice as well as make a valid search for a clarification, which the physician, in the scenario of an inaccessible system, would not be able to provide. Thus, opaque medical AI may be a barrier to the delivery of accurate information and so might put explicit consent in danger. Therefore, it is crucial to uphold ethically acceptable and explainability criteria in order to protect the informed consent's autonomy-preserving function.

The potential for opaque AI to encourage authoritarianism by limiting patients' ability to voice their interests and desires on medical practices should be taken into consideration [191]. Complete patient autonomy is a prerequisite for shared decision-making, Full autonomy, however, may only be attained if the patient is given a variety of worthwhile alternatives from which to choose [192]. In this way, as opaque AI becomes increasingly crucial to medical decision-making, patients' capabilities to exercise their autonomy about medical procedures are diminished. The difficulty with opaque MDSS, in particular, is that it is still unknown whether and how patient opinions and desires are taken into consideration by the model. The "Value-flexible" AI that offers the patient a variety of options could be used to address this situation [193]. In addition, we contend that explainability is a prerequisite for value-flexible AI. In order to assess if the goals have been achieved, the patient must be able to comprehend which variables are crucial to the AI system's inner workings, whether the AI system's values and weighting are in line with theirs. The assessment of patients, for whom a "lessening of distress" is more significant, may not be aligned, for instance, with AI systems that are designed with "survival" as the desired objective [193]. Last but not least, when a decision is reached, patients must have confidence in the AI system and freedom to heed its advice [194]. If the AI model is opaque, then this is not feasible. As a result, explainability is a moral requirement for schemes supportive essential medical decision making from both the perspective of the doctor and the patient.

Although there are connections between the concepts of kindness and nonmaleficence, they nevertheless shed light on various topics, including explainability. Physicians are urged by beneficence to optimize patient advantages. Clinicians are therefore expected to employ AI-based technologies in a way that encourages the best possible outcome for the particular patient. However, in order to give patients the best options for enhancing their health and wellness, the system's full capabilities must be available to doctors. This suggests that doctors are familiar with the technology outside of its use for robotic tasks in a particular clinical scenario, enabling them to consider the results of the system for doctors. Instead than having to rely solely on an automated output, explainability in the form of visuals or natural language explanations permits informed clinical decisions. They can evaluate thoroughly the results produced by the system, and they decide for themselves if the outcomes appear reliable or not. This enables them to, if needed, modify predictions and advice to account for specific situations. As a result, physicians can lessen the chance of inspiring false optimism or inspiring false despair but can also use their clinical decision making to indicate possibly improper procedures [195]. This is particularly crucial when we consider a scenario in which a doctor and an AI system disagree, a difficult position to resolve [194]. This is ultimately a matter of intellectual competence, and furthermore, it is not apparent how doctors should determine whether they can put their faith in a black box model's epistemic authority enough to accept its judgment [194]. According to authors in [194], there is little epistemic evidence for deference in the situation of opaque AI. Furthermore, they contend that when faced with a "black-box" system, medical decision assistance may actually work against rather than in favor of doctors' competence. In order to avoid being scrutinized or held responsible, doctors may be compelled to use "defensive medicine" in this situation [194]. The autonomy of doctors would be seriously threatened in such a circumstance. Furthermore, doctors will infrequently have the opportunity to thoroughly examine the reasons why their clinical judgment conflicts with the AI system. Therefore, in the clinical context, focusing solely on a performance outcome is insufficient. Only healthcare professionals who are capable of making well-informed decisions about when to use an intelligence MDSS and how to comprehend its outcomes can be expected to provide patients with the best possible outcomes. Thus, it is difficult to see how any "black box" application might achieve beneficence in the context of medical AI.

When considering the notion of non-maleficence in relation to medical AI, the requirement for explainability becomes even more clear. According to non-maleficence, doctors have a moral obligation to protect their patients from any purposeful or unintentional injury, or by using medical procedures excessively or inappropriately. Why does performance fall short? It has been suggested that a black-box medical AI that only considers validated optimal performance is morally acceptable, even if the physician cannot determine the exact causative mechanisms underlying a particular AI-recommended intervention [196]. In fact, it is still fairly usual in medicine to base conclusions about a treatment's effectiveness only on anecdotal or experiential data. However, this does not provide justification for avoiding explanations, which are a crucial component of effective clinical judgment when one is in fact feasible. Recent developments in defining at least the key characteristics of AI models, while avoiding comprehensive mechanical justifications for AI judgments, a fundamental ethical responsibility to improve the interpretability and transparency of medical AI. If this were not done, a doctor's ability to monitor for potential clinical case misclassifications would be purposefully undermined, for instance, to training datasets with significant bias or variation. Thus, we come to the conclusion that explainability is a required quality of clinically applied AI systems, including with regard to beneficence and non-maleficence.

According to the fairness principle, no one should be ethically acceptable in discriminating against any selected individuals or social group in order to get the benefits of medical advancement [197]. Unfortunately, several AI systems go against this rule. For instance, authors in [198] recently reported on a medical AI system that prejudiced people of color. Explainability can assist developers and medical professionals in identifying and eliminating these biases, which are a significant source of prospective injustice, preferable in the early stages of the creation and validation of AI by, for instance, identifying significant traits that point to a bias in the model. However, in order for explainability to achieve this goal, the necessary participant parties must be made aware of the bias risk and what effects it might have on people's health and happiness. Sometimes, it could be tempting to put accuracy first, and simply avoid devoting energy to creating understandable AI. However, developers and doctors must be aware of the possible drawbacks and restrictions of these new tools if they are to guarantee that AI-powered DSSs function effectively. Explainability thus becomes a moral need for the creation and use of AI-based healthcare decision assistance, including from a just standpoint.

2.5.1 XAI weaknesses in medicine

Despite the impressive performance that many of these AI technologies may achieve, it is frequently challenging for these technologies to be fully embraced in real-world clinical settings since some of these algorithms are difficult to understand. XAI is becoming more prevalent to help patients and healthcare providers communicate internal decisions, behaviour, and activities. Clinicians may learn how to use predictive modelling in real-world scenarios rather than just blindly following the forecasts if XAI explains the prediction consequences, earning their trust. Due to the complexity of medical knowledge, there are still numerous possibilities to investigate regarding ways to advance the productivity of XAI in medical settings [199]. Predictive modelling has extensively exploited data from numerous sources in the healthcare industry. While some healthcare data are produced by patients and caregivers, some are produced by drug testing and investigation, and some are collected by sensor technology.

2.6 Conclusion and future directions

It has been demonstrated that MDSS can help healthcare professionals with a range of decisions and patient care duties, and they now enthusiastically and widely advocate the provision of high-quality healthcare. Some MDSS applications have more supporting data, particularly those that use CPOE. As we move into the era of the EMR, support for MDSS is growing, and there is nevertheless room for development in interoperability, execution speed and convenience, and cost. While doing so, we must continue to watch out for MDSS's possible flaws, this can range from just failing and losing resources to wearing down healthcare professionals and lowering the standard of patient care. Construction, implementation, and maintenance of MDSS must be done with extra caution and thoughtful planning. Therefore, in this chapter, the function of XAI in clinical decision support systems was examined from the viewpoints of technology, ethics, medicine, and patients. Thus, we have demonstrated that explainability is a complex idea with wide-ranging ramifications for the many key stakeholders concerned. Because it calls for a rethinking of roles and duties, medical AI presents difficulties for researchers, physicians, and policymakers. Based on our investigation, we believe that explainability is a prerequisite for addressing these issues in a lasting way that is consistent with professional standards and values. The survey also revealed various difficulties and restrictions. First off, it is unrealistic to utilize accuracy as the sole ML evaluation parameter for assessing the performance of a techniques. It is impossible to evaluate the ML algorithm objectively by using just one assessment metric. Second, there are currently no standardized XAI evaluation techniques that are widely acknowledged by researchers in the field. Because evaluation still depends on human cognition, only qualitative evaluation of XAI techniques is possible. But the majority of the papers included in this survey merely gave XAI methods, with no XAI evaluation. Only a small number of researchers offered XAI assessments by doctors. Third, several experiments solely used XAI or existing ML techniques. These XAI medicinal applications are deficient in creativity and earlier information from the doctors because no medical professionals were involved in the design of these AI techniques. As a result, they may not satisfy the actual clinical needs of the doctors.

Acknowledgment

The work of Agbotiname Lucky Imoize is supported in part by the Nigerian Petroleum Technology Development Fund (PTDF) and in part by the German Academic Exchange Service (DAAD) through the Nigerian-German Postgraduate Program under Grant 57473408.

References

[1] Folorunso SO, Awotunde JB, Banjo OO, Ogundepo EA, and Adeboye NO. Comparison of active COVID-19 cases per population using time-series models. *International Journal of E-Health and Medical Communications (IJEHMC).* 2021;13(2):1–21.

[2] Knapič S, Malhi A, Saluja R, and Främling K. Explainable artificial intelligence for human decision support system in the medical domain. *Machine Learning and Knowledge Extraction.* 2021;3(3):740–70.

[3] Awotunde JB, Oluwabukonla S, Chakraborty C, Bhoi AK, and Ajamu GJ. Application of artificial intelligence and big data for fighting COVID-19 pandemic. In *Decision Sciences for COVID-19* (pp. 3–26); 2022.

[4] Meske C and Bunde E. Transparency and trust in human-AI-interaction: the role of model-agnostic explanations in computer vision-based decision support. In *International Conference on Human-Computer Interaction* (pp. 54–69). Cham: Springer; 2020.

[5] Awotunde JB, Adeniyi EA, Ajamu GJ, Balogun GB, and Taofeek-Ibrahim FA. Explainable artificial intelligence in genomic sequence for healthcare systems prediction. In *Connected e-Health* (pp. 417–37). Cham: Springer; 2022.

[6] Voigt P. and Von dem Bussche A. *The EU General Data Protection Regulation. A Practical Guide.* New York, NY: Springer; 2017.

[7] Holzinger A, Biemann C, Pattichis CS, and Kell DB. What do we need to build explainable AI systems for the medical domain?. arXiv preprint arXiv:1712.09923. 2017 Dec 28.

[8] Folorunso SO, Awotunde JB, Adeniyi EA, Abiodun KM, and Ayo FE. Heart disease classification using machine learning models. In *International Conference on Informatics and Intelligent Applications* 2021 November 25 (pp. 35–49). Cham: Springer; 2021.

[9] Ribeiro MT, Singh S, and Guestrin C. "Why should i trust you?" Explaining the predictions of any classifier. In *Proceedings of the 22nd ACM SIGKDD International Conference on Knowledge Discovery and Data Mining*, 2016 August 13 (pp. 1135–44); 2016.

[10] van Wijk Y, Halilaj I, van Limbergen E, *et al* . Decision support systems in prostate cancer treatment: an overview. *BioMed Research International.* 2019;2019:1–10.

[11] Vellido A. The importance of interpretability and visualization in machine learning for applications in medicine and health care. *Neural Computing and Applications.* 2020;32(24):18069–83.

[12] Christopoulou SC. Impacts on context aware systems in evidence-based health informatics: a review. In *Healthcare* (Vol. 10, No. 4, p. 685). Basel: MDPI; 2022.

[13] Yang Z and Dong S. HAGERec: hierarchical attention graph convolutional network incorporating knowledge graph for explainable recommendation. *Knowledge-Based Systems.* 2020;204:106194.

[14] Jiang J, Li X, Zhao C, Guan Y, and Yu Q. Learning and inference in knowledge-based probabilistic model for medical diagnosis. *Knowledge-Based Systems.* 2017;138:58–68.

[15] Uzoka FM, Osuji J, and Obot O. Clinical decision support system (DSS) in the diagnosis of malaria: a case comparison of two soft computing methodologies. *Expert Systems with Applications.* 2011;38(3):1537–53.

[16] Alexy T, Sangkatumvong S, Connes P, *et al.* Sickle cell disease: selected aspects of pathophysiology. *Clinical Hemorheology and Microcirculation.* 2010;44(3):155–66.

[17] Bodas P, Huang A, O'Riordan MA, Sedor JR, and Dell KM. The prevalence of hypertension and abnormal kidney function in children with sickle cell disease – a cross sectional review. *BMC Nephrology.* 2013;14(1):1–6.

[18] Kanter J, Walters MC, Krishnamurti L, *et al.* Biologic and clinical efficacy of LentiGlobin for sickle cell disease. *New England Journal of Medicine.* 2022;386(7):617–28.

[19] Goyal S, Tisdale J, Schmidt M, *et al.* Acute myeloid leukemia case after gene therapy for sickle cell disease. *New England Journal of Medicine.* 2022;386(2):138–47.

[20] Frangoul H, Altshuler D, Cappellini MD, *et al.* CRISPR-Cas9 gene editing for sickle cell disease and β-thalassemia. *New England Journal of Medicine.* 2021;384(3):252–60.

[21] Moore M and Loper KA. An introduction to clinical decision support systems. *Journal of Electronic Resources in Medical Libraries.* 2011;8(4): 348–66.

[22] Spänig S, Emberger-Klein A, Sowa JP, Canbay A, Menrad K, and Heider D. The virtual doctor: an interactive clinical-decision-support system based on deep learning for non-invasive prediction of diabetes. *Artificial Intelligence in Medicine.* 2019;100:101706.

[23] Légat L, Van Laere S, Nyssen M, Steurbaut S, Dupont AG, and Cornu P. Clinical decision support systems for drug allergy checking: systematic review. *Journal of Medical Internet Research.* 2018;20(9):e8206.

[24] Sim I, Gorman P, Greenes RA, *et al.* Clinical decision support systems for the practice of evidence-based medicine. *Journal of the American Medical Informatics Association.* 2001;8(6):527–34.

[25] Karnieli-Miller O, Artom TR, and Neufeld-Kroszynski G. Time to rise to the challenge of truly implementing patient-centered care and shared decision making in Israel: the educational and policy mission. In *Zeitschrift für Evidenz, Fortbildung und Qualität im Gesundheitswesen*; 2022.

[26] Weed LL and Weed L. Ending medicine's chronic dysfunction: tools and standards for medical decision making. In *Synthesis Lectures on Assistive, Rehabilitative, and Health-Preserving Technologies.* 2021;10(1):i–177.

[27] Middleton B, Sittig DF, and Wright A. Clinical decision support: a 25 year retrospective and a 25 year vision. *Yearbook of Medical Informatics.* 2016;25(S01):S103–106.

[28] Jung SY, Hwang H, Lee K, *et al.* Barriers and facilitators to implementation of medication decision support systems in electronic medical records: mixed methods approach based on structural equation modeling and qualitative analysis. *JMIR Medical Informatics.* 2020;8(7):e18758.

[29] Dias D and Paulo Silva Cunha J. Wearable health devices—vital sign monitoring, systems and technologies. *Sensors.* 2018;18(8):2414.

[30] Sutton RT, Pincock D, Baumgart DC, Sadowski DC, Fedorak RN, and Kroeker KI. An overview of clinical decision support systems: benefits, risks, and strategies for success. *NPJ Digital Medicine.* 2020;3(1):1–0.

[31] Wake DT, Smith DM, Kazi S, and Dunnenberger HM. Pharmacogenomic clinical decision support: a review, how-to guide, and future vision. *Clinical Pharmacology & Therapeutics.* 2022;112(1):44–57.

[32] Dowse R. Advantages and functions of clinical and decision support systems. *Journal of Biomedical and Sustainable Healthcare Applications.* 2022;22:43–50.

[33] Shortliffe EH and Sepúlveda MJ. Clinical decision support in the era of artificial intelligence. *Jama.* 2018;320(21):2199–200.

[34] Osheroff JA, Teich JM, Middleton B, Steen EB, Wright A, and Detmer DE. A roadmap for national action on clinical decision support. *Journal of the American Medical Informatics Association.* 2007;14(2):141–5.

[35] Kwok R, Dinh M, Dinh D, and Chu M. Improving adherence to asthma clinical guidelines and discharge documentation from emergency departments: implementation of a dynamic and integrated electronic decision support system. *Emergency Medicine Australasia.* 2009;21(1):31–7.

[36] Davis DA and Taylor-Vaisey A. Translating guidelines into practice: a systematic review of theoretic concepts, practical experience and research evidence in the adoption of clinical practice guidelines. *Cmaj.* 1997;157 (4):408–16.

[37] Rand C, Powe N, Wu AW, and Wilson MH. Why don't physicians follow clinical practice guidelines. *JAMA*. 1999;282:14581465.

[38] Alsop DC, Detre JA, Golay X, *et al.* Recommended implementation of arterial spin-labeled perfusion MRI for clinical applications: a consensus of the ISMRM perfusion study group and the European consortium for ASL in dementia. *Magnetic Resonance in Medicine*. 2015;73(1):102–16.

[39] Lutz W, Deisenhofer AK, Rubel J, *et al.* Prospective evaluation of a clinical decision support system in psychological therapy. *Journal of Consulting and Clinical Psychology*. 2022;90(1):90.

[40] Siu PK, Tang V, Choy KL, Lam HY, and Ho GT. An intelligent clinical decision support system for assessing the needs of a long-term care plan. In *Recent Advances in Digital System Diagnosis and Management of Healthcare*. London: IntechOpen; 2019.

[41] Wu G, Yang P, Xie Y, *et al.* Development of a clinical decision support system for severity risk prediction and triage of COVID-19 patients at hospital admission: an international multicentre study. *European Respiratory Journal*. 2020;56(2):2001104.

[42] Folorunso SO, Ogundepo EA, Awotunde JB, Ayo FE, Banjo OO, and Taiwo AI. A multi-step predictive model for COVID-19 cases in Nigeria using machine learning. In *Decision Sciences for COVID-19* (pp. 107–36). Cham: Springer; 2022.

[43] Embi PJ, Jain A, Clark J, and Harris CM. Development of an electronic health record-based Clinical Trial Alert system to enhance recruitment at the point of care. In *AMIA Annual Symposium Proceedings* (Vol. 2005, p. 231). Bethesda, MD: American Medical Informatics Association; 2005.

[44] Cricelli I, Marconi E, and Lapi F. Clinical Decision Support System (MDSS) in primary care: from pragmatic use to the best approach to assess their benefit/risk profile in clinical practice. *Current Medical Research and Opinion*. 2022;38(5):827–9.

[45] Zare S, Mobarak Z, Meidani Z, Nabovati E, and Nazemi Z. Effectiveness of clinical decision support systems on the appropriate use of imaging for central nervous system injuries: a systematic review. *Applied Clinical Informatics*. 2022;13(01):037–52.

[46] Algaze CA, Wood M, Pageler NM, Sharek PJ, Longhurst CA, and Shin AY. Use of a checklist and clinical decision support tool reduces laboratory use and improves cost. *Pediatrics*. 2016;137(1).

[47] Pruszydlo MG, Walk-Fritz SU, Hoppe-Tichy T, Kaltschmidt J, and Haefeli WE. Development and evaluation of a computerised clinical decision support system for switching drugs at the interface between primary and tertiary care. *BMC Medical Informatics and Decision Making*. 2012;12 (1):1–8.

[48] Bell C, Jalali A, and Mensah E. A decision support tool for using an ICD-10 anatomographer to address admission coding inaccuracies: a commentary. *Online Journal of Public Health Informatics*. 2013;5(2):222.

[49] Haberman S, Feldman J, Merhi ZO, Markenson G, Cohen W, and Minkoff H. Effect of clinical-decision support on documentation compliance in an electronic medical record. *Obstetrics & Gynecology*. 2009;114(2 Part 1): 311–7.

[50] Turchin A, Shubina M, and Gandhi T. NLP for patient safety: splenectomy and pneumovax. In *Proceedings of AMIA 2010 Annual Symposium*, 2010 November 13.

[51] McEvoy D, Gandhi TK, Turchin A, and Wright A. Enhancing problem list documentation in electronic health records using two methods: the example of prior splenectomy. *BMJ Quality & Safety*. 2018;27(1):40–7.

[52] Vonbach P, Dubied A, Krähenbühl S, and Beer JH. Prevalence of drug–drug interactions at hospital entry and during hospital stay of patients in internal medicine. *European Journal of Internal Medicine*. 2008;19(6):413–20.

[53] Helmons PJ, Suijkerbuijk BO, Nannan Panday PV, and Kosterink JG. Drug-drug interaction checking assisted by clinical decision support: a return on investment analysis. *Journal of the American Medical Informatics Association*. 2015;22(4):764–72.

[54] Koutkias V and Bouaud J (eds.) , Section editors for the IMIA Yearbook section on decision support. Contributions from the 2017 literature on clinical decision support. *Yearbook of Medical Informatics*. 2018;27(01):122–8.

[55] Cornu P, Phansalkar S, Seger DL, *et al.* High-priority and low-priority drug–drug interactions in different international electronic health record systems: a comparative study. *International Journal of Medical Informatics*. 2018;111:165–71.

[56] Phansalkar S, Desai AA, Bell D, *et al.* High-priority drug–drug interactions for use in electronic health records. *Journal of the American Medical Informatics Association*. 2012;19(5):735–43.

[57] McEvoy DS, Sittig DF, Hickman TT, *et al.* Variation in high-priority drug–drug interaction alerts across institutions and electronic health records. *Journal of the American Medical Informatics Association*. 2017;24(2): 331–8.

[58] Cho I, Lee JH, Choi J, Hwang H, and Bates DW. National rules for drug–drug interactions: are they appropriate for tertiary hospitals? *Journal of Korean Medical Science*. 2016;31(12):1887–96.

[59] Eslami S, de Keizer NF, Dongelmans DA, de Jonge E, Schultz MJ, and Abu-Hanna A. Effects of two different levels of computerized decision support on blood glucose regulation in critically ill patients. *International Journal of Medical Informatics*. 2012;81(1):53–60.

[60] Jia P, Zhang L, Chen J, Zhao P, and Zhang M. The effects of clinical decision support systems on medication safety: an overview. *PLoS One*. 2016; 11(12):e0167683.

[61] Mahoney CD, Berard-Collins CM, Coleman R, Amaral JF, and Cotter CM. Effects of an integrated clinical information system on medication safety in a multi-hospital setting. *American Journal of Health-System Pharmacy*. 2007;64(18):1969–77.

[62] Peris-Lopez P, Orfila A, Mitrokotsa A, and Van der Lubbe JC. A comprehensive RFID solution to enhance inpatient medication safety. *International Journal of Medical Informatics.* 2011;80(1):13–24.

[63] Poon EG, Keohane CA, Yoon CS, *et al.* Effect of bar-code technology on the safety of medication administration. *New England Journal of Medicine.* 2010;362(18):1698–707.

[64] Van Der Veen W, Van Den Bemt PM, Wouters H, *et al.* Association between workarounds and medication administration errors in bar-code-assisted medication administration in hospitals. *Journal of the American Medical Informatics Association.* 2018;25(4):385–92.

[65] Kwan JL, Lo L, Ferguson J, *et al.* Computerised clinical decision support systems and absolute improvements in care: meta-analysis of controlled clinical trials. *Bmj.* 2020;370:m3216.

[66] Berner ES. Diagnostic decision support systems: why aren't they used more and what can we do about it?. In *AMIA Annual Symposium Proceedings* (Vol. 2006, p. 1167). Bethesda, MD: American Medical Informatics Association; 2006.

[67] Segal MM, Rahm AK, Hulse NC, *et al.* Experience with integrating diagnostic decision support software with electronic health records: benefits versus risks of information sharing. *eGEMs.* 2017;5(1):23.

[68] Kunhimangalam R, Ovallath S, and Joseph PK. A clinical decision support system with an integrated EMR for diagnosis of peripheral neuropathy. *Journal of Medical Systems.* 2014;38(4):1–4.

[69] Martinez-Franco AI, Sanchez-Mendiola M, Mazon-Ramirez JJ, *et al.* Diagnostic accuracy in Family Medicine residents using a clinical decision support system (DXplain): a randomized-controlled trial. *Diagnosis.* 2018; 5(2):71–6.

[70] Singh H, Schiff GD, Graber ML, Onakpoya I, and Thompson MJ. The global burden of diagnostic errors in primary care. *BMJ Quality & Safety.* 2017;26 (6):484–94.

[71] Singh H, Meyer AN, and Thomas EJ. The frequency of diagnostic errors in outpatient care: estimations from three large observational studies involving US adult populations. *BMJ Quality & Safety.* 2014;23(9):727–31.

[72] Fraser H, Coiera E, and Wong D. Safety of patient-facing digital symptom checkers. *The Lancet.* 2018;392(10161):2263–4.

[73] Jaspers MW, Smeulers M, Vermeulen H, and Peute LW. Effects of clinical decision-support systems on practitioner performance and patient outcomes: a synthesis of high-quality systematic review findings. *Journal of the American Medical Informatics Association.* 2011;18(3):327–34.

[74] Jha AK, DesRoches CM, Campbell EG, *et al.* Use of electronic health records in US hospitals. *New England Journal of Medicine.* 2009; 360(16):1628–38.

[75] Bright TJ, Wong A, Dhurjati R, *et al.* Effect of clinical decision-support systems: a systematic review. *Annals of Internal Medicine.* 2012;157(1): 29–43.

[76] Castaneda C, Nalley K, Mannion C, *et al.* Clinical decision support systems for improving diagnostic accuracy and achieving precision medicine. *Journal of Clinical Bioinformatics.* 2015;5(1):1–6.

[77] Van Biesen W, Van Cauwenberge D, Decruyenaere J, Leune T, and Sterckx S. An exploration of expectations and perceptions of practicing physicians on the implementation of computerized clinical decision support systems using a Qsort approach. *BMC Medical Informatics and Decision Making.* 2022;22(1):1–0.

[78] Simões AS, Maia MR, Gregório J, *et al.* Participatory implementation of an antibiotic stewardship programme supported by an innovative surveillance and clinical decision-support system. *Journal of Hospital Infection.* 2018;100(3):257–64.

[79] Henry KE, Kornfield R, Sridharan A, *et al.* Human–machine teaming is key to AI adoption: clinicians' experiences with a deployed machine learning system. *npj Digital Medicine*. 2022;5(1):1–6.

[80] Awotunde JB, Ayoade OB, Ajamu GJ, AbdulRaheem M, and Oladipo ID. Internet of Things and cloud activity monitoring systems for elderly healthcare. In *Studies in Computational Intelligence*, 1011 (pp. 181–207). Singapore: Springer; 2022.

[81] Ayo FE, Misra S, Awotunde JB, Behera RK, Oluranti J, and Ahuja R. A mobile-based patient surgical appointment system using fuzzy logic. In *Proceedings of Third International Conference on Computing, Communications, and Cyber-Security* (pp. 193–207). Singapore: Springer; 2023.

[82] Berner ES and Lande TJ. Overview of clinical decision support systems. In *Clinical Decision Support Systems* (pp. 3–22). New York, NY: Springer; 2007.

[83] Shabot MM, LoBue M, and Chen J. Wireless clinical alerts for physiologic, laboratory and medication data. In *Proceedings of the AMIA Symposium* (p. 789). Bethesda, MD: American Medical Informatics Association; 2000.

[84] Galanter WL, DiDomenico RJ, and Polikaitis A. Preventing exacerbation of an ADE with automated decision support. *Journal of Healthcare Information Management.* 2002;16(4):44–9.

[85] Elchynski AL, Desai N, D'Silva D, *et al.* Utilizing a human–computer interaction approach to evaluate the design of current pharmacogenomics clinical decision support. *Journal of Personalized Medicine*. 2021;11 (11):1227.

[86] Dowding D, Mitchell N, Randell R, Foster R, Lattimer V, and Thompson C. Nurses' use of computerised clinical decision support systems: a case site analysis. *Journal of Clinical Nursing.* 2009;18(8):1159–67.

[87] Zadro JR, Traeger AC, Décary S, and O'Keeffe M. Problem with patient decision aids. *BMJ Evidence-Based Medicine.* 2021;26(4):180–3.

[88] Goddard K, Roudsari A, and Wyatt JC. Automation bias – a hidden issue for clinical decision support system use. *International Perspectives in Health Informatics.* 2011:17–22.

[89] Matheus S and Den VB. An analysis of artificial intelligence based clinical decision support systems. *Journal of Biomedical and Sustainable Healthcare Applications.* 2021;1:9–17.
[90] Ash JS, Sittig DF, Campbell EM, Guappone KP, and Dykstra RH. Some unintended consequences of clinical decision support systems. In *Amia annual Symposium Proceedings* (Vol. 2007, p. 26). Bethesda, MD: American Medical Informatics Association; 2007.
[91] Shahid N, Rappon T, and Berta W. Applications of artificial neural networks in health care organizational decision-making: a scoping review. *PLoS One.* 2019;14(2):e0212356.
[92] O'Reilly D, Tarride JE, Goeree R, Lokker C, and McKibbon KA. The economics of health information technology in medication management: a systematic review of economic evaluations. *Journal of the American Medical Informatics Association.* 2012;19(3):423–38.
[93] Jacob V, Thota AB, Chattopadhyay SK, *et al.* Cost and economic benefit of clinical decision support systems for cardiovascular disease prevention: a community guide systematic review. *Journal of the American Medical Informatics Association.* 2017;24(3):669–76.
[94] Gordon WJ and Catalini C. Blockchain technology for healthcare: facilitating the transition to patient-driven interoperability. *Computational and Structural Biotechnology Journal.* 2018;16:224–30.
[95] Awotunde JB, Chakraborty C, and Folorunso SO. A secured smart healthcare monitoring systems using blockchain technology. *In Intelligent Internet of Things for Healthcare and Industry* (pp. 127–43). Cham: Springer; 2022.
[96] Alves NF, Ferreira L, Lopes N, *et al.* FHIRbox, a cloud integration system for clinical observations. *Procedia Computer Science.* 2018;138:303–9.
[97] Katehakis DG, Kouroubali A, and Fundulaki I. Towards the Development of a National eHealth Interoperability Framework to Address Public Health Challenges in Greece. InSWH@ ISWC 2018 October 9.
[98] Bincoletto G. Data protection issues in cross-border interoperability of Electronic Health Record systems within the European Union. *Data & Policy.* 2020;2:1–11.
[99] Bhavaraju SR. From subconscious to conscious to artificial intelligence: a focus on electronic health records. *Neurology India.* 2018;66(5):1270.
[100] Fernández-Cardeñosa G, de la Torre-Díez I, López-Coronado M, and Rodrigues JJ. Analysis of cloud-based solutions on EHRs systems in different scenarios. *Journal of Medical Systems.* 2012;36(6):3777–82.
[101] Rodrigues JJ, de la Torre I, Fernández G, and López-Coronado M. Analysis of the security and privacy requirements of cloud-based electronic health records systems. *Journal of Medical Internet Research.* 2013;15(8):e2494.
[102] Sloane EB and Silva RJ. Artificial intelligence in medical devices and clinical decision support systems. In *Clinical Engineering Handbook* 2020 (pp. 556–68). London: Academic Press.
[103] Tyler NS and Jacobs PG. Artificial intelligence in decision support systems for type 1 diabetes. *Sensors.* 2020;20(11):3214.

[104] Sechopoulos I and Mann RM. Stand-alone artificial intelligence – the future of breast cancer screening?. *The Breast.* 2020;49:254–60.
[105] Ngiam KY and Khor W. Big data and machine learning algorithms for health-care delivery. *The Lancet Oncology.* 2019;20(5):e262–73.
[106] Strasinger SK and Di Lorenzo MS. *Urinalysis and Body Fluids.* Philadelphia, PA: FA Davis; 2014.
[107] Charlton CL, Babady E, Ginocchio CC, *et al.* Practical guidance for clinical microbiology laboratories: viruses causing acute respiratory tract infections. *Clinical Microbiology Reviews.* 2018;32(1):e00042–18.
[108] Ter Voert EE, Muehlematter UJ, Delso G, *et al.* Quantitative performance and optimal regularization parameter in block sequential regularized expectation maximization reconstructions in clinical 68Ga-PSMA PET/MR. *EJNMMI Research.* 2018;8(1):1–5.
[109] Giuliano AE, Edge SB, and Hortobagyi GN. Eighth edition of the AJCC cancer staging manual: breast cancer. *Annals of Surgical Oncology.* 2018;25(7):1783–5.
[110] Shickel B, Tighe PJ, Bihorac A, and Rashidi P. Deep EHR: a survey of recent advances in deep learning techniques for electronic health record (EHR) analysis. *IEEE Journal of Biomedical and Health Informatics.* 2017;22(5):1589–604.
[111] Lai CL, Chien SW, Chang LH, Chen SC, and Fang K. Enhancing medication safety and healthcare for inpatients using RFID. In *PICMET'07-2007 Portland International Conference on Management of Engineering & Technology*, 2007 August 5 (pp. 2783–90). New York, NY: IEEE; 2007.
[112] Vijai C and Wisetsri W. Rise of artificial intelligence in healthcare startups in India. *Advances in Management.* 2021;14(1):48–52.
[113] Zagabathuni Y. Applications, scope, and challenges for AI in healthcare. *International Journal.* 2022;10(4):195–9.
[114] Dash S, Shakyawar SK, Sharma M, and Kaushik S. Big data in healthcare: management, analysis and future prospects. *Journal of Big Data.* 2019;6(1):1–25.
[115] Suzuki K and Chen Y (eds.) , editors. *Artificial Intelligence in Decision Support Systems for Diagnosis in Medical Imaging.* Cham: Springer; 2018.
[116] Musen MA, Middleton B, and Greenes RA. Clinical decision-support systems. In *Biomedical Informatics* (pp. 795–840). Cham: Springer; 2021.
[117] Davis KD, Aghaeepour N, Ahn AH, *et al.* Discovery and validation of biomarkers to aid the development of safe and effective pain therapeutics: challenges and opportunities. *Nature Reviews Neurology.* 2020;16(7):381–400.
[118] Erol O, Unsar S, Yacan L, Pelin M, Kurt S, and Erdogan B. Pain experiences of patients with advanced cancer: a qualitative descriptive study. *European Journal of Oncology Nursing.* 2018 1;33:28–34.
[119] Pappas AF and Christodoulou DK. A new minimally invasive treatment of pilonidal sinus disease with the use of a diode laser: a prospective large series of patients. *Colorectal Disease.* 2018;20(8):O207–14.

[120] Gupta P. Applications of fuzzy logic in daily life. *International Journal of Advanced Research in Computer Science.* 2017;8(5):1795–800.

[121] Lötsch J and Ultsch A. Machine learning in pain research. *Pain.* 2018;159 (4):623.

[122] Sajadi NA, Borzouei S, Mahjub H, and Farhadian M. Diagnosis of hypothyroidism using a fuzzy rule-based expert system. *Clinical Epidemiology and Global Health.* 2019;7(4):519–24.

[123] Graban M and Toussaint J. *Lean Hospitals: Improving Quality, Patient Safety, and Employee Engagement.* London: Productivity Press; 2018.

[124] Yang H, Chen Y, Song K, and Yin Z. Multiscale feature-clustering-based fully convolutional autoencoder for fast accurate visual inspection of texture surface defects. *IEEE Transactions on Automation Science and Engineering.* 2019;16(3):1450–67.

[125] Jakeman AJ, Letcher RA, and Norton JP. Ten iterative steps in development and evaluation of environmental models. *Environmental Modelling & Software.* 2006;21(5):602–14.

[126] Aljaaf AJ, Al-Jumeily D, Hussain AJ, Fergus P, Al-Jumaily M, and Abdel-Aziz K. Toward an optimal use of artificial intelligence techniques within a clinical decision support system. In *2015 Science and Information Conference (SAI)* 2015 July 28 (pp. 548–54). New York, NY: IEEE;2015.

[127] Tyler NS, Mosquera-Lopez CM, Wilson LM, *et al.* An artificial intelligence decision support system for the management of type 1 diabetes. *Nature Metabolism.* 2020;2(7):612–9.

[128] Cresswell K, Callaghan M, Khan S, Sheikh Z, Mozaffar H, and Sheikh A. Investigating the use of data-driven artificial intelligence in computerised decision support systems for health and social care: a systematic review. *Health Informatics Journal.* 2020;26(3):2138–47.

[129] Gupta S, Modgil S, Bhattacharyya S, and Bose I. Artificial intelligence for decision support systems in the field of operations research: review and future scope of research. *Annals of Operations Research.* 2021;1–60: 215–74.

[130] Söllner M. Paving the Way for Medical AI: Consumer Response to Artificial Intelligence in Healthcare (Doctoral dissertation, Technische Universität München).

[131] Awotunde JB, Folorunso SO, Jimoh RG, Adeniyi EA, Abiodun KM, and Ajamu GJ. Application of artificial intelligence for COVID-19 epidemic: an exploratory study, opportunities, challenges, and future prospects. In *Artificial Intelligence for COVID-19* (pp. 47–61). New Tork, NY: Springer; 2021.

[132] Awotunde JB, Folorunso SO, Bhoi AK, Adebayo PO, and Ijaz MF. Disease diagnosis system for IoT-based wearable body sensors with machine learning algorithm. In *Hybrid Artificial Intelligence and IoT in Healthcare* (pp. 201–22). Singapore: Springer; 2021.

[133] Yu KH, Beam AL, and Kohane IS. Artificial intelligence in healthcare. *Nature Biomedical Engineering.* 2018;2(10):719–31.

[134] Ho SY, Phua K, Wong L, and Goh WW. Extensions of the external validation for checking learned model interpretability and generalizability. *Patterns.* 2020;1(8):100129.

[135] Finlayson SG, Subbaswamy A, Singh K, *et al.* The clinician and dataset shift in artificial intelligence. *The New England Journal of Medicine.* 2021;385(3):283.

[136] Yang G, Ye Q, and Xia J. Unbox the black-box for the medical explainable AI via multi-modal and multi-centre data fusion: a mini-review, two showcases and beyond. *Information Fusion.* 2022;77:29–52.

[137] Rajan SP and Paranthaman M. Artificial intelligence in healthcare: algorithms and decision support systems. In *Smart Systems for Industrial Applications.* 2022:173–97.

[138] Folorunso SO, Awotunde JB, Ayo FE, Abdullah KK. RADIoT: the unifying framework for iot, radiomics and deep learning modeling. In *Hybrid Artificial Intelligence and IoT in Healthcare* (pp. 109–28). Singapore: Springer; 2021.

[139] Abiodun KM, Awotunde JB, Aremu DR, and Adeniyi EA. Explainable AI for fighting COVID-19 pandemic: opportunities, challenges, and future prospects. In *Computational Intelligence for COVID-19 and Future Pandemics* (pp. 315–32); 2022.

[140] Dauda OI, Awotunde JB, AbdulRaheem M, and Salihu SA. Basic issues and challenges on explainable artificial intelligence (XAI) in healthcare systems. *Principles and Methods of Explainable Artificial Intelligence in Healthcare.* 2022:248–71.

[141] Hamon R, Junklewitz H, and Sanchez I. *Robustness and Explainability of Artificial Intelligence*. Publications Office of the European Union. 2020.

[142] Arrieta AB, Díaz-Rodríguez N, Del Ser J, *et al.* Explainable Artificial Intelligence (XAI): concepts, taxonomies, opportunities and challenges toward responsible AI. *Information Fusion.* 2020;58:82–115.

[143] Turek M. *Explainable Artificial Intelligence (XAI)*. Arlington County, VA: Defense Advanced Research Projects Agency; 2016.

[144] Pauly M, Winston F, Naylor M, *et al. Seemed Like a Good Idea: Alchemy versus Evidence-Based Approaches to Healthcare Management Innovation.* Cambridge: Cambridge University Press; 2022.

[145] Ajagbe SA, Awotunde JB, Oladipupo MA, and Oye OE. Prediction and forecasting of coronavirus cases using artificial intelligence algorithm. In *Machine Learning for Critical Internet of Medical Things Applications and Use Cases* (pp. 31–54). New York, NY: Springer; 2022.

[146] Goebel R, Chander A, Holzinger K, *et al.* Explainable AI: the new 42?. *In International Cross-Domain Conference for Machine Learning and Knowledge Extraction* 2018 August 27 (pp. 295–303). Cham: Springer; 2018.

[147] Bayes T. LII. An essay towards solving a problem in the doctrine of chances. By the late Rev. Mr. Bayes, FRS communicated by Mr. Price, in a letter to John Canton, AMFR S. *Philosophical Transactions of the Royal Society of London.* 1763;53:370–418.

[148] Kyburg Jr HE, Kyburg J, and Teng CM. *Uncertain Inference.* Cambridge: Cambridge University Press; 2001.

[149] Ashburner M, Ball CA, Blake JA, *et al.* Gene ontology: tool for the unification of biology. *Nature Genetics.* 2000;25(1):25–9.

[150] Aldersey-Williams J and Rubert T. Levelised cost of energy – a theoretical justification and critical assessment. *Energy Policy.* 2019;124:169–79.

[151] Liaw A and Wiener M. Classification and regression by randomForest. *R News.* 2002;2(3):18–22.

[152] Hryniewska W, Bombiński P, Szatkowski P, Tomaszewska P, Przelaskowski A, and Biecek P. Checklist for responsible deep learning modeling of medical images based on COVID-19 detection studies. *Pattern Recognition.* 2021;118:108035.

[153] Huybers P, Liautaud P, Proistosescu C, *et al.* Influence of late Pleistocene sea-level variations on midocean ridge spacing in faulting simulations and a global analysis of bathymetry. *Proceedings of the National Academy of Sciences.* 2022;119(28):e2204761119.

[154] Hilbert M, Smith WA, and Randall RB. The effect of signal propagation delay on the measured vibration in planetary gearboxes. *Journal of Dynamics, Monitoring and Diagnostics.* 2022;1(1):9–18.

[155] Zhang Z. Introduction to machine learning: k-nearest neighbors. *Annals of Translational Medicine.* 2016;4(11):7.

[156] Sendak M, Elish MC, Gao M, *et al.* "The human body is a black box" supporting clinical decision-making with deep learning. In *Proceedings of the 2020 Conference on Fairness, Accountability, and Transparency*, 2020 January 27 (pp. 99–109).

[157] Hleg AI. *High-Level Expert Group on Artificial Intelligence: Ethics Guidelines for Trustworthy AI.* European Commission 2019; 2019.

[158] Amann J, Blasimme A, Vayena E, Frey D, and Madai VI. Explainability for artificial intelligence in healthcare: a multidisciplinary perspective. *BMC Medical Informatics and Decision Making.* 2020;20(1):1–9.

[159] Markus AF, Kors JA, and Rijnbeek PR. The role of explainability in creating trustworthy artificial intelligence for health care: a comprehensive survey of the terminology, design choices, and evaluation strategies. *Journal of Biomedical Informatics.* 2021;113:103655.

[160] Amann J, Vetter D, Blomberg SN, *et al.* To explain or not to explain?—artificial intelligence explainability in clinical decision support systems. *PLOS Digital Health.* 2022;1(2):e0000016.

[161] Lipton ZC. The mythos of model interpretability: in machine learning, the concept of interpretability is both important and slippery. *Queue.* 2018;16(3):31–57.

[162] Olsen HP, Slosser JL, Hildebrandt TT, and Wiesener C. What's in the box? The legal requirement of explainability in computationally aided decision-making in public administration. University of Copenhagen Faculty of Law Research Paper No. 2019-84, 2019.

[163] Babic B, Gerke S, Evgeniou T, and Cohen IG. Beware explanations from AI in health care. *Science.* 2021;373(6552):284–6.

[164] Hickman E and Petrin M. Trustworthy AI and corporate governance: the EU's ethics guidelines for trustworthy artificial intelligence from a company law perspective. *European Business Organization Law Review.* 2021;22(4):593–625.

[165] Awotunde JB, Folorunso SO, Ajagbe SA, Garg J, and Ajamu GJ. AiIoMT: IoMT-based system-enabled artificial intelligence for enhanced smart healthcare systems. *Machine Learning for Critical Internet of Medical Things*. 2022:229–54.

[166] Ajagbe SA, Amuda KA, Oladipupo MA, Oluwaseyi FA, and Okesola KI. Multi-classification of Alzheimer disease on magnetic resonance images (MRI) using deep convolutional neural network (DCNN) approaches. *International Journal of Advanced Computer Research.* 2021;11(53):51.

[167] Kavya R, Christopher J, Panda S, and Lazarus YB. Machine learning and XAI approaches for allergy diagnosis. *Biomedical Signal Processing and Control.* 2021;69:102681.

[168] Amoroso N, Pomarico D, Fanizzi A, *et al.* A roadmap towards breast cancer therapies supported by explainable artificial intelligence. *Applied Sciences.* 2021;11(11):4881.

[169] Dindorf C, Konradi J, Wolf C, *et al.* Classification and automated interpretation of spinal posture data using a pathology-independent classifier and explainable artificial intelligence (Xai). *Sensors.* 2021;21(18): 6323.

[170] El-Sappagh S, Alonso JM, Islam SM, Sultan AM, and Kwak KS. A multilayer multimodal detection and prediction model based on explainable artificial intelligence for Alzheimer's disease. *Scientific Reports*. 2021; 11(1):1–26.

[171] Peng J, Zou K, Zhou M, *et al.* An explainable artificial intelligence framework for the deterioration risk prediction of hepatitis patients. *Journal of Medical Systems.* 2021;45(5):1–9.

[172] Sarp S, Kuzlu M, Wilson E, Cali U, and Guler O. The enlightening role of explainable artificial intelligence in chronic wound classification. *Electronics.* 2021;10(12):1406.

[173] Tan W, Guan P, Wu L, *et al.* The use of explainable artificial intelligence to explore types of fenestral otosclerosis misdiagnosed when using temporal bone high-resolution computed tomography. *Annals of Translational Medicine.* 2021;9(12):969.

[174] Wu H, Chen W, Xu S, and Xu B. Counterfactual supporting facts extraction for explainable medical record based diagnosis with graph network. In *Proceedings of the 2021 Conference of the North American Chapter of the Association for Computational Linguistics: Human Language Technologies*, 2021 June (pp. 1942–55).

[175] Chen J, Dai X, Yuan Q, Lu C, and Huang H. Towards interpretable clinical diagnosis with Bayesian network ensembles stacked on entity-aware

CNNS. In *Proceedings of the 58th Annual Meeting of the Association for Computational Linguistics*, 2020 July (pp. 3143–53).

[176] Rucco M, Viticchi G, and Falsetti L. Towards personalized diagnosis of glioblastoma in fluid-attenuated inversion recovery (FLAIR) by topological interpretable machine learning. *Mathematics.* 2020;8(5):770.

[177] Gu D, Li Y, Jiang F, *et al.* VINet: a visually interpretable image diagnosis network. *IEEE Transactions on Multimedia.* 2020;22(7):1720–29.

[178] Kröll JP, Eickhoff SB, Hoffstaedter F, and Patil KR. Evolving complex yet interpretable representations: application to Alzheimer's diagnosis and prognosis. In *2020 IEEE Congress on Evolutionary Computation (CEC), 2020 Jul 19* (pp. 1–8). New York, NY: IEEE.

[179] Meldo A, Utkin L, Kovalev M, and Kasimov E. The natural language explanation algorithms for the lung cancer computer-aided diagnosis system. *Artificial Intelligence in Medicine*. 2020;108:101952.

[180] Wang L, Lin ZQ, and Wong A. Covid-net: a tailored deep convolutional neural network design for detection of Covid-19 cases from chest x-ray images. *Scientific Reports.* 2020;10(1):1–2.

[181] Sabol P, Sinčák P, Hartono P, *et al.* Explainable classifier for improving the accountability in decision-making for colorectal cancer diagnosis from histopathological images. *Journal of Biomedical Informatics.* 2020;109:103523.

[182] Chang Y-W, Tsai S-J, Wu Y-F, and Yang A. Development of an Al-based web diagnostic system for phenotyping psychiatric disorders. *Frontier Psychiatry*. 2020;11:1–10.

[183] Cho J, Alharin A, Hu Z, Fell N, and Sartipi M. Predicting post-stroke hospital discharge disposition using interpretable machine learning approaches. In *2019 IEEE International Conference on Big Data (Big Data)*, Los Angeles, CA, USA (pp. 4817–22). New York, NY: IEEE; 2019.

[184] Lamy JB, Sekar B, Guezennec G, Bouaud J, and Séroussi B. Explainable artificial intelligence for breast cancer: a visual case-based reasoning approach. *Artificial Intelligence in Medicine.* 2019;94:42–53.

[185] Das D, Ito J, Kadowaki T, and Tsuda K. An interpretable machine learning model for diagnosis of Alzheimer's disease. *Peer Journal*. 2019;7:e6543.

[186] Kletz S, Schoeffmann K, and Husslein H. Learning the representation of instrument images in laparoscopy videos. *Healthcare Technology Letter.* 2019;6:197–203.

[187] Beauchamp TL, Beauchamp TA, and Childress JF. *Principles of Biomedical Ethics*. Edicoes Loyola; 1994.

[188] Gillon R. Defending the four principles approach as a good basis for good medical practice and therefore for good medical ethics. *Journal of Medical Ethics.* 2015;41(1):111–6.

[189] Mittelstadt B. Principles alone cannot guarantee ethical AI. *Nature Machine Intelligence.* 2019;1(11):501–7.

[190] Spencer-Bonilla G, Thota A, Organick P, *et al.* Normalization of a conversation tool to promote shared decision making about anticoagulation in

patients with atrial fibrillation within a practical randomized trial of its effectiveness: a cross-sectional study. *Trials.* 2020;21(1):1–0.

[191] Faden RR and Beauchamp TL. *A History and Theory of Informed Consent.* Oxford: Oxford University Press; 1986.

[192] Raz J. *The Morality of Freedom.* Oxford: Oxford University Press; 2020. https://doi.org/10.1093/0198248075.001.0001/acprof-9780198248071.

[193] McDougall RJ. Computer knows best? The need for value-flexibility in medical AI. *Journal of Medical Ethics.* 2019;45(3):156–60.

[194] Grote T and Berens P. On the ethics of algorithmic decision-making in healthcare. *Journal of Medical Ethics.* 2020;46(3):205–11.

[195] Beil M, Proft I, van Heerden D, Sviri S, and van Heerden PV. Ethical considerations about artificial intelligence for prognostication in intensive care. *Intensive Care Medicine Experimental.* 2019;7(1):1–3.

[196] London AJ. Artificial intelligence and black-box medical decisions: accuracy versus explainability. *Hastings Center Report.* 2019;49(1):15–21.

[197] Jöbges S, Vinay R, Luyckx VA, and Biller-Andorno N. Recommendations on COVID-19 triage: international comparison and ethical analysis. *Bioethics.* 2020;34(9):948–59.

[198] Obermeyer Z, Powers B, Vogeli C, and Mullainathan S. Dissecting racial bias in an algorithm used to manage the health of populations. *Science.* 2019;366(6464):447–53.

[199] Yang CC. Explainable artificial intelligence for predictive modeling in healthcare. *Journal of Healthcare Informatics Research.* 2022;6(2):228–39.

Chapter 3

Explainable Artificial Intelligence-based framework for medical decision support systems

Joseph Bamidele Awotunde[1], Oluwafisayo Babatope Ayoade[2], Panigrahi Ranjit[3], Amik Garg[4] and Akash Kumar Bhoi[4,5,6]

Abstract

The rise in death tolls due to increased infectious diseases has become one of the most severe health problems and the largest source of death globally. Artificial Intelligence (AI)-based models have emerged and developed to assist medical experts in decision-making, thus reducing the mortality and morbidity rate. However, the most prominent weakness of these algorithms is the lack of interpretations for their results. In other words, the end-user is unfamiliar with the fundamental logic that supports the prediction. Hence, due to their black-box nature, physicians struggle to understand these models; thus, they often do not attract the confidence of the medical practitioners and, in most cases, are not permitted in medical practice. Therefore, this chapter reviews the most substantial reasons for and against explainable AI (XAI) in medical Decision Support Systems (MDSS) with future prospects. The chapter proposes a framework to address the above-mentioned issue in AL-based models using a deep Shapley additive explanations (DeepSHAP) for predicting various diseases. The framework relies on deep neural network architecture enabled with a feature selection method for disease prediction with an explanation. The proposed framework will provide medical experts with more accurate and personalized results for disease prediction and facilitate improved decision-making.

[1]Department of Computer Science, Faculty of Information and Communication Sciences, University of Ilorin, Nigeria
[2]Department of Computing and Information Science, School of Pure and Applied Sciences, College of Science, Bamidele Olumilua University of Education, Science and Technology, Nigeria
[3]Department of Computer Applications, Sikkim Manipal Institute of Technology, Sikkim Manipal University, India
[4]KIET Group of Institutions, Delhi-NCR, India
[5]Directorate of Research, Sikkim Manipal University, India
[6]AB-Tech eResearch (ABTeR), India

Keywords: Explainable Artificial Intelligence; Medical decision support systems; Artificial intelligence; Deep learning; Personalized medicine; Deep neural network; Black-box

3.1 Introduction

The emergence of modern technologies like machine learning (ML) and Artificial Intelligence (AI) methods in medical decision support systems (MDSSs) has tremendously increased to assist healthcare professionals in the diagnosis and prediction of patients from various diseases [1,2]. The power of novel MDSSs has the inherent ability to make quick suggestions even with vast medical data [3,4]. These have created the prospective way for personalized treatments, reduced medical costs, and improved patient diagnosis and treatments. The proof-of-concept of MDSSs enabled with AI-based models shows promising performance in healthcare laboratory settings [5–7]; however, in actual medical practice, MDSSs aided with AI frequently resulted in limited or few improvements [8,9]. Therefore, the explanation ascertained, where the results of the AI models are different from standard clinical recommendations used to be rejected.

The famous AI algorithm artificial neural networks (ANN) among various AI-based models are known as black-boxes since their inner processes and working remain hidden and impervious to the user. Their results in this situation, not be trusted by the medical experts, thus creating serious barriers to MDSSs adoption in healthcare systems [10,11]. Hence, it is crucial to identify measures to enhance the trust of medical practitioners in MDSSs so the models can be widely used and adopted in medical practice [12–14].

Foster trust is often the most positioned approach to increase the transparency of the systems [15,16], and the application of explainability is a significant part of growing and rising systems' transparency [17,18]. For example, in order to avoid the development of unintended coincidences, it is essential to gather sufficient knowledge of the system's behavior in order to find previously unknown weaknesses and flaws. According to researchers, explainability is vital to evaluate true ability of MDSSs. Studying the internal dynamics of an AI-based system is essential, particularly from the design point of view of developers, and very crucial to improve the algorithm by looking at the results. In other words, explainability can help to enhance the results of the AI-based models. As a result, enhancement is a secondary objective that can be accomplished utilizing XAI techniques. Figure 3.1 depicts the general objectives of Explainable AI (XAI).

While ML-based models lack equivalent predefined understanding, the correlations with linkages in data can be discovered using the inbuilt AI-based explanations, for example. According to experts, the goal of XAI is to create deep knowledge by learning from the operation and output of the algorithm [19]. AI is being used more frequently in critical scenarios that could have disastrous consequences for people. While numerous techniques explain the internal dynamics of an AI system, each approach has its own set of benefits and drawbacks [20–22].

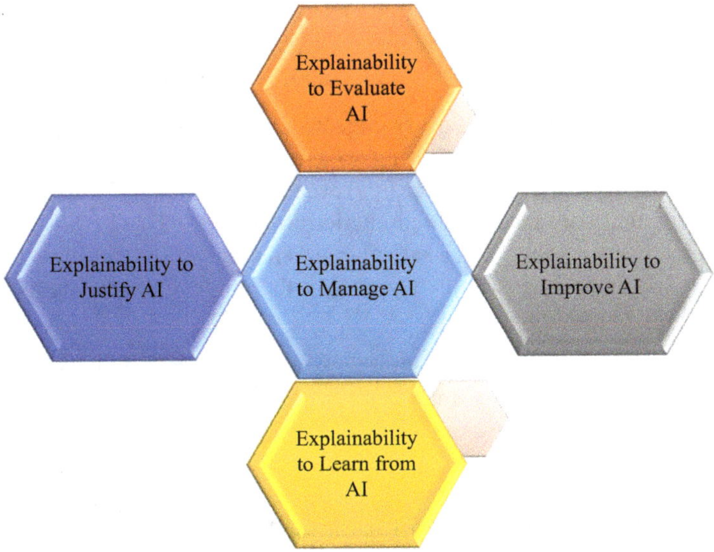

Figure 3.1 Comprehensive objectives of XAI-based model

Furthermore, what defines an excellent explanation depends on several elements, including the intended audience and use case [23–25]. Researchers have disagreed about whether explainability should be a requirement for AI-based models that enable MDSSs. There are several compelling arguments against explainability as a virtue that is useful, acceptable, desired, and even necessary [12,26–28], despite the overwhelming evidence to the contrary. Additionally, there are strong reasons against this prevailing viewpoint [29–31].

Transparency has been cited as one of the key requirements for reliable and trustworthy AI-based algorithms by the High-Level Expert Group on AI of the European Commission (AI HLEG). As a result, explainability is simply one of several approaches used to gauge how transparent an AI-based model is. Additional precautions include proper documentation of the datasets used and the algorithm codes used, as well as practical discussions of the systems' strengths and short-comings and vulnerabilities [23,32]. Although the AI HLEG values explainability, the AI-based experts do not believe that MDSSs should always be trusted. However, AI HLEG asserts that steps should be taken to integrate and include more transparency and accessibility elements into algorithms that lack this characteristic [33,34].

Humans generally do not support processes that are not clear-cut, simple to grasp, intuitive, interpretable, tractable, or trustworthy [35], which raises the need for ethical AI [28]. Although it is a frequent misconception that concentrating just on performance will produce better outcomes. Instead, this will only make the systems more complicated. It should be noted that performance and transparency of models are inversely correlated [36]. The shortcomings of a system might, nevertheless, be improved with a deeper understanding of it. The use of comprehensibility as a sophisticated design driver can improve the design and implementation of AL and ML-based models for three main reasons:

(a) Interpretability aids in ensuring objectivity in selection by detecting and correcting distortion in the training dataset.
(b) Interpretability aids resilience by revealing potentially antagonistic disturbances that could cause the prediction to shift.
(c) Interpretability might act as a safeguard to ensure that only pertinent variables are used to determine the outcome or that the model explanation has genuine underlying causation.

The system's interpretation must either give an understanding of the model's processes and assumptions, a visual representation of the model's discriminatory practises rules, or suggestions as to what might cause the model to be perturbed in order for the system to be considered practical for the reasons mentioned above [37]. Therefore, this chapter covers the multiple research opportunities and challenges identified, in addition to how explainable AI in MDSSs can be employed in medical procedures. Moreover, a framework has been proposed for the prediction of various diseases using explainable AI.

3.1.1 *Key contributions of the chapter*

The following are the significant contributions of this chapter:

(i) The chapter examines some of the key XAI application areas, evaluating the arguments for and against the explainability of MDSSs in the broader medical sector.
(ii) The significant challenges and the prospects of XAI for MDSSs and healthcare, in general, have been discussed.
(iii) For the early prediction of certain diseases, a framework based on explainable deep learning has been suggested, and a real-world medical dataset has been utilized to assess the effectiveness of the proposed framework.

3.1.2 *Chapter organization*

Section 3.2 presents the applicability of XA in MDSSs. Section 3.3 presents the challenges in the applicability of XAI in MDSSs. Section 3.4 presents proposed DeepSHAP enabled with DNN Framework. Section 3.5 discusses the experimental design for cancer prediction, Section 3.6 presents the experimental results and discussion. Section 3.7 explains the future research direction of XAI in healthcare systems, and finally, Section 3.8 concludes the chapter with future direction.

3.2 Applicability of XAI in MDSSs

Unfortunately, many terms are occasionally used without a detailed description when discussing the many strategies used to attain explainability, especially across fields [27,29,38]. As a result, it is essential to specify how to use these expressions in the MDSS application. The following section discusses the distinction between interpretable and explainable algorithms. Although both explainable and interpretable may share methodologies in common, as discussed by various experts in their general narrative, there are some differences, to foster innovation used, they are separate.

There are intrinsically interpretable methods embedded into the decision-making process of the model, similar to interpretable algorithms, which make it easier for the target user to comprehend. This category includes a variety of methods, but the most popular ones are logistic and linear regression techniques, which use the strengths of attributes to estimate the importance of features. Another example is decision trees, which are easy for people to comprehend. On the other hand, open-ended black-box models can be opened using explainable procedures. These strategies frequently use interpretable computational frameworks that mimic black-box methods. A black-box model can be given an interpretation by using an interpretable approximation. Examples of such models include the SHapley Additive exPlanations (SHAP) [39,40] and the Local Interpretable Model-Agnostic Explanations (LIME) [41,42]. Figure 3.2 displays the general concepts of XAI, which comprises of the concepts, methods and interactions.

When it comes to the acceptability of AI-based models in healthcare systems, the utilization of explainability methodologies for MDSSs methods is vital. A better understanding of the complexities of underlying AI systems can help detect probable errors and identify the root reasons of failure [43]. Furthermore, the explainability of the AI-based models would undoubtedly make it possible for medical professionals to evaluate the dependability of the expected result [44]. Also, these will help explain to patients why AI-based systems ensure action by the medical specialists when utilizing MDSSs techniques, which will develop trust in

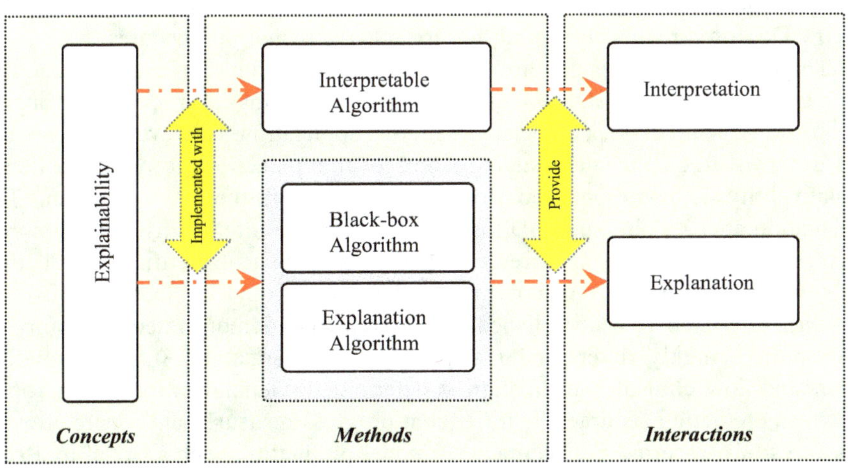

Figure 3.2 The general concepts of XAI

the connection between the physician and the patient [11,45]. Users can also use explainability to ensure that the system is not depending on abnormalities or interference in the learning algorithm. As a result, users can use explainability to identify whether or not a training data set includes incomplete or biased information in the dataset [46].

Additionally, AI-based systems can better be understandable using explanations that will improve their learnings, goals and swap [47]. It has also been demonstrated that offering medical experts a competent second opinion from an AI system can improve diagnostic accuracy over the AI system or the medical specialists on their own [48,49]. Some authors even go so far as to say that explainability is necessary for therapeutic diagnostics [50].

The first point to address from a medical standpoint is to distinguish between AI-based MDSS and existing diagnostic technologies, such as various laboratory testing. Especially since there are so many similarities: Both can produce results that can be used in MDSSs, prioritize performance, and their outcomes are verifiable. Understand how clinical laboratories work because it frequently affects a variety of other screening procedures. As a result, these should not be regarded as black box methods, such as imaging. Similarly, we cannot explain each test's outcomes using these methodologies. This demonstrates the importance of identifying two stages of explainability from a medical standpoint.

We can understand how the program arrives at plausible results thanks to the first degree of understandability. In the same way that we know which physiological and biochemical processes lead to the outcomes in laboratory tests, the same happens in AI-based MDSSs, where the model may provide correlation-based feature rankings that explain crucial inputs. We can ascertain exactly qualities essential in formulating a certain prognosis thanks to second-level explainability. Individual forecasts can be double-checked for similarities that could suggest a mistake, in the case of an out-of-sample scenario with anomalous feature dispersion, for example. This second degree of explainability will be accessible for AI-based CDSS on a regular basis, but not for other screening procedures.

This has implications for how explainability findings are communicated to clinicians and even patients. First-level explanations may suffice based on the therapeutic use and the risk associated with that application. However, alternate use scenarios will frequently necessitate second-level explanations to protect patients. Explainability is commonly considered only after further consideration. The explanation appears obvious; MDSSs, in particular, are intelligent Healthcare systems. To meet regulatory requirements and obtain medical certification, all products, whether AI-powered or not, must undergo a thorough evaluation [51,52].

Once this step is accomplished, the system has demonstrated its ability to perform in a highly diversified real-world clinical situation. It is essential to understand how clinical confirmation is determined. Generalization ability, often known as prediction accuracy, is a frequent progress measurement. There are various ways of measuring the accuracy of model prediction, each adapted to a specific use case, and however, they all have one thing in common: they indicate a model's prediction skill and, consequently its broad clinical use [53]. Therefore, improving predictive accuracy is one of the simulation model's main objectives and keeping the error rates low [54]. Likewise, the error rates of prediction have become lower using AI-based techniques and demonstrated to more useful in the area of reducing error rate of dataset prediction than conventional techniques [55,56].

Early intervention in disease diagnosis and prognosis reduces mortality and morbidity rates. Using countermeasures in every situation, on the other hand, can be costly and time-consuming. By providing comprehensive and accurate computational technology, a machine-learning-based MDSS can assist clinicians in lowering global mortality and morbidity rates. It would drastically reduce the amount of time and money spent if early prevention was provided. In clinical settings, the MDSS has a bright future and enormous potential, especially since many clinicians have used telemedicine to improve healthcare systems, particularly during the COVID-19 outbreak, in order to maintain social distance [57,58].

MDSSs hold great promise for improving healthcare delivery quality, but there is a paucity of literature on their effective implementation, particularly for AI and ML-based MDSSs. According to the authors of [12], for an MDSS to be accepted and integrated into healthcare workflow, its usability is very important in addition to the system being accurate, efficient, and well-accepted. An MDSS should be time-saving, straightforward, and simple to use in order to obtain system responses while navigating a demanding clinical schedule. At the same time, the authors stated that black-boxes should not be used in MDSSs. Their recommendations are consistent with the authors of [59], who state that explainability is a fundamental requirement for a CDSS to be successfully implemented for practical use. The authors' study in [60] demonstrates that while an ML-based model can detect a pattern in a dataset, it is incompatible with clinical experience and thus has no application in clinical settings. In a given population, asthmatic patients had a lower risk of succumbing to pneumonia, according to their analysis. This is because patients with allergies who develop pneumonia generally receive extensive treatment, which reduces their risk. Even though the system successfully gathered the training data, it would be troublesome if the model were to be used in clinical settings without first comprehending why it responded the way it did. Hence, XAI-based models can be used to easily resolve such problems in a medical setting. Numerous advantages have been found when using XAI in MDSS, like increasing decision confidence, creating causation hypotheses, and boosting the system's attractiveness, credibility, and acceptability. However, in the published research, there is a significant paucity of XAI applicability in MDSSs [59].

3.3 The challenges in the applicability of XAI in MDSSs

In defense of the concept of explainability in general, it was stated that healthcare practitioners prioritize accurate use of data from reliable sources over a thorough understanding of how the evidence was obtained [24,30]. In this situation, a system's actual usability and efficacy are valued more than its ability to describe its outputs, so long as these outcomes are shown to be somewhat trustworthy and verifiable [31,61]. In addition, research has shown that explainability does not improve consumers' predisposition to believe a judgment made by an intelligent agent, making it extremely difficult to uncover false predictions [62]. In rare instances, all healthcare specialists may not agree on a fundamental truth [63,64]. In such situations, the level of detail required of explanations is substantial, if not unattainable.

There is also the possibility that consumers will misinterpret reasonable links discovered by AI-based models. Despite the possibility that a ML MDSS may potentially justify its predictions, physicians may make erroneous causal assumptions, this is a critical issue to remember. However, because explanations are founded on correlations, due to random circumstances, they are prone to inaccuracy. As a result, clinical inference is investigated to establish causal factors, such as imbalanced datasets and misleading correlation [65]. But use a straightforward and understandable paradigm, contrary to popular belief, can reduce the possibility of consumers spotting model mistakes rather than employing tools that are not explainable. One explanation for this could be that the user is forced to excessively analyze data to focus on mistake identification [66]. It is also feasible to create deceptive explanations that appear reliable to the user and that, as a result, the user believes. Although they truly describe the decision-making procedure of the model on a technical level, elements of the explanation are omitted, leading to a decline in system trustworthiness [67–69].

Another technological difficulty is that current DL-based models are not intrinsically interpretable [70,71]. Approximations from explanation algorithms are frequently used in existing methods for providing interpretations for these DL-based models. As a result, there is a substantial risk that these approximations will be incorrect, hence, the explanations provided for some inputs do not adequately reflect the model. Additional reason for an erroneous portrayal of the initial formulation is that approximation-based explanations are not guaranteed to employ the same attributes as the initial formulation. The real risk of incorrect explanations makes it more difficult for professionals to believe them. As a result, the model they're explaining is not what these explanations are supposed to be about [71,72]. What defines a good explanation also varies depending on who is communicating with it [11,73] and there is no quantifiable way of predicting the most beneficial encounter ahead of time [74–76].

Due to the limitations of current explanation generation systems, various researchers [71,77] recommend using models that are transparent by nature, such as decision trees, rule lists, or regression. Deep neural network (DNN) models, for example, are more interpretable for end-users than conventional models, but those representations can only be considered interpretable if they follow a set of rules. For example, the number and nature of attribute values, or the model's complexity, makes one wonder whether any models are intrinsically transparent [78].

Others advocate using hypothetical explanations [79,80], which are descriptions of the kind "if the input was this current input rather than this old input, a different decision would have been made by the system" [81]. These explanations are easily understandable because they mimic the often metadiscourse style of human explanations [82,83]. Finding significant and usable hypothetical instances, on the other hand, can be challenging and costly to implement in reality [84,85]. Explainability, particularly in the medical field, might be deceptive and is not always required. The inadequacy of human affection to comprehend XAI decision-making description maps and the lack of a quantitative assessment of the explanation map's reliability and consistency are other problems of XAI in MDSSs in healthcare systems. As a result, the employment of visualization methods for purpose applications may need to be revisited in the future. It's also worth thinking about new ways to describe and communicate explanations. Figure 3.3 shows the basic issues of XAI in healthcare systems generally.

Privacy concerns are explicitly addressed throughout the life cycle of AI-based models, particularly when working with multiple data sources. This is especially important when dealing with personal information because individuals' privacy rights must always be respected. The importance of data governance in terms of privacy has been emphasized, as has ensuring the accuracy and consistency of the data used [86]. It should also include a contextual specification as well as the ability to process data securely.

Despite regulatory agencies' clear concerns, DL approaches have been found to harm privacy when no data fusion is performed. Even with image obfuscation, a few photos can compromise users' privacy [87], and a DNN's explanatory variables can be accessed by running input requests on the model [88,89]. One method for explaining lack of privacy is to use subjective privacy loss and intentional loss scores. Based on the function of a face in the photograph, the former provides a subjective assessment of the gravity of the invasion of privacy, while the latter depicts the spectators' desire to be included in the photograph. Such explanations have prompted the development of trustworthy complementing cryptographic algorithms to protect the privacy of photographers and witnesses [90–92]. Researchers strongly advocate for increased efforts in this area, particularly to ensure that XAI techniques do not jeopardize the privacy of the data used to train the AI-based model under consideration.

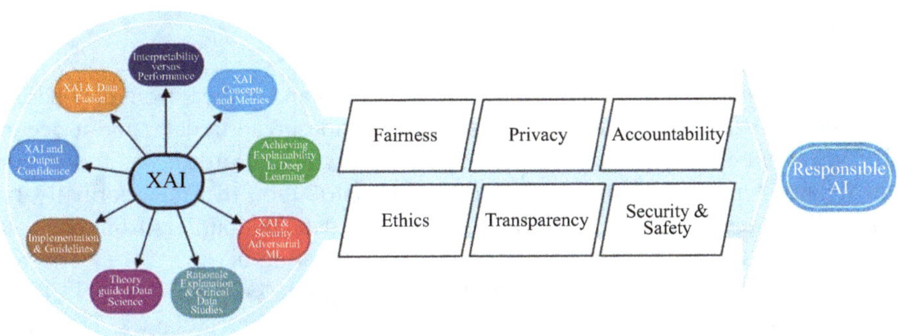

Figure 3.3 Basic challenges of XAI challenges in MDSSs

End-to-end concepts have also been demonstrated for Trustworthy AI [93], but those of Telefonica (a major Spanish ICT corporation with a global presence) are more AI-focused [94]. For example, an AI-based system can ensure the privacy and security of any interconnected IT system [95]. The same is true for privacy, even though privacy is probably even more important in developing AI systems than in normal IT processes. Because ML models require massive amounts of data and, more importantly, maintaining the privacy of protected material is becoming more difficult due to XAI capabilities and data fusion approaches.

3.4 The proposed DeepSHAP enabled with DNN framework

This chapter proposes a DeepSHAP-enabled DNN framework model for performance accuracy and explainable MDSS. The proposed framework consists of three stages: (i) the data pre-processing, (ii) DNN prediction model, and (iii) the explanation model for better understanding of the proposed model.

3.4.1 *The pre-processing stage*

Data pre-processing is an important step that prepares the dataset for training before building prediction model. Various pre-processing techniques were performed on the dataset to provide relevant data for the suggested model framework for model optimization. The following are the actions that were taken to restructure the dataset utilized in this study:

The normalization: The feature values are rescaled in the range of 0 to 1 using normalizing in this pre-processing stage. Maximum and minimum values between 0 and 1 were used to standardize all predictive features. To convert the desired feature, which is categorical by nature, to numerical, the transformation was used. The normalization equation is generated by subtracting the minimal value from the parameter to be standardized first [96]. After subtracting the minimum from the maximum, the prior result is divided by the latter as follows in Eq. (3.1):

$$X' = \frac{X - X_{\min}}{X_{\max} - X_{\min}} \tag{3.1}$$

Data cleaning: This eliminates incomplete values and outliers from data before it can be analyzed. The issue of lost variables is widespread in data, and it occurs when data values for attributes in an occurrence are not recorded. An outlier is a statistician's term for a value on the tails of a distribution that is abnormally far away from other values. As a result, we reject lost data and outliers from our data analysis process because they can create model estimation bias.

Dataset balanced: Synthetic minority oversampling technique (SMOTE) is a mechanism for artificially producing instances for outnumbered classes. This class is oversampled by assigning false instances to each outnumbered class instance along line segments connecting any/all of the k overcrowded class's nearest neighbors. This method eliminates overfitting and expands the decision zone of the outnumbered class's instances.

3.4.2 The hyper-parameters and DNN

The DNNs differ from traditional neural networks (NNs) in that they have far better generalization features and can model any non-linear function or connection by comparison [97]. This is partly due to the usage of many layers, which allows various function classes to be approximated in diverse ways. A NN is a network of interrelated neurons that work together to learn how to complete a specific task. Three or more layers, namely input, hidden, and output, are commonly used in NN configurations. The weighted nodes are contained in at minimum one hidden layer, and interconnection exists between nodes in surrounding levels, but not between nodes in the same layer. Therefore, the input and output mappings cannot learn successfully if a network would not have enough hidden nodes. Eventually, the output of hidden nodes is determined by the number of classes, and calculated by the use of an activation function. Hence, make final judgments Y using optimal weights to minimize the difference between predicted and actual values [98].

The selection of hyper-parameters has a strong influence on the DNN's effectiveness. An effective one-hidden-layer MLP can be used to learn NNs, and learning with two or more hidden layers of perceptrons, on the other hand, can generate significantly better results than learning with only one. Hence, the DNNs are used to convert the {2-10} hidden layers in the suggested framework. The performance of three activation functions was compared by varying the number of nodes per hidden layer {3, 5, 10, 15, 20, 25}, namely: Sigmoid (sigm), Rectified linear units (ReLU), and Tanh (tanh). The ideal hidden layers and their nodes in terms of the highest classification performance were presented in the experimental results. Additionally, Adam was used to optimize the models, and the learning rate is set at 0.001 while keeping the rest of the hyper-parameters constant. To limit the likelihood of over-fitting, regularization procedures are used.

3.4.3 The Shapley additive explainable (SHAP)

The SHAP as a post hoc explainability method to generate explanations is one of the most widely utilized human evaluation user studies. Deep SHAP explainer was investigated in the proposed approach, which incorporates concepts from the integrated gradients (consequently, integrating requires the usage of a unique standard value.), the SHAP and SmoothGrad (which receives an input image's gradient susceptibility maps combining them into a single equation for predicted value, then averaging them to identify pixels of interest. The proposed framework used the Kernel SHAP algorithm, it can be used to forecast SHAP quantities in any scenario without regard to the model. Each and every approach is consistent with the SHAP KernelExplainer. Nevertheless, it takes longer than other model kind of approaches, since it does not assume anything about the kind of model. Although slower than other explanations, it delivered the best performance, and giving an estimation of SHAP scores rather than actual values. Based on the open-source Shap library developed by Lundberg and Lee in 2017 [99], the Kernel SHAP algorithm was created. By analyzing the contribution of each characteristic to the prediction, SHAP is used to explain the predictions of an instance x.

SHAP uses the game-theoretic Shapley value to explain individual predictions [99,100]. To compute (as given in Eq. (3.2)), this method employs the concept of coalitions, the black-box model's features' Shapley value (f) prediction of instance (x). The average marginal contribution is known as the Shapley value, and given as $\left(\varnothing_j^m\right)$ of feature (j) in any imaginable coalition. Eq. (3.3) is used to compute the marginal contribution, where $\check{f}\left(\varnothing_{+j}^m\right)$ and $\check{f}\left(\varnothing_{-j}^m\right)$ are predictions made using black-box f without and with the replacement of the jth characteristic of sample instance x:

$$\varnothing_j(x) = \frac{1}{M}\sum_{m=1}^{M}\varnothing_j^m \tag{3.2}$$

$$\varnothing_j^m = \check{f}\left(\varnothing_{+j}^m\right) - \check{f}\left(\varnothing_{-j}^m\right) \tag{3.3}$$

3.5 Experimental design for cancer prediction

All tests were carried out on a PC with a 3.50 GHz Intel Core i7-4500K processor and 64GB RAM running Microsoft Windows 10 Operating Systems. The open libraries and packages of the Python programming language are used in the experimental analysis. The programming language have libraries and packages like DNN package, DeepSHAP [99], Statsmodels, Pandas [101–103] among others.

3.5.1 The Wisconsin breast cancer (WBCD) dataset

The suggested framework is based on the WBCD dataset (original), which contains 699 occurrences, each of which is linked to a set of nine attributes. The WBCD dataset contains about 66% benign and 34% cancerous records. The WBCD dataset recorded nine basic features that related to diabetes mellitus illness including clump thickness, cell size uniformity, normal nucleoli, size of single epithelial cells, Bare nucleus, marginal adhesion, Bland chromatin, mitoses, and normal nucleoli. The quantitative measure is given a value within 1 and 10 as an integer. The most anaplastic cases are given a number of 10, while the most benign ones are given a number of one. This WBCD dataset is the benchmark and is accessible in the repository for ML at UCI because it is important in assessing a variety of breast cancer tendencies. This WBCD information is also useful for accurately diagnosing breast tumors that fall into the malignant and benign categories. The threshold population compression factor and other parameters employed in the performance of the recommended technique are decided via trial and error in order to enhance effectiveness. The proposed DeepSHAP-enabled DNN architecture is displayed in Figure 3.4.

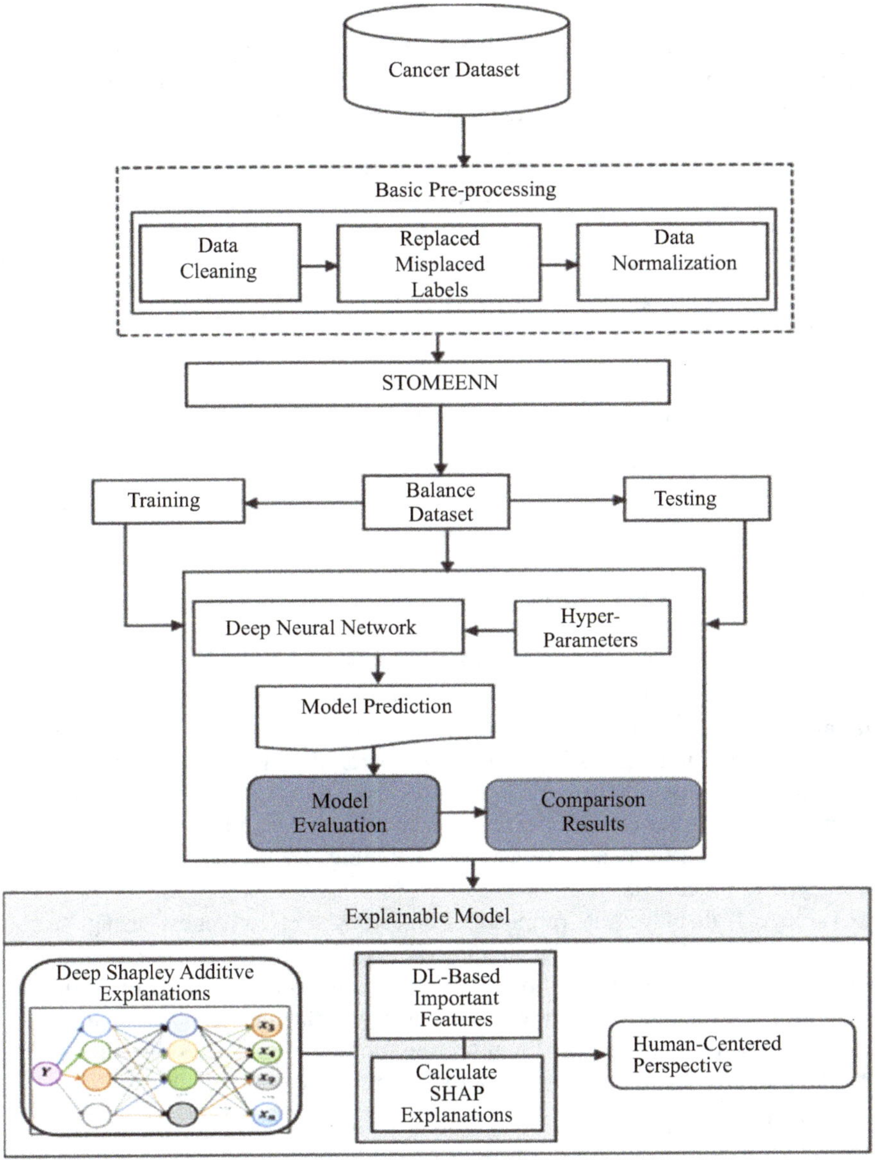

Figure 3.4 The proposed framework for XAI model in healthcare systems

3.5.2 The performance evaluation metrics

For evaluating the proposed framework for the prediction of various diseases, the commonly performance metrics are defined using four values to arrive at a conclusion, namely: true positive (TP), true negative (TN), false positive (FP), and false negative (FN). The total number of (TP+TN) that may understand correct and wrong predictions generated by the model is (FN+FP). The performance metrics used for the evaluation of the proposed model are accuracy, recall, precision, specificity, F1-score, and, the area under curve, respectively. Eqs (3.4)–(3.8) represent the measuring performance metrics using for the proposed system [104]:

$$Accuracy = \frac{TP + TN}{TP + FP + FN + TN} \tag{3.4}$$

$$Recall = \frac{TP}{TP + FN} \tag{3.5}$$

$$Precision = \frac{TP}{TP + FP} \tag{3.6}$$

$$Specificity = \frac{TN}{TN + FP} \tag{3.7}$$

$$F1 - Score = \frac{precision.recall}{precision + recall} \tag{3.8}$$

3.6 Experimental results

To verify the proposed model, the WDBC breast cancer dataset was predicted using the DNN classifier. The dataset contains 341 benign and 228 malignant patients totaling 569 instances. To verify the application of the suggested model, the dataset was split into training and testing portions, each comprising 70% of the dataset. The DNN was used to identify breast cancer cases from the dataset, which validated the implementation of the classifier on the dataset. For the validation of the proposed model, several performance measures, including F1-score, recall, sensitivity, accuracy, and precision.

Table 3.1 displays the proposed framework's effectiveness using various metrics and WBCD datasets. The anticipated model is crucial and relevant in breast cancer prediction and classification, as evidenced by the outcomes from numerous measures. For example, among the 569 occurrences, the model accurately classifies 564 of them. This demonstrates the proposed model's high level of accuracy. On a bar graph, Figure 3.4 depicts the model's performance.

Table 3.1 Proposed method evaluation

Dataset	Accuracy (%)	Precision (%)	Recall (%)	F-Measure (%)	Support	Class
WBCD	97.7	96.4	97.5	98.9	43	Benign
	96.8	98.8	97.2	96.3	95	Malignant
Mean/total	97.25	97.60	97.35	97.60		

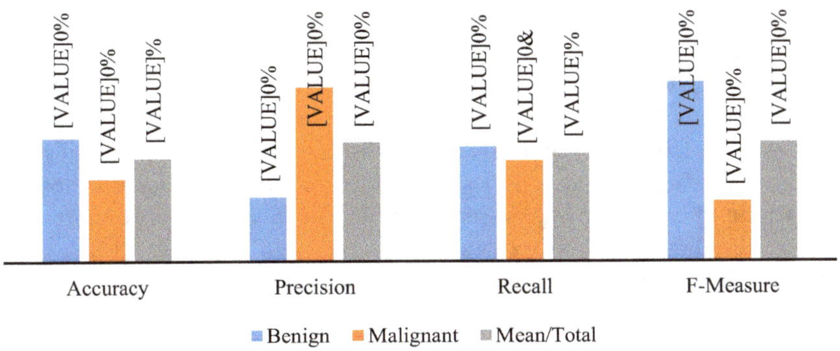

Figure 3.5 The performance evaluation of the proposed model

Table 3.2 The comparison of proposed model with the state-of-the-art classifiers

Authors	**Model**	**Accuracy (%)**
[105]	SVM, Naïve Bayesian, KNN	97.13
[106]	Decision trees	90.00
[107]	Naïve Bayesian	90.41
[108]	Association rules AND neural network	95.60
[109]	Ensemble method	89.20
[110]	Random Forest	95.71
[111]	Naive Bayes	81.30
[112]	Bayes belief network	91.70
[113]	RBF network	96.77

The performance of the proposed model is detailed in Table 3.1 after been measured using various performance metrics. The result of the performance metrics shows that the proposed DNN model generates an accuracy of 97.25%, recall of 97.35%, a specificity of 96.7%, the precision of 97.60%, and a F1-score of 97.2%, respectively. These results show that the proposed classifier can diagnose the breast cancer dataset correctly. The bar chart in Figure 3.5 displays the performance of the proposed classifier.

3.6.1 The comparison of the proposed model with existing methods

Table 3.2 examines the suggested strategy with several well-known strategies to demonstrate how effective the presented model is. The accuracy results for the proposed model and other algorithms utilizing the same WBCD dataset, are shown in Table 3.2. The proposed method is more accurate than previous methods. For example, the suggested method for diagnosing breast cancer is approximately given 97.25% accuracy. Compared to other approaches employing the compressed WBCD dataset, the suggested method outperformed other systems in terms of accuracy measures.

The proposed model differs from others classifiers used for breast cancer prediction since it uses the DNN with a hyper-parameter for classification efficiency and effectiveness. Furthermore, because of the reduced hidden layer, the model understands and investigates high-level functioning, effectively reducing the dimensionality of the data, and effectively displays significant properties. As a result, the suggested approach is best suited for real-time disease diagnosis and categorization in healthcare businesses that deal with large amounts of unregulated and unstructured data, such as medical data.

The model by the authors in [111] found four features very relevant for the classification of breast cancer using PSO algorithm with NB, K-NNs, and RepTree has classifiers, and the results reveal an accuracy of 81.3% for NB, 80% for RepTree, and 75% for k-NNs, respectively. Furthermore, the model recorded accuracy of 70% for NB, 96.3% for RepTree, and 66.3% for k-NNs, respectively with the implementation of PSO algorithm. The author in [112] applied the gradient boosting with some classifiers to improve the accuracy of the breast cancer prediction, and the results revealed an accuracy of 91.7% for BBN, 91.7% for BAN, and 94.11% for TAN, respectively. When comparing the accuracy with various existing classifiers, it was discovered that the proposed model performed significantly better than some of the start-of-the-art models.

3.6.2 The local explanation results

The model explanation is required to understand the predictive methods' rationale better. In the proposed model's second layer, a conventional forward pass is applied to DNN, and activation at each layer is integrated for the prediction model. The DNN output score is then transmitted backwards in the DNN in the third layer. The DeepSHAP approach's propagation rule was used to improve the interpretability of the breast cancer prediction model. The DeepSHAP technique allows for a human-centered viewpoint. From a human-centered perspective, the model considers a single individual's conditional relationship between features and classes. As a result, medical experts will be able to describe how effective DNNs work internally, and understand the key factors that contribute to the development of breast cancer in the overall population and in each person.

Moreover, to give individualized healthcare recommendations, local explanations are required. The XAI aids us in comprehending the key characteristics that prompted the algorithm to reach the appropriate judgment and forecast. The pictorial-based XAI also aids in visualizing the most relevant elements of a particular iteration, which aids in classification and decision-making.

3.7 The future research direction of XAI in healthcare systems

One of the goals of this chapter was to demonstrate the unresolved challenges surrounding the development of explainable models and make recommendations for further research in various application areas and responsibilities. According to the assessment mentioned above, the fundamental shortcoming of the presented approaches is the evaluation of the explanations. The studies used a variety of user research and experimentation strategies to resolve this problem, hence, a generic approach for evaluating explanations is still urgently needed. Additional obstacle that has been identified is algorithm-specific techniques to increasing explainability. It impedes explaining how entrenched systems are designed. A few potential research directions are highlighted below, based on the practical limitations of current explainable models:

(i) The dataset's impact (especially the effect of dataset imbalance, attribute complexity, and other factors), studies can be used to analyze the impact of various forms of bias problems (in data gathering and datasets, for example) on constructing an explainable model.

(ii) It was discovered that the majority of the work was done on NNs, and explanations were provided at the local level using post hoc approaches. Other models like ensemble and support vector machine (SVM) models have shown similar results, since its interpretation method is still a mystery to users. While various research has demonstrated methods for generating global explanations by emulating model behavior, they are not very good at performing. more studies can be conducted to generate a universal explanation without jeopardizing the models' effectiveness in the base task;

(iii) User studies were entreated to authenticate explanations constructed on natural language, textual explanations in a nutshell. The use of autonomous evaluation criteria for textual explanations is not yet widely used in research;

(iv) The most difficult aspect of appraising an explanation is devising a mechanism that can handle users' various degrees of competence and knowledge. These two qualities of users, in general, differ between individuals. To build a good model for measuring explanations focused on the competence and aptitude of the intended users, extensive study is required.

3.8 Conclusion and future scopes

The population growth globally has make disease diagnosis, prediction, monitoring, and treatment an issue in healthcare systems, especially in developing countries. Hence, medical system automation has become a necessity and a primary priority recently. The application of MDSSs and intelligent disease classification and prediction can help medical experts diagnose various diseases in real-time even in remote areas. Global population growth has made disease diagnosis, prediction, monitoring, and treatment more difficult in healthcare systems, particularly in developing countries. As a result, automation of medical systems has recently become a necessity and a top priority. MDSSs and intelligent disease classification and prediction can assist medical experts in diagnosing various diseases in real-time, even in remote locations. This can lead to lower mortality rates, lower healthcare costs, and more robust decision-making with high accuracy. The emergence of modern technologies such as the Internet of Things, Cloud computing, edge computing, ML, and AI-based models has increased the automation of human manual tasks, particularly in healthcare systems. The use of ML and AI-based algorithms in healthcare has greatly aided in processing massive amounts of data generated by various smart sensors and devices. However, the technologies are not fully accepted because they are black-box in nature; medical experts and end-users must well understand the outcomes of AI-based models. Especially in healthcare systems where decisions must be made carefully so that they do not negatively impact the patient. Several recent efforts have been made to explain the outcome of using DL complexity models in MDSSs in the process and analysis of various data involving disease diagnosis, classification, and monitoring. In this regard, explainable AI has been useful in explaining the justification for the results of AL-based models in various fields. As a result, this chapter discusses the use of XAI in healthcare systems, particularly in MDSS applications. The problems and prospects of XAI in MDSSs are also discussed. The chapter proposed a XAI-based framework for predicting and classifying various diseases, with breast cancer as a case study to demonstrate the model's efficacy. For the human-centered perspective of the DNN model, the model used SHAP, and the performance of the proposed system was measured using various evaluation metrics. When the model's results were compared to recent state-of-the-art classifiers, the results performed reasonably well, with an accuracy of 97.23%. This chapter demonstrates how XAI approaches improve model results comprehension in breast cancer prediction. Future work will investigate the application of other XAI models, such as LIME, to the explainer execution time to improve the model's reproducibility. In the future, various ensemble approaches, such as stacking and bagging, could be used and developed to increase the significance of the proposed model. DNN cross-validation could be used in future studies to obtain multiple training and testing data folds. It can assist with imbalanced datasets and class instability problems and increase computational costs. Although the current study focuses on breast cancer, it could be expanded to other diseases in the future.

References

[1] Abiodun MK, Misra S, Awotunde JB, Adewole S, Joshua A, and Oluranti J. Comparing the performance of various supervised machine learning techniques for early detection of breast cancer. In *International Conference on Hybrid Intelligent Systems* (pp. 473–82). Cham: Springer; 2021.

[2] Folorunso SO, Awotunde JB, Adeniyi EA, Abiodun KM, and Ayo FE Heart disease classification using machine learning models. In *International Conference on Informatics and Intelligent Applications* (pp. 35–49). Cham: Springer; 2021.

[3] Sutton RT, Pincock D, Baumgart DC, Sadowski DC, Fedorak RN, and Kroeker KI. An overview of clinical decision support systems: benefits, risks, and strategies for success. *NPJ Digital Medicine*. 2020;3(1):17.

[4] Spänig S, Emberger-Klein A, Sowa JP, Canbay A, Menrad K, and Heider D. The virtual doctor: an interactive clinical-decision-support system based on deep learning for non-invasive prediction of diabetes. *Artificial Intelligence in Medicine*. 2019;100:101706.

[5] De Fauw J, Ledsam JR, Romera-Paredes B, et al. Clinically applicable deep learning for diagnosis and referral in retinal disease. *Nature Medicine*. 2018;24(9):1342–50.

[6] Awotunde JB, Ajagbe SA, Oladipupo MA, *et al.* An improved machine learnings diagnosis technique for COVID-19 pandemic using chest X-ray images. In *International Conference on Applied Informatics* (pp. 319–30). Cham: Springer.

[7] Liu X, Faes L, Kale AU, *et al.* A comparison of deep learning performance against healthcare professionals in detecting diseases from medical imaging: a systematic review and meta-analysis. *The Lancet Digital Health*. 2019; 1(6):e271–97.

[8] Miotto R, Wang F, Wang S, Jiang X, and Dudley JT. Deep learning for healthcare: review, opportunities and challenges. *Briefings in Bioinformatics*. 2018;19(6):1236–46.

[9] Musen MA, Middleton B, and Greenes RA. Clinical decision-support systems. *In Biomedical Informatics* (pp. 795–840). Cham: Springer; 2021.

[10] Folorunso SO, Awotunde JB, Ayo FE, Abdullah KK. RADIoT: the unifying framework for iot, radiomics and deep learning modeling. *In Hybrid Artificial Intelligence and IoT in Healthcare* (pp. 109–28). Singapore: Springer; 2021.

[11] Amann J, Blasimme A, Vayena E, Frey D, and Madai VI. Explainability for artificial intelligence in healthcare: a multidisciplinary perspective. *BMC Medical Informatics and Decision Making*. 2020;20(1):1–9.

[12] Shortliffe EH and Sepúlveda MJ. Clinical decision support in the era of artificial intelligence. *Jama*. 2018;320(21):2199–200.

[13] Kawamoto K, Houlihan CA, Balas EA, and Lobach DF. Improving clinical practice using clinical decision support systems: a systematic review of trials to identify features critical to success. *BMJ*. 2005;330(7494):765.

[14] Mahadevaiah G, Rv P, Bermejo I, Jaffray D, Dekker A, and Wee L. Artificial intelligence-based clinical decision support in modern medical physics: selection, acceptance, commissioning, and quality assurance. *Medical Physics*. 2020;47(5):e228–35.

[15] Abiodun KM, Awotunde JB, Aremu DR, and Adeniyi EA. Explainable AI for fighting COVID-19 pandemic: opportunities, challenges, and future prospects. In *Computational Intelligence for COVID-19 and Future Pandemics* (pp. 315–32); 2022.

[16] Ehsan U, Liao QV, Muller M, Riedl MO, and Weisz JD. Expanding explainability: towards social transparency in ai systems. In *Proceedings of the 2021 CHI Conference on Human Factors in Computing Systems* (pp. 1–19).

[17] Shin D. The effects of explainability and causability on perception, trust, and acceptance: implications for explainable AI. *International Journal of Human-Computer Studies*. 2021;146:102551.

[18] Shin D. User perceptions of algorithmic decisions in the personalized AI system: perceptual evaluation of fairness, accountability, transparency, and explainability. *Journal of Broadcasting & Electronic Media*. 2020;64(4): 541–65.

[19] Meske C, Bunde E, Schneider J, and Gersch M. Explainable artificial intelligence: objectives, stakeholders, and future research opportunities. *Information Systems Management*. 2022;39(1):53–63.

[20] Pimenov DY, Bustillo A, Wojciechowski S, Sharma VS, Gupta MK, and Kuntoğlu M. Artificial intelligence systems for tool condition monitoring in machining: analysis and critical review. *Journal of Intelligent Manufacturing*. 2022:1–43.

[21] Schmidt P, Biessmann F, and Teubner T. Transparency and trust in artificial intelligence systems. *Journal of Decision Systems*. 2020;29(4):260–78.

[22] Asatiani A, Malo P, Nagbøl PR, Penttinen E, Rinta-Kahila T, and Salovaara A. Sociotechnical envelopment of artificial intelligence: an approach to organizational deployment of inscrutable artificial intelligence systems. *Journal of the Association for Information Systems*. 2021;22(2):8.

[23] Amann J, Vetter D, Blomberg SN, *et al.* To explain or not to explain?—artificial intelligence explainability in clinical decision support systems. *PLoS Digital Health*. 2022;1(2):e0000016.

[24] Sendak M, Elish MC, Gao M, *et al.* "The human body is a black box" supporting clinical decision-making with deep learning. In *Proceedings of the 2020 Conference on Fairness, Accountability, and Transparency* 2020 Jan 27 (pp. 99–109).

[25] Ploug T and Holm S. The four dimensions of contestable AI diagnostics – a patient-centric approach to explainable AI. *Artificial Intelligence in Medicine*. 2020;107:101901.

[26] Adadi A and Berrada M. Peeking inside the black-box: a survey on explainable artificial intelligence (XAI). *IEEE Access*. 2018;6:52138–60.

[27] Arrieta AB, Díaz-Rodríguez N, Del Ser J, *et al.* Explainable Artificial Intelligence (XAI): concepts, taxonomies, opportunities and challenges toward responsible AI. *Information Fusion*. 2020;58:82–115.

[28] Goodman B and Flaxman S. European Union regulations on algorithmic decision-making and a "right to explanation". *AI Magazine*. 2017;38(3): 50–7.

[29] Lipton ZC. The mythos of model interpretability: in machine learning, the concept of interpretability is both important and slippery. *Queue*. 2018; 16(3):31–57.

[30] London AJ. Artificial intelligence and black-box medical decisions: accuracy versus explainability. *Hastings Center Report*. 2019;49(1):15–21.

[31] Robbins S. A misdirected principle with a catch: explicability for AI. *Minds and Machines*. 2019;29(4):495–514.

[32] Awotunde JB, Oluwabukonla S, Chakraborty C, Bhoi AK, and Ajamu GJ. Application of artificial intelligence and big data for fighting COVID-19 pandemic. In *Decision Sciences for COVID-19* (pp. 3–26); 2022.

[33] AI H. High-level expert group on artificial intelligence. *Ethics Guidelines for Trustworthy AI*. 2019:6.

[34] Koszegi ST. High-level expert group on artificial intelligence. Brussels: European Commission; 2019.

[35] Zhu J, Liapis A, Risi S, Bidarra R, and Youngblood GM. Explainable AI for designers: a human-centered perspective on mixed-initiative co-creation. In *2018 IEEE Conference on Computational Intelligence and Games (CIG)* 2018 Aug 14 (pp. 1–8). New York, NY: IEEE.

[36] Došilović FK, Brčić M, and Hlupić N. Explainable artificial intelligence: a survey. In *2018 41st International Convention on Information and Communication Technology, Electronics and Microelectronics (MIPRO)*. New York, NY: IEEE; 2018.

[37] Hall P. On the art and science of machine learning explanations. arXiv preprint arXiv:1810.02909; 2018.

[38] Phillips PJ, Hahn CA, Fontana PC, Broniatowski DA, and Przybocki MA. *Four Principles of Explainable Artificial*. NIST Interagency/Internal Report (NISTIR). Gaithersburg, MD: National Institute of Standards and Technology.

[39] Kim J, Lee J, and Park M. Identification of smartwatch-collected lifelog variables affecting body mass index in middle-aged people using regression machine learning algorithms and SHapley additive explanations. *Applied Sciences*. 2022;12(8):3819.

[40] Mei F, Li X, Zheng J, Sha H, and Li D. A data-driven approach to state assessment of the converter valve based on oversampling and Shapley additive explanations. *IET Generation, Transmission & Distribution*. 2022;16(8):1607–19.

[41] Dauda OI, Awotunde JB, AbdulRaheem M, and Salihu SA. Basic issues and challenges on explainable artificial intelligence (XAI) in healthcare systems. In *Principles and Methods of Explainable Artificial Intelligence in Healthcare* (pp. 248–71). IGI-Global Publisher; 2022.

[42] Palatnik de Sousa I, Maria Bernardes Rebuzzi Vellasco M, and Costa da Silva E. Local interpretable model-agnostic explanations for classification of lymph node metastases. *Sensors*. 2019;19(13):2969.

[43] Folorunso SO, Ogundepo EA, Awotunde JB, Ayo FE, Banjo OO, and Taiwo AI. A multi-step predictive model for COVID-19 cases in nigeria using machine learning. In *Decision Sciences for COVID-19* (pp. 107–36). Cham: Springer; *2022*.

[44] Tonekaboni S, Joshi S, McCradden MD, and Goldenberg A. What clinicians want: contextualizing explainable machine learning for clinical end use. In *Machine Learning for Healthcare Conference* (pp. 359–80). PMLR; 2019.

[45] Awotunde JB, Misra S, Ayeni F, Maskeliunas R, and Damasevicius R. Artificial intelligence based system for bank loan fraud prediction. In *International Conference on Hybrid Intelligent Systems* (pp. 463–72). Cham: Springer; 2021.

[46] Lage I, Chen E, He J, *et al.* Human evaluation of models built for interpretability. In *Proceedings of the AAAI Conference on Human Computation and Crowdsourcing* (Vol. 7, pp. 59–67); 2019.

[47] Molnar C, Casalicchio G, and Bischl B. Interpretable machine learning – a brief history, state-of-the-art and challenges. In *Joint European Conference on Machine Learning and Knowledge Discovery in Databases* (pp. 417–31). Cham: Springer; 2020.

[48] Tschandl P, Rinner C, Apalla Z, *et al.* Human–computer collaboration for skin cancer recognition. *Nature Medicine*. 2020;26(8):1229–34.

[49] Oladipo ID, AbdulRaheem M, Awotunde JB, Bhoi AK, Adeniyi EA, and Abiodun MK. Machine learning and deep learning algorithms for smart cities: a start-of-the-art review. In *IoT and IoE Driven Smart Cities* (pp. 143–62); 2022.

[50] Awotunde JB, Folorunso SO, Jimoh RG, Adeniyi EA, Abiodun KM, and Ajamu GJ. Application of artificial intelligence for COVID-19 epidemic: an exploratory study, opportunities, challenges, and future prospects. In *Artificial Intelligence for COVID-19*. 2021:47–61.

[51] He J, Baxter SL, Xu J, Xu J, Zhou X, and Zhang K. The practical implementation of artificial intelligence technologies in medicine. *Nature Medicine*. 2019;25(1):30–6.

[52] Price W *and* Nicholson II. Regulating black-box medicine. *Michigan Law Review* 2017;116:421.

[53] Amasyali K and El-Gohary N. Machine learning for occupant-behavior-sensitive cooling energy consumption prediction in office buildings. *Renewable and Sustainable Energy Reviews*. 2021;142:110714.

[54] Awotunde JB, Chakraborty C, and Adeniyi AE. Intrusion detection in industrial internet of things network-based on deep learning model with rule-based feature selection. *Wireless Communications and Mobile Computing*. 2021;2021;1–17.

[55] Weng SF, Reps J, Kai J, Garibaldi JM, and Qureshi N. Can machine-learning improve cardiovascular risk prediction using routine clinical data?. *PLoS One*. 2017;12(4):e0174944.

[56] Zhao Y, Zhang H, Li Y, *et al.* AI powered electrochemical multi-component detection of insulin and glucose in serum. *Biosensors and Bioelectronics*. 2021;186:113291.

[57] Awotunde JB, Jimoh RG, AbdulRaheem M, Oladipo ID, Folorunso SO, and Ajamu GJ. IoT-based wearable body sensor network for COVID-19 pandemic. In *Advances in Data Science and Intelligent Data Communication Technologies for COVID-19* (p. 253–75); 2022.

[58] Awotunde JB, Jimoh RG, Oladipo ID, Abdulraheem M, Jimoh TB, and Ajamu GJ. Big data and data analytics for an enhanced COVID-19 epidemic

management. *In Artificial Intelligence for COVID-19 2021* (pp. 11–29). Cham: Springer.

[59] Antoniadi AM, Du Y, Guendouz Y, *et al.* Current challenges and future opportunities for XAI in machine learning-based clinical decision support systems: a systematic review. *Applied Sciences*. 2021;11(11):5088.

[60] Caruana R, Lou Y, Gehrke J, Koch P, Sturm M, and Elhadad N. Intelligible models for healthcare: Predicting pneumonia risk and hospital 30-day readmission. In *Proceedings of the 21th ACM SIGKDD International Conference on Knowledge Discovery and Data Mining* (pp. 1721–30); 2015.

[61] Durán JM and Jongsma KR. Who is afraid of black box algorithms? On the epistemological and ethical basis of trust in medical AI. *Journal of Medical Ethics*. 2021;47(5):329–35.

[62] Buçinca Z, Lin P, Gajos KZ, and Glassman EL. Proxy tasks and subjective measures can be misleading in evaluating explainable AI systems. In *Proceedings of the 25th International Conference on Intelligent User Interfaces* (pp. 454–64); 2020.

[63] Trocin C, Mikalef P, Papamitsiou Z, and Conboy K. Responsible AI for digital health: a synthesis and a research agenda. *Information Systems Frontiers*. 2021:1–9.

[64] Rosenfeld A and Richardson A. Explainability in human–agent systems. *Autonomous Agents and Multi-Agent Systems*. 2019;33(6):673–705.

[65] Topuz K and Delen D. A probabilistic Bayesian inference model to investigate injury severity in automobile crashes. *Decision Support Systems*. 2021;150:113557.

[66] Ma X, Shao Y, Tian L, *et al.* Analysis of error profiles in deep next-generation sequencing data. *Genome Biology*. 2019;20(1):1–5.

[67] Stiglic G, Kocbek P, Fijacko N, Zitnik M, Verbert K, and Cilar L. Interpretability of machine learning-based prediction models in healthcare. *Wiley Interdisciplinary Reviews: Data Mining and Knowledge Discovery*. 2020;10(5):e1379.

[68] Lakkaraju H and Bastani O. "How do I fool you?" Manipulating user trust via misleading black box explanations. In *Proceedings of the AAAI/ACM Conference on AI, Ethics, and Society* (pp. 79–85); 2020.

[69] Wanner J, Herm LV, Heinrich K, Janiesch C, and Zschech P. White, grey, black: effects of XAI augmentation on the confidence in AI-based decision support systems. In *ICIS* 2020 Sep.

[70] Zhavoronkov A, Vanhaelen Q, and Oprea TI. Will artificial intelligence for drug discovery impact clinical pharmacology? *Clinical Pharmacology & Therapeutics*. 2020;107(4):780–5.

[71] Rudin C. Stop explaining black box machine learning models for high stakes decisions and use interpretable models instead. *Nature Machine Intelligence*. 2019;1(5):206–15.

[72] Sharma S, Rawal YS, Pal S, and Dani R. Fairness, Accountability, Sustainability, Transparency (FAST) of artificial intelligence in terms of

hospitality industry. *In ICT Analysis and Applications* (pp. 495–504). Singapore: Springer; 2022.

[73] Miller T. Explanation in artificial intelligence: insights from the social sciences. *Artificial Intelligence*. 2019;267:1–38.

[74] Bhatt U, Andrus M, Weller A, and Xiang A. Machine learning explainability for external stakeholders. arXiv preprint arXiv:2007.05408; 2020.

[75] Holzinger A, Langs G, Denk H, Zatloukal K, and Müller H. Causability and explainability of artificial intelligence in medicine. *Wiley Interdisciplinary Reviews: Data Mining and Knowledge Discovery*. 2019;9(4):e1312.

[76] Holzinger A, Carrington A, and Müller H. Measuring the quality of explanations: the system causability scale (SCS). *Künstliche Intelligenz*. 2020; 34(2):193–8.

[77] Rudin C and Radin J. Why are we using black box models in AI when we don't need to? A lesson from an explainable AI competition. *Harvard Data Science Review*. 2019;1(2).

[78] Alaa AM and van der Schaar M. Demystifying black-box models with symbolic metamodels. *Advances in Neural Information Processing Systems*. 2019;32:11301–11.

[79] Karimi AH, Barthe G, Balle B, and Valera I. Model-agnostic counterfactual explanations for consequential decisions. In *International Conference on Artificial Intelligence and Statistics* (pp. 895–905). PMLR; 2020.

[80] Mothilal RK, Sharma A, and Tan C. Explaining machine learning classifiers through diverse counterfactual explanations. In *Proceedings of the 2020 Conference on Fairness, Accountability, and Transparency* (pp. 607–17); 2020.

[81] Gomez O, Holter S, Yuan J, and Bertini E. ViCE: visual counterfactual explanations for machine learning models. In *Proceedings of the 25th International Conference on Intelligent User Interfaces* (pp. 531–5); 2020.

[82] Miller T, Howe P, and Sonenberg L. Explainable AI: Beware of inmates running the asylum or: How I learnt to stop worrying and love the social and behavioural sciences. arXiv preprint arXiv:1712.00547; 2017.

[83] Folorunso SO, Awotunde JB, Banjo OO, Ogundepo EA, and Adeboye NO. Comparison of active COVID-19 cases per population using time-series models. *International Journal of E-Health and Medical Communications*. 2021;13(2):1–21.

[84] Russell C. Efficient search for diverse coherent explanations. In *Proceedings of the Conference on Fairness, Accountability, and Transparency* (pp. 20–8); 2019.

[85] Awotunde JB, Abiodun KM, Adeniyi EA, Folorunso SO, and Jimoh RG. A deep learning-based intrusion detection technique for a secured IoMT system. In *International Conference on Informatics and Intelligent Applications* (pp. 50–62). Cham: Springer; 2021.

[86] Smuha NA. The EU approach to ethics guidelines for trustworthy artificial intelligence. *Computer Law Review International*. 2019;20(4):97–106.

[87] Oh SJ, Benenson R, Fritz M, and Schiele B. Faceless person recognition: privacy implications in social media. In *European Conference on Computer Vision* (pp. 19–35). Cham: Springer; 2016.

[88] Orekondy T, Schiele B, and Fritz M. Knockoff nets: Stealing functionality of black-box models. In *Proceedings of the IEEE/CVF Conference on Computer Vision and Pattern Recognition* (pp. 4954–63); 2019.

[89] Oh SJ, Schiele B, and Fritz M. Towards reverse-engineering black-box neural networks. In *Explainable AI: Interpreting, Explaining and Visualizing Deep Learning* (pp. 121–44). Cham: Springer; 2019.

[90] Aditya P, Sen R, Druschel P, *et al.* I-pic: a platform for privacy-compliant image capture. In *Proceedings of the 14th Annual International Conference on Mobile Systems, Applications, and Services* (pp. 235–48); 2016.

[91] Awotunde JB, Ogundokun RO, Ayo FE, and Matiluko OE. Speech segregation in background noise based on deep learning. *IEEE Access.* 2020;8:169568–75.

[92] Abdulraheem M, Awotunde JB, Jimoh RG, and Oladipo ID. An efficient lightweight cryptographic algorithm for IoT security. In *International Conference on Information and Communication Technology and Applications* (pp. 444–56). Cham: Springer; 2020.

[93] Kaissis G, Ziller A, Passerat-Palmbach J, *et al.* End-to-end privacy preserving deep learning on multi-institutional medical imaging. *Nature Machine Intelligence.* 2021;3(6):473–84.

[94] Horowitz MC, Allen GC, Saravalle E, Cho A, Frederick K, and Scharre P. *Artificial Intelligence and International Security*. Washington, DC: Center for a New American Security; 2018.

[95] Awotunde JB, Jimoh RG, Folorunso SO, Adeniyi EA, Abiodun KM, and Banjo OO. Privacy and security concerns in IoT-based healthcare systems. In *The Fusion of Internet of Things, Artificial Intelligence, and Cloud Computing in Health Care* (pp. 105–134). Cham: Springer; 2021.

[96] Tan PN, Steinbach M, and Kumar V. *Introduction to Data Mining*. India: Pearson Education; 2016.

[97] Bodapati JD, Shaik NS, and Naralasetti V. Composite deep neural network with gated-attention mechanism for diabetic retinopathy severity classification. *Journal of Ambient Intelligence and Humanized Computing.* 2021;12(10):9825–39.

[98] Liu W, Wang Z, Liu X, Zeng N, Liu Y, and Alsaadi FE. A survey of deep neural network architectures and their applications. *Neurocomputing.* 2017;234:11–26.

[99] Lundberg SM and Lee SI. A unified approach to interpreting model predictions. *Advances in Neural Information Processing Systems.* 2017;30: 4765–74.

[100] Shapley L. A value for n-person games. contributions to the theory of games II (1953) 307–317. In *Classics in Game Theory* (pp. 69–79). Princeton, NJ: Princeton University Press; 2020.

[101] McKinney W. Pandas: a foundational Python library for data analysis and statistics. *Python for High Performance and Scientific Computing*. 2011; 14(9):1–9.

[102] Hunter JD. Matplotlib: a 2D graphics environment. *Computing in Science & Engineering*. 2007;9(03):90–5.

[103] Oliphant TE. *A guide to NumPy*. USA: Trelgol Publishing; 2006.

[104] Van Stralen KJ, Stel VS, Reitsma JB, Dekker FW, Zoccali C, and Jager KJ. Diagnostic methods I: sensitivity, specificity, and other measures of accuracy. *Kidney International*. 2009;75(12):1257–63.

[105] Asri H, Mousannif H, Al Moatassime H, and Noel T. Using machine learning algorithms for breast cancer risk prediction and diagnosis. *Procedia Computer Science*. 2016;83:1064–9.

[106] Huo D, Ikpatt F, Khramtsov A, *et al.* Population differences in breast cancer: survey in indigenous African women reveals over-representation of triple-negative breast cancer. *Journal of Clinical Oncology*. 2009; 27(27):4515.

[107] Kharya S and Soni S. Weighted naive bayes classifier: a predictive model for breast cancer detection. *International Journal of Computer Applications*. 2016;133(9):32–7.

[108] Karabatak M and Ince MC. An expert system for detection of breast cancer based on association rules and neural network. *Expert Systems with Applications*. 2009;36(2):3465–9.

[109] Mohebian MR, Marateb HR, Mansourian M, Mañanas MA, and Mokarian F. A hybrid computer-aided-diagnosis system for prediction of breast cancer recurrence (HPBCR) using optimized ensemble learning. *Computational and Structural Biotechnology Journal*. 2017;15:75–85.

[110] Islam M, Haque M, Iqbal H, Hasan M, Hasan M, and Kabir MN. Breast cancer prediction: a comparative study using machine learning techniques. *SN Computer Science*. 2020;1(5):1–4.

[111] Sakri SB, Rashid NB, and Zain ZM. Particle swarm optimization feature selection for breast cancer recurrence prediction. *IEEE Access*. 2018;6:29637–47.

[112] Thirumalaikolundusubramanian P. Comparison of Bayes classifiers for breast cancer classification. *Asian Pacific Journal of Cancer Prevention*. 2018;19(10):2917.

[113] Chaurasia V, Pal S, and Tiwari BB. Prediction of benign and malignant breast cancer using data mining techniques. *Journal of Algorithms & Computational Technology*. 2018;12(2):119–26.

Chapter 4

Prototype interface for detecting mental fatigue with EEG and XAI frameworks in Industry 4.0

Martín Montes Rivera[1], Luciano Martinez[1], Alberto Ochoa Zezzatti[2], Alan Navarro[2], Jesús Rodarte[2] and Néstor López[2]

Abstract

Mental fatigue correlates to prolonged cognitive activity. It stresses the brain of so many ideas or thoughts that can translate into commitments, jobs, and to-do at home—leaving a person exhausted and hindering productivity and overall cognitive function. Moreover, extracranial electroencephalogram (EEG) signals are an excellent indicator of the brain conditions of a person. Besides, mental fatigue increases power in frontal theta (θ) and parietal alpha (α) EEG rhythms. On the other hand, artificial intelligence (AI) and EEG signals improved the classification and regression results in different applications with new convolutional neural networks (CNNs), including EEGNet. Some of these results in the literature are applications for disabled persons, detection of mental fatigue driving, mental workload, and schizophrenia. Despite the benefits of applying CNNs to interpret EEG signals, the final products' applications are still limited due to the expertise required for working with this model. Alternatively, explainable AI (XAI) refers to the principle of AI operation and the presentation of the results obtained in the most user-friendly way possible. Explainable models must provide a clear description of their results without having to forget high learning efficiency. It must also be possible for users to understand the emerging generation of artificial intelligence mechanisms, place a certain degree of trust in it, and work with it and manage it efficiently. The present chapter proposes a new application that brings the advantages of using EEG signals together with the EEGNet structure, adding explainable intelligent models simplifying the detection of mental fatigue and preventing accidents in Industry 4.0. A study of various activities as a stimulus under a workstation scenario is analyzed to determine criteria associated with preventing accidents in the physical plant of an industrial building. Typically, it is difficult for

[1]Universidad Politécnica de Aguascalientes, Mexico
[2]Universidad Autónoma de Ciudad Juárez, Instituto de Ingeniería y Tecnología, Departamento de Ingenieria Electrica y Computación, México

the device to provide us with high-quality signals; there are invasive systems that allow greater precision. In this project, we use a non-invasive device for this purpose.

Keywords: Mental fatigue; EEG signals; Explained Artificial Intelligence

4.1 Introduction

Mental fatigue is associated with prolonged cognitive activity. It stresses the brain of so many ideas or thoughts that can translate into commitments, jobs, and to-do at home—leaving a person exhausted and reducing productivity and cognitive functions. Symptoms of mental fatigue include mental block, lack of motivation, irritability, stress from eating or losing appetite, and insomnia. Mental exhaustion can affect the short and long term [1,2].

In the industry, employees must perform manufacturing operations constantly manipulating objects in the processes of some production line; this causes mental fatigue or fatigue, which is the cause of human errors that manage to delay production times and schedules already established by the company's managers [3]. For example, the works in [4,5] claim that we have finite concentration periods and must schedule breaks to improve retention of information, motivation, and effectiveness even in sports, driving, and studying, among others.

These studies show that long working hours can lead to increased stress levels, poor eating habits, lack of physical activity, and illness. Thus, it is essential to recognize the mental fatigue symptoms of workers and their potential impact on the safety and health of each worker and their co-workers. Moreover, some research suggests that constant mental exhaustion can affect physical endurance [4,5].

Extracranial electroencephalogram (EEG) signals are an excellent indicator of the brain conditions of a person. For example, when one is in the process of drowsiness and tends to sleep, neural waves are measurable; some research demonstrates and decodes images based on dreams trying to discover sleep patterns [2]. Moreover, mental fatigue produces increasing power in frontal theta (θ) and parietal alpha (α) EEG rhythms [6]. Figure 4.1 shows the phases of an EEG signal and its associated condition. Thus, with a wireless brain-machine interface, one can perform measurements of a person's fatigue and analyze them to make decisions that help reduce human errors in the production processes caused by fatigue.

Psychological care is another branch from which this proposal benefits because it is functional for personalized attention to the operator. For example, suppose it is possible for metal fatigue detection from the cranial cortex in a non-invasive way with the analysis of the evoked potentials (EEG signals). In that case, the production supervisor can estimate failures due to fatigue behaviors in his staff. Then, he can adapt activities for keeping a person operating with consecutive linear series, reducing errors in the process, thus adding a layer to the quality of the products.

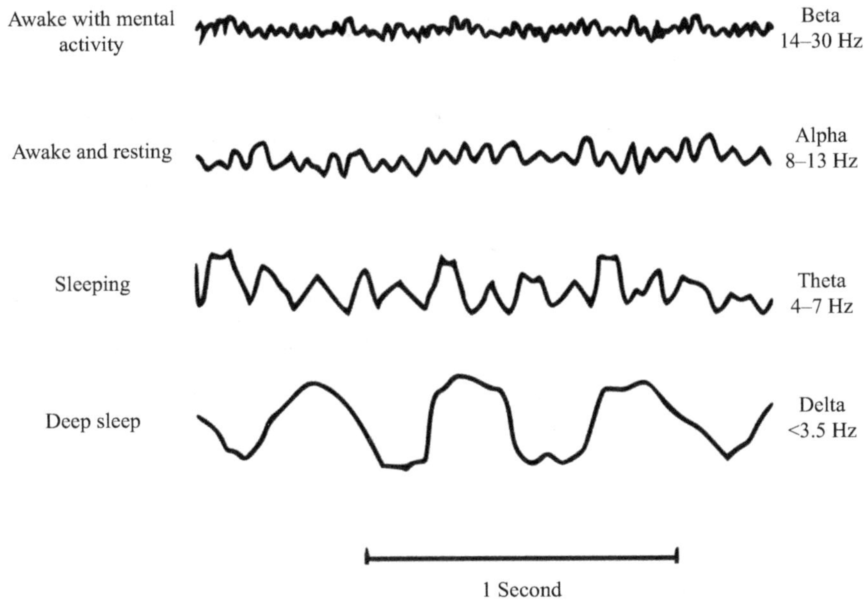

Figure 4.1 Standard frequencies of EEG acquisitions

4.1.1 Measurement of mental fatigue

Research on mental fatigue in the work environment considers different physiological factors. For example, in [1], the construction industry followed an induction process where the evaluated employees had to move loads between two places and unload them after meditating and relaxing. Furthermore, as shown in Figure 4.2, they conducted four trials, including activation of physical and mental fatigue with personnel (nine subjects) previously evaluated in health factors to give reliability to the EEG measurements.

4.1.2 EEG in mental fatigue

EEG consists of the electrical activity of neuronal populations that can be recorded from the scalp or cranial cortex invasively by acquisition with tiny potential processing devices. There are different frequency bands, designated as delta (0–4 Hz), theta (5–8 Hz), alpha (9–12 Hz), beta (13–30 Hz), and gamma (31–100 Hz) as far as neuronal signals are concerned, these are classified into these five groups because they are of interest characteristics. The neural activity recorded depends on the location of the electrode. Some examples of these recorders include inter-cortical local field potential (LFP), intracranial electrocorticogram (ECoG), or EEG [1,7].

The EEG signals recorded under a specific neuronal stimulation event are known as ERP, from the acronym (event-related potentials), which indicates any

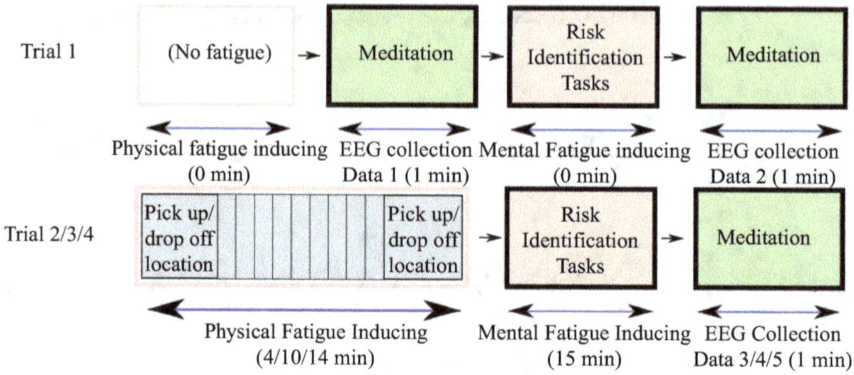

Figure 4.2 Mental and physical fatigue experiment designed with four independent trials in [1]

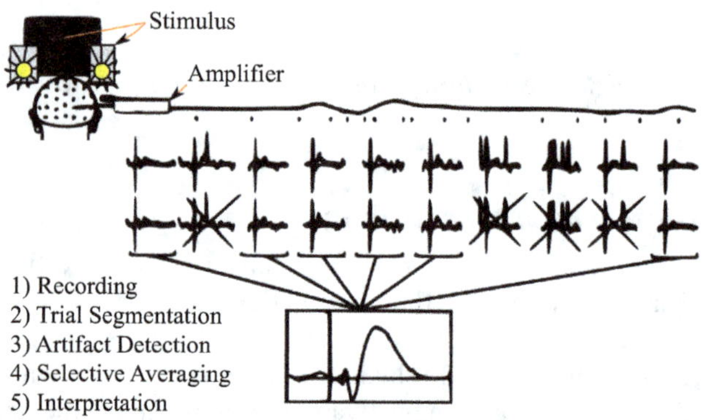

Figure 4.3 ERP reading procedure with its stages as described in [7]

electrophysiological response to an internal or external stimulus. In other words, any measured brain response directly results from a thought process or perception [7].

The ERP allows the researcher to know patterns and discover what is happening in the human brain under stimulus based on their signal recordings according to the general characteristics of signal processing, for example, frequency, amplitude, and areas of most significant activity according to the positioning of selected electrodes. The ERP reading is detailed in [7], which mentions the basic procedure of an EEG system with ERP reading, as shown in Figure 4.3.

The classification of signals is part of a pattern recognition system, which for EEG signals as inputs corresponds to the procedure described in [8] and, in our proposal, performs the classification of two classes of signals belonging to or not to

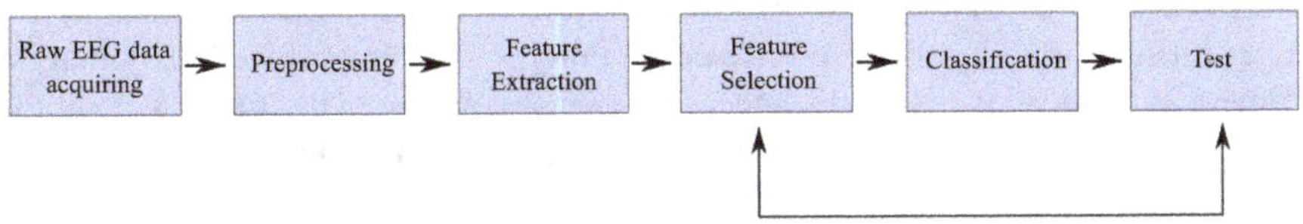

Figure 4.4 Block diagram of the proposed system for mental fatigue detection

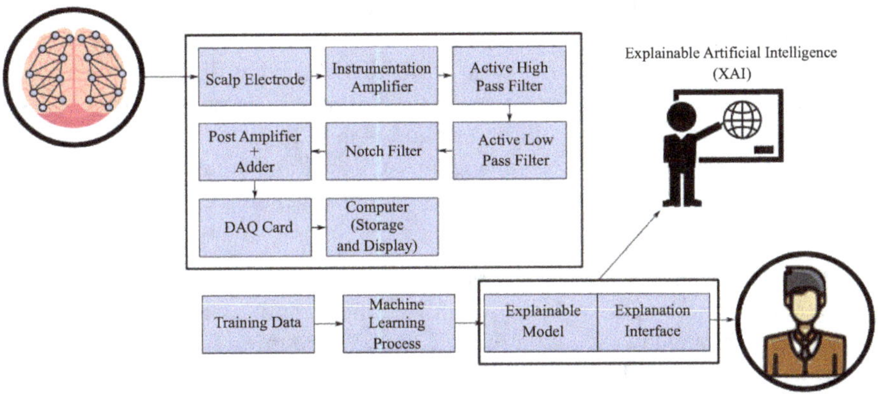

Figure 4.5 Block diagram of an EEG signal acquisition system, including an external explainable AI (XAI) module

mental fatigue. First, we acquire de EEG data. Then the preprocessing stage includes attenuation of the signal and reduction of noise. After that, we select characteristics based on the input signals and the classification quality obtained, as shown in Figure 4.4.

The block diagram of a pattern recognition system in [8] is sufficient to implement a classifier of EEG signals corresponding or not to a person with mental fatigue.

4.1.3 Acquisition with brain–machine interface (BCI)

A BCI must have specifications for reading clear neural signals. In other words, an acquisition only occurs when signals are readable and omit pure reading noise. OpenBCI already provides the hardware with software integration to make acquisition and transmission by WiFi of the signals. According to [9], the design of a BCI system must include the elements shown in Figure 4.5 as blocks.

Many packages perform acquisition and preprocessing with amplification, filtration, and noise elimination in EEG signals. For example, OpenBCI uses the ADS1299 microchip package. Once recorded the signals, there are many ways to perform neural analysis with the ADS1299. Moreover, many BCI libraries are free of use and supported by the scientific community. For example, Ref. [10] analyses commercial BCI devices that detect drowsiness states. The OpenBCI library implementation for working with the Cyton platform and LabVIEW used in this work is in [11].

4.1.4 EEGNET

This artificial neural network architecture in [12] is a compact convolutional neural network (CNN) compatible with different BCI paradigms, which can be entered with minimal data and reproduce neurophysiologically interpretable attributes. Figure 4.6 shows the visualization of the EEGNET model. To summarize, it starts with a convolution so that the network learns the frequency filters, then performs another convolution, but deeper to learn the special frequency-specific filters – called parity convolution. The full description is in [12].

Each phase has a series of parameters established and adjustable according to the needs and requirements in its use, Ref. [12] fundamentally explains the parts of the deep neural network as follows:

- Block 1: Performs two convolutional steps in sequence. First, 2D F1 convolutional filters of size (1, 64), with the filter length assigned to half of the sampling frequency of the data (here, 128 Hz), outputting F1 feature maps with the EEG signal at different frequencies. Then, an average pooling layer of size (1, 4) reduces the signal's sampling frequency to 32 Hz. Finally, there is a normalizer spatial filter using a maximum norm constraint of 1 on their weights |w|2< 1.

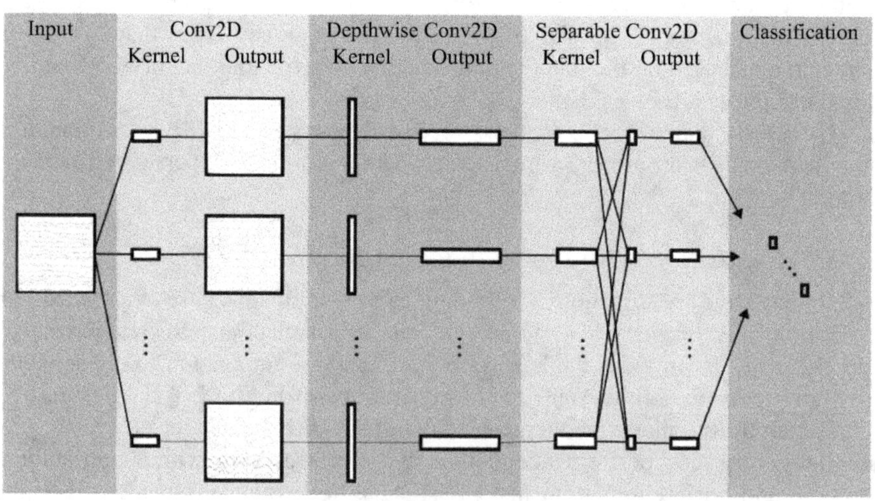

Figure 4.6 EEGNet architecture [12]

- Block 2: Uses a separable convolution with sizes (1, 16), representing 500 ms of EEG activity at 32 Hz) followed by F2 (1, 1) convolutions. The authors first learn a 500 ms "summary" of each feature map and then combine the results. Finally includes a dimensionality reduction, with a layer of pooling averaging size (1, 8).
- Classification block: The features raw pass to a SoftMax classification with *N* units, where *N* is the number of classes in the data.

The present chapter introduces a solution for detecting mental fatigue in Industry 4.0 with EEG signals and the EEGNet–CNNs designed for working with brain activity inputs.

The novelty of our proposal is on three specific elements:

- Application of EEGNet with training data specifically for each user-generated using non-invasive sensors so that workers can continue with their activities while detecting mental fatigue.
- Include an XAI user interface for inspection and identifying the level of fatigue so that relaxing activities be scheduled with mentally fatigued employees.
- We also generated a mental fatigue dataset, and here we introduce the schema for adding new datasets, allowing training of new users for mental fatigue detection.

Here we described the general introduction to EEG signals and their use with artificial intelligence (AI) techniques like the EEGNet CNNs. In this section, we also describe the schema of our proposal and the more relevant novelties of our work. The rest of this chapter is organized as follows: Section 4.2 shows the more recent work in state of the art using AI techniques to interpret EEG signals. Section 4.3 describes the methods and materials used to implement our proposal. Section 4.4 shows the results of this research and a brief about them. Finally, Section 4.5 describes the conclusions from the research described in this chapter.

4.2 Related work

Several works with EEG signals for different proposes have appeared in recent years with remarkable results due to the improvements in AI algorithms, computer power, and the new algorithms for working with them.

Ref. [13] claims that the EEG signals demand massive datasets for training because of the variability of the characteristics of those signals. Nevertheless, they found that working with smaller datasets focused on a specific task with specific individuals is possible.

Previous works showed that conventional neural networks have difficulties mapping the characteristics of EEG signals with time and space as inputs. Moreover, this produces a negative effect on the accuracy of classification tasks. However, new neural networks with structures designed explicitly for EEG signals successfully tackle this situation and allow them to work directly with time and space inputs. Examples of these proposed in [12,14] new EEG neural networks are EEGNet and S-EEGNet, based on CNNs.

Additionally, the new structures of CNNs are expanding approaches and applications. For example, Ref. [15] applies the EEGNet structure with EEG signals captured from the ear, which has a low length in the visual potentials. Similarly, Ref. [16] applies different basic algorithms with characteristics including wavelet transforms and artificial neural networks to identify mental fatigue in drivers. Likewise, the work in [17] also analyzes mental fatigue, but it does with specific hardware that produces a signal they call a mental wave. Also, Ref. [18] concentrates information of the several electrodes with autoencoders in a compressed signal.

EEG and CNNs work together well in analyzing brain information, as described in [19]. For example, Ref. [20] shows an application. Another application with EEG and CNNs designed in [19] uses brain activity to detect alert levels, avoiding accidents in drivers. Similarly, brain activity detected with EEG signals and CNNs allows identifying schizophrenia [21]. Likewise, Ref. [22] allows detecting mental workload with EEG signals and CNNs with adaptative learning.

4.3 Materials and methods

The hardware we decided to use for reading the EEG signals is the Cyton card, but before purchasing this device, we evaluated other platforms considering the details described in [10]. Then, we simulated the signal filtration phase. Figure 4.7 is the structure used for acquiring the output of a single channel with the Cyton

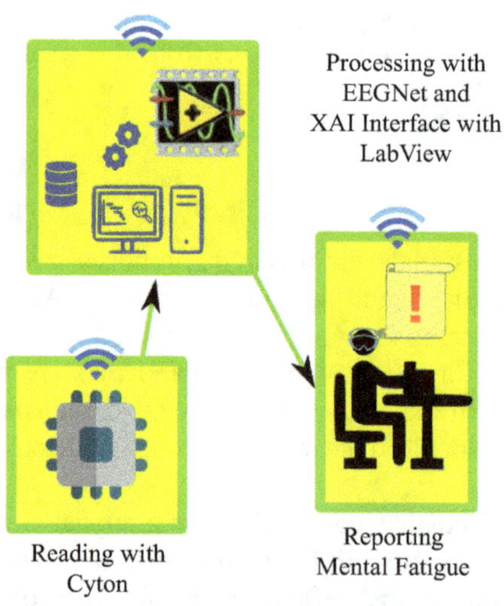

Figure 4.7 Diagram of the EEG proposed system with XAI for EEGNet

acquisition card, sending the preprocessed signals to the EEGNet classifier, and finally presenting the XAI interface to the user reporting mental fatigue detection.

4.3.1 Selection of computer equipment for mental fatigue detection

In the part of the deep neural network's classification and training, GPUs in processing accelerate the training process. Additionally, fine adjustments require laboratory equipment with the LabVIEW software for preprocessing stages. According to Ref. [12], the hardware they used for training was an NVIDIA Quadro M6000 GPU graphics card with CUDA 9 and cuDNN v7 in Tensorflow, using the Keras API.

However, the neural network could also train on the Google Colab platform by executing the EEGNet code with the information acquired in CSV; the files recorded the EEG signals with a time–voltage electrode structure (time, voltage). After that, the python code using the TensorFlow library runs to train the model. Then, we export the trained model to include it later in the XAI interface. Thus, the design of the XAI interface and the reading of signals imply a computer with LabVIEW to develop the graphical interface and Python programming language to execute the EEGNet neural network and TensorFlow used in its architecture to train and execute it.

Therefore, the minimum requirement for processing is a PC with LabView, Python, Tensorflow, and Nvidia video card. Alternatively, one could use Google Colab, an instance in the cloud that allows Google servers to train models faster. The general process is in Figure 4.8.

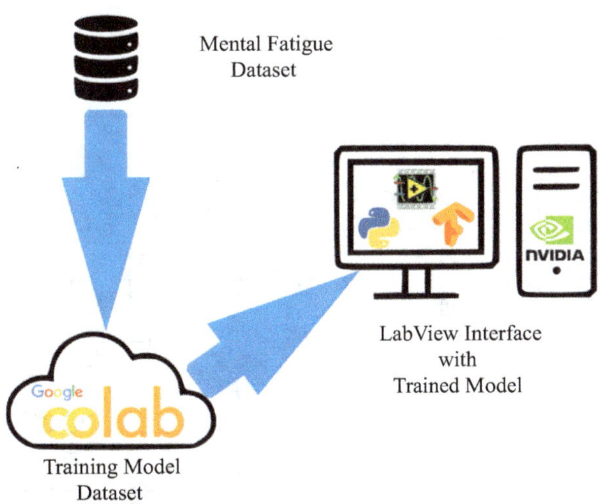

Figure 4.8 Schema for training EEGNet and using it in mental fatigue detection with LabView XAI interface

4.3.2 Generation of the dataset for training

The generation of the dataset starts by assigning repetitive tasks to a person. Then the person puts on the EEG device to record signals of mental fatigue. After that, the person relaxes to record the signals of not mental fatigue, like in Figure 4.9.

The fatigue induction process (15 min) includes matrix filling in a printed document, making this repetitive activity as it occurs in the industry. Figure 4.10 shows the pattern the subjects should follow, in which they must write a cross in one box, not the next. The entire matrix has 31 by 32 squares. On the other hand, the meditation process (15 min) changes the user to comfortable poses with continuous respiration. Finally, the measurement (15 min) includes putting on the Cyton headset and recording the EEG signals, as stated in Figure 4.9.

After recording the signals, we preprocessed them to avoid noise and extract interest characteristics for BCIs. The preprocessing includes a second-order digital Butterworth filter high pass with a limit of 98 Hz, then a second-order digital Butterworth filter low pass with a limit of 9.4 Hz. After that, we normalize the information and generate three virtual channels performing independent component

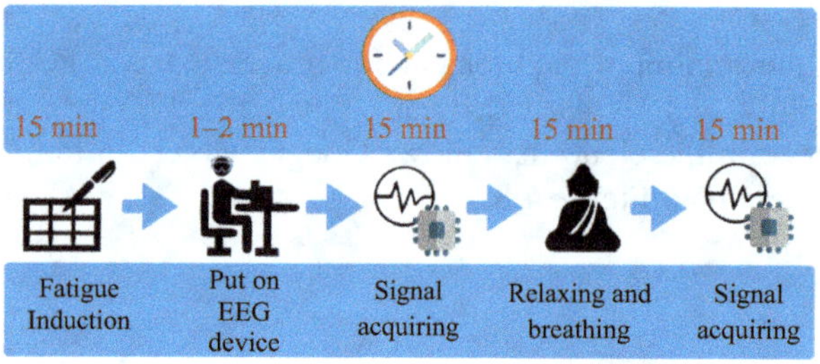

Figure 4.9 Timeline of the procedure

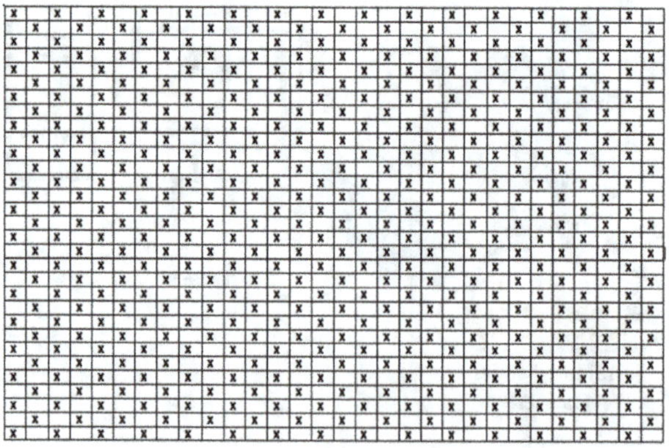

Figure 4.10 Pattern with squares to fill for induction of mental fatigue

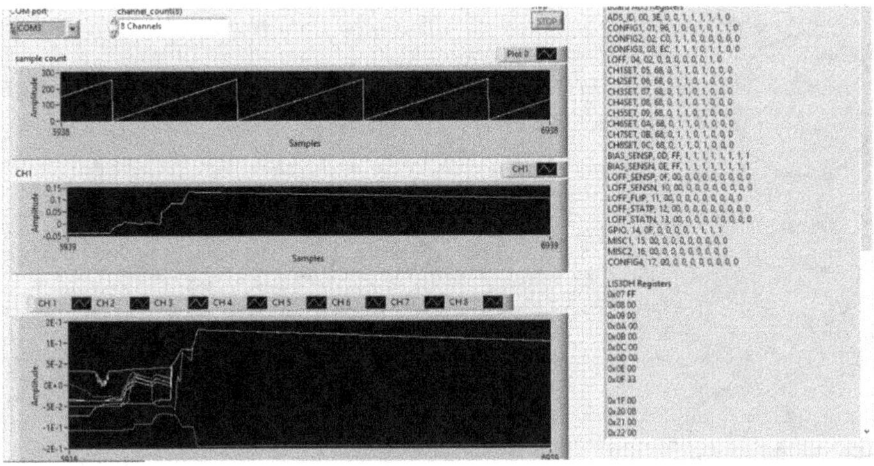

Figure 4.11 Open BCI interface for LabVIEW

analysis (ICA). Finally, we applied discrete wavelet transform (DWT) in 6 levels with the type of wavelet Haar-Daubechies with order fourth (db04 in LabVIEW Advanced Signal Processing Toolkit), as recommended and included in OpenBCI for LabVIEW in [11]. The interface for OpenBCI in LabVIEW is in Figure 4.11.

4.3.3 Training of EEGNet

Training EEGNet, as in any other artificial neural network, determines the numerical parameters for success in performing its goal. However, before training, the structure and training parameters must be defined. In this work's case of the EEGNet, we used the training algorithm and parameters recommended in [12] and described in Section 4.1.4.

However, we changed the original EEGNet model that includes four classes in the softmax activation function: LA: left auditive stimulation; RA: right auditive stimulation; LV: left visual stimulation; and RV: proper visual stimulation; to work with two classes, mental fatigue and not mental fatigue.

The input variables are the EEG virtual channel signals collected after the preprocessing stage from the information recorded with a 16 kHz sampling frequency in the 15 min acquisition stage with eight channels.

The inputs include the raw signal without filters, the processed signal with high pass and low pass filters; the ICA filter; the alfa waves; and the potential in the alpha waves—these last two are obtained with the wavelet transform described in Section 4.3.2.

After that, we recommend training an EEG neural network specifically for the user because training a single model for anyone implies collecting a massive amount of data. Alternatively, one can train the system for a specific person and then continually read to warn about metal fatigue.

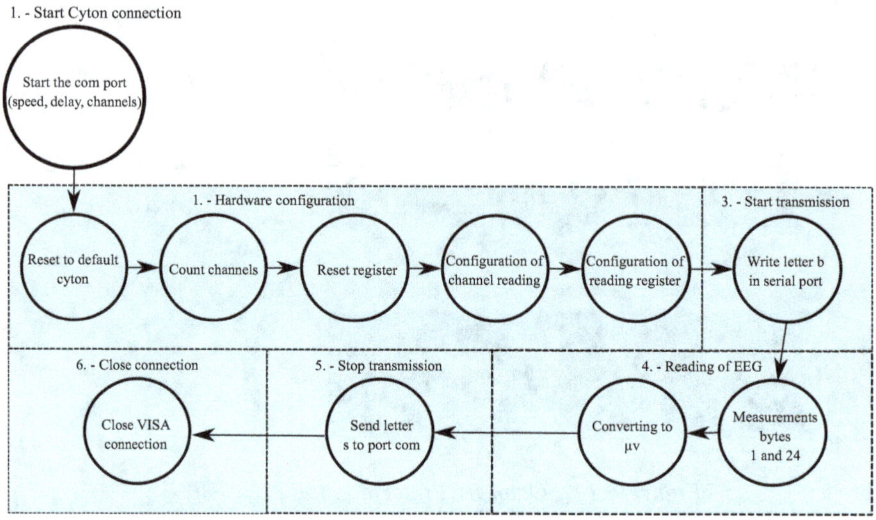

Figure 4.12 Communication process controlled by the graphical interface in LabVIEW

Therefore, in this proposal, we recommend training the model for each user using the Adam optimizer as in [12], a method extensively used for training CNN proposed and described in [23].

4.3.4 Graphical interface and control of communication with the trained model

Developing a graphical interface with LabVIEW achieves a breakdown of the results in a manageable and user-friendly user interface. The graphical interface also controls the communication process according to the state diagram in Figure 4.12.

The EEGNet output is the percentage of security that the neuron has that a pattern belongs to a class, in this work, to classes mental fatigue or not. In this research, we propose a LabVIEW interface showing a percentage from 0% to 100% labeled as the quantity of mental fatigue, linking that value to the prediction accuracy obtained in the output of the Softmax activation function. We recommend this change because it is more straightforward to interpret a level of mental fatigue for a regular user than security belonging to a class, as is commonly reported in the output of a Softmax activation function for machine learning experts.

4.4 Results and discussions

4.4.1 Results

This section shows the obtained results after following the methods described in Section 4.3.

On the part of the software, we developed a LabVIEW program that allows acquiring signals for 15 min directly from the Cyton with a sampling frequency of 128 kHz and saves them after preprocessing in a separate file per comma, giving us eight columns dedicated to each channel, and each line is a sample, sampling 200 times per second for the CSV file (200 Hz).

The preprocessing results take the raw signals with a sampling frequency of 128 kHz and process them as described in Section 4.3.2. Figure 4.13 shows, for a user of the device after induction of mental fatigue, the raw signals, the output of the Butterworth filters, the outputs of virtual channels obtained with ICA, and the alpha and power of alpha outputs obtained with the DWT of fourth-order and six levels using Haar-Daubechies waves.

Similarly, Figure 4.14 shows, for a user of the device after relaxing to reduce mental fatigue, the raw signals sampled at 128 kHz, the output of the Butterworth filters, the outputs of virtual channels obtained with ICA, and the alpha and power of alpha outputs obtained with the DWT of fourth-order and six levels using Haar-Daubechies waves.

After that, we tested the EEGNet training and classifying data from event-related potentials (ERP) from a four-class classification task (Figure 4.15) using the sample dataset provided in the MNE package. The information used by the network as input for training uses the file FIF (Fractal Image Format Bitmap file), but in future work, we will implement another type of file with the raw signals after preprocessing using the information we generated for the dataset. This test aims to read the data and verify compatibility and the framework's configuration for working with the EEGNet. We proof this concept locally using the purchased computer equipment described above. Although we do not show the results of training the model with our dataset, this is not the goal of this work but to show an XAI interface that delivers to the user a result easier to interpret associated with detecting metal fatigue.

The graphical XAI interface we generated is in Figures 4.16–4.18. The interaction with the user begins in the main window (Figure 4.16), giving the selection options in the communication port, selection of acquisition channels, and the person's name to classify or make a database for training, allowing the generation of a model for each user (like we recommended in Section 4.3.3). Next, the acquisition histogram window (Figure 4.17) provides the signal acquired by Cyton when recording the database in real-time. Finally, the fatigue measurement window (Figure 4.18) shows the spectra in time and frequency of the preprocessing and the final result of the fatigue percentage based on the neural network's accuracy.

4.4.2 Discussions of results

The first result we obtained is the communication and measuring system which allows us to configure the essential characteristics of the EEG device and acquire the signals for being shown and processed in the developed application. The system produced accomplishes all the requirements stated in Section 4.3. In addition, it allows controlling the communication port and the number of channels for

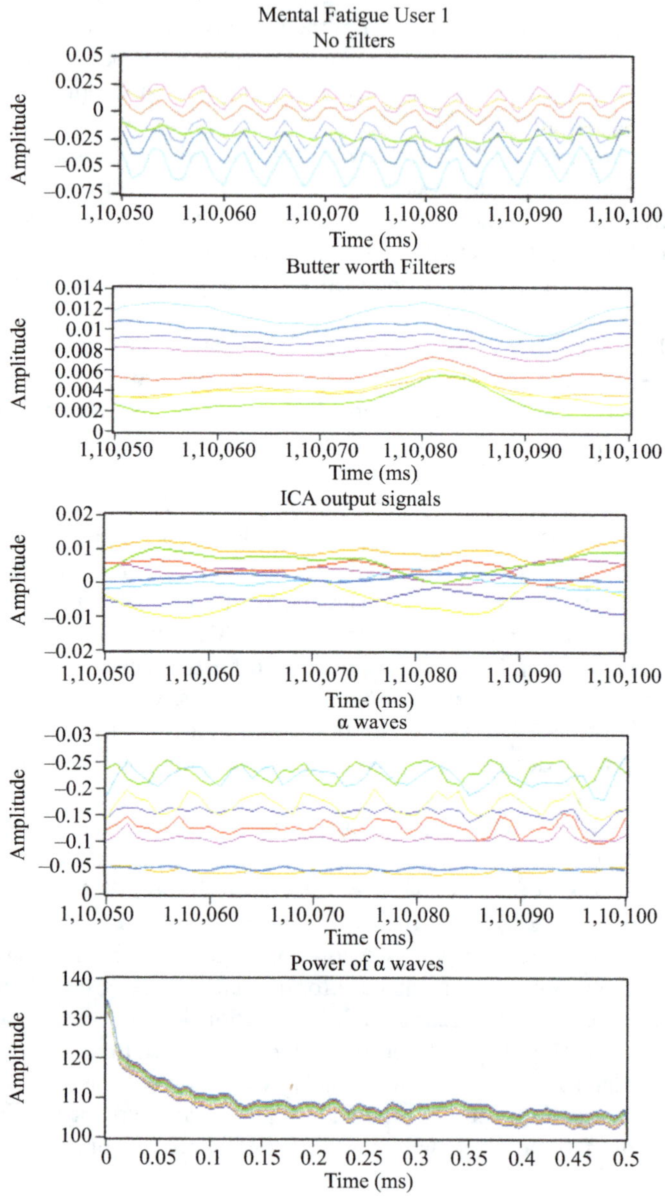

Figure 4.13 EEG signals with preprocessing after inducting mental fatigue

measuring, including the name of whose the signals belong for later grouping in the dataset generated. Moreover, it allows two modalities, one for dataset generation and the other for measuring and classification with the trained system, as shown in Figure 4.16 in Section 4.4.1.

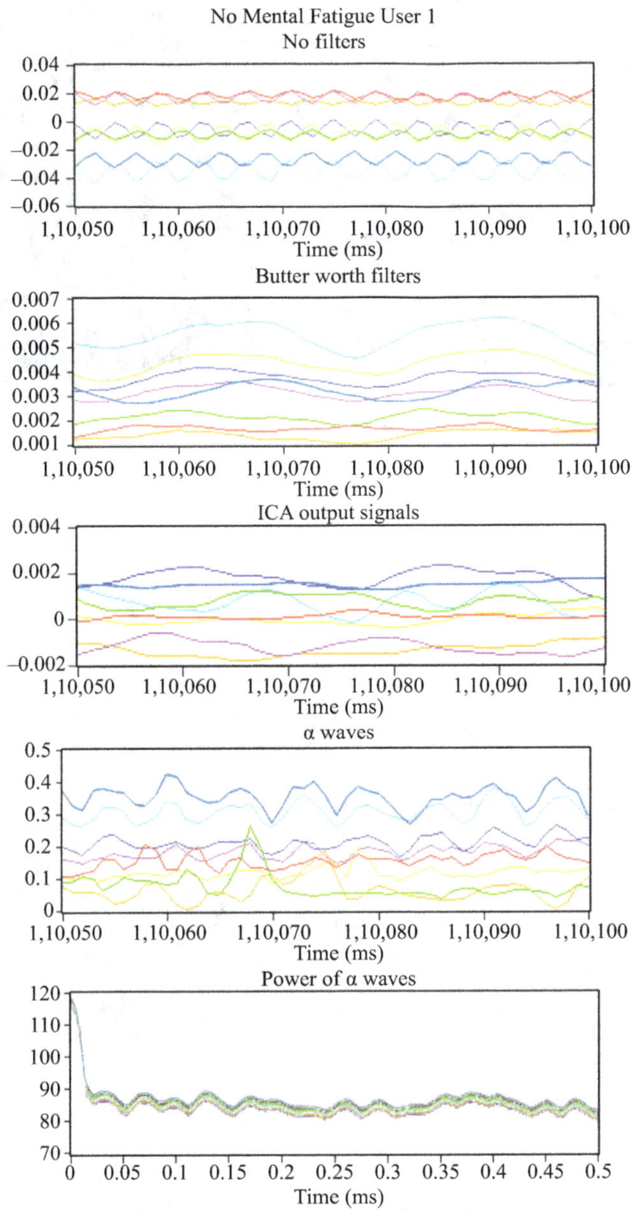

Figure 4.14 EEG signals with preprocessing after relaxing to reduce mental fatigue

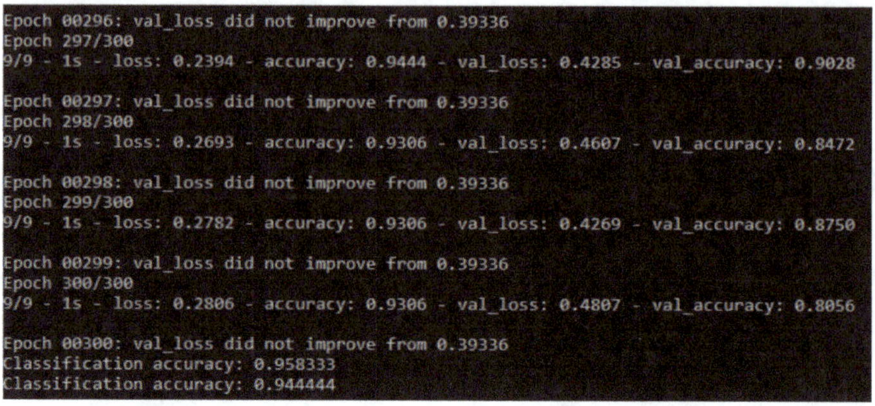

Figure 4.15 Test of EEGNet framework with the MNE dataset of ERP signals

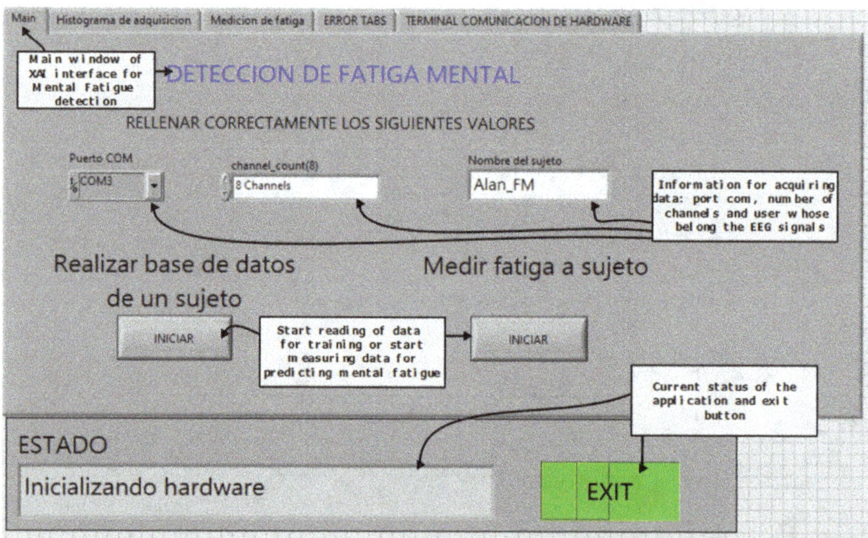

Figure 4.16 Main window of the proposed XAI mental fatigue interface

The system also includes plotting tools to see the acquired signals and their different characteristics depending on the level of the processing—for example, the raw acquired signals in the histogram acquisition tab in Figure 4.17, which we detailed from the dataset saved in Figures 4.13 and 4.14. Similarly, after preprocessing the filtered signals, the level of fatigue is perceived in the signals, as shown in Figure 4.18 for the mental fatigue tab, where the EEGNet process the filtered signals with the TensorFlow trained model with the results shown in Figure 4.15 with the concept proof. Finally, Figure 4.18 shows the tab with the

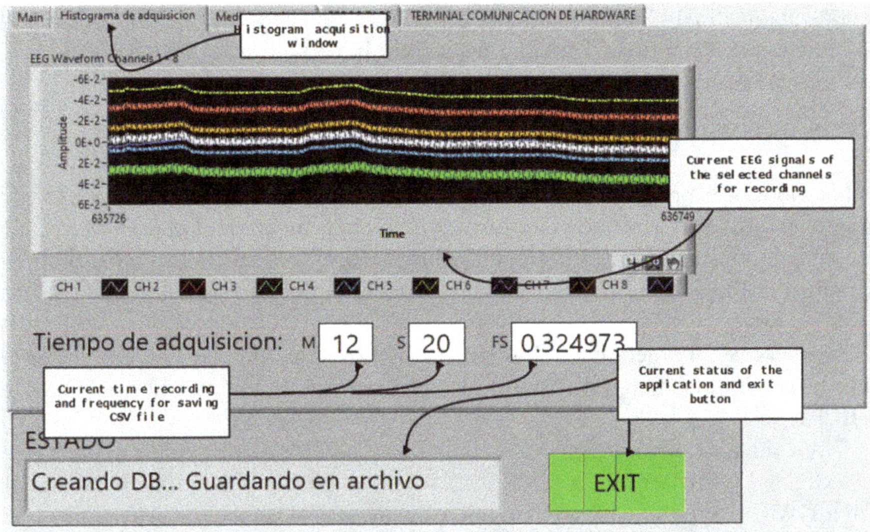

Figure 4.17 Histogram of acquisition window of the proposed XAI mental fatigue interface

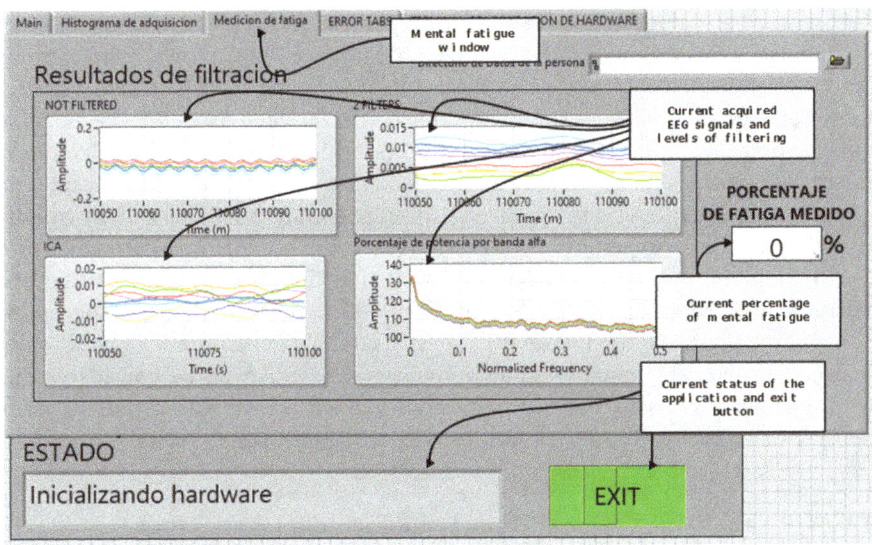

Figure 4.18 Mental fatigue measurement window of the proposed XAI mental fatigue interface

mental fatigue final result, which the supervisor can verify intuitively due to the XAI processing. The mental fatigue results are in a range from 0% to 100%.

4.5 Conclusions

In this work, we propose the use of an XAI interface to help inexpert users to comprehend the output value of a Softmax activation function of an EEGNet–CNN specialized in interpreting EEG signals. The proposal focuses on solving a problem of interest in Industry 4.0, which measures mental fatigue in workers who perform repetitive tasks.

We successfully acquired EEG signals at 128 kHz with the Cython platform. Moreover, the device complied with the requirements to obtain information from each channel.

After that, we preprocessed the signals in LabVIEW with Butterworth filters and extracted virtual channels with ICA. Then, we obtained the alpha band signals with a DWT of fourth-order and six levels using Haar-Daubechies waves for users with mental fatigue induction and after relaxing from it.

Induction of mental fatigue was achieved with the proposed activity of filing squares while changing to comfortable positions with continuous breathing allow reduce mental fatigue.

The selected EEGNet architecture worked as a proof of concept because we only tested the model and the framework for training it.

Our proposal is novel since it applies the EEGNet to a new application for detecting mental fatigue in the presence of repetitive activities in industry 4.0. For this purpose, we designed and acquired an interface that allows characteristic extraction for the generation of datasets for subject training models to detect mental fatigue for repetitive tasks in Industry 4.0 specific for each user. We did not find similar approaches in the literature to our proposal for mental fatigue detection in Industry 4.0. Moreover, we include an interface that simplifies the final result with an XAI model that allows any supervisor to set and understand the system without artificial intelligence experience or knowledge in convolutional neural networks.

Additionally, we generate a dataset and models for working with one user at a time for a selected group of people who fulfilled the function of providing neural information with the Cython interface. These datasets can be further applied to test other models detecting mental fatigue in Industry 4.0. However, we considered that the group of people was not representative enough; other works present more extensive databases with information from up to 100 people.

References

[1] Xing X, Zhong B, Luo H, Rose T, Li J, and Antwi-Afari MF. Effects of physical fatigue on the induction of mental fatigue of construction workers: a pilot study based on a neurophysiological approach. *Autom Constr.* 2020;120:103381.

[2] Liu Y, Lan Z, Cui J, Sourina O, and Müller-Wittig W. Inter-subject transfer learning for EEG-based mental fatigue recognition. *Adv Eng Inform*. 2020;46:101157.

[3] Mahmud, N. Study the impact of fatigue and optimizing productivity of an assembly line of garment industry. *Int J Sci Eng Res*. 2011;2(11) [cited 2022 May 17]. Available from: http://www.ijser.org

[4] Weerakkody NS, Taylor CJ, Bulmer CL, *et al.* The effect of mental fatigue on the performance of Australian football specific skills amongst amateur athletes. *J Sci Med Sport*. 2021;24(6):592–6.

[5] Russell S, Jenkins DG, Halson SL, Juliff LE, Connick MJ, and Kelly VG. Mental fatigue over 2 elite Netball Seasons: a case for mental fatigue to be included in athlete self-report measures. *Int J Sports Physiol Perform*. 2021;17(2):160–9 [cited2022 May 17]. Available from: https://journals.humankinetics.com/view/journals/ijspp/17/2/article-p160.xml

[6] Trejo LJ, Kubitz K, Rosipal R, Kochavi RL, and Montgomery LD. EEG-based estimation and classification of mental fatigue. *Psychology*. 2015;6 (05):572–89 [cited 2022 May 17]. Available from: http://file.scirp.org/Html/7-6901416_55452.htm

[7] Cong F, Ristaniemi T, and Lyytinen H. Advanced signal processing on brain event-related potentials: filtering ERPs in time, frequency and space domains sequentially and simultaneously. In *Advanced Signal Processing on Brain Event-Related Potentials: Filtering ERPs in Time, Frequency and Space Domains Sequentially and Simultaneously*. Singapore: World Scientific Publishing Co.; 2015, pp. 1–202.

[8] Simao M, Mendes N, Gibaru O, and Neto P. A review on electromyography decoding and pattern recognition for human-machine interaction. *IEEE Access*. 2019;7:39564–82.

[9] Bhagawati AJ and Chutia R. Design of single channel portable EEG signal acquisition system for brain computer interface application. *Int J Biomed Eng Sci*. 2016;3(1).

[10] LaRocco J, Le MD, and Paeng DG. A systemic review of available low-cost EEG headsets used for drowsiness detection. *Front Neuroinform*. 2020;14:42.

[11] OpenBCI. GitHub – rcassani/OpenBCI-Toolkit-LabVIEW: OpenBCI toolkit for LabVIEW [Internet]. [cited 2022 May 24]. Available from: https://github.com/rcassani/OpenBCI-Toolkit-LabVIEW

[12] Lawhern VJ, Solon AJ, Waytowich NR, Gordon SM, Hung CP, and Lance BJ. *EEGNet: A Compact Convolutional Neural Network for EEG-Based Brain-Computer Interfaces*. 2018 [cited 2022 May 18]. Available from: https://github.com/vlawhern/arl-eegmodels.

[13] Wan Z, Yang R, Huang M, Zeng N, and Liu X. A review on transfer learning in EEG signal analysis. *Neurocomputing*. 2021;421:1–14.

[14] Huang W, Xue Y, Hu L, and Liuli H. S-EEGNet: electroencephalogram signal classification based on a separable convolution neural network with bilinear interpolation. *IEEE Access*. 2020;8:131636–46.

[15] Zhu Y, Li Y, Lu J, and Li P. EEGNet with ensemble learning to improve the cross-session classification of SSVEP based BCI from ear-EEG. *IEEE Access*. 2021;9:15295–303.

[16] Ettahiri H, Vicente JMF, and Fechtali T. EEG signals in mental fatigue detection: a comparing study of machine learning technics VS deep learning. In: Ferrández Vicente JM, Álvarez-Sánchez JR, de la Paz López F, and Adeli, H editors, *Artificial Intelligence in Neuroscience: Affective Analysis and Health Applications*. Cham: Springer; 2022, pp. 625–33.

[17] Dahmani A Al, Goncharenko I, Gu Y. E-Worker mental fatigue detection through mindwave EEG data and deep neural networks. In *2022 IEEE 4th Global Conference on Life Sciences and Technologies (LifeTech)*. 2022, pp. 501–2.

[18] Riyad M, Khalil M, and Adib A. Dimensionality reduction of MI-EEG data via convolutional autoencoders with a low size dataset. In: Fakir M, Baslam M, and El Ayachi R, editors. *Business Intelligence*. Cham: Springer; 2022, pp. 263–78.

[19] Craik A, He Y, and Contreras-Vidal JL. Deep learning for electro-encephalogram (EEG) classification tasks: a review. *J Neural Eng*. 2019;16 (3):31001.

[20] Amin SU, Alsulaiman M, Muhammad G, Mekhtiche MA, and Shamim Hossain M. Deep Learning for EEG motor imagery classification based on multi-layer CNNs feature fusion. *Futur Gener Comput Syst*. 2019;101: 542–54.

[21] Korda AI, Ventouras E, Asvestas P, Toumaian M, Matsopoulos GK, and Smyrnis N. Convolutional neural network propagation on electro-encephalographic scalograms for detection of schizophrenia. *Clin Neurophysiol*. 2022;139:90–105.

[22] Yin Z and Zhang J. Cross-session classification of mental workload levels using EEG and an adaptive deep learning model. *Biomed Signal Process Control*. 2017;33:30–47.

[23] Kingma DP and Ba JL. Adam: a method for stochastic optimization. In *3rd Int Conf Learn Represent ICLR 2015 – Conf Track Proc*, 2014 [cited 2022 May 24]. Available from: https://arxiv.org/abs/1412.6980v9

Chapter 5

XAI for medical image segmentation in medical decision support systems

Abasiama Godwin Akpan[1], *Flavious Bobuin Nkubli*[2], *Victoria Nnaemeka Ezeano*[1], *Anayo Christian Okwor*[3], *Mabel Chikodili Ugwuja*[3] *and Udeme Offiong*[4]

Abstract

Medical image segmentation has contributed immensely to medical care delivery. With the speedy development of deep learning (DL), medical image segmentation processing based on deep convolutional neutral networks (CNNs) has become a research interest. Explainable artificial intelligence (XAI) provides pathways for useful MDSSs. The necessity for XAI in MDSSs is largely based on ethical, fair decision making, strengthening means of chronological procedures, and unfairness that should be revealed during medical image segmentation processes. Studies have shown that an inaccurate diagnosis is as a result of not identifying the limits of a pathological lesion or organ. It is clear that the likelihood of survival can be improved if the tumor is identified and classified properly at its early stage. In this study, we provide an enhanced application of fuzzy C-means and Artificial Neural Network Algorithm for medical image segmentation. The paper intends to review and contrast the techniques of automatic detection of brain tumor through magnetic resonance imaging (MRI) by the application of fuzzy C-mean and artificial neural network (ANN). Explanation given to these AI processes creates medical decision confidence, trustworthiness, acceptability, and potentials for its incorporation in the medical image segmentation workflow. Based on the discussions on human pathological tissues and organs, the specificity between them and their classic segmentation algorithms is revealed.

Keywords: Explainable artificial intelligence (XAI); Artificial neural network (ANN); Medical decision support systems (MDSSs); Convolutional neutral networks (CNNs); Magnetic resonance image (MRI)

[1]Department of Computer Science & Mathematics, Evangel University, Nigeria
[2]Department of Medical Radiography, University of Maiduguri, Nigeria
[3]Department of Radiography & Radiation Sciences, Evangel University, Nigeria
[4]Department of Psychology, Chukwuemeka Odumegu University, Nigeria

5.1 Introduction

Explaining medical decisions to patients or users in artificial intelligence (AI)-based predictive models is a necessity [1]. Segmentation processes with artificial neural network (ANN) is the foundation of machine learning (ML) and deep learning (DL). The algorithm is composed of levels of related points [2]. The input data may be radiomic characteristics imported from the image files. The real carrier of automated medical imaging is the visual recognition based on AI to derive lower error rates than the human observation [3]. The aim of ML in medical imaging includes detecting and classifying lesions, automated image segmentation, data analysis, extraction of radiomic features, prioritizing reporting, studying triage, and reconstructing medical images. ANN runs parallel with conventional statistical analysis that provides insights into clinical data [4]. The resultant of the segmentation affects medical image analysis processes, for example, image representation, object definition, measuring object features, and classifying of medical objects. Hence, image segmentation becomes a necessary process for assisting the delineation, characterization, and visualization of regions of interest in any medical image [5]. The physical segmentation is labor intensive, tiresome, lingering, and imprecise, particularly with the modalities made up of a large amount of images that need to be checked. Cheng *et al.* [6] posited that convolutional neutral networks (CNNs) are the basis of DL means for medical imaging, with multifaceted ANNs made up of prejudiced links involving neurons that are accustomed and with repetitive spotlight to training data. Modern AI systems have witnessed an increase of obscure (black-box) decision systems, for example, deep neural networks (DNNs) [7]. Regrettably, the majority of the AI representations designed for ML and DL are tagged "black-box" by researchers since the fundamental composition is difficult, non-linear, exceptionally complicated, and elucidated to medical experts. This vagueness has made explainable AI (XAI) architectures to be a necessity based on three reasons, as revealed by Ref. [8]:

- Claims to build interpretable models.
- Demand for methods that create human relationship.
- Requisite for reliability of assumptions.

However, there have been a lot of problems that are associated with mistrust as a result of the cloudiness with un-XAI models. Some of these problems being addressed are listed below:

- Un-explainable and incorrect diagnosis of images (in the case of brain tumor) where loss of life may occur.
- The life span of patients affected by wrong medical image segmentation techniques that are not good enough to enhance the image received from magnetic resonance imaging (MRI), producing wrong result.
- Non-detection of brain tumor as its initial stage of early formation may result to the failure of the brain.

5.1.1 Contributions of the current study

The current study contributes to the followings:

- Integration of other segmentation techniques with fuzzy C-means to achieve a better explainable results in the detection of brain tumor.
- An enhanced filtering technique that improves the quality of the MRI brain image with specific explanation.
- The integration of fuzzy C-means with the ANN that combines with other operation like morphological operation to produce a more accurate and improved quality of MRI of the brain tumor detected and also inclusion of three performance evaluation in percentages to determine accuracy, specificity, and precision of the brain images.

5.1.2 Chapter organization

This chapter is presented in the following sections: Section 5.1 focuses on the introduction of the XAI for medical image segmentation in medical decision support systems (MDSSs). Section 5.2 presents literatures of related works in XAI for medical image segmentation in MDSSs. In Section 5.3, the methodology will be discussed including system architecture, program flowcharts used for the medical image detection purposes; comprising feature extraction, dimensionality reduction, detection, segmentation, and classification. Section 5.4 will be for discussion.

5.2 Related work

5.2.1 Concept of XAI

An unclear method of AI does not promote confidence and acceptance in the midst of medical experts. Transparency of AI methods becomes the best means in AI-driven processes [9]. Explainability is a vital aspect of trust given that it depends fully on the expert understanding of the algorithms or the outcomes. Hence, AI (DNNs) processes should be human centered for justifications of its outcomes [10]. Human-centered results clarifying a definite prediction on the patient, by revealing the behavior of the clinical outcomes. In this study, we will use terms such as, *understandability, intelligibility, comprehensibility, interpretability and transparency* in discussing XAI [10]. Schoenborn and Althoff [11] argued that XAI assist experts understand transparent, relevant, and justified clinical outcomes. In the other hand, ML algorithms function by training raw clinical data and present them as new data [12]. Barredo Arrieta [7,13,14] opined that ML methods are linked with outcomes from biased clinical decisions. In this study, we will use the definition of XAI as posited by [15], which means that XAI is the collection of novel ML systems whose outcomes are explicable to enhance experts understanding, definite trust, and efficient management of usable AI systems. The definition adopted shows the exact aim of the interpretability from the expert viewpoint.

5.2.2 Framework for XAI

XAI methods can be categorized in two ways: *Transparent explainability*—this method of interpretability is through transparent methods and *Post-hoc explainability*—its explanation employs exterior XAI methods. Additionally, the former can be splitted into *pre-modeling* and *during-modeling*. Elshawi *et al.* [16] listed the objective of the latter to be included to understand the methods and also describe clinical data used in developing the representations. But the objectives for modeling interpretability create intrinsic representations. The means for the classification of post-hoc interpretability scheme is interpreting the global or local interpretations [7,16]. In this chapter, we include XAI methods in the outcome as shown in Figure 5.1.

5.2.3 Explainability in healthcare

In healthcare delivery, models developed through AI methods will intermingle with healthcare experts and patients [17]. Patients have a significant part to play in AI applications; the patients' data are affected by the realization of AI. According to Ploug and Holm [18], patient-centered explanation involves the right to question clinical outcomes based on the request for proper interpretation. This is known as contestability. Hence, it is practical to adapt explainability based on suitable clinical explanations. In such a medium, experts find ways of interpreting outcomes to patients [19]. Therapeutic assessment trail-related process is adapted where preeminent examination is chosen as a result of existing records. Hence, clinicians in regular examination spotlight on the preeminent result that matches the data,

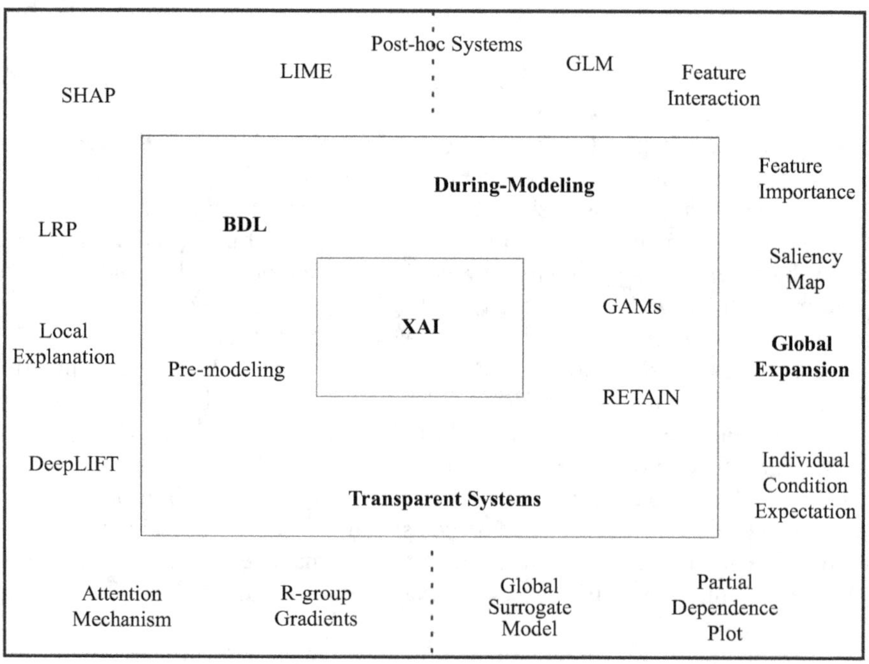

Figure 5.1 Adapted classification of XAI methods [16]

highlighting the usefulness of the situation. In many situations, full interpretations may not occur in the process [20]. A typical illustration is an anesthesia precise process revealing trouncing of consciousness that was unidentified despite its frequent usage [21]. Still, clinicians appreciate its usage and paybacks. Also, conservative antidepressants, like serotonin reuptake inhibitors, are regularly given, even though accurate methods are not revealed [22]. Lack of full underlying understanding seems to portray some of the clinician's practice [23]. Hence, the best part is to have enough to practically understand methods [24], as such, a method to test malignancy traits of ovarian cysts applying imaging information that discards feature like menopausal condition and family record [25]. Furthermore, ML platform non-replicating of the local situation might be prejudiced [26]. Also, elucidating the dangers of unfairness will result in the fundamental retention of patients and dealing with the moral necessities of justice [27]. Wrist wearables are used for determining heart rate but inaccurate on dim skin and give a phony result on black patients [28]. Hence, in making explanation useful to clinicians, it becomes doubtful that a sole clarification offers sufficient data to seize insinuations of using AI in medical decision. Acceptability of lower levels of interpretation is part of the vast medical examination for medical decision [29]. A typical example is via AI supporting examination decisions [30]. Inference of the likelihood stating how the method is understandable needs the sum of the collective standards as shown in Figure 5.2.

In addition, a ML algorithm for forecasting pregnancy result that goes after in vitro fertilization could accept satisfying less explainability standards than a ML algorithm for breast cancer analysis.

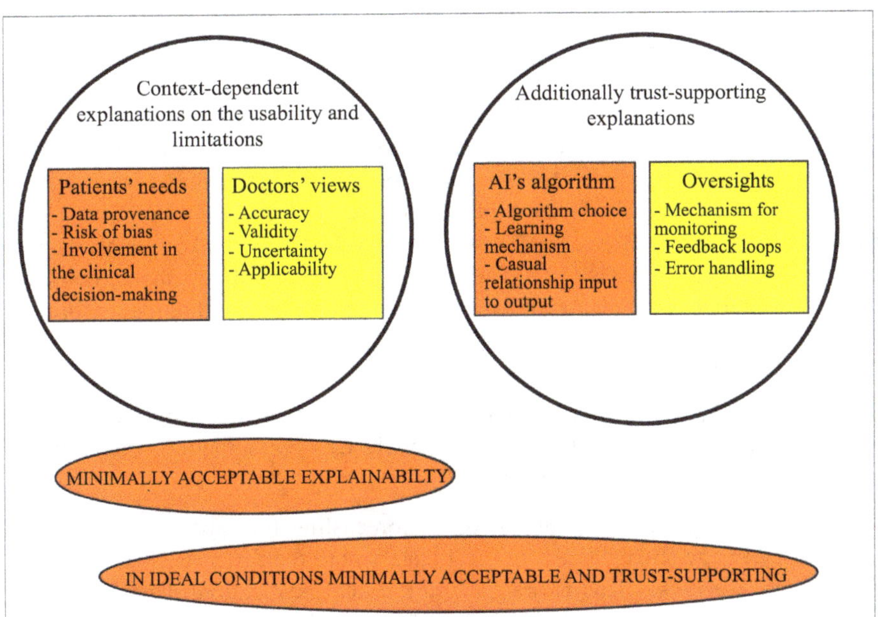

Figure 5.2 Adapted standard for explainability [30]

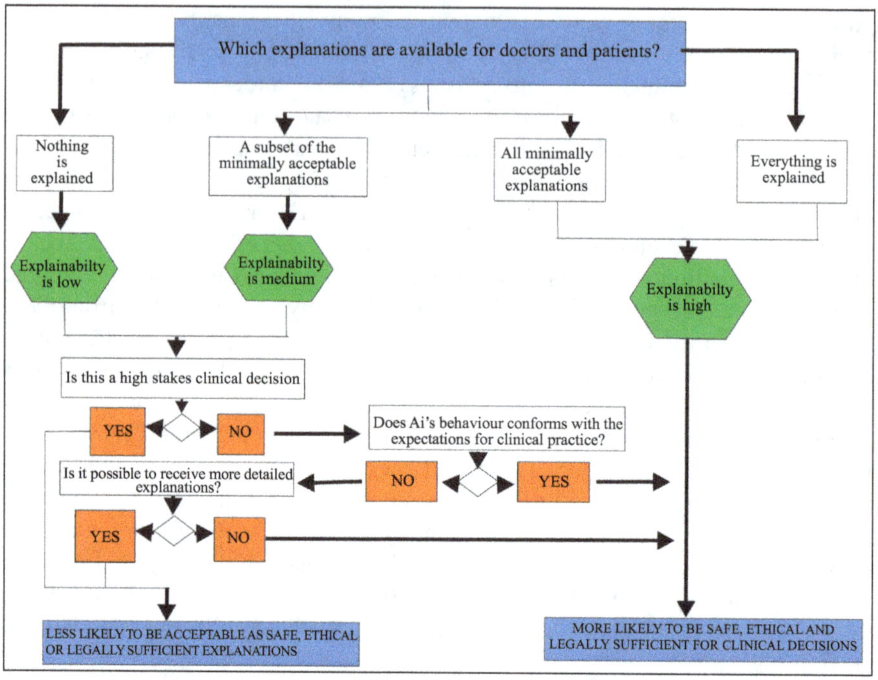

Figure 5.3 Adapted evaluation flow showing explainability in clinical practice [30]

As shown in Figure 5.3, the explainability for healthcare is focused on the clinical processes. In distinction, higher explanation is needed in risk forecasting systems because of the high priority rating [31].

5.2.4 DL concept and applications

Currently, complex ANN-based structures exist in clinical task and categorization issues are diverse. Here, DNNs are used to denote all models. The observed achievement of DL models, for example, DNNs, result in a blend of proficient algorithms with vast statistical space. This space-composed layer makes DNNs as complicated "black-box" models [32,33]. Based on the use of ML methods, the lesion detection techniques are automated with convincing precision and human effort [34].

5.2.5 Computer vision tasks

Cheng *et al.* [6] in Figure 5.4 listed various clinical tasks performed by a computerized-based vision for which DL models are used in medical imaging as in the following:

(i) *Image categorization:* This involves forecasting the labeling of the entire binary image (two classes) or multiclass (more than two).
(ii) *Object detection:* This is the recognition and localization of precise unit of the element of image.

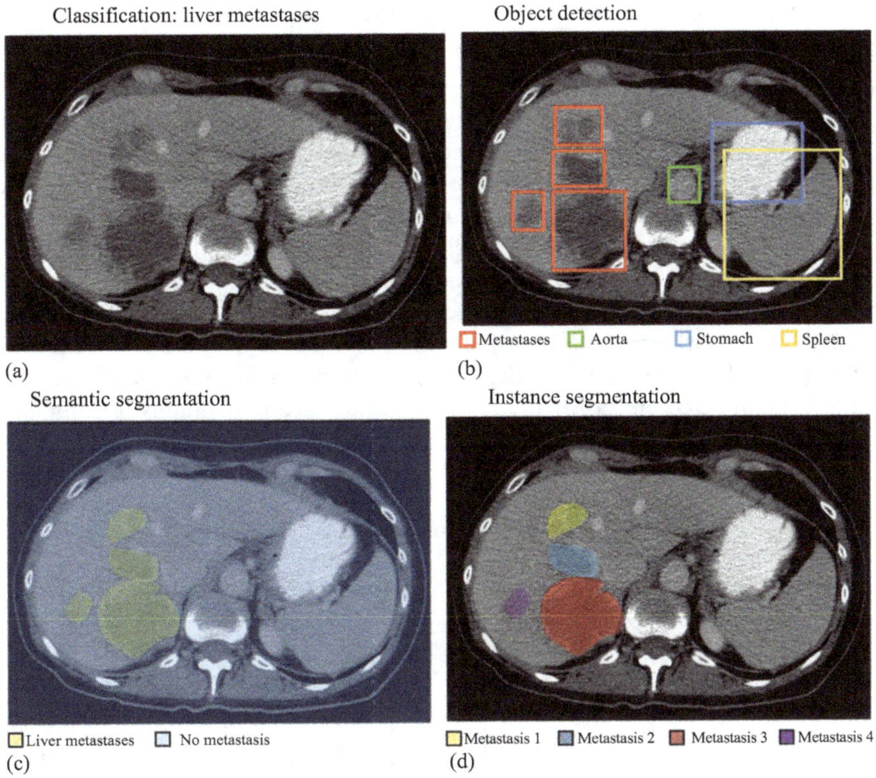

Figure 5.4 Contrast-enhanced CT images showing computer vision tasks [6]

(iii) *Semantic segmentation:* It allocates individual picture element to exact classes. A typical illustration is that individual picture element in lever can be allocated to parenchyma, tumor, or blood vessel.

(iv) *Instance segmentation:* Refers to the picture element level recognition and demarcation of several objects in the same class, for example, lung nodules independently differentiated on a chest radiograph.

From Figure 5.4, Cheng *et al.* [6] argued that:

(a) The categorization of image is to allocate a label from a structure to a known image.

(b) The detection of an object aims to identify human organs, lesions area, structures like metastases that are in red, aorta in green, stomach in blue, and the spleen in a yellow (all square size).

(c) The semantic segmentation allocates an object classification label to an individual picture image, for example, liver metastases which have yellow color.

(d) The instance segmentation allocates labels to every picture element, for example, each of the liver metastases is in sections (i.e., red, blue, purple, and yellow).

5.2.6 Convolutional neural networks (CNNs)

CNN is a made up of heap of layers, individually handling a definite operation, for example, convolution, pooling, and calculating loss. Each intermediate layer gets the output of the former layer as its input. Layers get the input layer from the past layer as the input. The starting layer becomes the input layer linked to the input image with neurons that are equipment to the pixel numbers in the input image. The next stage is the CNN layers showing outcomes of convolving quantities of filters with input data to present feature extraction. The filters are called kernels, and of smaller dimensions, based on the essential part. The neuron reacts to the exact region of the former layer, termed the receptive field. The yield of each CNN layer is called the activation map, emphasizing the consequence of using a precise filter on the input. These stretched the idea by applying various imaging modalities, for example, brain MRI, breast MRI, and cardiac computed tomography angiography for segmentation process [38].

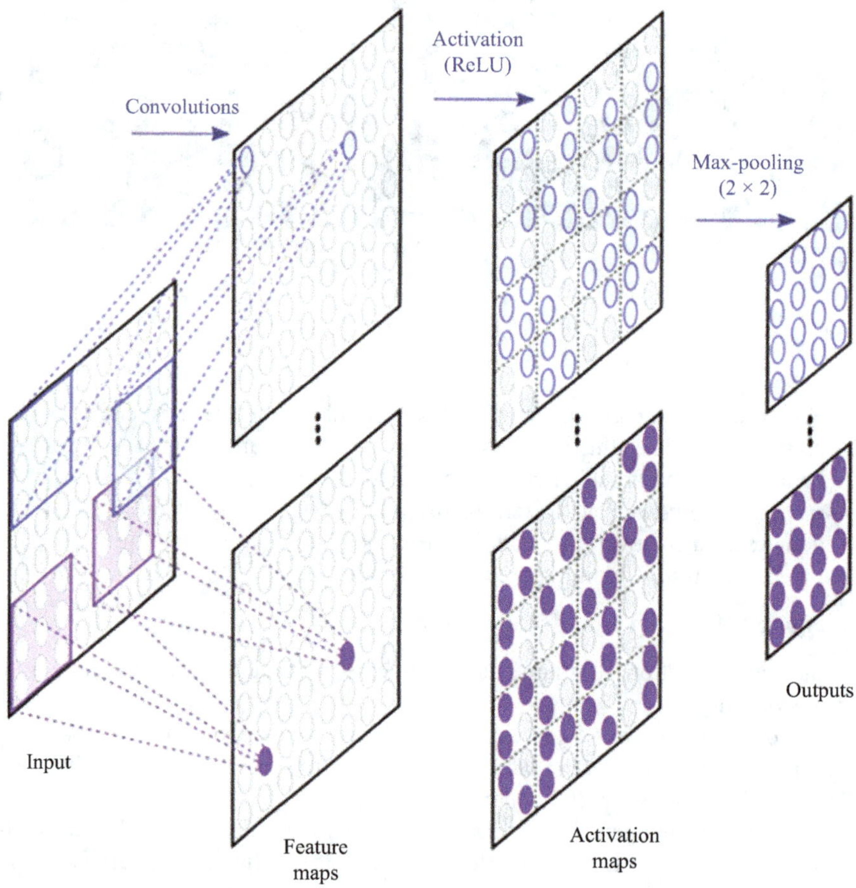

Figure 5.5 The structure of CNNs [6]

However, showing 2D convolutions with an isotropic kernel on anisotropic 3D images can be difficult [39]. In detecting, the information represents images denoted with leaping box match up, demarcating traits of interest. Organizing object data in ML algorithm process is difficult [34,35]. Even with inadequate information, the model can be trained to forecast labels exactly [36]. The means to increase the training dataset to avert over fitting is image augmentation. Figure 5.6 shows data augmentation process. This involves:

(i) *Classic data augmentation*: This involves applying a range of transformations, for example, translating randomly, rotating, flipping, scaling, cropping, bighting, and contrasting adjustments to original CT images.
(ii) *Artificial data augmentation:* This employs a generative adversarial network (GAN) creating extra artificial images with a numerical distribution.

The illustration in Figure 5.6 shows the CycleGAN training to switch contrast-enhanced CT images to non-contrast images. The generator afterward supplements early dataset for training on a task segmenting non-contrast images [37].

A similar process of raising the amount of preparing images involves translating randomly, rotating, flipping, scaling, cropping, brighting, and contrasting adjustments. GANs can produce irregular images resembling exact images [21].

5.2.7 Medical image segmentation

Segmentation is an important aspect of processing image. Aarish and Devanand [5] posited medical image segmentation as the practice of facilitating the delineation, characterization, and visualization of regions of interest in any medical image. Previous to denoising an image, it is segmented to recuperate the original image. The major reason for segmentation is to decrease the information for effortless

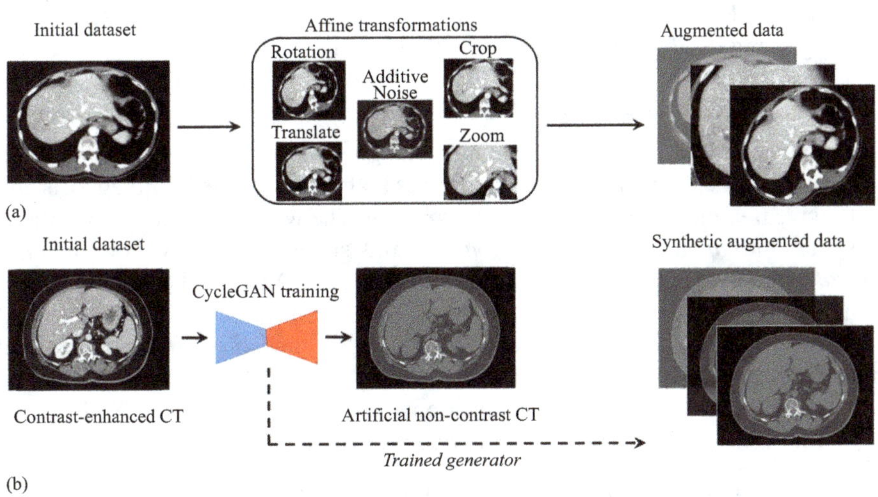

Figure 5.6 Demonstration of data augmentation [6]

analysis. Numerous techniques for automated segmentation of computed tomography (CT) and MRIs are used in transforming medical practices. Imaging experts are engaged in image elucidation for patients with cancer, obesity, cardiovascular disease, neurodegeneration, osteoporosis, arthritis, etc. These approaches will aid in disease diagnosis, determining the prognosis, selecting the patients for therapy, and to observe responses to treatment. Bensalah *et al.* [40] argued that in classifying medical image segmentation, the following approaches should be mentioned. These approaches are the following:

(i) *Semi-automatic vs automatic segmentation*: The object, for example, organ, tissue, pathological lesion, or other structure, is used for the examination or treatment of a particular disease.
(ii) *Supervised and unsupervised segmentation*: Supervised segmentation needs prior training, such as intensity normalization and classification. While unsupervised does not need training and is less accurate.

5.2.8 Medical image segmentation techniques

Dar and Padha [5] listed segmentation methodology to include the following:

(i) *Thresholding*: Local thresholding, Otsu's method, Gaussian mixture approach.
(ii) *Region based*: Region merging and splitting.
(iii) *Edge based/boundary based:* Edge detection, Prewitt filter, Sobel filter, Canny filter, Laplacian of Gaussian LOG, Watershed.
(iv) *Clustering methods:* k-means/iso-data algorithm, Fuzzy C-means algorithm, expectation maximization (EM) algorithm.
(v) *Other methods:* Level set method (LSM), ANNs, Atlas-guided approach, generic algorithms.

Dar and Padha [5] compared segmentation methodologies by specifying advantages and disadvantages as given in Table 5.1.

5.2.9 Medical imaging modality

In the diagnosis and treatment of patients, imaging assists radiologists to perform diagnosis. A wide range of imaging modalities that are being used for diagnosis and in effective treatment planning currently is in use. The widely used main modalities are spitted into *anatomical* and *functional*. In this study, we discuss about the *anatomical* modality. Digital images can be represented in 2D, 3D, and 4D systems. The elements of an image in 2D are referred to as pixels, while as in 4D, they are referred to as Voxels. In this study, specific medical imaging modalities are presented and emphasis is mainly focused on ultrasound, MRI, CT, and X-ray. Table 5.2 shows various imaging modalities, their application areas, and recommended methods.

(i) CT:
CT helps in capturing different sectional planes (tomography) which are difficult to process otherwise. It visualizes small density gradients i.e., in

Table 5.1 Difference in segmentation techniques

Methodologies	Advantages	Disadvantages
Local thresholding	Ease of implementation No need of prior information	Produces noisy and blurred edges
Otsu's method	Minimizes inter-class and intra-class variations. No particular histogram shape considered prior. Extendable to multi-level thresholding	Formation of binary classes in gray-level images. Enhancement in density with increase in levels of threshold. Regions might get fused or varied
Gaussian mixture approach	Used for histogram-based problems Decreases categorization error probability Favored for small size-classes Iterative model	All histograms do not follow Gaussian model Resultant intensities are fixed and non-negative
Region growing	Its support is on similarity and immune to noise	Costly method
Region merging and splitting	Dividing an image on demand resolution and calculating mean, variance of segment pixel value	May result in blocky segments
Edge detection	Choose a huge region in an image. It uses images with irregular elucidation	Applicability for basic backgrounds
Prewitt filter	Calculates edges and their orientations in 8 directions of pixel	Less accuracy Sensitive to noise
Sobel filter	Calculates edges in horizontal and vertical orientations Better noise suppression Isotropical results	Expensive
Canny filter	Calculates wide range of edges and orientations Adaptive	Difficulty in working effectively at curves, corners
Laplacian of Gaussian (LoG)	Detection of blurry edges and sharp detail	Complexity in working at corners
Watershed	Decreases over-segmentation Division of overlapping objects. Quick and dependable output.	Time consuming and gradient based
K-Means/iso-data algorithm	Quick and easier to used	Sensitive to variety and initialization of centroids

(Continues)

Table 5.1 (Continued)

Methodologies	Advantages	Disadvantages
Fuzzy C-means algorithm	Unsupervised and considers vagueness, uncertainty in an image	Best solution is undefined Initialization is susceptible Slightest compatible for noisy images
EM algorithm	Unsupervised Iterative and reduced sensitivity	Results in noise generation and intensity-inhomogeneity. Slow convergence rate Gets stuck into local optima High-computational cost
Level set method (LSM)	It is efficient, versatile, robust and accurate	Sensitive, requires considerable design planning for level set function
ANN	Ease of implementation Applicable to diverse problems	Selection of architecture Black-box problem
Atlas-guided approach	Computationally fast Suited for structures that are constant over populace of study Labels are transferred during segmentation	Difficulty in accurate segmentation of complex structures with non-linear registration methods
Genetic algorithms	Incremental segmentation Adaptive to user access patterns Computationally fast	Choosing number of generations, population size It does not always result in optimal solution

Table 5.2 Medical imaging modalities and their application areas

Technique	Recommended methods	Application area
Thresholding and region-based segmentation	Abdomen, appendix, bladder, brain, breast, chest, cervix, kidney, lungs, pancreas, esophagus	CT scan
Watershed and region-growing (3D) Clustering (2D)	Neuro-imaging, cardiovascular, musculoskeletal, liver, gastro-intestinal, functional, oncology, phase contrast	MRI
Thresholding based	Transrectal, breast, Doppler, abdominal, transabdominal, cranial, gall-bladder, spleen	Ultrasound
Edge-based and watershed	Radiography, mammography, fluoroscopy, contrast-radiography, anthography, discography, dexa-scan, upper GI	X-ray

case of the brain, it distinguishes between gray-matter, white-matter, and cerebro-spinal-fluid (CSF).

(ii) Ultrasound:
It is a non-invasive imaging process using sound waves to produce computerized images reflected by organs and interior organs of body. It can also be used for interventional procedures. It does not have any known harmful effects on human body in clinical imaging. It is inexpensive technique but it cannot visualize the anatomical regions (i.e., brain).

(iii) MRI:
MRI uses magnetic fields and radio-frequencies to generate visualization area of different body organs. The variation in reflected frequencies helps in localization of different body organs with the help of magnetic field. This method is employed to get fine details of organs i.e., brain, liver, chest, pelvis, and abdomen.

(iv) X-ray:
X-ray is also a non-invasive imaging method and one of the oldest imaging techniques that use ionizing radiations that are rapid and of shorter duration. This imaging technique is inexpensive as compared to others. It can also be used in interventional procedures for detecting fractures in bones.

5.2.10 Summary of related works

Table 5.3 shows related works on XAI models; the purpose, model used and findings by various authors.

Table 5.3 Related XAI models

S. no.	Citations	Purpose	Model used	Findings
1.	[41]	The authors argued on the determinant of biochemical explainable ML	Metabolic allele classifier	Post-hoc and visual XAI methods and techniques were used
2.	[42]	This paper applied interpretable ensemble AI methods to classify hemodialysis-patient	K-means	R-group gradients XAI method was used
3.	[8]	This paper built an explainable QSAR methods with ML algorithms	SVM	Visual XAI and post-hoc methods were used
4.	[43]	The authors presented clinical explainable DL method for detecting glaucoma	CNN	Visual XAI and post-hoc methods were used
5.	[44]	Authors used deep affinity for explainable DL and CNN	RNN and CNN	Post-hoc, attention mechanism and visual methods and techniques were used

(Continues)

Table 5.3 (*Continued*)

S. no.	Citations	Purpose	Model used	Findings
6.	[45]	Authors used deep affinity for seriously ill patients	RNN	Post-hoc, attention mechanism and visual methods and techniques were used
7.	[46]	This paper assessed and validated an explainable DL structure for Alzheimer's disease classification	FCN	Post-hoc and visual XAI methods and techniques were used
8.	[47]	This paper classified genetic patterns	Sequential rule mapping (SRM)	Transparent and visual XAI methods and techniques were used
9.	[48]	This study interpreted AI for breast cancer	Weighted K-nearest neighbor (WkNN)	Transparent and visualization XAI methods and techniques were used
10.	[49]	The authors assessed DL for RNA reports	Neural network (NN)	Post-hoc, DeepLIFT and visualization XAI methods and techniques were used
11.	[8]	The authors predicted avoidance of hypoxaemia	XGBoost	Post-hoc, local, SHAP, and visualization XAI methods and techniques were used
12.	[8]	This article discusses Gaussian Process Regression for ML interpretability	GPR	Transparent and global were XAI methods and techniques were used
13.	[50]	This paper discusses interpretable analysis for COVID-19 from chest CT	Decision tree (DT)	Transparent and visual XAI methods and techniques were used
14.	[51]	This paper agrees on causal inference interpretable ML	Generalized linear regression model (GLM)	Transparent and rule-based XAI methods and techniques were used
15.	[52]	The authors presented visually interpretable DL for the prediction of mortality	Multi-scale CNN	Post-hoc and visual XAI methods and techniques were used
16.	[16]	This paper discusses explainable ML for hypertension	Random forests ensemble	Post-hoc, local-LIME, and SHAP were used

(Continues)

Table 5.3 (Continued)

S. no.	Citations	Purpose	Model used	Findings
17.	[53]	This paper discusses interpretable DL for survival analysis	DNN	Post-hoc was used
18.	[54]	The authors argued on 3D DL on hyperspectral images	Deep CNN (DCNN)	Post-hoc was used
19.	[8]	This paper presented a study on adverse drug reactions with interpretable DL structure	DNN	Post-hoc was used
20.	[8]	This paper presented diabetic retinopathy with ML algorithm	SVM	Post-hoc and decision tree XAI methods and techniques were used
21.	[55]	This paper presented stimulated-based structure in surgery	Linear SVM	Transparent XAI methods and techniques were used
22.	[56]	The authors argued on interpretable skin lesion classification with DL models	CNN	Post-hoc, LIME, and local were used
23.	[57]	The authors evaluated XAI on medical imaging tasks	The modality-specific feature importance (MSFI) metric	The outcomes indicated that recent XAI algorithms are inadequate
24.	[58]	The authors evaluate XAI for X-ray image analysis	Review was done on Kitchenham and Charters	No confidence in the explanation
25.	[59]	The authors argued on the application of XAI in diagnosis and surgery	XAI trends in diagnosis and surgery	Summarizing the XAI methods
26.	[60]	This paper presented XAI for CNN-based prostate tumor segmentation in multi-parametric MRI correlated to whole mount histopathology	An explainable DL model to understand the predictions of a CNN for prostate tumor segmentation	The CNN achieved a mean Dice Sorensen Coefficient 0.62 and 0.31 for the prostate gland and the tumor lesions
27.	[61]	The paper proposed a framework for eXplainable DL for prediction of brain tumor	CNN, LIME, and SHAP	Higher interpretability

5.3 Analysis of the proposed system

The proposed system is the hybridization of enhanced fuzzy C-means and ANN in medical image segmentation process to detect brain tumor at its initial stage of formation. This system creates pathways for useful decisions. The system achieved the following:

(i) Detection of tumor of the brain using enhanced fuzzy C-means and artificial neural network.
(ii) Incorporate an enhanced image processing operators with a 5×5 template used for digital image processing.
(iii) Incorporate high-pass filtering technique and Gaussian convolution on image from MRI machine.

Figure 5.7 shows the single use case diagram for a user, while Figure 5.8 shows the single use case diagram for the system using enhanced fuzzy C-means.

In Figure 5.7, the user logins before the access is granted.

In Figure 5.8, the system verifies and authenticates login detail supply by the user before the access is granted for the system to invoke the enhanced fuzzy C-means on selection by the user which will perform the following operations:

(i) Denoise image that was uploaded
(ii) Segment the image
(iii) Extract features
(iv) Detect tumor and indicate the type if it is mass or malignant type of tumor

The study further unravels the algorithmic operation as shown in Figure 5.9.

(i) *Image preprocessing stage*: Images of the brain are captured via the MRI machine scanned, as a result, these images contain noise, so we will first of

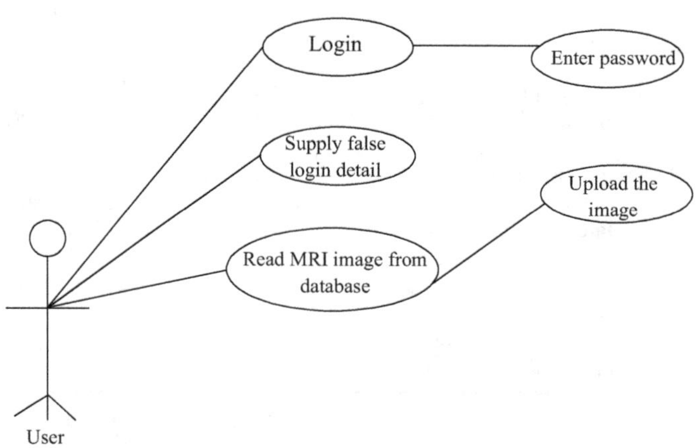

Figure 5.7 Single use case diagram for a user

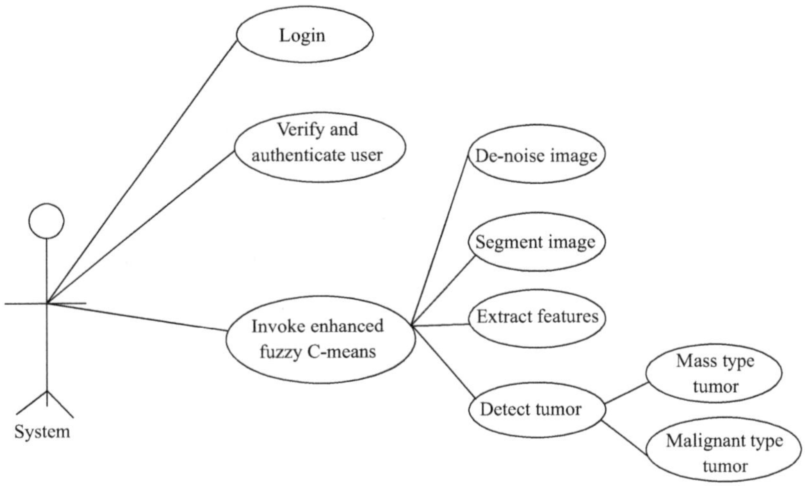

Figure 5.8 Single use case diagram for a system using enhanced fuzzy C-means

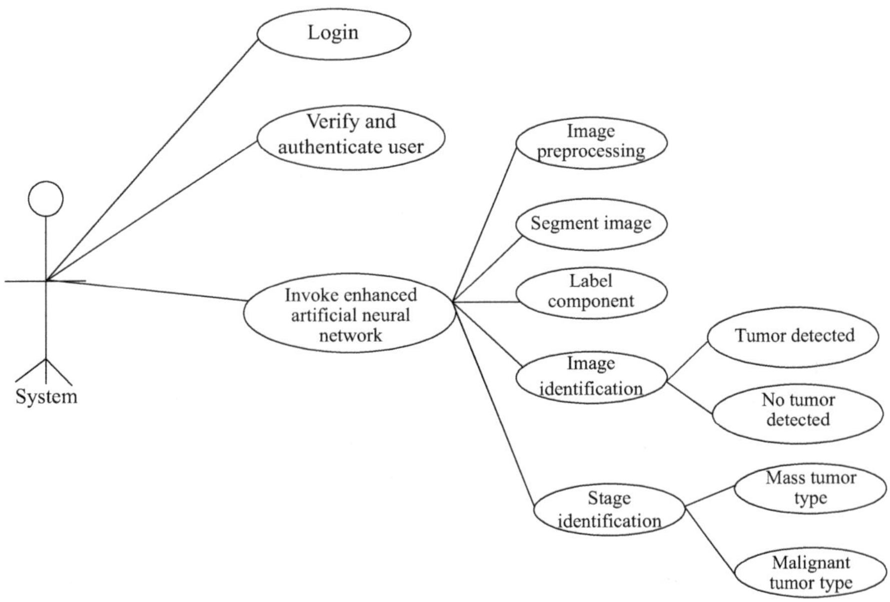

Figure 5.9 Single use case diagram for a system using enhanced ANN (EnANN)

all denoise the image by process filtering technique using Gaussian convolution.

(ii) *Segmentation stage*: In this stage, we applied the region growing technique which is an easy region-based image segmentation technique. It can be

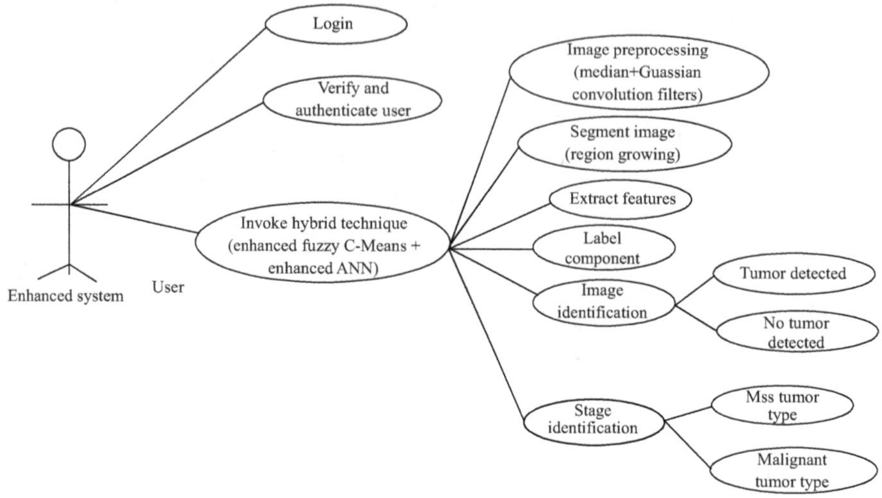

Figure 5.10 Single use case diagram for enhanced system using hybrid technique

referred to as pixel-based image segmentation due to its name as it engage the process of selecting initial seed point.

(iii) *Label of connected component stage*: This is the third stage after the recognition of connected component of all the images, every set of connected pixel having the same gray-level assigned the same unique region label.

(iv) *Tumor identification*: In this stage, we are having the dataset, previously collected from the brain MRIs and it is in 3D representation and used of 3D analyzer from which we are extracting features. A knowledge based will be created for comparison.

(v) *Stage identification:* In this step, we will identify that patients who are suffering from brain tumor, it is also necessary for us to find out which type of tumor the patients is suffering from if it is mass or malignant types of tumor.

Figure 5.10 shows the hybridization of the enhanced fuzzy C-means and ANN that is more robust used in detecting brain tumor.

5.3.1 Analysis of algorithm for proposed system

The below is the algorithm for an enhanced fuzzy C-means with Laplacian operators of a 5 × 5 image and ANN that is used in the detection and segmentation of brain tumors in patients:

Proposed algorithm for multiple kernel fuzzy C-means (MKFCM) with spatial biasing

Step 1: Open the folder; load 3D representation from MRI machine scanned (JPEG format)

Step 2: Ensure image is RGB, else change the image to gray image

Step 3: Change 3D representation twice pixels value
Step 4: For MKFCM, predefine the clusters center C_i (c= 3 clusters)
Step 5: Get the size of the whole image
Step 6: Perform feature extraction from segmented image of the brain
Step 7: To determine TP (true positive), TN (true negative), FP (false positive), FN (false negative)
Step 8: Perform the calculation using the formula below:

$$Accuracy = \frac{TP + TN}{TP + FP + FN + FP} \times 100, \quad Sensitivity = \frac{TP}{TP + FN} \times 100$$

$$Specificity = \frac{TN}{FP + TN} \times 100,$$

Step 9: Convert the input matrix to a vector
Step 10: Compute the membership value by using

$$U_{ij} = \frac{\left((1 - K_M(x_j, C_i)) + n_i(1 - K_M(x_j, c_i))\right)^{-1/(m-1)}}{\sum_{i=1}^{c} \left((1 - K_M(x_j, C_i)) + n_i(\bar{x}_j, C_i)\right)^{-1/(m-1)}}; i = 1, 2, \ldots\ldots C$$

Step 11: Update the cluster center by using:

$$C_i = \frac{\sum_{j=1}^{n} U_{ij}^m (K_M(x_j, C_i)x_j + n_i K_M(\bar{x}_j, C_i)\bar{x}_j)}{\sum_{j=1}^{n} U_{ij}^m (K_M(x_j, C_i)x_j + n_i K_M(\bar{x}_j, C_i))}; i = 1, 2 \ldots\ldots C$$

Iteration Process Start:
Step 12: Update the membership value U_{ij} by using:

$$O_m^G(U, C) = \sum_{i=1}^{c} \sum_{j=1}^{n} U_{ij}^m (1 - K_M(x_j, C_i) + \sum_{i=1}^{c} \sum_{j=1}^{n} n_i U_{ij}^m (1 - K_M(\bar{x_j}, C_i)$$

where $K_M(x_j, C_i) = K_1(x_j, C_i)\ x\ K_2(x_j, C_i)$, $K_1(x_j, C_i) = \exp\left(\frac{-\|x_j - C_i\|^2}{\sigma_1^2}\right)$

$$K_2(x_j, C_i) = \exp\left(\frac{-\|x_j - C_i\|^2}{\sigma_2^2}\right)$$

x is the mean for MKFCM_S1 and the median for MKFCM_S2 of the neighbor pixels σ_1^2, σ_2^2 that are the variances.
Step 13: Update the cluster center C_i by using:

$$U_{ij} = \frac{\left((1 - K_M(x_j, C_i)) + n_i(1 - K_M(x_j, c_i))\right)^{-1/(m-1)}}{\sum_{i=1}^{c} \left((1 - K_M(x_j, C_i)) + n_i(\bar{x}_j, C_i)\right)^{-1/(m-1)}}; i = 1, 2, \ldots\ldots C$$

Step 11: f | Cnew – Cold| $> \varepsilon; (\varepsilon = 0.001)$ then go to Step 1
Else stop Assign each pixel to a specific cluster for which the membership is maximal
Step 12: Display result: accuracy, precision, specificity, and sensitivity.

Detection stage

In this processing stage, the image under segmentation can be achieved by means of binarization method (i.e., either 1 or 0). We denote the binary image by summing up the total number of black and white pixels as given in the formula below:

$$\text{Image}, I = \sum_{W=0}^{m} \sum_{H=0}^{m} [f(0) + f(1)] \tag{5.1}$$

Pixels = Height (H) × Width (W)
f (l) = black pixel (digit 1)
f (0) = white pixel (digit 0)

$$\text{No\_of\_White}, \quad P = \sum_{w=0}^{m} \sum_{H=0}^{m} [f(0)] \tag{5.2}$$

where m is the maximum image size; P is the total number of white pixels (height × width)

1 Pixel = 0.264

The formula for area of tumor size

$$\text{Size\_of\_Tumor S} = \left[\left(\sqrt{P}\right) \times 0.264\right] \text{mm}^2 \tag{5.3}$$

A. Algorithm for detection

The steps used for brain tumor detection are shown:

Step 1: Apply .JPEG MRI images from a database.

Step 2: Confirm image format is specified, go to Step 3.

Step 3: Authenticate color is gray, then change to gray-scale using rgb to gray ( ).

Step 4: Locate the edge of the gray-scale image using binarization and thresholding techniques.

Step 5: Compute the sum digit of white pixels (digit 0) in the image using:

$$\text{No\_of\_White}, \quad P = \sum_{w=0}^{m} \sum_{H=0}^{m} [f(0)]$$

Step 6: Calculate the size of the tumor using:

$$\text{Size\_of\_tumor S} = \left[\left(\sqrt{P}\right) \times 0.264\right] \text{mm}^2$$

Step 7: Test if tumor area > 6 mm^2, then display message "Abnormal" else display message "Normal":

Test if Tumor = "Mass" then display message "Mass Type"
Else display message "Malignant Type"

Step 8: Stop the process.

B. Proposed algorithm for ANN
Step 1: Read image from database obtained from MRI-scanned machine
Step 2: Verify image is RGB else change the image to gray image
Step 3: Change 3D image representation twice the pixels value
Step 4: Denoise the image by applying improve high-pass filtering
Step 5: Perform segmentation of the image by using region-based growing
Step 6: Perform feature extraction from segmented image of the brain
Step 7: To determine TP, TN, FP, FN
Step 7: Perform classification by using EnANN
Step 8: Compute the sum of white pixels (digit 0) and size of the tumor using:

$$\text{No\_of\_White,} \quad P = \sum\nolimits_{w=0}^{m} \sum\nolimits_{H=0}^{m} [f(0)]$$

$$\text{Size\_of\_tumor S} = \left[\left(\sqrt{P}\right) \times 0.264\right] \text{mm}^2$$

Step 9: Perform calculation using the formula below:

$$Accuracy \leftarrow \frac{TP + TN}{TP + FP + FN + FP} \times 100$$

$$Specificity \leftarrow \frac{TN}{FP + TN} \times 100, \quad Sensitivity \leftarrow \frac{TP}{TP + FN}$$

Step 10: Test if tumor area > 6 mm^2, then display message "Abnormal" else display message "Normal:"

Test if tumor = "Mass" then display message "Mass Type"
Else display message "Malignant Type"

Step 11: Display result: accuracy, precision, specificity, and sensitivity

C. Proposed hybrid technique algorithm
Step 1: Create the database of image in MATLAB©
Step 2: Browse and upload the image from the created database
Step 3: Convert the image to binarization

If image $\neq$ gray_image then
Perform binarization and thresholding

Else
Compute: P$\leftarrow \sum\nolimits_{w=0}^{m} \sum\nolimits_{H=0}^{m} [f(0)]$, S $\leftarrow \left[\left(\sqrt{P}\right) \times 0.264\right] \text{mm}^2$

Step 4: Apply the filtering techniques on the image using:

high-pass + Gaussian convolution + median filters

Step 5: Apply segmentation using the hybrid technique

Region growing + fuzzy C-means + ANN

Step 6: Extract the segmented image of tumor area
Step 7: To determine TP, TN, FP, FN
Step 8: Perform image classification by EnANN
Step 9: Perform calculation using the formula below:

$$Accuracy = \frac{TP + TN}{TP + FP + FN + FP},$$

$$Specificity = \frac{TN}{FP + TN} \times 100, \quad Sensitivity \leftarrow \frac{TP}{TP + FN}$$

Step 10: Test if tumor area > 6 mm^2, then display message “Abnormal” else display message “Normal”:

Test if Tumor = “Mass” then display message “Mass Type”
Else display message “Malignant Type”

Step 11: Display result: accuracy, precision, specificity, and sensitivity.

D. Analysis of output

The system output is designed to include the user login details, the user enter user-name, and password which serves as a means of data validation. Output back-end of the system is the database that contains the records of images of the brain. Due to this fact, the output interface is made interactive, very simple, and most of importantly the ease of usability. The output model of the system is illustrated in Figure 5.11.

5.3.2 Advantages of the hybrid system

The anticipated hybrid technique which combines fuzzy C-means and ANN algorithm has the following advantages:

(i) It is less expensive as it does not require any backup.
(ii) It gives better peak signal-to-noise ratio in medical image.
(iii) It has better mean square error.

5.3.3 Disadvantages of the system

The disadvantages of the hybrid system that combines the fuzzy C-means and ANN are stated below:

(i) It can be more complex than other segmentation.
(ii) It potentially increased the complexity.
(iii) Applications and languages are not available after image deployment is complete.

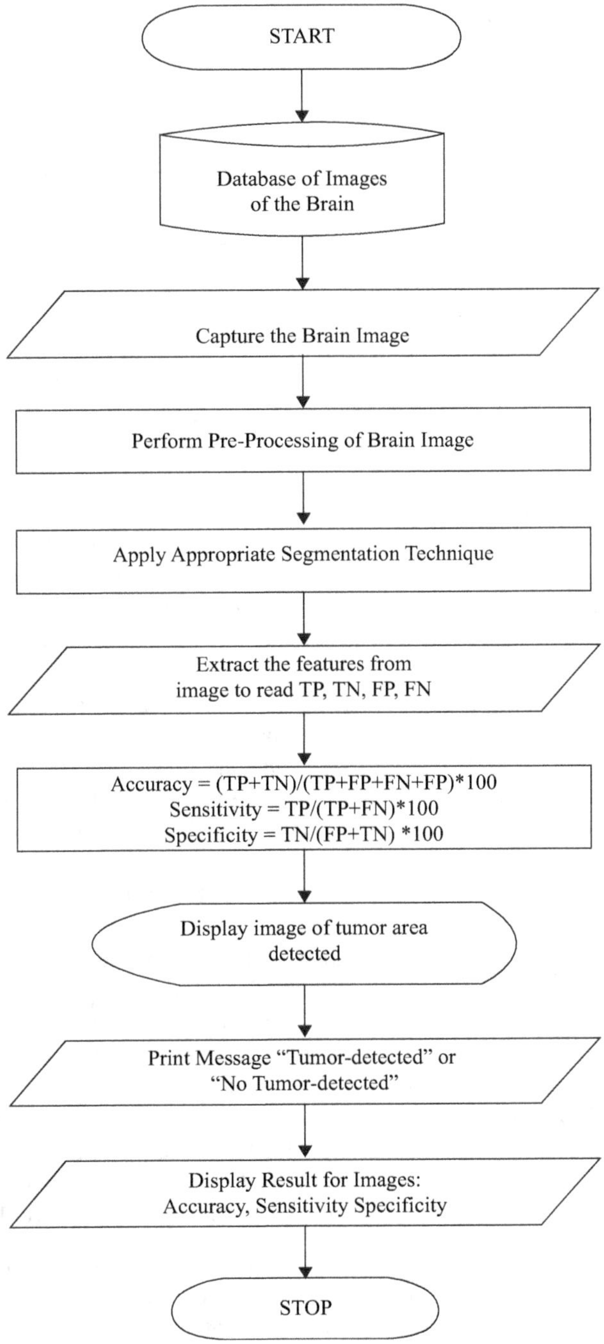

Figure 5.11 Output model of the proposed hybrid system

5.3.4 Justification of the system

The anticipated system combines fuzzy C-means and ANN justifying the following:

(i) The system will be able to detect brain tumor at any location in the brain. This is made possible by the application of fuzzy C-means with spatial biasing and ANN to classify images to obtain a more accurate result.
(ii) The system use hybrid filtering technique. This is made possible by the combination of Gaussian convolution and median filters to denoise images.

5.4 Conclusion

This paper describes the XAI for medical image segmentation in medical decision support systems, the algorithm for various stages, and the analysis of the resulting data. This is based on the quantitative variables applied on the structures to give a clear interpretation of tasks performed by the AI-driven models to the patients or experts. The quantitative approach reveals the need for the design of AI algorithm to be clear, precise, and interpretable. Explanation of medical decisions to patients or experts in AI-based predictive models is important. In healthcare delivery, models developed through AI methods must interact with healthcare experts and patients for seamless flow of healthcare delivery [17]. In this paper, we have developed XAI model approach by using a hybrid analysis (fuzzy C-means and ANN) to denoise medical images (brain tumor) at its initial stage of formation with an interactive module to aid explanation. Our approach was demonstrated using case diagrams and flowcharts showing image segmentation stages with better peak signal-to-noise ratio and a better mean square error. We find complexity and delays forming part of its weaknesses. Nonetheless, in justifying the model based on the hybrid analysis approach, the system was able to detect brain tumor at location with clearer algorithmic interpretation for a better medical discussion support. Also, the mechanisms for image filtering had high explainability rating, hence, creating a means for useful medical decision support. The image segmentation techniques, such as pre-existing, emerging techniques, and their applicability, were part of the techniques used. Since the accuracy of segmentation remains a concern issue for patients with complications, there is the need for explainability.

Conclusively, performing brain tumor detection using image segmentation models was discovered to be a difficult task due to the fact that most images are noisy. This is one of the major challenges which medical expert faces when it comes to segmentation processes. The developed model can be used to achieve the following:

(i) A platform for fuzzy C-means use for medical image segmentation to detect brain tumor and also do a performance evaluation to determine the percentage in terms of accuracy, specificity, and precision.
(ii) A platform for ANN use for medical image segmentation to detect brain tumor and also do a performance evaluation to determine the percentage in terms of accuracy, specificity, and precision.

(iii) A hybrid platform that integrates the fuzzy C-means with the ANN use for medical image segmentation to detect brain tumor and also do a performance evaluation to determine the percentage in terms of accuracy, specificity, and precision.

References

[1] Jin, W., Fatehi, M., Abhishek, K., Mallya, M., Toyota, B., and Hamarneh, G. 'Artificial intelligence in glioma imaging: challenges and advances'. *J Neural Eng*. 2020; 17(2): 210–212.

[2] Geoff, C., Hawk, K., and Vial, A. 'Machine learning and deep learning in medical imaging: intelligent imaging'. *J Med Imaging Radiat Sci*. 2019; 50 (1): 477–487.

[3] Langlotz, C., Allen, B., and Erickson, B. 'A roadmap for foundational research on artificial intelligence in medical imaging'. In *NIH/RSNA/ACR/ The Academy Radiology Workshop 1906 Report*, 2019.

[4] Currie, G. 'Intelligent imaging: radiomics and artificial neural networks in heart failure'. *J Med Imaging Radiat Sci*. 2019; 50(4): 571–574.

[5] Dar, A. and Padha, D. 'Medical image segmentation: a review of recent techniques, advancements and a comprehensive comparison'. *Int J Comput Sci Eng*. 2019; 7(7): 114–124.

[6] Cheng P., Emmanuel, M., Rikiya, Y., *et al.* 'Deep learning: an update for radiologists'. *RadioGraphics*. 2021; 41(5): 1427–1445.

[7] Barredo Arrieta, A., Diaz-Rodriguez, N., Del Ser, J., *et al.* 'Explainable artificial intelligence (XAI): concepts, taxonomies, opportunities and challenges toward responsible AI'. *Inf Fusion*. 2020; 58(1): 82–115.

[8] Chakrobartty, S. and El-Gayar, O. 'Explainable artificial intelligence in the medical domain: a systematic review'. In *Proceedings of AMCIS*, USA, 2021.

[9] Holzinger, A., Langs, G., Denk, H., Zatloukal, K., and Müller, H. 'Causability and explainability of artificial intelligence in medicine'. *Wiley Interdisciplin Rev DMKD*. 2019; 9(4): 13–12.

[10] Arrieta, A., Díaz-Rodríguez, N., Del Ser, J., *et al.* 'Explainable artificial intelligence (XAI): concepts, taxonomies, opportunities and challenges toward responsible AI'. *Inf Fusion*. 2020; 58(1): 82–115.

[11] Schoenborn, J. and Althoff, K. 'Recent trends in XAI: a broad overview on current approaches, methodologies and interactions'. In *Presented at 27th International Conference on Case-Based Reasoning*, USA, 2019.

[12] Suresh, H. and Guttag, J. 'A Framework for Understanding Unintended Consequences of Machine Learning'. ArXiv: 1901.10002.

[13] Bhatt, U., Xiang, A., Sharma, S., *et al.* 'Explainable machine learning in deployment'. In *Proceedings of the 2020 Conference on Fairness, Accountability, and Transparency, FAT* '20*. New York, NY: ACM, pp. 648–657.

[14] Gill, N., Hall, P., Montgomery, K., and Schmidt, N. 'A responsible machine learning workflow with focus on interpretable models, post-hoc explanation, and discrimination testing'. *Information*. 2020; 11(3): 137–141.

[15] Gunning, D. and Aha, D. 'DARPA's explainable artificial intelligence program'. *AI Mag*. 2019; 40(2): 44–58.

[16] Elshawi, R., Al-Mallah, M., and Sakr, S. 'On the interpretability of machine learning-based model for predicting hypertension'. *BMC Med Inf Decision Mak*. 2019; 19(1): 146–150.

[17] Topol, J. 'High-performance medicine: the convergence of human and artificial intelligence'. *Nat Med*. 2019; 25(1): 44–56.

[18] Ploug, T. and Holm, S. 'The four dimensions of contestable AI diagnostics – a patient-centric approach to explainable AI'. *Artif Intell Med*. 2020; 10(7): 112–114.

[19] Trimble, M and Hamilton, P. 'The thinking doctor: clinical decision making in contemporary medicine'. *Clin Med*. 2016; 16: 343–346.

[20] London, J. 'Artificial intelligence and black-box medical decisions: accuracy versus explainability'. *Hastings Cent Rep*. 2019; 49(1):15–21.

[21] He J, Baxter, S., Xu J., *et al.* 'The practical implementation of artificial intelligence technologies in medicine'. *Nat Med*. 2019; 25(1): 30–36.

[22] Malhi, S., Morris G., Bell E., *et al.* 'A new paradigm for achieving a rapid antidepressant response'. *Drugs*. 2020; 80(1): 755–764.

[23] Starke, G., De Clercq, E., and Elger, S. 'Towards a pragmatist dealing with algorithmic bias in medical machine learning'. *Med Health Care Philos*. 2021; 24(1): 341–349.

[24] Páez, A. 'The pragmatic turn in explainable artificial intelligence (XAI)'. *Minds Mach*. 2019; 29(1): 441–459.

[25] Drukker, L., Noble, J., and Papageorghiou, A. 'Introduction to artificial intelligence in ultrasound imaging in obstetrics and gynecology'. *Ultrasound Obstet Gynecol*. 2020; 56(1): 498–505.

[26] Alami, H., Rivard, L., and Lehoux, P. 'Artificial intelligence in health care: laying the foundation for responsible, sustainable, and inclusive innovation in low- and middle-income countries'. *Glob Health*. 2020; 16(1): 52–58.

[27] Vollmer, S., Mateen, B., Bohner, G., *et al.* 'Machine learning and artificial intelligence research for patient benefit: 20 critical questions on transparency, replicability, ethics, and effectiveness'. *Br Med J*. 2020; 36(8): 16–27.

[28] Colvonen, P., DeYoung, P., Bosompra, N., *et al.* 'Limiting racial disparities and bias for wearable devices in health science research'. *Sleep*. 2020; 43(1): 159–162

[29] Wang, F., Kaushal, R., and Khullar, D. 'Should health care demand interpretable artificial intelligence or accept "black box" medicine?' *Ann Intern Med*. 2019; 17(2): 59–60.

[30] Hassan, M., Al-Insaif, S., and Hossain, M. 'A machine learning approach for prediction of pregnancy outcome following IVF treatment'. *Neural Comput Appl*. 2020; 32(1): 2283–2297.

[31] Arbelaez, O., Georg, S., Giorgia, L., Julia, V., David, S., and Bernice, E. 'Re-focusing explainability in medicine. Review article.' *Digital Heath*. 2022; 8(1): 1–9.

[32] Malhi, A., Kampik, T., Pannu, H., Madhikermi, M., and Fr·amling, K. 'Explaining machine learning-based classifications of in-vivo gastral images'. In *Proceedings of International Workshop on Explainable, Transparent Autonomous Agents and Multi-Agent Systems; Netherlands*, September 2019. The Netherlands: IEEE; 2019. pp. 1–7.

[33] Meske, C. and Bunde, E. 'Transparency and trust in human-AI-interaction: the role of model-agnostic explanations in computer vision-based decision support'. In *Proceedings of International Conference on Human-Computer Interaction*, London, July 2020. London: Springer; 2020. pp. 54–69.

[34] Montagnon, E., Cerny, M., Cadrin-Chênevert, A., *et al.* 'Deep learning workflow in radiology: a primer'. *Insights Imaging*. 2020; 11(1): 22–25.

[35] Willemink, M., Koszek, W., Hardell, C., *et al.* 'Preparing medical imaging data for machine learning'. *Radiology*. 2020; 295(1): 4–15.

[36] Salman, S. and Liu, X. 'Overfitting mechanism and avoidance in deep neural networks'. *CoRR*. 2019; 50(2): 24–29.

[37] Sandfort, V., Yan, K., Pickhardt, P., and Summers, R. 'Data augmentation using generative adversarial networks (Cycle-GAN) to improve generalizability in CT segmentation tasks'. *Sci Rep*. 2019; 9(1): 68–84.

[38] Tan, M. and Le, V. 'EfficientNet: rethinking model scaling for convolutional neural networks'. In Chaudhuri, K. and Salakhutdinov, R. eds. *Proceedings of the 36th International Conference on Machine Learning*. Cambridge, MA: PMLR, 2019. pp. 6105–6114.

[39] Mohammad, H., Wenjing, J., Xiangjian, H., and Paul, K. 'Deep learning techniques for medical image segmentation: achievements and challenges'. *J Digital Imag*. 2019; 32(1): 582–596.

[40] Bensleh, M., Boujelben, A., Baklouti, M., and Abid, M. 'A comparative study of medical image segmentation methods for tumor detection'. *Int J Comput Inf Eng*. 2021; 15(4): 285–290.

[41] Kavvas, E., Yang, L., Monk, J., Heckmann, D., and Palsson, B. 'A biochemically-interpretable machine learning classifier for microbial GWAS'. *Nat Commun*. 2020; 11(1): 2580. https://doi.org/10.1038/s41467-020-16310-9.

[42] Kanda, E., Epureanu, B., Adachi, T., *et al.* 'Application of explainable ensemble artificial intelligence model to categorization of hemodialysis-patient and treatment using nationwide-real-world data in Japan'. *PLoS One*. 2020; 15(5): e0233491.

[43] Liao, W., Zou, B., Zhao, R., Chen, Y., He, Z., Zhou, M. 'Clinical interpretable deep learning model for glaucoma diagnosis'. *IEEE J Biomed Health Inf*. 2020; 24(5): 1405–1412. https://doi.org/10.1109/JBHI.2019.2949075.

[44] Karimi, M., Wu, D., Wang, Z., and Shen, Y. 'DeepAffinity: interpretable deep learning of compound-protein affinity through unified

recurrent and convolutional neural networks'. *Bioinformatics (Oxford, England)*. 2019; 35(18): 3329–3338.

[45] Shickel, B., Loftus, T. J., Adhikari, L., Ozrazgat-Baslanti, T., Bihorac, A., and Rashidi, P. 'DeepSOFA: a continuous acuity score for critically ill patients using clinically interpretable deep learning'. *Sci Rep*. 2019; 9(1): 1879.

[46] Qiu, S., Joshi, P., Miller, M., *et al.* 'Development and validation of an interpretable deep learning framework for Alzheimer's disease classification'. *Brain: J Neurol* (*England*). 2020; 143(6): 1920–1933.

[47] Anguita-Ruiz, A., Segura-Delgado, A., Alcalá, R., Aguilera, C.M., and Alcalá-Fdez, J. "EXplainable artificial intelligence (XAI) for the identification of biologically relevant gene expression patterns in longitudinal human studies, insights from obesity research'. *PLoS Comput Biol*. 2020; 16(4): e1007792. https://doi.org/10.1371/journal.pcbi.1007792.

[48] Lamy, J.-B., Sekar, B., Guezennec, G., Bouaud, J., and Séroussi, B. 'Explainable artificial intelligence for breast cancer: a visual case-based reasoning approach'. *Artif Intell Med* (*Netherlands*). 2019; 94: 42–53. https://doi.org/10.1016/j.artmed.2019.01.001.

[49] Fiosina, J., Fiosins, M., and Bonn, S. 'Explainable deep learning for augmentation of small RNA expression profiles'. *J Comput Biol*. 2019; 27(2): 234–247.

[50] Warman, A., Warman, P., Sharma, A., *et al.* 'Interpretable artificial intelligence for COVID-19 diagnosis from chest CT reveals specificity of ground-glass opacities. *MedRxiv: The Preprint Server for Health Sciences*. 2020.

[51] Markus A., Kors, J., and Rijnbeek, P. 'The role of explainability in creating trustworthy artificial intelligence for health care: a comprehensive survey of the terminology, design choices, and evaluation strategies'. *J Biomed Inform Elsevier*. 2021; 113: 1–11.

[52] Caicedo-Torres, W. and Gutierrez, J. 'ISeeU: visually interpretable deep learning for mortality prediction inside the ICU'. *J Biomed Inform*. 2019; 98: 103269.

[53] Hao, J., Kosaraju, S.C., Tsaku, N.Z., Song, D.H., and Kang, M. 'PAGE-Net: interpretable and integrative deep learning for survival analysis using histopathological images and genomic data'. *Pacific Sympos Biocomput*. 2020; 25: 355–366.

[54] Nagasubramanian, K., Jones, S., Singh, A., Sarkar, S., Singh, A., and Ganapathysubramanian, B. 'Plant disease identification using explainable 3D deep learning on hyperspectral images'. *Plant Methods*. 2019; 15: 98. https://doi.org/10.1186/s13007-019-0479-8.

[55] Mirchi, N., Bissonnette, V., Yilmaz, R., Ledwos, N., Winkler-Schwartz, A., and Del Maestro, R.F. 'The virtual operative assistant: an explainable artificial intelligence tool for simulation-based training in surgery and medicine'. *PloS One*. 2020; 15(2): e0229596.

[56] Xiang, A. and Wang, F. 'Towards interpretable skin lesion classification with deep learning models'. In *AMIA Symposium*, 2019, pp. 1246–1255.

[57] Weina, J., Xiaoxiao, L., and Ghassan, H. *Evaluating Explainable AI on a Multi-Modal Medical Imaging Task: Can Existing Algorithms Fulfill Clinical Requirements?* Association for the Advancement of Artificial Intelligence (www.aaai.org).

[58] Miró-Nicolau, M., Moyà-Alcover, G., and Jaume-i-Capó, A. 'Evaluating explainable artificial intelligence for X-ray image analysis'. *Appl Sci.* 2022; 12(4459): 1–28. https:// doi.org/10.3390/app12094459.

[59] Yiming, Z., Ying, W., and Lund, J. 'Applications of explainable artificial intelligence in diagnosis and surgery'. *Diagnos Rev MDPI.* 2022; 12(237): 1–18.

[60] Gunashekar, D.D., Bielak, L., Hägele, L., *et al.* 'Explainable AI for CNN-based prostate tumor segmentation in multi-parametric MRI correlated to whole mount histopathology'. *Radiat Oncol.* 2022; 17: 65. https://doi.org/10.1186/s13014-022-02035-0.

[61] Gaur, L., Bhandari, M., Razdan, T., Mallik, S., and Zhao, S. 'Explanation-driven deep learning model for prediction of brain tumour status using MRI image data'. *Front Genet.* (*Comput Genom*). 2022; 13: 8226666. https://doi.org/10.3389/fgene.2022.822666.

Chapter 6

XAI robot-assisted surgeries in future medical decision support systems

Aishat Titilola Rufai[1], Kenechi Franklin Dukor[2], Opeyemi Michael Ageh[3] and Agbotiname Lucky Imoize[4,5]

Abstract

Artificial intelligence (AI) models are gaining widespread applications in various areas such as the healthcare system, especially robotic surgeries. The output of these models needs to be easily explained to surgeons and other stakeholders. These explanations assist stakeholders or end-users of these AI models in establishing trust and understanding the output of the model. However, there are identified limitations in fully implementing these AI models, particularly in critical areas such as robotic surgeries. This is mainly due to the complexity of its results, patient safety, and growing security concerns. Thus, explainable AI (XAI) aims to bridge the gap in understanding the results of AI models. Toward this end, this chapter provides an overview of the current applications, importance, and limitations of XAI robotic-assisted surgeries in the medical decision support system (MDSS). The chapter discusses the privacy and security concerns of patients while utilizing XAI techniques in robotic surgeries. The chapter also explores current trends and issues regarding the future deployment of XAI robotic-assisted surgeries in supporting medical decision-making systems. Finally, the chapter addresses the limitations of machine learning (ML) tools used for robotic surgeries.

Keywords: Explainable Artificial Intelligence (XAI); Medical decision support system (MDSS); robotic-assisted surgery; Artificial Intelligence (AI) system; Machine learning (ML); Interpretability; AI models; Electronic Health Records

[1]Department of Electrical and Electronics Engineering, Faculty of Engineering, University of Lagos, Nigeria
[2]Department of Mechanical Engineering, Faculty of Engineering, University of Lagos, Nigeria
[3]Department of Agriculture and Production Engineering, Doala Ageh Nigeria Limited, Nigeria
[4]Department of Electrical and Electronics Engineering, Faculty of Engineering, University of Lagos, Nigeria
[5]Department of Electrical Engineering and Information Technology, Institute of Digital Communication, Ruhr University, Germany

6.1 Introduction

In recent times, there has been substantial growth in the use of artificial intelligence (AI) systems, as evidenced by Ref. [1], where the global corporate investment increased from $14.99 billion in 2015 to $176.47 billion in 2021. Also, these AI models have been adopted in various fields such as the medical, financial, and agricultural sectors. AI applications range from increasing productivity in businesses to improving healthcare performance. The prevalence of these models has recently led to the use of complex AI algorithms, particularly machine learning (ML) models, to solve problems. Although these complex models generate ground-breaking results, they use the "black-box model" concept where the decision-making process is unclear to both AI users and developers [2]. This is a major concern, especially in healthcare and the legal system, where human lives depend on AI results [3].

For this reason, AI systems have not been widely integrated into most healthcare domains. In order to fully utilize the potential of these AI models, it is crucial to understand the reasons behind the decisions of these AI systems to detect irregularities or faults in the system [4]. The lack of interpretability of suitable AI models led to a renewed interest in the explainable AI (XAI) field.

Scientists have always debated the explainability of AI decisions even though XAI field has existed for over 40 years. Recently, AI researchers have discovered various methods for achieving transparency in black-box models [5]. The concept of explainability covers the ability of humans—specifically AI users to understand the decision-making process of AI models [6]. While these AI algorithms may yield accurate results, there is an offset between the performance(accuracy) and the transparency [4]. For example, the neural network (NN) model is highly accurate but lacks interpretability, whereas older models such as decision trees are interpretable but lack accuracy [3,7]. For applications in the medical field, medical practitioners must comprehend the processes of attaining the results of the system regardless of its accuracy. For example, a survey among surgeons in 2021 indicated that the major issues and concerns relating to implementing AI are lack of trust, risk of bias, and loss of autonomy [8]. Other concerns in the healthcare sector include the privacy and security of patients. Since healthcare professionals are saddled with the responsibility of providing the best care for patients, the decision support system in a clinical setting must be interpretable and explainable [9]. Therefore, the increasing demand for explainability has resulted in developing rules and regulations for AI models to achieve secure and reliable applications in the healthcare domain and other applicable sectors.

In order to achieve the full potential of XAI, there are regulations in place that require the implementation of XAI, such as the general data protection regulation (GDPR) for European Union (EU) citizens and residents. The GDPR is a regulation for data protection and privacy [10], stating that any model inferred from data must be accountable. In other words, the GDPR, commonly acknowledged as "rights to an explanation" allows users to acquire explanations about the decisions of a model that can impact them physically, mentally, legally, and financially [11–13]. Thus, the

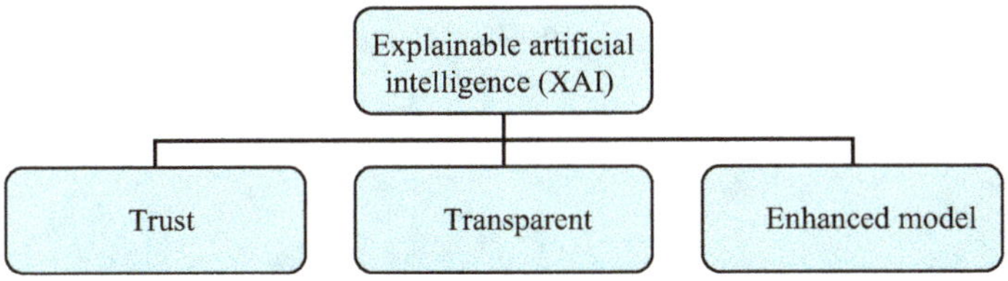

Figure 6.1 XAI system and its possible outcomes

privacy and security of users are also considered while using AI models. With these regulations, more decision-making results will be governed by such regulations. Current studies have shown that a model has to be explainable to its users. However, these regulations are yet to attain their prospects since there are no specific requirements to be considered for explanations [14]. This explainability gives users assurance and security, specifically regarding the medical decision support system (MDSS).

XAI could be a possible solution to the problems in the AI system. One of its main goals is the model's capability to explain itself so that humans can comprehend these algorithms. These explanations would foster research and lead to justification, management, and improvement of these AI models [15]. The goals of XAI can be summarized in Figure 6.1.

There are a considerable number of studies on the measurement and evaluation of XAI methods [11,14,16,17]. These studies highlighted different XAI approaches and metrics used, but there are no standard ways to select and evaluate the appropriate XAI methods to use for a system.

We will discuss the following topics in the next section: current application of AI in the healthcare system, concepts and related terms of explainability, history of medical robots, applications of XAI in the medical field, application of XAI, and different types of explanation methods for MDSS in robot-assisted surgeries.

6.2 Related work

In this section, we will review various literature on the current application and limitations of AI in the healthcare sector-robotic surgery, XAI applications in the medical field and robotic surgeries.

6.2.1 Current applications of AI in the healthcare systems

Artificial intelligence (AI) applications in healthcare have gradually increased within the past 7 years. This is shown in Figure 6.2 from Google trends.

AI-based systems can be used in various areas in the healthcare system, such as disease diagnostics and prediction, clinical decision making, drug interaction and discovery, patient care, and robotic surgery [18–20]. Thus, the use of AI can help transform various areas in the healthcare sector [21]. Some AI applications in the medical field are discussed below.

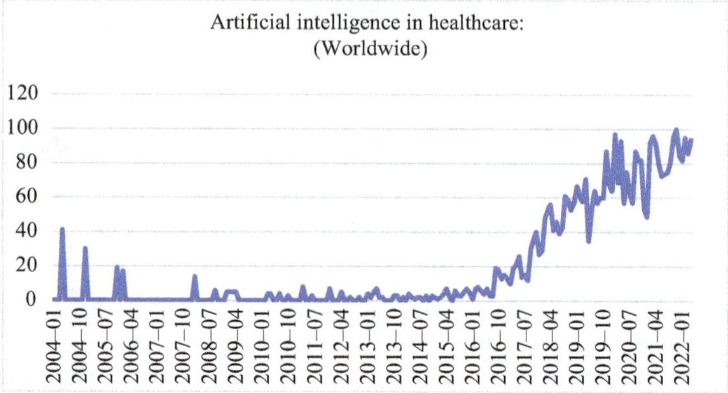

Figure 6.2 AI in healthcare from 2004 till date. Source: Google Trends, search term: AI in healthcare

6.2.1.1 Drug interaction and discovery

Since drug discovery and development are time-consuming, AI assists pharmaceutical companies in expediting the drug discovery process [18,19]. Although some researchers question the application of AI in the medical field, most agree that AI in healthcare will be critical in the future [19]. Currently, researchers encounter both benefits and challenges of AI, specifically when the methods are integrated with automation [22].

6.2.1.2 Clinical decision-making

One of the most significant aspects of AI application in the healthcare system is the MDSS. MDSS is synonymous with the clinical decision support system (CDSS). MDSS is an AI-aided technology used to assist medical practitioners and/or patients in improving decision-making in the medical field [23,24]. Studies indicate that AI for MDSS can be mainly categorized into the knowledge-based system, which is a rule-based expert system and the data-driven system [24–26]. For the knowledge-based system, programming is used to determine the decisions.

In contrast, the data-driven system derives its decision by using ML techniques to obtain insights from a large amount of data [24]. In addition, MDSS has various uses in the medical domain, which are not limited to the following: diagnostics, image interpretation, patient monitoring, postoperative care, outpatient care surgical devices, etc. Therefore, MDSS is intended to make the decision-making process for medical practitioners faster, less prone to error, and more empirical [27].

6.2.1.3 Disease diagnostic and prediction

AI obtains insights from past patients' data to diagnose diseases. In addition, AI analyses medical images like X-rays, magnetic resonance imaging (MRI), and ultrasounds by recognizing patterns for the early and accurate detection of diseases such as cancer. The AI algorithms used for such analysis are NNs, decision trees, support vector machines (SVMs), and Artificial NNs (ANNs) [18]. Another application of AI for disease prediction is IBM's "Watson for Oncology" where the patient's data is inputted and assists medical staff in predicting the best treatment based on various past data [28]. However, it has been reported that IBM's Watson has sparked criticisms from doctors globally, stating that its patient recommendations are unsuitable [29].

6.2.1.4 Robotic surgery

Recently, robot-assisted surgery has found applications in the following surgeries: urological, colorectal, cardiothoracic, orthopedic, maxillofacial, and neurosurgery [30]. Surgeons prefer robot-assisted surgeries because of their increased efficiency, flexibility, and control [28]. Also, robot-assisted surgery is mostly used in complex surgeries that surgeons find difficult to perform. A typical example is a smart tissue autonomous robot (STAR) used to perform intestinal surgery in 2016. It makes use of ML algorithms. Its features include real-time communication support and can also be controlled by a distance monitoring system.

6.2.2 Limitations of AI in the medical field

The following are the problems that AI is currently facing in the medical field:

- *Regulatory compliance*: Regulations must be incorporated into AI algorithms that will ensure patient's privacy while being compatible with technical innovations [21].
- *Human–computer interaction* (*HCI*): Studies have shown that medical researchers have suggested that the most likely reason for AI's failure is the lack of HCI considerations [23, 31]. Also, most AI-enabled technologies are not user-friendly. That is, clinicians find it difficult to relate to and use such technologies.
- *Automation bias*: Research has indicated that most AI users, such as medical professionals, rarely question the decision-making process of these models [23]. Also, due to the black-box nature of these models for MDSS, it is difficult for medical practitioners to question these models. This makes it challenging to evaluate these models leading to automation bias. The implications of automation bias are the limits it imposes in the healthcare sector, which are the inability to detect errors in the model and determine who is responsible (medical practitioner or AI developer) for such errors.
- *Ethical issues*: Although the ethical implications of AI in the medical industry have not received much attention. Few studies highlight the importance of ethical principles for AI. According to [32], there is a need for the implementation of these principles in health care, which should comprise of the following: the need for human autonomy, explainability, patient privacy, fairness, and safeguarding patients from harm. For human autonomy, humans must be able to choose and evaluate AI decisions. Human autonomy has to do with humans having control over AI decisions to prevent or reduce automation bias and being able to choose other alternatives model [33]. The lack of proper implementation of these principles has led to the current constraints of AI-based technologies in the healthcare system. In other words, the proper implementation and integration of these principles could lead to the successful adoption of AI in healthcare.

In summary, the data quality of the trained model can also impact the accuracy of its results and the unavailability of clinical data, thus making it difficult to obtain accurate information from the model. For example, AI finds it challenging to adapt the trained data to real-life clinical data [23]. Finally, the most prevalent problem amongst medical practitioners is AI models' lack of interpretability. This is because it is crucial to understand the process for achieving particular results for medical decision-making. For this reason, there has been a renewed interest in the XAI field. Thus, further research and evaluation of XAI techniques are needed to adopt AI in the medical field fully.

6.2.3 XAI

Although researchers are yet to agree on the definition of "XAI," XAI can be defined as a system that provides the reasons for a particular outcome by giving understandable explanations to end-users and accurately reflecting the system's process by adhering to the specifications the system intends to meet [34,35]. These explanations can primarily help to improve the interaction between the ML models and humans, such as the clinicians, patients, and other stakeholders in the health-care settings. Consequently, factors should be considered for an explanation, and the AI system should be able to distinguish what to explain to an end-user and an expert. For example, for medical diagnoses, XAI systems should be able to inform the patient about their diagnosis. However, the AI models (i.e., processes) would not be explained to the patients; the decision-making process will be explained to the medical practitioner(s). Since the XAI field is still relatively young, there is a need for more research on these methods and evaluation measures to justify the results and determine their scope [5,36].

6.2.3.1 Concepts and terms of explainability

Although there is no specific definition for the term XAI, it encompasses the concept of making AI models interpretable and trustworthy. In some cases, XAI can also be referred to the following terms: responsible AI, interpretable AI, and transparent AI, just to mention a few [2]. All these terms are closely related but have different definitions. Also, interpretability and explainability are frequently used interchangeably, but these terms have subtle differences. Some of these related ML terms and their definitions are described in Table 6.1.

6.2.4 XAI in healthcare

AI methods can be one of the critical factors in determining the future of MDSS [38]. AI applications in the healthcare systems range from prognosis and diagnosis to patient-facing application, robotic surgery, drug development, and other ground-breaking outcomes. AI models have become more complex, making it difficult to interpret the results; the ambiguity of the AI decisions has led to physicians' and users' distrust of the system [39]. This has increased interest in XAI techniques in the medical sector.

Table 6.1 Terms of explainability

Terms	Meaning
Interpretability	The ability to convey meaning in an understandable human form [4]
Explainability	It is the model's ability to explain its internal process by providing an accurate representation of the model and is still understandable by humans [4,37]
Transparency	A model is transparent if it is interpretable [4,16]
Understandability	The ability of a model to make humans understand its function [4,11]
Responsible AI	The application of AI models ensures that appropriate stakeholders are held responsible for any error emanating from the model

The history of XAI can be traced back to the origins of the AI system, where a knowledge-based experts' system was developed [40]. However, as time progressed, there was more emphasis on building more ML models based on the performance of the models rather than the model's interpretability. As a result, this has led to several limitations for AI in different fields. For example, in robotic surgeries, recent studies show that explainability improves pre-emptive management of adverse effects of surgery [17,41,42]. The role of explainability can be beneficial in surgery if the risks are discovered before surgery (such that the necessary interventions can be done to reduce or prevent complications during and after surgery) and the reasons behind the risks to prevent or mitigate them. For example, ML techniques can be used to predict the occurrence of hypoxemia before the surgery. However, this information is insufficient for a typical clinical setting due to its lack of interpretability [43]. This means that the factors leading to possible hypoxemia have not been stated clearly. With the application of ML algorithms before, during and after surgeries, it is important that the clinicians and all other stakeholders understand the reasons for the model's result to achieve an improved outcome. Due to a lack of interpretability, ML techniques have not been incorporated in most medical domains. Examples of these models are shown in Figure 6.3 according to their interpretability:

The most interpretable model in Figure 6.2 is the classification model, while the least interpretable model is the NN. In addition, the model's accuracy is inversely proportional to its interpretability, where the least interpretable model is the most accurate and vice versa for the most interpretable model.

While several studies agree that XAI improves trust and understanding of AI models for medical professionals [38,40,44,45], it is apparent that XAI hopes to achieve other functions. These functions include fairness, security, privacy, ethics, and confidence [40,44]. As discussed earlier, AI poses concerns about its lack of interpretability in the medical sector. This is because the healthcare system is a critical area where decisions or predictions made by an AI must be clear and understandable. Also, in the medical field, the following are problems faced while utilizing AI: lack of interpretability, bad quality of data, automation bias, poor human–computer interaction, non-homogeneity, unpredictability, and high

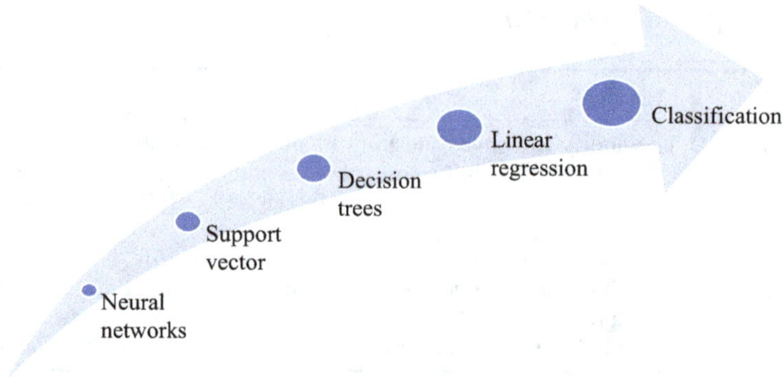

Figure 6.3 AI models according to their interpretability

dimension data which leads to uncertainty [46–49]. Thus, XAI aims to bridge these gaps by providing explanations to models. In this section, existing literature on XAI in clinical settings is discussed.

6.2.5 How explainability works—bridging the AI gap

Since explainability is one of the major limitations of AI, this section explains the reasons behind the explanation. In order to obtain an explanation, the following questions need to be considered: who is the explanation intended for? Why is an explanation needed? The answer to this former question is the users, which could range from clinicians to AI developers to regulatory agencies, just to mention a few [4]. Thus, such users will require various explanations depending on the context and users' needs [4].

In order to explain a model, the explanation model and explanation interface are required. The explanation interface is where the human–computer interaction (HCI) takes place, while the model is the technique used for explanation. As mentioned earlier, healthcare scholars suggested that lack of HCI is one of the major reasons for AI failure for MDSS [31]. As a result, there is a need for a proper understanding of the user's workflow to aid HCI which gives rise to the need for XAI. Therefore, the human-centered approach to XAI is concerned with discovering various approaches for explanation to humans (end-users), by iteratively involving the users in the development process (e.g., through interviews, hypothetical scenarios, focus groups, and questionnaires) [50]. For example, some clinicians might require a type of explanation while others might not. Thus, this subjective nature of explanations must be considered while designing the model. Thus, the proper integration of HCI in a clinical environment can assist clinicians by reducing tasks that could otherwise be difficult for humans to perform.

The successful integration of AI into MDSS will be useful in most medical domains such as diagnosis and drug discovery [51]. However, Ref. [52] highlights the need for a clear governance framework for MDSS to safeguard people from harm, particularly from unethical practices. This government framework determines the system's objectives, values, policies, culture, and accountabilities [53]. In other words, XAI might be the vehicle for achieving these properties. Consequently, XAI methods need to be evaluated to determine the accuracy and effects of their results on human rights. Ref. [54] raised the following questions "Are there specific conditions that all XAI must fulfill"? If true, are there limits for discrepancies between the measures (e.g., accountability and accuracy), and how can it be assessed? Thus, the ability to carry out a proper assessment (i.e., both technical and ethical assessments) of XAI for MDSS will ensure patient safety, privacy, and assurance of the system. The ethical aspect of XAI has covered most of the limitations of AI for MDSS.

6.2.6 Benefits of XAI for the medical field

This section focuses on the characteristics of XAI and how they impact the users (clinicians, patients, and others). This section discusses the applications and key importance of XAI for clinical systems.

The following are the benefits of XAI for MDSS,

6.2.6.1 Obtaining insights from the system

AI algorithms are applied to a large amount of data to obtain meaningful insights from the data. For example, the application of AI to electronic health records (EHRs) will lead to obtaining predictions from this data. However, this prediction lacks transparency, leading to limited use of these models in healthcare [55]. XAI is recently used to get new scientific information from the data [56]. Therefore, XAI methods can be applied to patients' data to achieve transparency and interpretability of the models' predictions, unlike AI black box model. In addition, XAI allows medical experts to review the performance of these AI models to improve them. In addition, medical experts and scientists must discover new information in these AI models [57]. Thus, XAI allows scientists to make new informed discoveries and learn new information about the input data.

6.2.6.2 For evaluating the system

While there has been a significant progress in the adoption of AI systems in various fields, the authors of [7] suggested that the application of AI methods will be most likely limited in the healthcare systems provided that all the issues regarding the full implementation of AI systems in the medical field are not resolved. These issues are not limited to explainability, interpretability, etc. For example, one of the difficulties researchers face—particularly in the medical sector, is the accessibility of quality data. The available data might not fit the algorithms. Using such data could lead to bias in the results. Some examples of possible bias in EHRs are differences in patient populations, types of equipment, omission of some data, imaging parameters used, and the lack of representation of rare diseases [5]. Hence, it is important to assess the accuracy of AI algorithms based on the source of the data in the medical field [57–59]. Since XAI allows experts to evaluate AI results by verifying if the predictions are correct, this process will lead to the improvement of the model.

6.2.6.3 Following rules and regulations

Adopting AI models calls for serious safety concerns in the medical field, such as patients' privacy and security. The continuous accumulation of different data types—such as patients' information and medical history—makes it challenging to ensure patient privacy and data security [60]. Therefore, patients' data can be anonymized to protect patient's privacy, which causes the loss of some data leading to a trade-off between privacy and usage [61].

Recently, there are security and privacy safeguards for patients' such as Health Insurance Portability and Accountability Act (HIPAA) and the Health Information Technology for Economic and Clinical Health (HITECH) Act, compliance to these acts is important to safeguard health information privacy, security, and confidentiality [62]. Furthermore, XAI can assist the justice system in identifying who(person) or what (AI system) is responsible for the decisions of AI-enabled technology [63]. This is because the application of XAI makes the decision-making process more interpretable, making it easier to figure out who is liable for errors.

Since cybersecurity threats pose a risk to patients' privacy and data security, there is a need to address these threats faster and look for ways to mitigate such threats. This calls for the expertise of a cybersecurity professional who must meet the growing demands of the healthcare industry [64]. Therefore, the healthcare system will need AI developers and/or cybersecurity experts to mitigate cyber threats. Since XAI models provide the rationale for a particular decision, these can help clinicians, cybersecurity experts, and AI developers detect possible threats to patients' privacy and safeguard patient data. Therefore, the AI model needs to be explainable so that medical experts, computer scientists, cybersecurity experts, and AI developers collaborate to protect patients' data and privacy. In addition, for AI to be considered in the medical field (e.g. AI-based MDSS), a clinical evaluation is carried out where the model's prediction performance is measured to determine its application in a real-world scenario [65]. Since these models produce a low error, XAI application is an absolute necessity for MDSS to understand the inner workings of the models [14]. With the application of XAI, patients' security and privacy concerns are considered [66].

6.3 Medical robots

Recently, robots have supported humans in various settings such as schools, manufacturing industries, businesses, and other sectors [67]. It has also seen moderate application in the medical field. The concept of robots in the medical field can be traced back to the late 1960s. While authors argue about the timeframe of the first robotic surgery, the development of the first medical robots came into reality in 1978, and the first robotic-assisted surgery was implemented in 1985. Medical robots have found various applications in the medical field and are as

Table 6.2 Difference between traditional and robotic surgery

Features	Traditional surgery	Robotic surgery
Accessibility	• More flexible	• Designed to work in difficult positions
Surgical procedure	• Prone to tremor • Surgeons perform surgery with hand	• No tremor • Lacks sense of touch • Can experience equipment failure
Surgical instruments	• Limited sterilization • Limited vision	• Can be sterilized and reused • High-quality 3D vision
Radiation risk for surgeons	• Surgeons are prone to radiation risk	• Rare
Precision	• Precision depends on surgeon's skills	• High precision
Risk of infection (patients)	• High	• Low
Risk of infection (surgeons)	• Prone to infection	• Rare
Dexterity	• Prone to fatigue and tremor	• Improved dexterity, flexibility
Decision-making	• Ability to make good judgments	• Robotic technology can be improved
Cost	• Easy to get humans unlike expensive machines that can only be in place	• More expensive to set up
Real-time data usage	• Unable to use quantitative data	• Unable to use qualitative data
Size	• Surgeons handle surgical instruments	• Heavy and difficult to handle e.g. robotic arm

follows: patient care, rehabilitation, assistive, robotic surgery, and more. This chapter was majorly concerned with robotic surgery or robot-assisted surgery.

Surgical robots can be used to assist surgeons and perform specific tasks that are difficult and risky for surgeons. For example, in orthopedic surgery, surgeons are exposed to radiation [68] and fatigue during surgery [69]. All these concerns in traditional surgeries (i.e., surgeries performed without the help of robots) can be limited with the aid of robotic surgeries. On the other hand, robotic surgery has its limitations. For example, the size of the robotic arms makes it difficult to handle by the surgeon, and since the robot lacks touch sensation, it is usually challenging for the surgeons to detect how much depth to cut [52]. There is a need for a proper evaluation of surgical skills for both traditional and robot-assisted surgeries.

Therefore, using surgical robots will aid clinicians and surgeons under challenging tasks and could lead to better surgical outcomes. The adoption of these technologies will lead to an efficient healthcare system. Table 6.2 summarizes the benefits and limitations of robotic surgery and traditional surgery.

6.3.1 History of robotic surgery

Robotic surgery can also be known as robot-assisted surgery. Robotic surgery can be used to perform surgeries with the robotic arm, and they are majorly used for minimally invasive surgeries (surgeries with little incisions). The concept of robots in the medical field can be traced back to the late 1960s. However, the development of the first medical robots came into reality in 1978 and was first implemented in 1985.

In 1978, the Programmable Universal Manipulation Arm (PUMA) was developed by Victor Scheinman and was the first robot used for surgery; the PUMA 560 had six degrees of freedom, and it was adaptable compared to human hands [55]. The adaptability and flexibility of these medical robots can assist surgeons during surgeries to perform challenging tasks for human—surgeons. In 1985, the PUMA robot was first used to perform stereotactic brain biopsy surgery and by 1988, it was converted to a surgeon-assistant robot for prostatectomy [55,70]. The SARP was used for prostate surgeries [71]. Therefore, the PUMA 560 gave rise to the application of medical robots for surgeries in subsequent years.

In 1992, ROBODOC was designed to aid total hip replacement surgery [55]. This is the process where the hip joints are replaced by prosthesis [72]. This procedure was a far-reaching discovery in orthopedic surgery. The robodoc surgical robot was the first approved by the food and drug administration (FDA) [73].

In 1993, Yulin Wang, founder of Computer Motion Inc., developed an automated endoscopic system for optimal positioning (AESOP) [71]. AESOP is widely used in the following areas: laparoscopic cholecystectomy, hernioplasty, fundoplication, and colectomy and was approved by FDA [55,74,75]. The operation of the AESOP is described below: the robotic arm is controlled by the surgeon's voice command by manipulating the endoscopic camera during surgery [76]. The use of AESOP gave rise to several advancements in robots, leading to the development of ZEUS.

In 1998, ZEUS was developed by Defense Advanced Research Projects Agency (DARPA), ZEUS's first application in a robotic surgical system was a fallopian tube anastomosis, and was also implemented in 2001 as the first transcontinental telesurgery [75,77]. Also, ZEUS was designed to replicate the surgeon's arms movements, thereby incorporating arms and surgical instruments in the design, which is operated and regulated by the surgeon [71,78].

The computer Motion Inc. company was obtained by Intuitive Surgical Inc., which stopped the production of the ZEUS system, replacing it with the da Vinci surgical system. However, some parts of the ZEUS system were incorporated into the da Vinci system, using the master–slave system [79]. The master–slave system is a system where the surgeon controls the robot. The da Vinci robot is the prevalent surgical robot and was approved by the FDA in 2001 [73,75]. It has several applications in surgical operations [73]. In addition, the Da Vinci robot eliminated the problems of laparoscopic surgery. With technological advancements in these robots, this robot is preferable for surgeons. These advancements are as follows: improved 3D vision, precisely controlled endo wrist instruments, the seven degrees of freedom, and the preservation of natural eye and instruments alignment made of robotic platforms [73].

With the use of da Vinci robots in different surgeries, other robots were also developed, such as the Mako Rio robots, which were created in 2019 to perform total knee surgery. The Mako Rio is the first approved robot for knee surgeries, and it has sophisticated 3D devices which can be used for total or partial knee replacement. Over the years, the da Vinci robot has been transformed for different applications in surgery. Thus, the development of the da Vinci robot gave rise to different technological innovations in robotic surgery.

6.3.2 Current and future use of medical robots and devices

With the increasing production of medical devices, these devices need to be examined to be safe for users and patients to reduce errors, particularly in high-risk devices such as robotic surgical systems. This brought about the need for regulating medical devices. For example, in the USA, the FDA is responsible for approving these medical devices. These devices are subdivided into different classes, Class I–Class III. Class I is considered low risk, while Class III is a high risk. In such circumstances, different processes are used for approval depending on the medical device classes [80–82]. Thus, Class III medical devices go through a thorough examination to ascertain their safety for users. Also, European regulations go through a rigorous process regarding regulating medical robots, such as all healthcare robots must undergo the confirming European mark according to the European medical device's directive. In this case, the robots are classified into different classes, and regulations apply to each. Although the general data protection regulation (GDPR) has some regulations regarding the application of AI systems in Europe, these regulations have been limited [81]. Therefore, regulations and implementation are important factors for robotic surgical systems' future application and development.

Some have suggested that the size of surgical robots be reduced. This is to make handling of such systems in the operating theatre easy for the surgeon and the surgical team. This has led to the growing research and development of smaller surgical robots, making the robots easy to use [83]. For example, laparoscopic fiber was used for checking the internal body parts. Currently, micro-robots are built for this same purpose. In addition, this micro-robot can be used for internal repair without the aid of external equipment [80]. Thus, technological innovations such as AI in the medical sector will pave the way for future research and implementation of medical robots, particularly for surgical robotic systems.

6.3.3 Robotic surgery and AI

While AI has significantly improved healthcare processes for physicians, it has yet to achieve a similar breakthrough in MDSS or clinical results [84]. According to a Harvard business review (HBR) analysis, robot-assisted surgery was among the leading AI applications that are likely to change healthcare. Consequently, for orthopedic surgery, AI-enabled robotic surgery performs real-time analysis of patients' preoperative data to assist physicians during surgery. For example, in a review of 379 orthopedic patients, Mazor Robotics' AI-aided robotic surgery method decreased surgery complications by five compared to the surgeon performing the procedure alone [84,85]. This is because an AI-enabled robot surgery performs real-time analysis of patient's preoperative data to assist the physician during surgery leading to a 21% decrease in patient's post-op hospital stay and reduced cost [84]. Thus, HBR predicted that AI robot-assisted surgery will reduce annual spending by $40 billion by 2026.

In the following subsection, we will discuss the current use of AI in surgery, limitations, roles of surgeons, ethical considerations, human-robot interaction, various AI approaches, and future directions of robotic surgery.

6.3.4 Current application of AI in robotic surgery

AI consists of multiple fields for application in the healthcare sector. This section will briefly discuss the most common subcategories of AI for the medical field.

The subdivisions are discussed as follows:

ML

A subset of AI that allows machines to learn from large data and discover patterns humans cannot see from data. ML can also be categorized into the following learning algorithms: supervised, unsupervised, semi-supervised, and reinforcement learning [75,86].

For supervised learning, the computer learns from labeled data, and the outcome of the data is known [87]. For instance, you can teach the computer how to identify parts of a body such as eyes, hands, or stomach, this data is fed into the algorithm, and the result of the prediction will either be eyes, hands, or stomach. Thus the computer is trained to identify body parts, a new data set can be fed into the algorithm, and the algorithm learns from experience to identify the body parts.

The computer discovers patterns or clusters from unstructured or unlabeled data for unsupervised learning. At the same time, semi-supervised learning is a combination of both labeled and large unlabeled data—a combination of both supervised and unsupervised learning. Reinforcement learning is used to perform specific tasks where decision-making is essential while learning from its successes and failures. Such tasks include driving cars, and robotics [86,87].

Computer vision (CV)

CV is an engineering field that allows the machine to see [88]. It has applications in various fields, from robotics to medical imaging and self-driving cars. CV has also led to the advancement of various sectors in the medical arena—particularly in the operating room. Examples of such applications include surveillance of the operating room, endoscopic equipment, and surgical team activities (which can be used to measure the performance and skills of the surgeon) [88]. Therefore, accumulating all this information can help enhance the surgical robotic system, aid research, and improve the development of autonomous surgical systems. The aid of AI and CV in the medical sector has allowed for accurate detection and treatment of such diseases [89]. In addition, most successful CV techniques are developed with ML techniques such as SVMs, k-nearest neighbor (KNN), and convolutional NNs (CNN) [89].

Natural language processing (NLP)
NLP is the computer's capacity to comprehend human language [75]. NLP extracts information from large unstructured text data [90]. For instance, NLP has been used to identify critical words and phrases in operative and progress reports that predict postoperative complications (e.g., anastomotic leak after colorectal surgery) [84]. However, these predictions displayed simple clinical knowledge. Still, the algorithm could modify the predictive factors of the phrases that depict patients' emotions like irritated and tired with respect to the postoperative day to predict the surgical complications (e.g. anastomotic leak) [91].

ANN and deep learning
The main concepts in ANNs research can be traced to the brain [92]. ANN is used to accomplish tasks by detecting patterns through one or two layers of NNs. Deep learning also recognizes patterns using multiple layers of NNs, identifying more complex patterns, unlike ANN [84].

6.3.5 Current application of AI in emerging robotic systems

AI has various uses in surgery and can be used in both intraoperative and postoperative stages. In this section, we discuss the current applications of AI in surgery. For surgeries, studies have shown that AI has helped predict intraoperative, postoperative complications and postoperative care for patients [93–95]. For robot-assisted surgery (RAS), the subfields of AI discussed earlier form the basic building blocks for surgical procedures. In addition, there are three main types of RAS, namely [96].

The active system, also known as an autonomous system, performs tasks independently (while remaining under the surgeon's control). Examples of such systems are PROBOT and ROBODOC platforms. A semi-active system is a surgeon-driven element to complement the pre-programmed element of these robot systems [96]. Master–slave or haptic system is a surgeon-guided procedure [97], where the surgeon's hand movement is replicated on the laparoscopic surgical instruments, which imitates the surgeon's hand activity. Examples include the da Vinci and ZEUS systems [96]. As discussed earlier, ML can help RAS during surgery by using computer vision to monitor and learn from the surgeons.

From studies in Table 6.3, ML techniques can assist RAS with collecting information during surgery through computer vision, aiding MDSS [99]. Collection and analysis of data using ML algorithms can be done efficiently with the help of MDSS. On the other hand, AI also faces numerous challenges in surgery, including poor data quality, human–robot interaction, etc. Thus, ML and RAS techniques have a long way to go before incorporation into surgeries.

In the past, experts who observed various surgeries carried out skill assessments for surgery. This method is prone to error and time-consuming, making it

Table 6.3 Summary of AI applications in surgery

Surgery type	Surgery stage	AI algorithms mentioned used	Focus/results	Limitations	Ref.
Abdominal surgery	Post-op	SVM, KNN, Logistic regression	Prediction of post-operative complications after abdominal surgery	Data imbalance	[94]
Various surgeries	Post-op	SVM and logistic regression, random forest, gradient boosting tree (GBT), and deep NN (DNN)	Predicting the risk of postoperative complications related to pneumonia, acute kidney injury, deep vein thrombosis, pulmonary embolism, and delirium	Data imbalance	[98]
Orthopedic robotic surgery	–	CNN	• Use of (AI) for MDSS for diagnosis and treatment for orthopedic surgery • Use of robotic surgery in surgical treatment	• AI and robots' helplessness during complications • AI and robots' non-liability • Difficult to incorporate due to its Complex technology	[97]
Urologic surgery	Intra-op		Highlights recent findings and applications of ML in robotic-assisted urologic surgery	• ML techniques not well-established in surgery • Security of high-volume surgical data	[99]
Wide application of surgeries	Intra-op	• SVM • ANN • k-NN • Recurrent NN (RNN)	Gives surgeons additional information, accelerating intraoperative pathology, and recommending surgical steps	Methodological shortcoming	[100]

Note: Intra-op: intra-operative; Post-op: post-operative.

unreliable for assessment [99]. The aid of ML for surgeries allows for faster and more accurate means of evaluating surgical skills. ML has also been used to recognize surgical tasks (i.e., knot tying, suturing, and needle passing) in a simulated lab setting [99].

ML can also be used in autonomous robotic surgery. Thus, the following tasks need to be accomplished: autonomous camera positioning and other autonomous surgical tasks such as suturing, knot tying, and tissue dissection using ML techniques.

A robot requires the use of computer vision to "see," and ML algorithms to think (task planning) and to do (task execution) to complete an autonomous task [99].

Therefore, digital surgery is one emerging field that could transform surgical robotics systems for the future of surgery. However, it is still in its infancy. Digital surgery (DS) hopes to introduce novel scientific practices and transparency to the surgical system through machines that assist surgeons with better understanding and good decision-making. DS hopes to contribute to the technical development of surgeries, including robot-assisted and computer-assisted surgeries. Furthermore, DS deals with several advanced technologies, which are not limited to the following: robotics, advanced instrumentation, connectivity, enhanced vision, data analytics and ML algorithms. Such technologies are the building blocks of digital surgery and will assist the surgical team in achieving a better outcome for surgical procedures.

6.3.6 XAI robot-assisted surgeries for MDSS

One major problem AI and robotics face for MDSS is lack of transparency. Although different technologies are underway to solve this issue, there is yet to be a consensus between the stakeholders involved. Also, the lack of human–robot interaction has made it difficult for surgical teams to adopt AI and robotic technologies into their procedures fully.

Studies have suggested that the prediction of complications is not sufficient. Thus, the cause of the complication should be known. Researchers agree that XAI in the operating room can help improve MDSS by detecting and preventing surgical complications in real-time. For example, by monitoring vital signs, an XAI tool, "Prescience" detects hypoxemia during surgery up to 5 min prior to occurrence [41]. XAI can assist the surgical team and predict in advance possible surgical adverse events by updating surgeons periodically and giving reasons for its predictions [41].

Thus, surgeons need to work with computer scientists to extract relevant surgical data so that the computer scientist can obtain the right data and provide accurate solutions for the surgical members.

6.3.7 Current limitations of XAI and robotic surgery for MDSS

The current difficulties in applying XAI for robotic surgeries for MDSS include data quality, surgeon trust, and a well-established framework for explanations.

During robotic surgery, a large accumulation of data is generated. This makes real-time analysis of such data difficult, and this can result in using irrelevant data. Thus, it is important to identify important data for analysis during surgery. However, an efficient MDSS can aid in collecting and analyzing user data. Also, surgeons find it difficult to trust an AI system since the decision-making process lacks transparency. Medical practitioners will be more likely to trust a system with interpretable decisions. In addition, incorporating these novel techniques (robotic surgery and XAI) in the operating room can lead to a complicated work setting and increased operation time for surgical teams. Furthermore, there are no well-defined techniques for XAI methods for robotic surgery. Therefore, there is need for more research to be done to determine standard techniques for the healthcare sector.

6.4 Explanation methods

Several XAI methods are available for AI models, the most common methods will be discussed in this section. This has brought about the classification of explanations into two methods as follows [39,66,101]:

- *Post-hoc explanations*: explanations are provided after the results. Deep learning models such as NNs are examples of models with such explanations.
- *Ante-hoc explanations*: the explanations are already embedded in the model that is the AI model is self-interpretable. Examples of models with such explanations are decision trees, and linear regression. This can also be called the model-based explanation.

The post-hoc explanations can be further divided into:

- *Model agnostic*: this type of explanation can be applied to all types of models.
- *Model specific*: can only be applied to certain models.

Also, the scope of the explanation methods can be categorized into:

- *Global explanations*: here, the whole model is explained, these explanations can apply to population-based decisions such as epidemic outbreak [44].
- *Local explanations*: individual predictions are explained. This type of explanation can be used in surgery to obtain risk factors for complications in surgery. Example of local explanation is the local interpretable model agnostic explanations (LIME). LIME can describe a prediction by measuring the contributing factors associated with obtaining the predictions [102].

In addition, there are other common methods of explanations used for deep NNs (DNN) and are as follows, these methods have been used in the medical domain:

- *Sensitivity analysis*: the most relevant input features are those to which the output is sensitive [57].
- Layer wise relevance propagation explains predictions relative to the state of maximum uncertainty [5].

It is noteworthy to mention that different explanation results can be in a numerical, textual, pictorial, or hybrid format making it easy for stakeholders (clinicians) to understand the results. Furthermore, the input data for the algorithms can also be in any format (numerical, text, image). However, there is a need to evaluate the various available methods to obtain suitable explanations for healthcare domains. Table 6.4 summarizes the different classifications of XAI methods and different input and output data types.

Table 6.4 A summary of all XAI methods and different input and output types

Methods	Explanation types
1. Stage	Ante-hoc: explanation within the model Post-hoc: can further be classified into: • Model agnostic: • Model specific
2. Scope	Global, local
3. Problem type	Classification, regression
4. Input data	Numerical/categorical, Image, Text, Time series
5. Output	Numerical, image, text, hybrid

Different explainability methods can be applied to different types of AI models. Thus, it is essential to determine the accurate evaluation for different XAI methods and check if the explanations are relevant to their various applications. However, these evaluations are still under development [48].

6.4.1 Explanation methods in robotics

Robotics consists of different technologies working together to achieve an objective goal. The robot's actions are sometimes informed by a ML model, some planning algorithms or a reinforcement learning model. As such, for explainability to be achieved, research areas of XAI, explainable AI planning (XAIP), and explainable reinforcement learning (XRL) are important.

Our focus, XAI, focuses on systems finding interpretations for complex pattern recognition models such as deep NNs. Such a model includes a "judgment basis" factor that is not readable by a human, and this factor must be presented in a format that is readable by a human. Such systems may play a role in close human contact work only if humans accept the presented bases of their judgments. XAI research raises questions about how humans perceive such models' reliability and expands the scope of ML applications by presenting the bases of judgments [103].

In day-to-day human activities, there has been the need to simplify the job and other needs to make work easier in our day-to-day journey and routine. Owing to the development of Al techniques, surgical robots can achieve superhuman performance. Al helps boost the capability of surgical robotic systems in perceiving complex in vivo environments, making decisions, and performing the desired tasks with increased precision, safety, and efficiency. Common Al techniques used for robotic systems include the following: perception and human–robot interactions.

Various methods have been used to achieve model explainability; however, this chapter highlights three popular methods used in the industry:

1. Local interpretable model-agnostic explanations (LIME)
2. Layer-wise relevance propagation (LRP)
3. Shapley additive explanations (SHAP)

6.4.2 SHAPs

Shapley values are the concept of the cooperative game theory field where the objective is to measure each player's contribution to the game.

Shapley values emerge from the context where "n" players participate collectively, obtaining a reward "p," which is intended to be fairly distributed to each one of the "n" players according to the individual contribution, and such a contribution is a Shapley value [104]

The SHAP method was introduced for ML model predictions interpretability, and results are obtained through Shapely values. The key idea of SHAP is to calculate the Shapley values for each feature of the sample to be interpreted, where each Shapley value represents the impact that the feature to which it is associated generates in the prediction.

For a model with a prediction function f(x) and M features, we can obtain Shapley values as:

$$\phi_i = \sum_{S \subseteq N \setminus \{i\}} \frac{|S|!(M - |S| - 1)!}{M!} [f_x(S \cup \{i\}) - f_x(S)] \tag{6.1}$$

We sum all possible subsets (S) of feature values in the formula, excluding the ith feature value. |S|! represents the number of permutations of feature values that appear before the ith feature value. (|M|−|S|−1)! represents the number of permutations of feature values that appear after the ith feature value and ∪ is the union of sets.

The different term is the marginal contribution of adding ith feature value to S.

There are different variants of SHAP applied today, and their applications are dependent on the kind of problem posed. Some of the prevalent ones are:

1. *Kernel SHAP*: Kernel Shap is based on a weighted linear regression where the coefficients of the solution are the Shapley values. It utilizes LIME for the calculation of the shapley values.
2. *Tree SHAP*: The algorithm allows the computation of exact SHAP values for decision trees-based models [105,106].
3. *Deep SHAP*: It is a version of the DeepLIFT algorithm (deep SHAP) similar to Kernel SHAP, where the conditional expectations of SHAP values are approximated using a selection of background samples [107].

6.4.3 Layer-wise relevance propagation

Layer-wise relevance propagation (LRP) aims to explain any NN's output in its input domain. This method does not interact with the network's training, so you can easily apply it to already trained classifiers. An example is if your network is aimed at supporting a robot to predict objects captured by the attached cameras, then the explanation given by LRP would be a map of which pixels in the original image contribute to the output or decision.

Algorithmically, LRP uses the network weights and activations created during the forward-pass to propagate the output back through the network up until the input layer. This makes it possible to visualize the pixels that really contributed to the output. The magnitude of the contribution of each intermediate neuron (pixel) "relevance" values is represented as R in the equation below.

$$R_j = \sum\nolimits_k \frac{a_j w_{jk}}{\sum\nolimits_{0,j} a_j w_{jk}} R_j \tag{6.2}$$

From the equation above, j and k are two neurons of any consecutive layers in the NN. a is the activation for the neurons and w denotes the weight between two neurons. With the relevance R in the output layer, new R's are calculated backwards and iteratively for every neuron in the previous layers [104].

6.4.4 LIMEs

The LIMEs approach to model interpretability generates an explanation for a prediction from the components of an interpretable model (e.g., the coefficients in a linear regression model), which is similar to the black-box model at the area of the point of interest and which is trained over a new data to ensure interpretability [101,108].

There are three main ideas of LIME, which are as follows:

(a) *Model-agnosticism*: The LIME model is agnostic (model-independent) because it can explain any model without making assumptions while providing explanations. It treats the model as a black box, so it only has to understand its behavior by perturbing the input and seeing how the predictions change.
(b) *Interpretability*: Explanations must be easy and intuitive to understand by the user, which is not usually the case because the feature space used by the model may use too many input variables or have very complex or artificial variables. LIME can explain those classifiers in terms of interpretable representations (words), even if that is not the representation used by the classifier.
(c) *Locality*: It produces an explanation by approximating the black-box model by an interpretable model in the neighborhood of the instance to be explained [109–111].

6.5 Conclusion

The utilization of computer vision, robotics, and ML technologies in the operating room contributes significantly to improved healthcare delivery. By incorporating ethical considerations in AI and robotic surgery, the need for interpretability, explainability, patient security and safety become imperative. Interestingly, the XAI field addresses all these issues. Thus, XAI in surgery aids surgeons' work by providing real-time explanations to predictions. Since XAI is relatively young, there are relatively few studies on XAI related to robotic surgery. Therefore, there is a need for further studies on the future of XAI in robotic surgery. In addition, scientists must develop proper evaluation methods suitable for different medical domains to integrate XAI methods into robotic surgeries successfully. One of the limitations of the current study is that the physical activity of the robotic system, surgical skill assessment, and simulation studies were not covered. Simulation studies of the explanation methods must be conducted before clinical studies are implemented. However, the chapter focuses more on AI algorithms and explanation methods. Future work will determine surgical skills assessment and the proper incorporation of all aspects of AI. Finally, since XAI and surgical robotics technologies require interdisciplinary knowledge, there is a need for collaboration between surgeons and other stakeholders in medicine, engineering, and computer science.

Acknowledgment

The work of Agbotiname Lucky Imoize is supported in part by the Nigerian Petroleum Technology Development Fund (PTDF) and in part by the German Academic Exchange Service (DAAD) through the Nigerian-German Postgraduate Program under Grant 57473408.

References

[1] Zhang D, Maslej N, Brynjolfsson E, *et al.* "The AI Index 2022 Annual Report," AI Index Steering Committee, Stanford Institute for Human-Centered AI, Stanford University, March 2022.

[2] Adadi A and Berrada M. Peeking inside the black-box: a survey on explainable artificial intelligence (XAI). *IEEE Access*. 2018;6:52138–60.

[3] Matarese M, Rea F, and Sciutti A. A User-Centred Framework for Explainable Artificial Intelligence in Human-Robot Interaction; 2021. arXiv preprint arXiv:2109.12912.

[4] Arrieta AB, Díaz-Rodríguez N, Del Ser J, *et al.* Explainable Artificial Intelligence (XAI): concepts, taxonomies, opportunities and challenges toward responsible AI. *Information Fusion.* 2020;58:82–115. https://doi.org/10.1016/j.inffus.2019.12.012.

[5] Xu F, Uszkoreit H, Du Y, Fan W, Zhao D, and Zhu J. Explainable AI: a brief survey on history, research areas, approaches and challenges. In *CCF International Conference on Natural Language Processing and Chinese Computing* (pp. 563–74). Cham: Springer; 2019.

[6] Royal Society. *Explainable AI: The Basics*; 2019. https://royalsociety.org/-/media/policy/projects/explainable-ai/AIand-interpretability-policy-briefing.pdf

[7] Meske C, Bunde E, Schneider J, and Gersch M. Explainable artificial intelligence: objectives, stakeholders, and future research opportunities. *Information Systems Management*. 2022;39(1):53–63.

[8] Voskens FJ Abbing JR, Ruys AT, Ruurda JP, and Broeders IA. A nationwide survey on the perceptions of general surgeons on artificial intelligence. *Artificial Intelligence Surgery*. 2022;2(1):8–17.

[9] Rajabi E and Etminani K. Towards a knowledge graph-based explainable decision support system in healthcare. *Studies in Health Technology and Informatics*. 2021;281:502–3. doi:10.3233/SHTI210215.

[10] [Online]: https://gdpr-info.eu/art-22-gdpr/

[11] Vilone G and Longo L. Notions of explainability and evaluation approaches for explainable artificial intelligence. *Information Fusion*. 2021;76:89–106.

[12] Sheh R and Monteath I. Introspectively assessing failures through explainable artificial intelligence. In *IROS Workshop on Introspective Methods for Reliable Autonomy* 2017, Vancouver, Canada (pp. 40–7); 2017.

[13] Montavon G, Samek W, and Mller K-R. Methods for interpreting and understanding deep neural networks. *Digital Signal Processing*. 2018;73: 1–15. doi:10.1016/j.dsp.2017.10.011.

[14] Payrovnaziri SN, Chen Z, Rengifo-Moreno P, *et al.* Explainable artificial intelligence models using real-world electronic health record data: a systematic scoping review. *Journal of the American Medical Informatics Association*. 2020;27(7):1173–85. doi:10.1093/jamia/ocaa053.

[15] Dazeley R, Vamplew P, Foale C, Young C, Aryal S, and Cruz F. Levels of explainable artificial intelligence for human-aligned conversational explanations. *Artificial Intelligence*. 2021;299:103525.

[16] Markus AF, Kors JA, and Rijnbeek PR. The role of explainability in creating trustworthy artificial intelligence for health care: a comprehensive survey of the terminology, design choices, and evaluation strategies. *Journal of Biomedical Informatics*. 2021;113:103655.

[17] Rudzicz F and Joshi S. Explainable AI for the operating theater. In Atallah S, ed., *Digital Surgery* (pp. 339–50). Orlando, FL: Springer; 2021.

[18] Manne R and Kantheti SC. Application of artificial intelligence in healthcare: chances and challenges. *Current Journal of Applied Science and Technology*. 2021;40(6):78–89.

[19] Shaheen MY. *Applications of Artificial Intelligence (AI) in Healthcare: A Review*. Berlin: ScienceOpen, Preprints; 2021.

[20] Secinaro S, Calandra D, Secinaro A, Muthurangu V, and Biancone P. The role of artificial intelligence in healthcare: a structured literature review. *BMC Medical Informatics and Decision Making*. 2021;21(1):125. doi:10.1186/s12911-021-01488-9.

[21] Hazarika I. Artificial intelligence: opportunities and implications for the health workforce. *International Health*. 2020;12(4):241–5. doi:10.1093/inthealth/ihaa007.

[22] Chan HCS, Shan H, Dahoun T, Vogel H, and Yuan S. Advancing drug discovery via artificial intelligence. *Trends in Pharmacological Sciences*. 2019;40(8):592–604.

[23] Magrabi F, Ammenwerth E, McNair JB, *et al.* Artificial intelligence in clinical decision support: challenges for evaluating AI and practical implications. *Yearbook of Medical Informatics*. 2019;28(1):128–34. doi:10.1055/s-0039-1677903.

[24] van Baalen S, Boon M, and Verhoef P. From clinical decision support to clinical reasoning support systems. *Journal of Evaluation in Clinical Practice*. 2021;27(3):520–8. doi:10.1111/jep.13541.

[25] Montani S and Striani M. Artificial intelligence in clinical decision support: a focused literature survey. *Yearbook of Medical Informatics*. 2019;28 (1):120–7. https://doi.org/10.1055/s-0039-1677911.

[26] Steels L and Lopez de Mantaras R. The Barcelona declaration for the proper development and usage of artificial intelligence in Europe. *AI Communication*. 2018;31(6):485–94. https://doi.org/10.3233/AIC-180607.

[27] Sikma T, Edelenbosch R, and Verhoef P. The use of AI in healthcare: a focus on clinical decision support system. 2020 [Case-Study in the RECIPES Project: https://recipes-project.eu/sites/default/files/2020-11/D2_3_AI_In_Healthcare%28CDSS%29_HarvardStyle.pdf] Google Scholar.

[28] Lee D and Yoon SN. Application of artificial intelligence-based technologies in the healthcare industry: opportunities and challenges. *International Journal of Environmental Research and Public Health*. 2021;18(1):271.

[29] Ross C and Swetlitz I. IBM's Watson Supercomputer Recommended 'Unsafe and Incorrect' Cancer Treatments, Internal Documents Show. STAT +. 25 July 2018. https://www.statnews.com/wp-content/uploads/2018/09/IBMs-Watsonrecommended-unsafe-and-incorrect-cancer-treatments-STAT.pdf (accessed on 26 June 2022).

[30] Connelly TM, Malik Z, Sehgal R, Byrnes G, Cofey JC, and Peirce C. The 100 most influential manuscripts in robotic surgery: a bibliometric analysis. *Journal of Robotic Surgery*. 2020;14(1):155–65.

[31] Steinfeld A and Zimmerman J. Unremarkable AI: fitting intelligent decision support into critical, clinical decision-making processes. In *Proceedings of the 2019 CHI Conference on Human Factors in Computing Systems* 2019 May 2 (pp. 1–11).

[32] Karimian G, Petelos E, and Evers SM. The ethical issues of the application of artificial intelligence in healthcare: a systematic scoping review. *AI and Ethics*, 2022:1–13.

[33] Morley J, Floridi L, Kinsey L, and Elhalal A. From what to how: an initial review of publicly available AI ethics tools, methods and research to translate principles into practices. *Science and Engineering Ethics*. 2020;26 (4):2141–68. https://doi.org/10.1007/ s11948-019-00165-5

[34] Holzinger A, Langs G, Denk H, Zatloukal K, and Müller H. Causability and explainability of artificial intelligence in medicine. *WIREs Data Mining and Knowledge Discovery*. 2019;9:e1312. https://doi.org/10.1002/widm.1312

[35] Phillips PJ, Hahn CA, Fontana PC, Broniatowski DA, and Przybocki MA. *Four Principles of Explainable Artificial Intelligence*. Gaithersburg, MD; 2020.

[36] Zhang Y, Weng Y, and Lund J. Applications of explainable artificial intelligence in diagnosis and surgery. *Diagnostics (*Basel*)*. 2022;12(2):237. doi:10.3390/diagnostics12020237.

[37] Guidotti R, Monreale A, Ruggieri S, Turini F, Giannotti F, and Pedreschi D. A survey of methods for explaining black box models. *ACM Computing Surveys (CSUR)*. 2018;51(5):1–42.

[38] Rajabi E and Etminani K. Towards a knowledge graph-based explainable decision support system in healthcare. *Studies in Health Technology and Informatics*. 2021;281:502–3. doi:10.3233/SHTI210215.

[39] Khedkar S, Subramanian V, Shinde G, and Gandhi P. Explainable AI in healthcare. In *Healthcare* (April 8, 2019). *2nd International Conference on Advances in Science & Technology* (*ICAST*), 2019.

[40] Holzinger A, Langs G, Denk H, Zatloukal K, and Müller H. Causability and explainability of artificial intelligence in medicine. *Wiley Interdisciplinary Reviews: Data Mining and Knowledge Discovery*. 2019;9(4):e1312.

[41] Gordon L, Grantcharov T, and Rudzicz F. Explainable artificial intelligence for safe intraoperative decision support. *JAMA Surgery*. 2019;154 (11):1064–5. https://doi.org/10.1001/jamasurg.2019.2821

[42] Chen D, Afzal N, Sohn S, *et al.* Postoperative bleeding risk prediction for patients undergoing colorectal surgery. *Surgery*. 2018;164:1209–16. https://doi.org/10.1016/j.surg.2018.05.043.

[43] Lundberg SM, Nair B, Vavilala MS, *et al.* Explainable machine-learning predictions for the prevention of hypoxaemia during surgery. *Nature Biomedical Engineering*. 2018;2(10):749–60.

[44] Adadi A and Berrada M. Explainable AI for healthcare: from black box to interpretable models. In *Embedded Systems and Artificial Intelligence* (pp. 327–37). Singapore: Springer; *2020*.

[45] Fellous JM, Sapiro G, Rossi A, Mayberg H, and Ferrante M. Explainable artificial intelligence for neuroscience: behavioral neurostimulation. *Frontiers in Neuroscience*. 2019;13:1346.

[46] Holzinger A. From machine learning to explainable AI. In *2018 World Symposium on Digital Intelligence for Systems and Machines* (*DISA*) (pp. 55–66); 2018, doi:10.1109/DISA.2018.8490530.

[47] Holzinger A, Stocker C, and Dehmer M. Big complex biomedical data: towards a taxonomy of data. In MS Obaidat and J Filipe, eds, *Communications in Computer and Information Science CCIS 455* (pp. 3–18). Berlin, Heidelberg: Springer; 2014. doi:10.1007/978-3-662-44791-8 1.

[48] Friedman JH. On bias, variance, 0/1loss, and the curse-of-dimensionality. *Data Mining and Knowledge Discovery*. 1997;1(1):55–77. doi:10.1023/A:1009778005914.

[49] Plsek PE and Greenhalgh T. Complexity science: the challenge of complexity in health care. *British Medical Journal*. 2001;323(7313):625–8.

[50] Schoonderwoerd TA, Jorritsma W, Neerincx MA, and Van Den Bosch K. Human-centered XAI: developing design patterns for explanations of clinical decision support systems. *International Journal of Human–Computer Studies*. 2021;154:102684.

[51] Morley J, Machado CC, Burr C, *et al.* The ethics of AI in health care: a mapping review. *Social Science & Medicine*. 2020;260:113172.

[52] Ngiam KY and Khor IW. Big data and machine learning algorithms for healthcare delivery. *Lancet Oncology*. 2019;20(5):e262–73. https://doi.org/10.1016/S1470-2045(19) 30149-4.

[53] Smith C and Brooks DJ. *Security Science: The Theory and Practice of Security*. Oxford: Butterworth-Heinemann; 2012.

[54] Izumo T and Weng YH. Coarse ethics: how to ethically assess explainable artificial intelligence. *AI and Ethics*. 2022:2(3):449–461.

[55] Kalan S, Chauhan S, Coelho RF, *et al.* History of robotic surgery. *Journal of Robotic Surgery*. 2010;4(3):141–7.

[56] Roscher R, Bohn B, Duarte MF, and Garcke J. Explainable machine learning for scientific insights and discoveries. *IEEE Access*. 2020;8:42200–16.

[57] Samek W, Wiegand T, and Müller KR. Explainable Artificial Intelligence: Understanding, Visualizing and Interpreting Deep Learning Models; 2017. arXiv preprint arXiv:1708.08296.

[58] Zeng F, Liang X, and Chen Z. New roles for clinicians in the age of artificial intelligence. *BIO Integration*. 2020;1(3):113–7.

[59] Rubin DL. Artificial intelligence in imaging: the radiologist's role. *Journal of the American College of Radiology*. 2019;16(9):1309–17.

[60] Rickert J. On patient safety: the lure of artificial intelligence – are we jeopardizing our patients' privacy? *Clinical Orthopaedics and Related Research*. 2020;478(4):712–4. doi:10.1097/CORR.0000000000001189.

[61] Puiu A, Vizitiu A, Nita C, Itu L, Sharma P, and Comaniciu D. Privacy-preserving and explainable AI for cardiovascular imaging. *Studies in Informatics and Control*. 2021;30(2):21–32.

[62] Moore W and Frye S. Review of HIPAA, Part 1: history, protected health information, and privacy and security rules. *Journal of Nuclear Medicine Technology*. 2019;47(4):269–72.

[63] Lupton M. Some ethical and legal consequences of the application of artificial intelligence in the field of medicine. *Trends in Medicine*. 2018;18 (4):100147.

[64] Kruse CS, Smith B, Vanderlinden H, and Nealand A. Security techniques for the electronic health records. *Journal of Medical Systems*. 2017;41(8):1–9.

[65] Amann J, Blasimme A, Vayena E, Frey D, and Madai VI. Precise4Q consortium. Explainability for artificial intelligence in healthcare: a multidisciplinary perspective. *BMC Medical Informatics and Decision Making*. 2020;20(1):310. doi: 10.1186/s12911-020-01332-6. PMID: 33256715; PMCID: PMC7706019.

[66] Holzinger A, Biemann C, Pattichis CS, and Kell DB. What do we need to build explainable AI systems for the medical domain?; 2017. arXiv preprint arXiv:1712.09923.

[67] Kyrarini M, Lygerakis F, Rajavenkatanarayanan A, *et al.* A survey of robots in healthcare. *Technologies*. 2021;9(1):8.

[68] Hayda RA, Hsu RY, DePasse JM, and Gil JA. Radiation exposure and health risks for orthopaedic surgeons. *Journal of the American Academy of Orthopaedic Surgeons*. 2018;26(8):268–77.

[69] Yang L, Wang T, Weidner TK, Madura JA, Morrow MM, and Hallbeck MS. Intraoperative musculoskeletal discomfort and risk for surgeons during open and laparoscopic surgery. *Surgical Endoscopy*. 2021;35(11):6335–43.

[70] Hockstein NG, Gourin CG, Faust RA, and Terris DJ. A history of robots: from science fiction to surgical robotics. *Journal of Robotic Surgery*. 2007;1(2):113–8.

[71] Leal Ghezzi T and Campos Corleta O. 30 years of robotic surgery. *World Journal of Surgery*. 2016;40(10):2550–7.

[72] [Online]: https://www.hopkinsmedicine.org/health/treatment-tests-and-therapies/hip-replacement-surgery (accessed 19 March 2022).

[73] Lanfranco AR, Castellanos AE, Desai JP, and Meyers WC. Robotic surgery: a current perspective. *Annals of Surgery*. 2004;239(1):14.

[74] Bacá I, Schultz C, Grzybowski L, and Götzen V. Voice-controlled robotic arm in laparoscopic surgery. *Croatian Medical Journal*. 1999;40:409–12.

[75] Ozmen MM, Ozmen A, and Koç ÇK. Artificial intelligence for next-generation medical robotics. In *Digital Surgery* (pp. 25–36). Cham: Springer; *2021*.

[76] Abdul-Muhsin H and Patel V. History of robotic surgery. In *Robotics in General Surgery* (pp. 3–8). New York, NY: Springer; *2014*.

[77] Falcone T, Goldberg J, Garcia-Ruiz A, Margossian H, and Stevens L. Full robotic assistance for laparoscopic tubal anastomosis: a case report. *Journal of Laparoendoscopic & Advanced Surgical Techniques*. 1999;9(1):107–13.

[78] Satava RM. Robotic surgery: from past to future—a personal journey. *Surgical Clinics*. 2003;83(6):1491–500.

[79] George EI, Brand CT, and Marescaux J. Origins of robotic surgery: from skepticism to standard of care. *Journal of the Society of Laparoendoscopic Surgeons*. 2018;22(4):e2018.00039.

[80] Kasina H, Bahubalendruni MR, and Botcha R. Robots in medicine: past, present and future. *International Journal of Manufacturing, Materials, and Mechanical Engineering*. 2017;7(4):44–64.

[81] [Online]: https://www.medicaldevice-network.com/comment/robotics-in-medical-2021-regulatory-trends/ (accessed 23 March 2022).

[82] [Online]: https://www.greenlight.guru/blog/fda-clearance-approval-granted (accessed 23 March 2022)

[83] Beasley RA. Medical robots: current systems and research directions. *Journal of Robotics*. 2012;2012:1–14. 1-s2.0-S2405456921001127-main.pdf

[84] Kalis B, Collier M, and Fu R. 10 promising AI applications in health care. *Harvard Business Review*. 2018:1–5.

[85] Schroerlucke SR, Wang MY, Cannestra AF, *et al.* Complication rate in robotic-guided vs fluoro-guided minimally invasive spinal fusion surgery: report from MIS refresh prospective comparative study. *The Spine Journal*. 2017;17(10):S254–5.

[86] Hashimoto DA, Rosman G, Rus D, and Meireles OR. Artificial intelligence in surgery: promises and perils. *Annals of Surgery*. 2018;268(1):70.

[87] Burkov A. *The Hundred-Page Machine Learning Book*. Quebec City, QC: Andriy Burkov; 2019.

[88] Kennedy-Metz LR, Mascagni P, Torralba A, *et al.* Computer vision in the operating room: opportunities and caveats. *IEEE Transactions on Medical Robotics and Bionics*. 2021;3(1):2–10. doi:10.1109/tmrb.2020.3040002.

[89] Olveres J, González G, Torres F, *et al.* What is new in computer vision and artificial intelligence in medical image analysis applications. *Quantitative Imaging in Medicine and Surgery*. 2021;11(8):3830–53. doi:10.21037/qims-20-1151.

[90] Hao T, Huang Z, Liang L, Weng H, and Tang B. Health natural language processing: methodology development and applications. *JMIR Medical Informatics*. 2021;9(10):e23898. doi:10.2196/23898.

[91] Soguero-Ruiz C, Hindberg K, Rojo-Alvarez JL, *et al.* Support vector feature selection for early detection of anastomosis leakage from bag-of-words in electronic health records. *IEEE Journal of Biomedical and Health Informatics*. 2016;20(5):1404–15. doi:10.1109/JBHI.2014.2361688.

[92] Yang GR and Wang XJ. Artificial neural networks for neuroscientists: a primer. *Neuron*. 2020;107(6):1048–70.

[93] Fritz BA, Chen Y, Murray-Torres TM, *et al.* Using machine learning techniques to develop forecasting algorithms for postoperative complications: protocol for a retrospective study. *BMJ Open*. 2018;8(4):e020124.

[94] Stam WT, Goedknegt LK, Ingwersen EW, Schoonmade LJ, Bruns ER, and Daams F. The prediction of surgical complications using artificial intelligence in patients undergoing major abdominal surgery: a systematic review. *Surgery*. 2021;171:1014–21.

[95] Hindin D. Artificial intelligence and machine learning: implications for surgery. In *Digital Surgery* (pp. 311–7). Cham: Springer; *2021*.

[96] Lane T. A short history of robotic surgery. *Annals of the Royal College of Surgeons of England*. 2018;100(6_sup):5–7. doi: 10.1308/rcsann.supp1.5.

[97] Beyaz S. A brief history of artificial intelligence and robotic surgery in orthopedics & traumatology and future expectations. *Joint Diseases and Related Surgery*. 2020;31(3):653.

[98] Xue B, Li D, Lu C, *et al.* Use of machine learning to develop and evaluate models using preoperative and intraoperative data to identify risks of postoperative complications. *JAMA Network Open*. 2021;4(3):e212240.

[99] Ma R, Vanstrum EB, Lee R, Chen J, and Hung AJ. Machine learning in the optimization of robotics in the operative field. *Current Opinion in Urology*. 2020;30(6):808.

[100] Navarrete-Welton AJ and Hashimoto DA. Current applications of artificial intelligence for intraoperative decision support in surgery. *Frontiers of Medicine*. 2020;14(4):369–81.

[101] Ribeiro MT, Singh S, and Guestrin C. "Why should I trust you?" Explaining the predictions of any classifier. In *Proceedings of the 22nd ACM SIGKDD International Conference on Knowledge Discovery and Data Mining*, 2016 August 13 (pp. 1135–44); 2016.

[102] Pawar U, O'Shea D, Rea S, and O'Reilly R. Explainable AI in healthcare. In *2020 International Conference on Cyber Situational Awareness, Data*

Analytics and Assessment (*CyberSA*) 2020 June 15 (pp. 1–2). New York, NY: IEEE; 2020.

[103] Sakai T and Nagai T. Explainable Autonomous Robots: A Survey and Perspective (Version 1); 2021. arXiv. https://doi.org/10.48550/ARXIV.2105.02658

[104] Binder A, Montavon G, Bach S, Müller K-R, and Samek, W. Layer-wise relevance propagation for neural networks with local renormalization layers (Version 1); 2016. arXiv. https://doi.org/10.48550/ARXIV.1604.00825

[105] Shapley, LS A value for n-person games. In *Contributions to the Theory of Games* (*AM-28*), Volume II (pp. 307–18). Princeton: Princeton University Press; 1953. https://doi.org/10.1515/9781400881970-018

[106] Lundberg SM, Erion GG, and Lee S-I. Consistent Individualized Feature Attribution for Tree Ensembles (Version 3); 2018. arXiv. https://doi.org/10.48550/ARXIV.1802.03888

[107] Yang J. Fast TreeSHAP: Accelerating SHAP Value Computation for Trees (Version 2); 2021. arXiv. https://doi.org/10.48550/ARXIV.2109.09847

[108] Shrikumar A, Greenside P, and Kundaje A. Learning Important Features Through Propagating Activation Differences; 2017. arXiv. https://doi.org/10.48550/ARXIV.1704.02685

[109] Lundberg S and Lee, S-I A Unified Approach to Interpreting Model Predictions (Version 2); 2017. arXiv. https://doi.org/10.48550/ARXIV.1705.07874

[110] Kumar RL, Wang Y, Poongodi T, and Imoize AL, eds. *Internet of Things, Artificial Intelligence and Blockchain Technology*, 1st ed. Switzerland AG: Springer Nature; 2021.

[111] Ramasamy LK, Khan KPF, Imoize AL, Ogbebor JO, Kadry S, and Rho S. Blockchain-based wireless sensor networks for malicious node detection: a survey. *IEEE Access*. 2021;9:128765–85, doi: 10.1109/ACCESS.2021.3111923

Chapter 7

Prediction of erythemato squamous-disease using ensemble learning framework

Efosa Charles Igodan[1], Olumide Olayinka Obe[2], Aderonke Favour-Bethy Thompson[3] and Otasowie Owolafe[3]

Abstract

The erythemto-squamous (skin) disease is characterized by redundant and noisy features. One of the biggest challenges in the artificial intelligence field has been finding relevant features for the target concept. This is a result of similarities among the six classes in the dataset. From the literature, most studies focus mainly on building models on one-phase combined feature selection methods. This paper assesses the performance of models derived from machine learning techniques experimentally using ensemble feature selection techniques. The skin dataset was evaluated using chi-squared, information gain, gain ratio, and relief F as the filter-based features selection methods and RFE-, PRIFEB-, and MIFEB-based on SVMs as the embedded feature selection methods in determining distinctive feature subsets. Then, a variety of classification algorithms have been used to create models that are then compared to seek the optimal feature combinations that produce model performance. The experimenter results show that our proposed stacking models outperform other models in terms of accuracy and applicability.

Keywords: Filter method; Embedded method; Ensemble method; SVMs, Stacking ensemble, Skin disease

7.1 Introduction

The skin on a human body measures 1.85806 square meters (20 sq. ft. area). When and where it is necessary, the skin controls body temperature and shields the body from cold, heat, and sundry diseases. Skin disease is a condition that can affect the

[1]Department of Computer Science, Faculty of Physical Sciences, University of Benin, Nigeria
[2]Department of Computer Science, Faculty of Computing, Federal University of Technology, Nigeria
[3]Department of Cyber Security, Faculty of Computing, Federal University of Technology, Nigeria

skin due to environmental or genetic factors. Though classified into six classes, its identification and diagnosis can be challenging due to shared clinical characteristics with minor variations inherent in all the classes [1–3]. It is important to recognize skin disease at early stage [4] whether it is fungal, allergic, or viral in order to stop the spread. To determine the parameters of skin disease, first, a dermatologist examines 12 clinical features and if a symptom is present, 22 histopathological features are then microscopically analyzed. Particularly, dermatologists have a difficult time treating and diagnosing skin diseases because the symptoms may overlap with each other [3]. Many times, different combinations of these techniques are used for classification, prediction, and diagnosis of various medical diseases since the advent of artificial intelligence and its subfields were introduced to the field of medicine [5]. These learning algorithms have been used to treat a variety of conditions irrespective of the data size, including skin disease, kidney disease, lung cancer, breast cancer, and many more [6–11]. However, high latitude data with redundant, irrelevant, and noisy features is one of the factors limiting the quality of the data. Due to the fact that instances with numerous irrelevant features provide very little information [10], these features can negatively impact the performance of prediction models [12]. The need to choose the optimal feature subsets for high performance is therefore driven by these constraints.

Ensemble and feature selection methods are the two most recent machine learning research areas frequently adopted to enhance the generalization performance of single machine learning [7,9,13,14]. The concept is that combining the output of multiple experts is superior to the output of an expert [15]; also it increases the accuracy and diversity of the base model. Bol'on-Canedo *et al.* [15,16] used feature selection as a technique for creating diversity in classification ensembles. Diversity was integrated into this case as an objective in the search for the best feature subsets. Techniques for feature selection are categorized using two different criteria: first categorized as supervised, unsupervised, or semi-supervised. Second, divided into the filter, wrapper, embedded method [17–19], hybrid method [12,20], and ensemble method [1] depending on how the modeling and selection algorithms is combined. The ensemble multiple-filter-multiple-embedded feature selection (EMFMEFS) method is presented as an ensemble framework in our work. The technique combines the four filter-based methods; information gain (IG), gain ratio (GR), Chi-square, and Relief F; and three embedded methods; Recursive Feature Elimination for Support Vector Machine (RFE-SVM) [21], Prediction Risk-based Feature Selection for Bagging (PRIFEB) of SVMs, and Mutual Information-based Feature Selection for Bagging (MIFEB), to choose both descriptive and informative features. While the filter methods leverage the descriptive information of the features and are autonomous of the classifier, the embedded method uses the SVM classifier to evaluate the feature subset importance. Based on distance, dependency, and information measurements, the intrinsic attributes of features are examined and graded [6,7,22]. In explainable artificial intelligence (XAI), the methods used to explain these models are either local or global in scope. The global method seeks to explain overall model predictions comprehensively from a top-to-bottom approach, i.e., it provides an understanding of how the structures and parameters of the model make predictions. The local method explains how a specific sample is mapped to its output by providing an understanding of how the model arrived at its prediction. Furthermore, these methods are categorized in stages as pre-model, intrinsic, post-hoc, and Specific/Model Agnostic when applied before, during, or after predictions. The pre-models only apply to data, i.e., independent of the model. It follows that they can only take place prior to model selection because it is crucial to investigate and comprehend the data before considering a model. However, accuracy suffers as a result of natural explainability. The intrinsic interpretability methods, such as decision trees, generalized linear, logistic, and clustering models, are self-explanatory models that take advantage of internal structure to provide natural explainability. The post-hoc interpretability methods are a group of methods that can be used with any trained black-box model without having to comprehend its internal workings. By resolving relationships between input samples and predictions, they offer explanations for the global or local behavior of models. They work with intrinsic models as well. The majority of pre- and post-hoc (models) explainability techniques are model-agnostic in the sense that they can be used with a diverse range of models. Some, particularly deep neural networks, are model-specific and only apply to a particular group of models (e.g., CNNs). Model-specific methods are superior to model-agnostic methods because they make use of the model's unique properties or architecture to increase explainability in ways that model-agnostic methods might not be able to. Some of the explainability techniques used include principal component analysis (PCA), Shapley Adaptive Explanations (SHAP), Local Interpretable Model-Agnostic Explanations (LIME), and Partial Dependence Plots (PDP) [23,24].

However, this study does focus on some of the models for XAI applications but on the adoption of some of the intrinsic and complex models [23] only. The objectives of feature selection are to decrease analysis time, eliminate redundant features, decrease dimensionality, improve model interpretability, and increase predictive accuracy [22,25–28]. In this study, Naive Bayes (NB), support vector machines (SVMs), K-nearest neighbor (KNN), Decision trees (DT), and Multilayer Perceptron (MLP), as well as a meta-classifier (logistic regression) as some of the well-known machine learning algorithms [29] are adopted and stacked ensemble classifiers. The performance of our proposed approach is assessed using the benchmark dataset of erythemato-squamous (skin disease), with 34 features.

According to the literature, the combination of feature selection techniques aids in the discovery of a stable feature subset that enhances the predictive accuracy of the majority of classification models, making them easier to understand [30]. Few ensemble feature selection methods are used in the work of most authors, who generally support ensemble methods because they outperform single classifiers. This study aims to add to knowledge by developing an ensemble of multi-filter and embedded feature selection methods (EMFE-FS). Following are the remaining details of our study. Related work is presented in the second part. The third part describes the proposed EMFMEFS method, and part four presents the classification algorithms and benchmark dataset. The fifth part discusses the results of our experiments while part six summarizes the work.

7.2 Related literature review

Artificial Intelligence (AI) has lately involved abundant considerations and increasingly being adopted to predict various medical-related diseases. Verma and Pal [2] developed a predictive model for skin disease with three features selection techniques using stacking ensemble methods to obtain 98.64% accuracy. Verma *et al.* [3] built a classification model for skin disease using ensemble methods. In Ref. [12], a diagnosis system for skin cancer using CNN-ensemble with random forest was designed by applying three filtering methods to achieve 90.7% accuracy. Skin disease detection using association rule-based a priori algorithms, fuzzy logic, and ensemble methods was developed in Bose *et al.* [31]. Other related works are shown in Table 7.1.

Regardless of the methodology, the survey's executive summary identifies gaps in the methodologies used as feature selection approaches. Majority of the literature applied either a single selection approach using filter-based, wrapper-based, embedded-based methods, or combination of filter-embedded or filter-wrapper approaches. However, to the best of our knowledge, none has applied the filter-embedded ensemble of feature selection methods in layers as done in our study. Also, the issue of overfitting in most of the works was not considered and addressed. Lastly, some of the features selected were descriptive but not informative while some selected were informative but not descriptive. These limitations noticed affected the performances of their works. This study combines the advantages of both filter and embedded feature selection approach on six base classifiers and stacking ensemble and applied the logistic regression to address the overfitting problem in the ensemble learning framework.

7.3 Materials and methods

A description of the research's methodology in four steps is contained in this section. The block diagram in Figure 7.1 describes the proposed methodology. Data gathering is the first step while attribute selection process in step two. The feature selection method is broken down in two steps: the multi-embedded-based feature selection method and the multi-filtering-based attribute selection [44]. Following that, ensemble classifiers by bagging, boosting, and stacking are used to create classification models using SVM, KNN, DT, NB, and MLP. The Theorem, "No Free Lunch," states that every algorithm performs well equally when its performance is averaged across all potential problems [9]. Nevertheless, this does not imply that all hope is lost because understanding the underlying issue, the available data, and the surrounding circumstances can guide the development of more effective solutions [24]. An ensemble model, which outperforms individual base classifiers, reduces the variance of error estimation by combining multiple sets of classifiers [45]. The models are evaluated using a dataset of skin diseases in the final step, and the top ensembles are determined by comparing bagging, boosting, and stacking methods. The best ensemble is selected based on prediction accuracy after comparing the results of the ensemble methods. As a result, in this research, we employed the feature ensemble and classifier ensemble learning techniques referred to as the ensemble learning framework in this study.

Table 7.1 Related literature review

S/N#	Year	Classifiers	Feature selection methods	Model accuracy (%)	Limitation
[32]	2000	P A C, L D A, R N C, BNB, GNB, ETC and Bagging, AdaBoost, and GBC.	Feature importance method (+DT)	99.68	Time complexity No feature selection method.
[33]	2011	SVM, EHO	–	98.61–99.07	No feature selection
[4]	2014	Bagged tree ensemble, KNN, SVM, ResNet50, VGG16, and GoogleNet	GLCM and its statistical information	ML classifier: 83–93, DL classifier: 58–69	No feature selection
[34]	2016	ABC, FELM	–	99.57	Time complexity
[35]	2017	ResNet, InceptionV3, DenseNet, Inception ResNetV2 and VGG-19 using TL	–	98–98.6 majority and weighted voting	No feature selection
[5]	2017	C5.0, CDT, CART, RF, Forest-PA RF, RoF, GBC, XGB, C-Forest, AdaBoost. DL, MLP, DNN with a SAEDNN, LSVM. RIP, PART & OneR	Multi-round cross-validation, sub-sampling, and cross-validation	Improved performance	No feature selection
[36]	2019	SVM, GNB, DT, LR, and stacking ensemble	Chi-squared, decision importance, Heat map	SVM: 97.3 Stacked ensemble 99.8	Increased time
[3]	2019	GNB, KNN, DT, SVM, RF and MLP	Chi-squared, information gain, and a principal component	Single classifier—97.3–96.0, Ensemble: 95.94, 97.70 and 99.67	Filter-based feature selection methods affected modeling
[2]	2020	CART, SVM, DT, RF, and GBC (+DT), and ensemble method	–	98.64	No feature selection
[37]	2021	Optimal path forest classifiers	Pearson correlation coefficient, GR, information gain, re-lief F, PCA (+greedy-stepwise, best-first search)	94.3–96.7	Computational complexity

(Continues)

Table 7.1 (Continued)

S/N#	Year	Classifiers	Feature selection methods	Model accuracy (%)	Limitation
[31]	2021	Association rule-based a priori algorithm, FL, Real, Gentle, AdaBoost, SVM, and modest boosting variants		99.3	Overfitting problem
[12]	2021	RF	Chi-squared, information gain and Pearson correlation coefficient	91	Only considered filter-based feature selection methods
[38]	2021	SVM-RFE, DT, KNN, NB, and neural networks	Chi-square	96.70–100	One filter technique No EA considered.
[39]	2021	SVM, RF, stacking ensemble, KNN and NB	Info gain and NB as wrapper approach (+Best first, Greedy stepwise, and Ran k search)	93.8–100	One filter method
[40]	2021	Res Net50, Res NeXt50, Res NeXt101, EfficientNet-B4, Mo bileNetV2, Mo bileNetV3-Large and Minas Net	Image balancing, augmentation, and normalization	Improved models	Time complexity
[41]	2022	S V M, K NN, L R, R F, ET, G B DT, X-GBoost, LightGB M, C atBoost, and M L P. Stack ensemble	Embedded GBDT + Pearson correlation + SHAP	89.86	Few dataset Poor performance
[42]	2022	KNN, DT, LR, NB, SVM	–	92.30	No feature selection Poor performance
[43]	2022	NN, SVM and KNN	Pearson correlation matrix	93	Only one feature selection method adopted

S/N#: serial number; PAC, passive-aggressive classifier; LDA, linear discriminant analysis; RNC, radius neighbor classifier; BNB, Bernoulli Naïve Bayes; GNB, Gaussian Naïve Bayesian; ETC, extra tree classifier; GBC, gradient boosting classifier; SVM, support vector machine; EHO: elephant herding optimization algorithm; KNN, k-nearest neighbor; DT, decision tree; GLCM, grey level co-occurrence matrix; ABC, artificial bee colony; FELM, fuzzy extreme learning machine; TL, transfer learning; CDT, credal decision tree; RF, Random Forest; RoF, Rotation Forest; DNN, deep neural network; XGB: extreme gradient boosting machine; SAEDNN: stacked auto-encoder deep neural network; RIP, requested incremental pruning; PART, partial decision tree; LR, logistic regression; MLP, multilayer perceptron; FL, fuzzy logic; RFE, recursive feature elimination; NB, Naïve Bayes; SHAP: SHarpley Additive exPlanation.

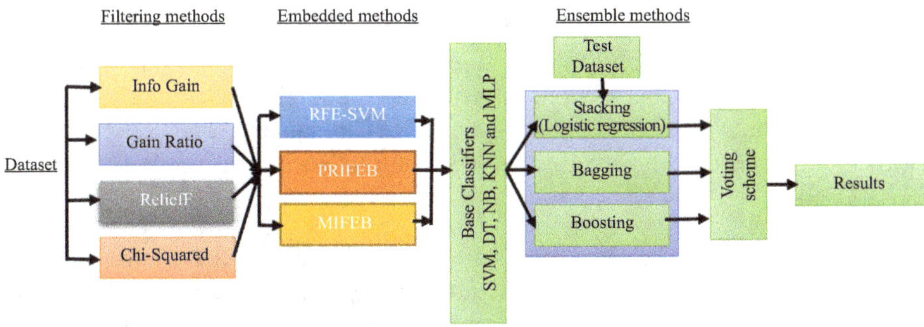

Figure 7.1 A proposed methodology for ensemble data mining and feature selection method

7.3.1 Data collection

The collection of the data from the UCI repository dataset [46] is the first step of this study. The data description is given in [2,3,36,46]. The experimenter platform configurations were done on a processor with 64 bit Operating System, Windows 10 Pro, Core^{TM} i5, 1.9 GHz Intel®. Python 3.7 notebook was used as the implementation language.

7.3.2 Dataset analysis

According to [2,3], the dataset for skin diseases contains 6 classes, 22 histopathological features, and 12 clinical features. Age is a nominal attribute used in clinical terminology. The following terms define the feature set:

$$\textit{Family history}\ (f11) = \left\{ \begin{matrix} 1 \text{ if disease is in family} \\ 0 \qquad\qquad \text{otherwise} \end{matrix} \right\} \tag{7.1}$$

$$\textit{Other attributes} = \left\{ \begin{matrix} 0 & \text{no disease found} \\ 1,2 & \text{disease within limit} \\ 3 & \text{high value} \end{matrix} \right\} \tag{7.2}$$

7.3.3 Feature selection

The attribute selection process, which is a crucial part of machine learning, is typically a critical preprocessing activity toward high performance in modeling. Because the optimal attributes aid inductive learners in enhancing their capacity for better generalization, improved learning speed and induced model simplicity. If a model is simple, it can be interpretable and explainable too [23]. However, while only a few researchers focused on ensemble and hybrid methods, many others concentrated their efforts either on the filter, wrapper, or embedded methods only. In [25,47], the justification for using the hybrid and ensemble feature selection methods is emphasized. Ensemble of multi-filter and embedded-based feature selection methods is adopted in this study.

7.3.4 Multi-filter-based feature selection approach

Using Algorithm 7.1, our proposed ensemble of multi-filter feature selection methods is described in this section. These methods combine the output of the four filter selection methods in formation gain, GR, Chi-squared, and Relief F. The justification for using the four filter-based feature selection methods is found in [7,10,47–49].

- *Information gain*: The top features are chosen by ranking based on the information theory. When the feature is present, the entropy decreases. When the feature is absent, the entropy increases. The information gained can be calculated as follows:

$$IG(X|Y) = H(X) \quad - \quad H(X|Y) \tag{7.3}$$

$$H(X|Y) = \sum_j P(y_j) \sum_i P(x_i|y_i) log_2 P(x_i|y_i) \tag{7.4}$$

$$H(X) = -\sum_i P(x_i) log_2 (P(x_i)) \tag{7.5}$$

  Note $P(x_i)$ denotes the value of prior probabilities of X, $P(x_i|y_i)$ is the posterior probabilities of X given Y.

- *GR*: Due to the bias inherent in information gain toward attributes with larger number of different values, the GR is used instead to calculate the ratio of the information gain of a certain attribute to its split information (or intrinsic value) [50]. Eqs (7.6) and (7.7) are used to compute the gain ratio of a given feature x with class value of y:

$$GR(y,x) = \frac{information\ gain\ (y,x)}{intrinsic\ value\ (x)} \tag{7.6}$$

$$Intrinsic\ value(x) = -\sum \frac{|S_i|}{|S|} * \log_2 \frac{|S_i|}{S} \tag{7.7}$$

  Note $|S|$ represents the number of possible values attributes x can take, while $|S_i|$ is the number of actual values of attribute x.

- Relief F: Based on the differences in feature values and target values between neighboring instances, the score values of the features are calculated. The Relief F decreases the features' weight if a set of neighboring instances have different values for a feature but the same target value. But, Relief F increases the feature's weight, if neighboring instances have different values for a feature and different target value. This process is repeated for a set of sampled instances and their neighbors to calculate an overall score for each feature [51]. The weight is updated using E. (7.8):

$$W = \sum ((x - Miss) - (x - Hit)) \tag{7.8}$$

where x represents the feature value, Hit and Miss represents the feature value of a nearest neighbor with the same class and the feature value of a nearest neighbor with the opposite class value of x respectively.

- Chi-square (x^2): The Chi-square x^2 statistic measures feature significance between the feature and the class label. High Chi-square scores are selected in the new dataset. Eq. (7.9) is used to calculate the Chi-square score [52–55]:

$$x^2 = \sum_{i=1}^{x} \sum_{j=1}^{y} \frac{\left(A_{ij} - E_{ij}\right)^2}{E_{ij}} \tag{7.9}$$

The observed and the expected values are represented in A_{ij} and E_{ij} whilst y is the number of class labels, x is the attribute value.

However, each method has its limitations which can only be overcome by the combination of two or more methods to maximize their ability to achieve higher classification performance [56]. Algorithms 7.1 and 7.5 illustrate the logic flow of the combined methods. The combination of the embedded methods improves the diversity of individuals and creates an ensemble method that builds a more flexible and robust model exhibiting accurate results using few feature subsets [25,47,48,57]. Some of the merits and demerits of feature selection methods are included in [47,48,56,57].

Algorithm 7.1 Multiple filter-based feature selection

Input: Training set T, $nF = \{\text{Ing}, \text{Grat}, \text{Chs}, \text{ReF}\}$, Count Threshold $= cT \geq 3$, and *Rn* ranking

Output: Final features Subset, *sF*

Procedure:

1. For each n from 1 to *nF* do
2. Obtain ranking *Rn* using (feature selection) method n
3. End
4. For each n from 1 to *Rn* do
5. Select two-third split F_{split} of each method
6. End
7. cF = Combine all selected F_{split}
8. sF = using majority vote, select final subset of features with $cT \geq 3$

7.3.5 Multi-embedded-based feature selection approach

The PREFEB, RFE-SVM, and MIFEB are embedded feature selection models. The PREFEB and REF-SVM use the prediction risk criteria and recursive feature elimination criteria, which combine feature selection with the bagging of SVMs, to improve the overall performance of single learning algorithms. The MI-FEB uses the mutual information criteria employed to the bagged SVM in order to demonstrate that "many could be better than all."

- *RFE-SVMs*: In the SVM recursive feature selection method, the SVM classifier is used to rank the features recursively and discard the least informative ones. By reranking the features after each iteration according to their contribution to the SVM classifier, this method recursively removes features [21,27,58]. The classification boundary equation for linear SVMs is:

$$\widehat{y}(x)=w^T.x+b \tag{7.10}$$

where sign$[\widehat{y}(x)]$ is the expected class feature vector (x), w is the weight vector, and b is a constant.

The process of SVM training generates a set of (N_{train}) parameters, where (N_{train}) is the number of training points, from a training set of feature vectors and the class labels that correspond to them. The weight vector w in the classification boundary equation (7.11) can be calculated using the parameters $\{\alpha_i\} = \{\alpha_1, \ldots, \alpha_{N_{train}}\}$.

$$w=\sum_{i=1}^{N_{train}} \alpha_i y_i x_i \tag{7.11}$$

therefore, the ranking criterion is given as:

$$c_j=(w_i)^2 \tag{7.12}$$

After training an SVM with the entire set of features for each feature j, the SVM-RFE uses Eqs. (7.11) and (7.12) to determine the ranking criterion values $c\,j$. In the resulting ranked feature set, the feature with the lowest cj value is removed from the training set and positioned at the bottom. The process is repeated until the training list is devoid of any features, with the final training feature being used to retrain the SVM. The entire feature set is now organized logically in the ranked features list that follows. The top m features for classification are selected from an ordered feature set that is the output of the SVM-RFE algorithm. The general structure of the linear SVM-RFE is shown in Algorithm 7.2.

Algorithm 7.2 RFE-SVMs

Inputs: Training Samples $\{x_i y_i\}$
Output: Ranked features list R
Initialize: $S = \{1, 2, \ldots, D\}, \quad R = \varnothing$
While S is not e m p t y, do:

Restrict the features of X_j to the remaining S

Calculate weight vectors after training SVM

Compute the ranking criteria $c_k = w_k^2,\ k = 1, \ldots, |S|$

Obtain features with the least value of c_k, referred to as feature p

Add feature p into $R(R = (\{p\}UR)$

Remove feature p from $S\left(S = \frac{S}{p}\right)$

- Prediction risk-based feature selection for bagging (PRIFEB): The prediction risk criteria are used to rank and select features in the embedded feature selection model known as PRIFEB. The PRIFEB was proposed by the authors in [59], and it measures the prediction error of the datasets when the values of all instances of a feature are replaced by their average values [57]. That is, PRIFEB searches for the best subsets at each increasing cardinality using the sequential forward search approach [60]. Algorithm 7.3 provides an overview of the PRIFEB methodology.

$$S_i = Err_{Test}\left(\bar{x}^i\right) - Err_{train} \tag{7.13}$$

where the training and testing errors are represented as Err_{Train} and Err_{Test} and the ith feature is defined as:

$$Err_{Train}\left(\bar{x}^i\right) = \frac{1}{l}\sum_{j=1}^{1}\left(\bar{y}(x_j^1, \ldots, x_j^1, \ldots, x_j^D)\right) \neq y_j \tag{7.14}$$

where $\bar{y}(\ )$ is the prediction value of the jth example after the value of the ith feature is replaced by its mean value and $\bar{x}^i$ is the mean value of the ith feature. l and D are the numbers of examples S_i and feature respectively.

Algorithm 7.3 PRIFEB method

Input: Training set $T_r(x^1, x^2, \ldots, x^D, C)$, *number of individuals* T, T_{rk} is ¾ of T_r, Individual model L_k, prediction risk value = R_i, optimal features = $T_{rk-optimal}$

Output: Ensemble model N,

Procedure:

For k = 1: T

Bootstrap training set T_r to create subset T_{rk}

Train L_k on $\boldsymbol{T_{rk}}$ using Eq. (7.16) and compute Err_{Train}.

Compute R_i using Eq. (7.11)

If $R_i > 0$, then

  ith feature is selected as one of the optimal features.

Repeat until features in = T_{rk}.

Create $T_{rk-optimal}$ features from T_{rk}

Train the individual model N_k on the $T_{rk-optimal}$ using Eq. (7.16)

End

Apply majority voting to obtain Ensemble classifier.

- The mutual information-based feature for bagging (MIFEB): The MIFEB uses the mutual information (MI) criteria with the bagging of SVMs, in contrast to the previously mentioned embedded feature selection model, which relies on the learning machine using the prediction risk criteria and the recursive feature elimination criteria. The information-theoretical measure (MI) is used to describe how statistically two random features depend on how much information each feature knows about the other. The MI between two features R and S can be defined as follows:

$$I(R:S)=\sum_{r\in R}\sum_{s\in S}P\left\{r,s\right\}log\frac{P\{r,s\}}{P\{r\}.P\{s\}} \tag{7.15}$$

where $P\{r\}$ and $P\{s\}$ capture the individual probability distribution of intensities; $P\{r,s\}$ represents the combined probability distribution of intensities of two features R and S. Algorithmic MIFEB is depicted in Algorithm 7.4 as follows.

Algorithm 7.4 MI-FEB approach

Input: Training set $T_r(x^1,x^2,\ldots,x^D,C)$, *number of individuals* T, T_{rk} is ¾ of T_r.
Output: Ensemble model N
Procedure:
For k = 1: T
Bootstrap training set T_r to create subset T_{rk}
Apply Eq. (7.1 3) on the training subset T_{rk} and obtain the value vector
Rank the vector in descending order
Compute sum(MI)
Select all features greater than RMI* sum(MI), where RMI is a predefined ratio in the range {0,1}
Generate optimal features $T_{rk-optimal}$ features from T_{rk} according to the optimal features obtained.
Train model N_k on the $T_{rk-optimal}$ features using Eq. (7.16)
End
Apply majority voting to obtain Ensemble classifier.

7.3.6 *An ensemble multi-feature selection (EMFME-FS) approach*

The output of the multi-filter-based selection method described in Algorithm 7.1 is used with the individual ensemble methods in Algorithms 7.2, 7.3, and 7.4, respectively, in our suggested EMFME-FS approach. Before using the multi-embedded methods, which ultimately choose the best optimal feature subsets prior to using a learning algorithm, the multi-filter method is a first step pre-processing phase. Prior to choosing a two-thirds split of the ranked features, the four filter methods are used to rank the feature set of the original datasets to create a mutually exclusive subset (i.e., 22 features). At this point, these features are regarded as high descriptive features with respect to each filter method. Calculating the simple majority vote combining all filter methods yields the final single feature subset output of the multi-filter methods. Following each embedded feature selection algorithm, the feature subset obtained from the multi-filter methods is then used for classification using the bagging, boosting, and stacking ensemble methods as described in Algorithms 7.2, 7.3, and 7.4. Algorithm 7.5 provides an illustration of the EMFME-FS strategy.

Algorithm 7.5 Ensemble multi-feature selection approach (EMFME-FS)

Input: Training set $S_{rk}(x^1, x^2, \ldots, x^k, C)$, Em=REF$_S$VM, PRIFEB, and MIFEB
Output: Ensemble classifier N with metrics performance
Procedure:
From Algorithm 7.1
For Em=REF_SVM, PRIFEB, and MIFEB
    Perform ensemble learning
Compute performance evaluation
Determine the best model

7.3.7 Machine learning classifiers

Five different machine learning classifiers, including logistic regression as a m eta-classifier. Due to the various ensemble methods used in the study, the classifiers were chosen as a combination of homogeneous and heterogeneous classifiers.

- SVMs: The SVM identifies hyperplane that can maximize the margin between different classes in a case of binary (2) classification or multi-class (>2) scenario using one-to-one or one-to-many approach. The hyperplane is described in Eq. (7.8) above. Using the Lagrange multiplier techniques, the dual problem of the objective is shown in Eqs. (7.16) and (7.17).

$$\max_{a} \sum_{i=1}^{n} \alpha_i - \frac{1}{2} \sum_{i=1}^{n} \sum_{j=1}^{n} \alpha_i \alpha_j y_i y_j x_i^T x_j \tag{7.16}$$

$$\text{s.t.} \sum_{i=1}^{n} \alpha_i y_i = 0, \alpha_i \geq 0, i = 1, 2, \ldots, n \tag{7.17}$$

  Once α_i is calculated, w can be obtained using Eq. (7.9). The radial basis function is used as the kernel function and the regularization parameters as C and σ which are set to 1 0 0 and 1 0, respectively.
- DT: DTs are tree like structure for decision deduction at the nodes and reach some outcome at the leaf nodes. It is used for classification analysis where each path is a set of decisions leading to a class. The trees are constructed

based on entropy inputs that are high, constructed based on the divide and conquer approach. The approach is illustrated in [13,23].

- NB: Naïve Bayes learning model is a probabilistic classifier that works under the principle of Bayes theorem with naïve (strong) independent assumptions between features [23,61,62].
- K-nearest neighbor (KNN): KNN is a non-parametric technique that is used to classify data points based on the distance function stored by the algorithm on the training dataset. The Euclidean distance is adopted in this study because it is used by most instance-based learners [1,63,64]. To classify a new instance using k-NNN, Eq. (7.18) is adopted:

$$d(x,y) = \sqrt{\left(\sum_{i=1}^{N} (x_i - y_i)^2\right)} \tag{7.18}$$

where testing vector $x = x_1, x_2, \ldots, x_n$ and training vector $y = y_1, y_2, \ldots, y_n$ in $\mathbb{R}^2$ vector space.

- Multilayer perceptron (MLP): MLP is one popular "black box" neural network model that learns by applying a backpropagation algorithm to adjust propagated error to obtain an arbitrary level of accuracy [8,65]. In the MLP, the input vector x_i is multiplied by a weight vector w_i, and added to a bias b to produce an output $\widehat{y}$ using the following Eqs. (7.19)–(7.22):

$$y_i = f\left(\sum_{i=1}^{n} w_i x_i + b\right) \tag{7.19}$$

where n represents input–output pairs, f represents an activation function presented as:

$$f = \frac{1}{1 + exp^{-x_i}} \tag{7.20}$$

$$E(\widehat{y}, y) = \frac{1}{2}\sum_{i=1}^{n} (\widehat{y} - y)^2 \tag{7.21}$$

where E is the error function.

$$\delta_i = \frac{dE}{w} \tag{7.22}$$

where δ_i is the gradient descent, w represents weight. This study adopts one hidden layer MLP as it gives the best accuracy.

7.3.8 Ensemble methods

The main goal of the ensemble methodology is to create a composite global model with more accurate and reliable decision estimates than a single model can. Each model in the ensemble set solves the same initial problem. According to published research, combining the results of various classifiers lowers the generalization error. There are two types of ensemble methods:

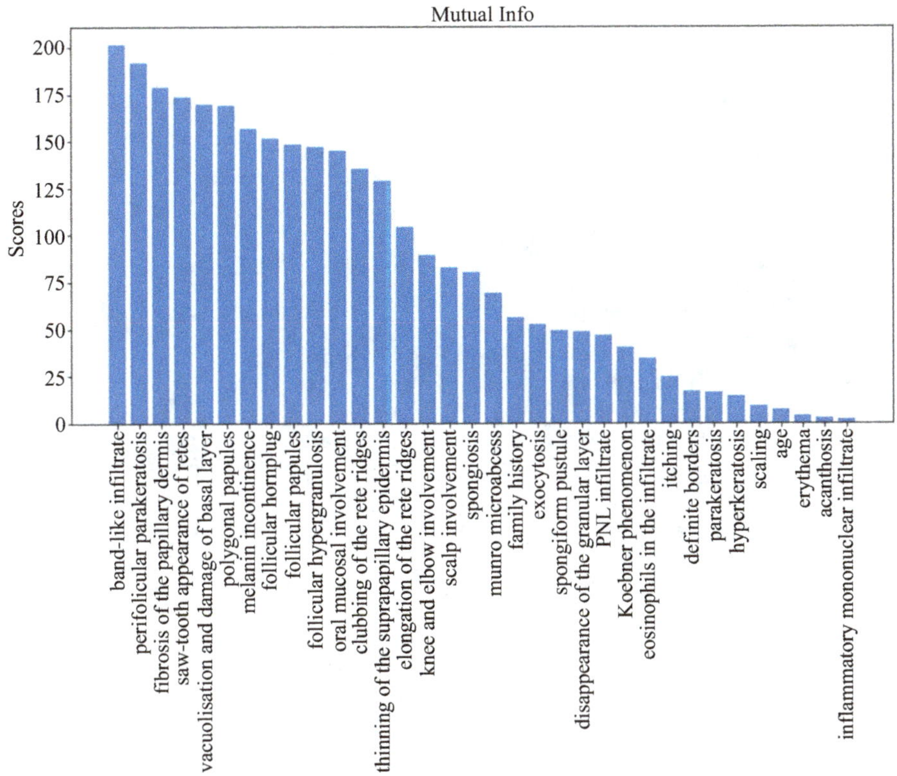

Figure 7.2 Information gain

homogeneous and heterogeneous. The boosting, bagging, and stacking ensemble methods are used in this study to combine both categories [36].

- *AdaBoost*: AdaBoost combines weak classifiers into an effective strong classifier by using the iterative ensemble method. The fundamental principle of AdaBoost is to train data samples to predict a class target of a given data instance with two classes by setting the classifier weights and using the average majority vote [66–68] as shown in Eq. (7.23):

$$\sum_{t=1}^{T} w_t d_{t,j}(x) = \max_{j=1}^{C} \sum_{t=1}^{T} w_t d_{t,j}\ (x) \tag{7.23}$$

where $d_{t,j}$ (x) represents support given by the *t*th classifier to the *j*th class for the instance x, w_t is the weight of classifier t and T is the total number of classifiers.

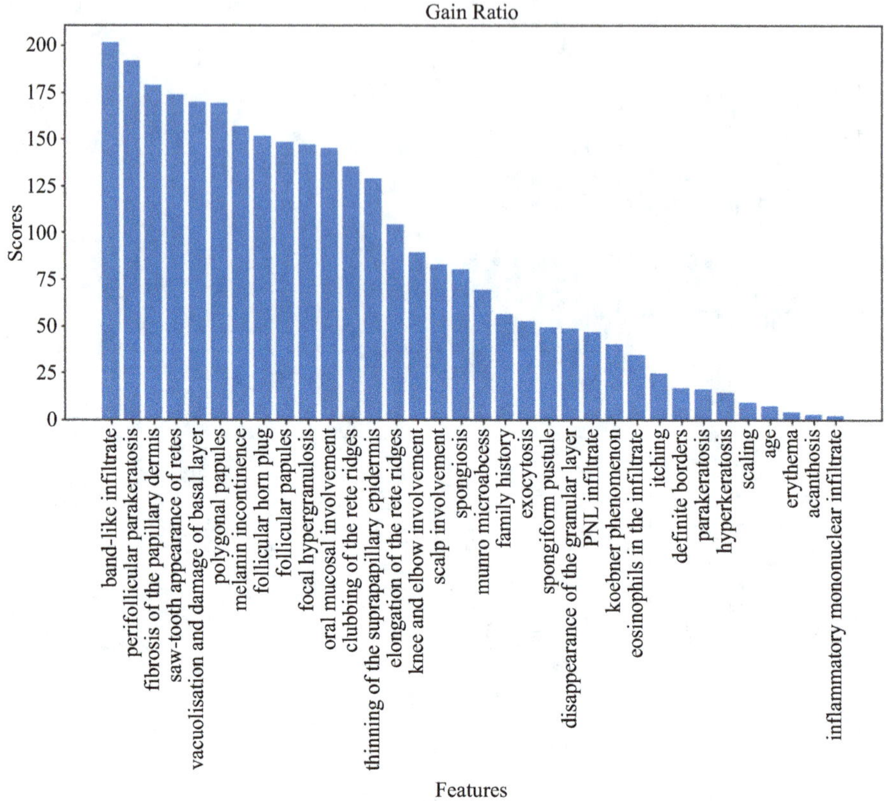

Figure 7.3 GR

- *Bagging*: The simplest but most effective independent ensemble method for enhancing the accuracy of unstable learning algorithms is bagging [67,68], which is derived from bootstrap aggregation. During the bagging process, datasets are divided among various bootstrap replicates. The original dataset is used to create each replicate, which contains, on average, 63.2% of the original data. The process entails putting the slow learner through several bootstraps on repeat. The weak learner's classifier is combined into a strong composite classifier with each iteration, producing higher accuracy than any individual component classifier could manage. The total of all base learners is then calculated using the majority voting system (or plurality voting) represented in Eq. (7.24):

$$\sum_{t=1}^{T} \mathrm{d}_{i,j} = \max_{j=1}^{C} \sum_{t=1}^{T} \mathrm{d}_{t,j}(\mathrm{x}) \tag{7.24}$$

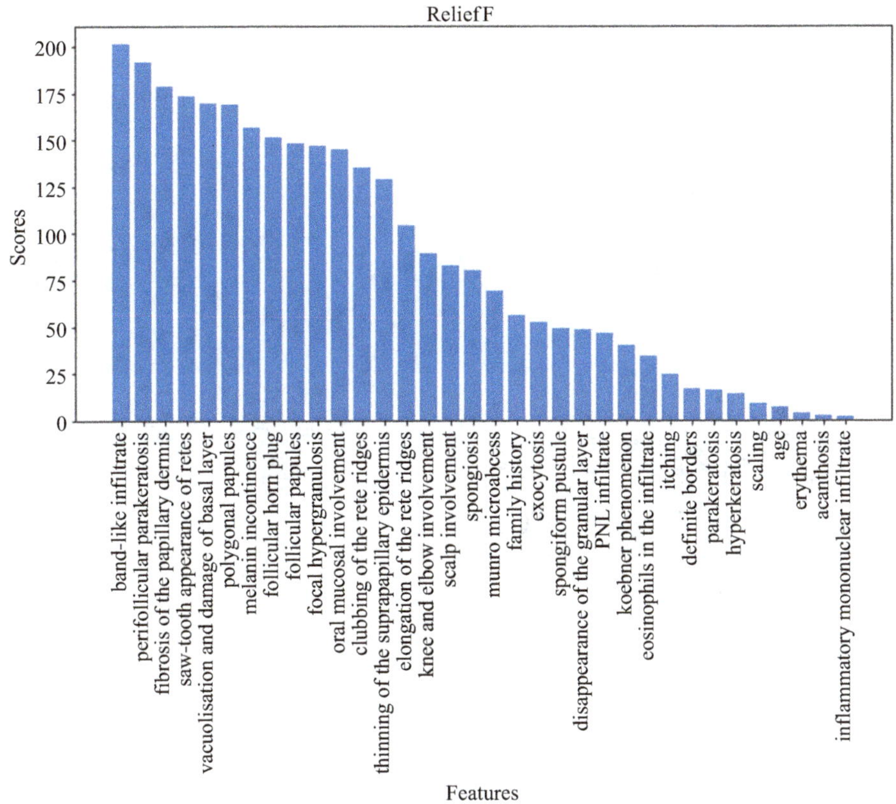

Figure 7.4 Relief F

where the decision of the *t*th classifier is defined as $d_{t,j} \in \{0, 1\}$, $t = 1, \ldots, T$ and $j = 1, \ldots, C$. T represents the size of the classifiers, and C represents the size of the classes. If *t*th chooses ω_j, then $d_{t,j} = 1$, otherwise 0.

- *Stacking*: With the aid of a meta-classifier, stacking [69] is used as an ensemble technique to combine heterogeneous models. In order to get the final outputs from the base classifier for prediction results, five base classifiers—SVM, DT, NB, K-NN, and MLP—are trained, while the logistic regression is then used as the meta-classifier to avoid the overfitting phenomenon generated by the base models in the ensemble [41].

7.4 Experimental results and discussion

By combining several filter-based feature selection models, including Info Gain, GR, Relief F, and Chi-square to produce various feature subsets, we created a novel methodology for feature selection in this work. In each subset, 22 features

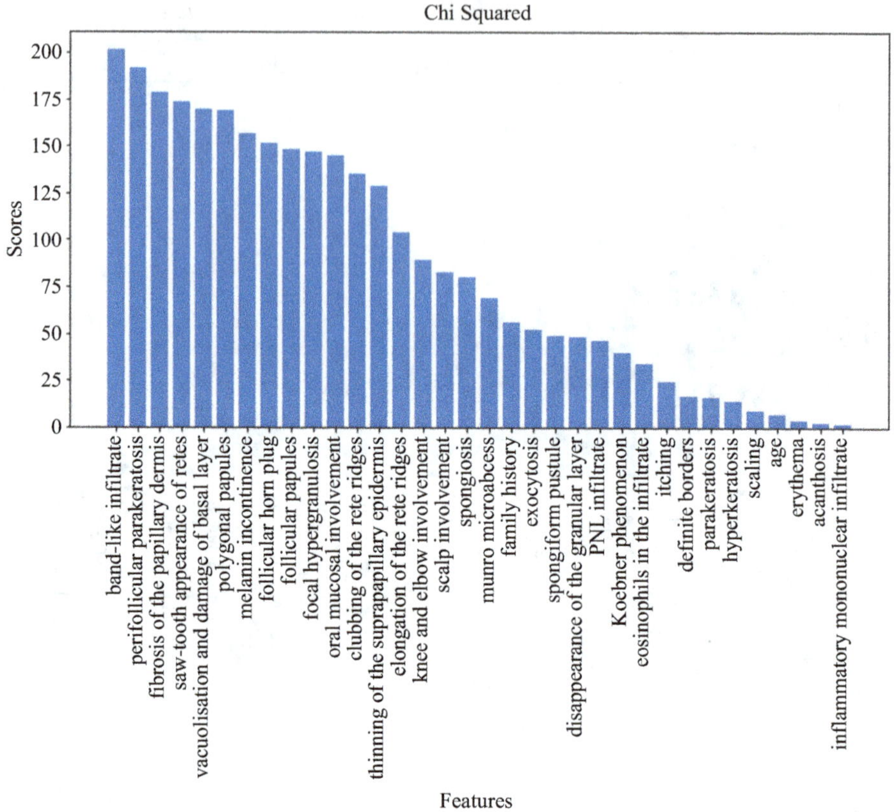

Figure 7.5 Chi-squared

Table 7.2 Metrics measurement.

Base classifiers	Accuracies	Sensitivity	Specificity	Precision	Recall	F1 score
SVM	**98.5**	85.1	98.2	83.9	85.0	88.8
DT	94.7	88.4	95.2	72.7	88.3	71.6
MLP	88.0	85.7	99.9	77.3	85.6	77.3
K-NN	94.7	86.3	**98.0**	84.6	86.3	85.4
NB	77.3	69.4	93.8	53.1	64.9	57.6
Bag (SVM)	80.9	85.1	**98.2**	88.9	85.0	80.8
Bag (DT)	**98.6**	83.3	97.8	73.2	83.3	80.2
Bag (MLP)	80.9	85.3	**98.1**	89.7	81.5	81.4
Bag (KNN)	94.7	88.7	97.0	89.9	88.6	80.1
Bag (NB)	71.7	72.8	95.5	79.0	72.7	65.0
Boost (SVM)	49.4	17.7	83.3	04.2	16.6	05.3
Boost (DT)	**98.6**	80.5	93.9	79.1	78.5	80.8
Boost (NB)	77.3	64.5	95.6	52.7	61.4	52.2
Stacking (Log Reg.)	**98.6**	**87.8**	**98.2**	**90.2**	**90.7**	**88.5**

Note: The bold values shown in Table 7.2 indicates that the ensemble learning methods (using both feature selection and classifiers) demonstrate competitive results than the individual classifiers in majorly all the metrics used. Therefore, it is important to note that the ensemble learning is suitable and has the advantage regarding freeing users from making decision in choosing the best possible feature selection method for any given problem.

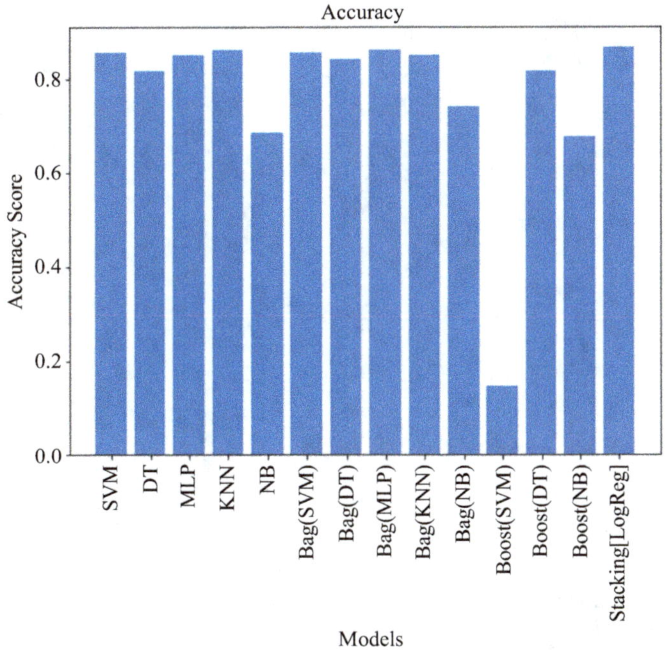

Figure 7.6 Models accuracies

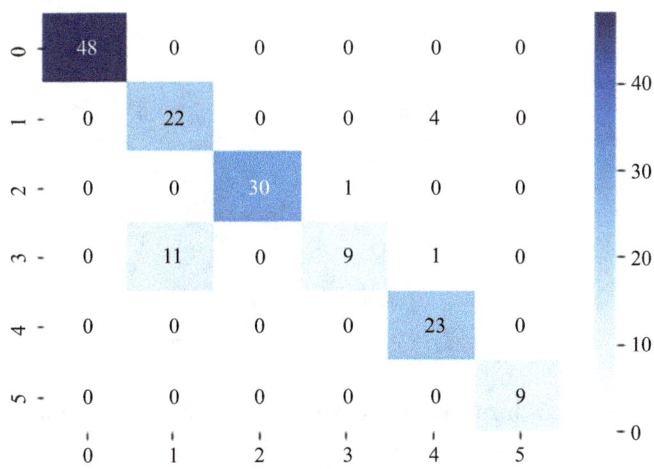

Figure 7.7 SVM

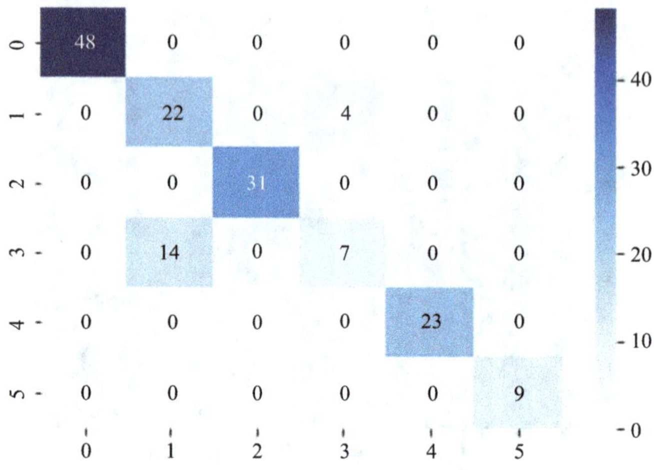

Figure 7.8 DT

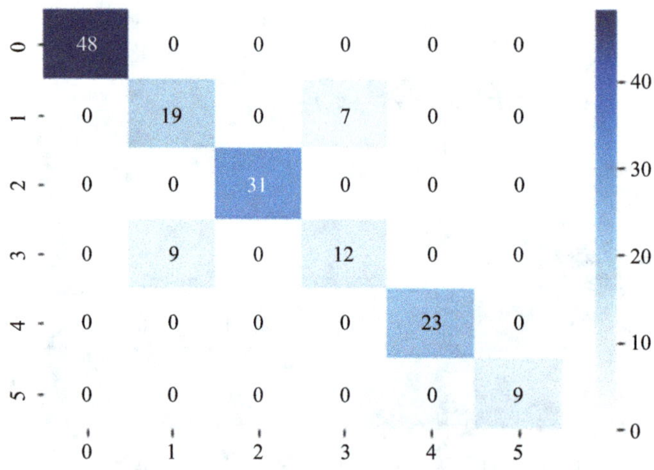

Figure 7.9 MLP

were chosen (i.e., two-thirds of the 34-original features). Then, only 13 features were chosen for the embedded feature selection methods using the simple majority voting technique. Finally, nine features for the base classifier ensemble learning method—boosting, bagging, and stacking—were obtained using the PRIFEB, MIFEB, and RFE based on the SVM model. The outcomes at each stage are listed below. The various features obtained by the filtering techniques are displayed in Figures 7.2–7.5. Two-thirds of the features with the highest

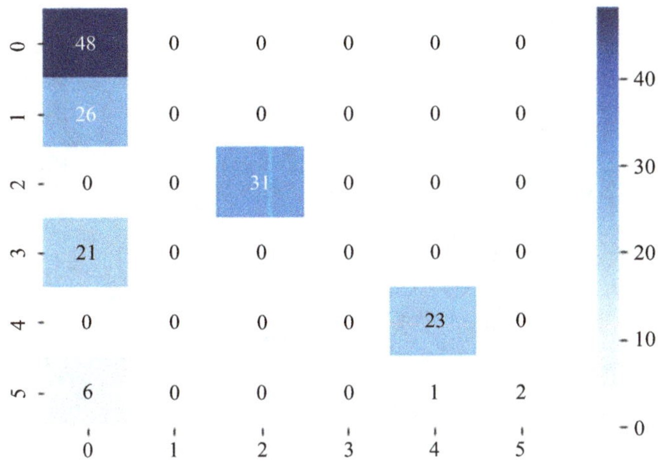

Figure 7.10 KNN

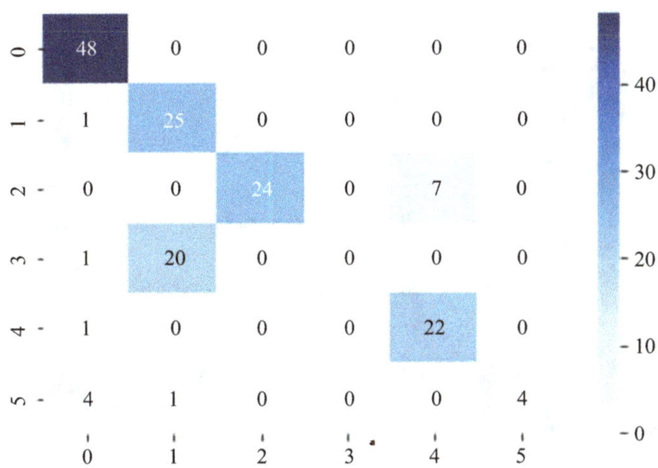

Figure 7.11 NB

scores were chosen. The accuracy results from the top features from the PRIFEB, MIFEB, and RFE-SVM embedded selection models are shown in Table 7.2.

The accuracy and other measure scores after five iterations of each base model are shown in Table 7.2. The individual and ensemble models perform well in terms of accuracy, with the exception of the Boosted SVM, which had an accuracy rate of 49.4%. The selection of the obtained hyper-parameter values was poor, which is why the boosted SVM performed poorly. The base SVM, bagged DT, boosted DT, which uses majority and weighted majority voting techniques, and stacking of all

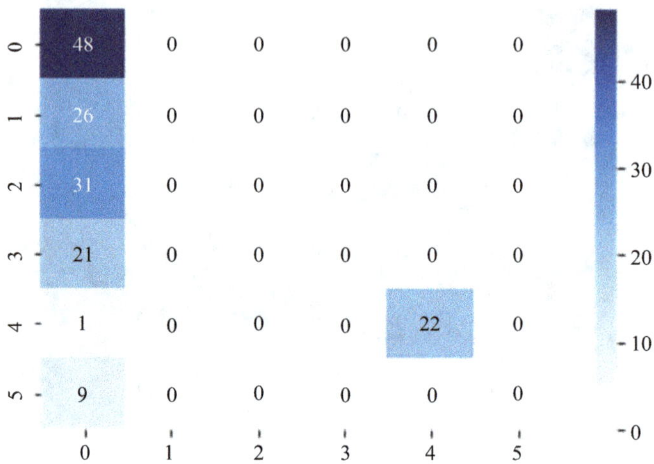

Figure 7.12 Bag (SVM)

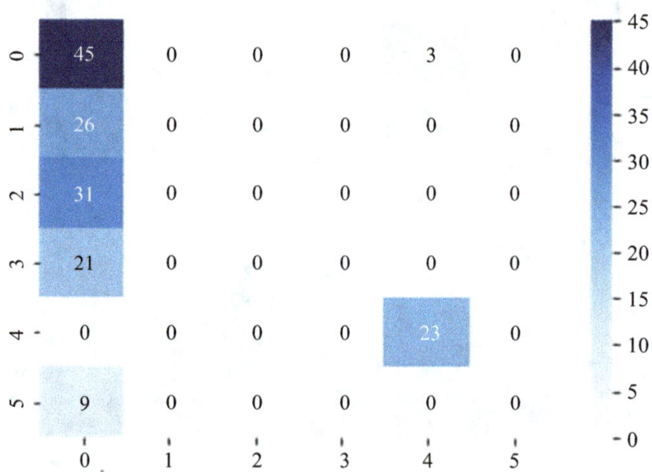

Figure 7.13 Bag (DT)

models using logistic regression as the meta classifier, are the models that perform the best in terms of accuracy. The accuracy of each model is shown in a single chart in Figure 7.6. The SVM, bagged DT, boosted DT, and stacked ensemble model generated the highest accuracies of 98.5–98.6%. The stacked ensemble has the highest performances still in sensitivity, specificity, precision, recall, and F1 score

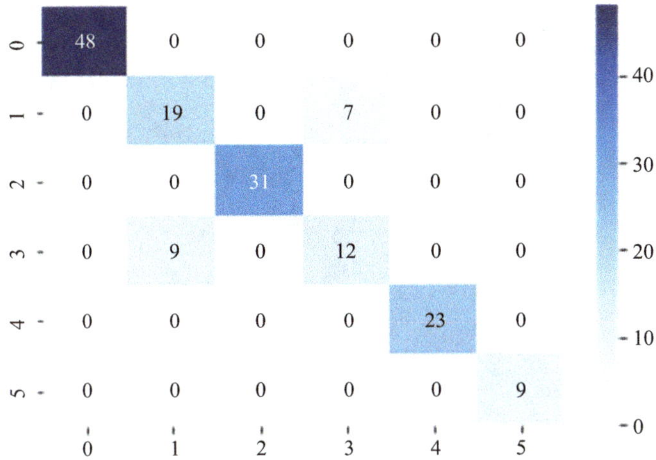

Figure 7.14 Bag (MLP)

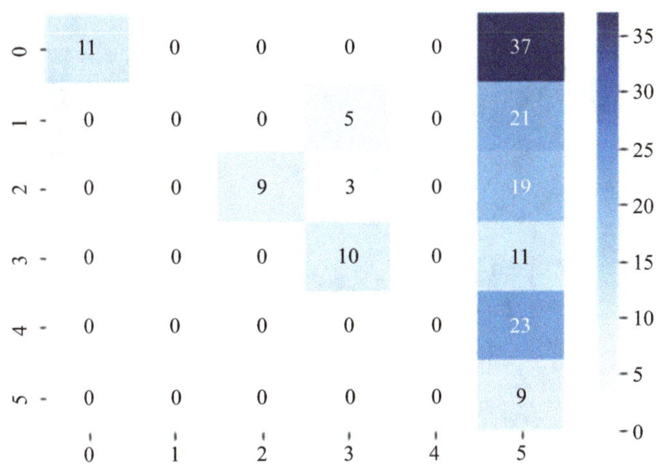

Figure 7.15 Bag (KNN)

of 87.8%, 98.2%, 90.2%, 90.7%, and 88,5% respectively. Table 7.2 demonstrates that heterogeneously staking models can result in models with significantly improved classification performance.

Figures 7.7–7.19 show the confusion matrices for the individual and ensemble models. Figure 7.20 depicts the box plots of the models. The accuracy of each

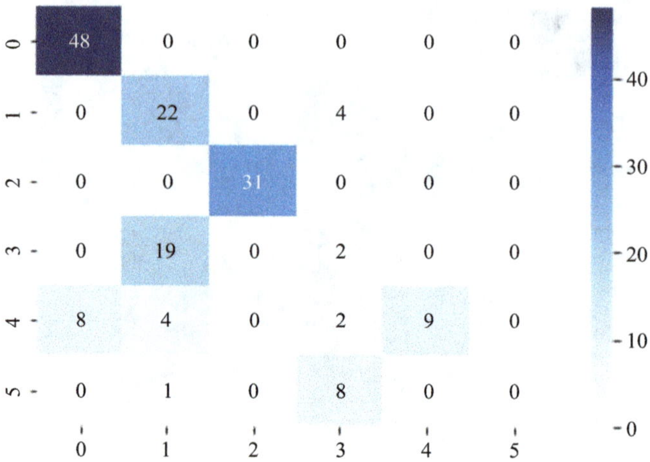

Figure 7.16 Bag (NB)

Figure 7.17 Boost (SVM)

model used in this paper is shown in the box plots. The stacked ensemble learning model, which combines all five models and a metaheuristic classifier to classify the dataset, is clearly the best performing model, according to the box plot, while the worst model is the boosted SVM, which exhibits signs of weak hyperparameter values used for the SVM classification. In order to reduce the size of the original features for the skin disease dataset obtained from the UCI machine learning

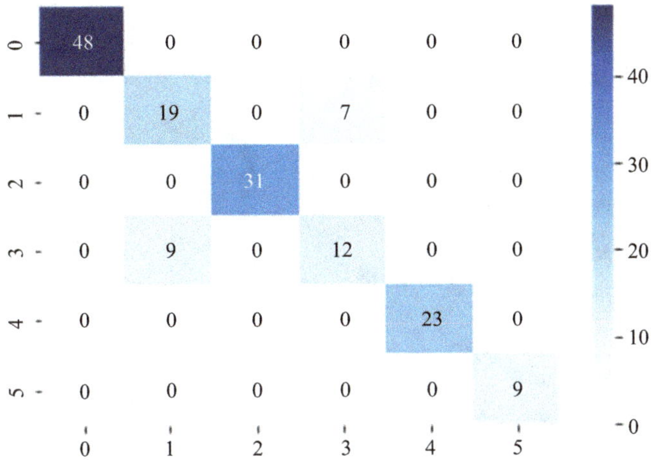

Figure 7.18 Boost (DT)

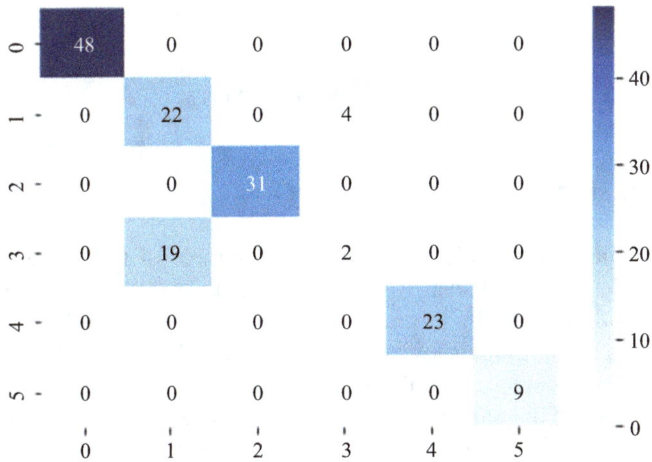

Figure 7.19 Stacking

repository, the study develops an ensemble of multi-feature selection approaches using both the filter feature selection techniques and the embedded feature selection methods. Any large dataset for classification can be used with this method. The focus of this study, however, is only the skin data set using three embedded models and filters to choose the features, along with a few chosen classification learning algorithms and an ensemble of them.

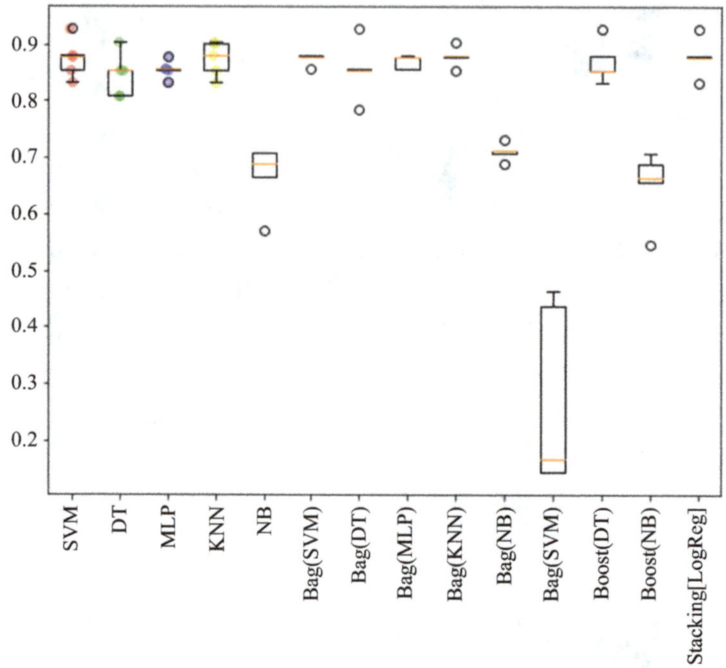

Figure 7.20 Box plot

7.5 Conclusion

The ability to classify skin diseases using an ensemble of multi-feature selection algorithms is demonstrated in this paper. The algorithms' time and space complexity were both reduced by the multi-feature selection method's feature reduction. The results of this study show the stacked ensemble of classifiers outperforms others in terms of accuracy, specificity, sensitivity, recall, precision, and F1 score. The black-box nature of some of the machine learning algorithms used in this study limits the explainability of these models and their practical deployment in medicine. This is due to the fact that the cardinal of AI deployment in the clinical environment is not only for model accuracy, but the explainability of the models hence the critical need for medical XAI for the diagnosis of erythemato-squamous disease [23]. The XAI deals with the transparency, comprehensibility, interpretability and understandability of the causality of the learned representations in the decision-making process of the classifiers. This is a current challenge that is critical to the medical personnel's acceptability and adoption. As fantastic as our current approaches seem, it is still considered black-box algorithms problems [11,23,70], in that, while some models like DTs can learn the mappings of inputs to outputs, others are next to impossible to how predictions are made [24]. In the future, we hope to address some of the challenges of explainability in AI applications.

References

[1] Kumari B. and Swarkar T. 'Filter versus wrapper feature subset selection in large dimensionality microarray: a review'. *International Journal of Computer Science and Information Technology*. 2011;2(3):1048–1053.

[2] Verma A.K. and Pal S. 'Prediction of skin disease with three different feature selection techniques using stacking ensemble method'. *Applied Biochemistry and Biotechnology*. 2020;191(2):637–656. doi: 10.1007/s12010-019-03222-8.

[3] Verma A.K., Pal S., and Kumar S. 'Classification of skin disease using ensemble data mining techniques'. *Asian Pacific Journal of Cancer Preview*. 2019;20(6):1887–1894.

[4] Bol´on-Canedo V., Porto-D´ıaz I., S´anchez-Maro˜no N., and Alonso-Betanzos A. 'A framework for cost-based feature selection'. *Pattern Recognition*. 2014;47(7):2481–2489.

[5] Tuba E., Ribic I., Capor-Hrosik R., and Tuba M. 'Support vector machine optimized by elephant herding algorithm for erythemato-squamous diseases detection'. In *5th International Conference on Information Technology and Quantitative Management ITQM*, vol. 122, 2017, pp. 916–923.

[6] Liang Y., Gharipour A., Kelemen E., and Kelemen A. 'Ensemble machine learning approaches for proteogenomic cancer studies'. 2020. https//doi.org/10/21203/rs-101902/v1.

[7] Akinloye F.O., Obe O., and Boyinbode O. 'Development of an affective-based e-healthcare system for autistic children Scientific African'. *Hypotheses*. 2011;9(1):1–9. https://doi.org/10.1016/j.sciaf.2020.e00514.

[8] Qaseem S.N. and Saeed F. 'Hybrid feature selection and ensemble learning methods for gene selection and cancer classification'. *International Journal of Advanced Computer Science and Application*. 2021;12(2):193–200. doi:10.14569/IJACSA.2021.0120225.

[9] Raina R. 'An investigation into cervical cancer using ensemble learning approach'. Unpublished master's thesis, School of Computing, National College of Ireland, Ireland, 2017.

[10] Rustam Z. and Kharis S.A.A. 'Comparison of support vector machine recursive feature elimination and kernel function as feature selection using support vector machine for lung cancer classification'. In *Basic and Applied Science Interdisciplinary Conference*, 2017, pp. 1–6. doi:10.1088/1742-6596/1442/1/012027.

[11] Hauser K., Kurt A., and Haggenmuller S. 'Explainable artificial intelligence in skin cancer recognition'. *European Journal of Cancer*. 2022;167:54–69. https://doi.org/10.1016/j.ejca.2022.02.025. Assessed 27 February 2022.

[12] Jiang P. 'CNN-based diagnosis system on skin cancer using ensemble method weighted by cubic precision'. TechRxiv. Preprint, 2021. https://doi.org/10.36227/techrxiv.16964893.v1. Assessed 27 March 2022.

[13] Li G-Z. and Yang J.Y. 'Feature selection for ensemble learning and its application'. In *Machine Learning in Bioinformatics*. New York, NY: John Wiley & Sons, 2008, pp. 135–155.
[14] Oduntan O.E., Adeyanju I.A., Falohun A.S., and Obe O.O. 'A comparative analysis of Euclidean distance and cosine similarity measure for automated essay-type grading'. *Journal of Engineering and Applied Sciences*. 2018;13 (11): 4198–4204.
[15] Bolón-Canedo V., Sánchez-Maroño N., and Alonso-Betanzos A. 'A review of feature selection methods on synthetic data'. *Knowledge and Information Systems*. 2013;34:483–519. https://doi.org/10.1007/s10115-012-0487-8.
[16] Bol´on-Canedo V., S´anchez-Maro˜no N. and Alonso-Betanzos. A. 'An ensemble of filters and classifiers for microarray data classification'. *Pattern Recognition*. 2011;45:531–539.
[17] Jia, L. 'A hybrid feature selection method for software defect prediction'. In *IOP Conference on Series: Materials Science and Engineering*. 2018, p. 394. doi: 10.1088/1757-899X/394/3/032035, 2018.
[18] Kausar N., Hameed A., Sattar M., et al. 'Multiclass skin cancer classification using an ensemble of fine-tuned deep learning models'. *Applied Sciences*. 2021;11:10593. https://doi.org/10.3390/app112210593.
[19] Obe O.O. and Dumitrache I. 'Adaptive neuro-fuzzy controller with genetic training for mobile robot control'. *International Journal of Computer Communications and Control*. 2021;7(1):145–156.
[20] Guyon I., Weston J., Barnhill S., and Vapnik V. 'Gene selection for cancer classification using support vector machines'. *Machine Learning*. 2002;46 (1–3):389–422.
[21] Hasanpour H., Meibodi R.G., Navi K., and Asadi S. 'Novel ensemble method for the prediction of response to fluvoxamine treatment of obsessive-compulsive disorder'. *Neuropsychiatric Disease and Treatment*. 2018;14:2027–2038. doi:10.2147/NDT.S173388.
[22] Igodan E.C., Obe O.O., Thompson A.F., and Owolafe O. 'A hybrid feature ensemble method for cervical cancer classification'. Unpublished manuscript.
[23] Zhang Y., Weng Y., and Lund J. 'Applications of explainable artificial intelligence in diagnosis and surgery'. *Diagnostics*. 2022;12:237. https://doi.org/10.3390/diagnostics12020237.
[24] Kamath U. and Liu J. *Explainable Artificial Intelligence: An Introduction to Interpretable Machine Learning*. Switzerland: Springer, 2021.
[25] Moody J. and Utans J. 'Principled architecture selection for neural networks: application to corporate bond rating prediction'. In J.E. Moody, S.J. Hanson, and R.P. Lippmann (eds.), *Advances on Neural Information Processing Systems*. Burlington, MA: Morgan Kaufmann Publishers, Inc., 1992, pp. 683–690.
[26] Osanaiye O., Cai H., Choo K-K., Dehghantanha R., Xu Z., and Dlodlo M. 'Ensemble-based multi-filter feature selection method for DDoS detection in cloud computing'. *EURASIP Journal on Wireless Communications and Networking*. 2016;130:1–10. doi:10.1186/s13638-016-0623-3.

[27] Schapire R.E. *Boosting: Foundations and Algorithms*. Cambridge, MA, London: The MIT Press, 2012.

[28] Witten I.H., Frank E., Hall M.A., and Pal C.J. *Data Mining: Practical Machine Learning Tools and Techniques*. Fourth ed., Cambridge, MA: Morgan Kaufmann, 2017.

[29] Fagbola T.M., Adejanju I.A., Oloyede A., et al. 'Development of mobile-interfaced machine learning-based predictive models for improving students' performance in programming courses'. *International Journal of Advanced Computer Science and Applications*. 2018;9(5):105–115.

[30] Liu H. and Setiono R. 'CHi2: feature selection and discretization of numeric attributes'. In *IEEE 7th International Conference of Tools with Artificial Intelligence*, 1995, pp. 388–392.

[31] Bose P., Bandyopadhyay S.K., Bhaumik A., and Poddar S. 'Skin disease detection: machine learning vs. deep learning'. Preprints. 2021; 2021090209. doi:10.20944/preprints202109.0209.v1.

[32] Weston J., Mukheriee S., Chapelle O., Pontil M., Poggio T., and Vapnik V. 'Feature selection for SVMs'. In *NIPS' 00: Proceedings of the 13th International Conference on Neural Information Processing Systems*, 2000, pp. 647–653.

[33] Thirey B. and Eastbuy C. 'Increasing accuracy through class detection: ensemble creation using optimized binary KNN classifiers'. *International Journal of Computer Science, Engineering and Application*. 2011;1(2):1–11.

[34] Badrinath N., Gopinath G., and Ravichandran K.S. 'Estimation of automatic detection of erythemato-squamous diseases through AdaBoost and its variants classifiers'. *Artificial Intelligence Review. Springer Nature*. 2016;45:471–488. doi:10.1007/s10462-015-9436-8.

[35] Oliveira O.R.B., Pereira A.S., Manuel J., and Tavares R.S. 'Computer methods and programs in biomedicine'. *Computer Methods and Programs in Biomedicine*. 2017;149:43–53.

[36] Verma A. K., Pal S., and Kumar S. 'Comparison of skin disease prediction by feature selection using ensemble data mining techniques'. *Informatics in Medicine Unlocked*. 2019;16:100202.

[37] Asogbon Y.A., Oluwarotimi W.S., Nsugbe E., *et al.* 'A deep learning based model for decoding motion intent of traumatic brain injured patients' using HD-sEMG recordings'. In *2021 IEEE International Workshop on Metrology for Industry 4.0 & IoT* (*MetroInd4.0&IoT*), 2021, pp. 609–614. doi:10.1109/MetroInd4.0IoT51437.2021.9488440.

[38] Karpagam S., Kaleeswari M., Kavitha K., and Priyadarsini S. 'Heart disease prediction using machine learning algorithm'. *International Journal of Scientific Development and Research*. 2021;5(8):334–337.

[39] Zhenya Q. and Zhang Z. 'A hybrid cost-sensitive ensemble for heart disease prediction'. *BMC Medical Informatics and Decision Making*. 2021;21 (73):1472–6947. https://doi.org/10.1186/s12911-021-01436-7.

[40] Kang J., Ullah Z., and Gwak J. 'MRI-based brain tumor classification using ensemble of deep features and machine learning classifiers'. *Sensors (Basel)*. 2021;21(6)2222. https//doi.org/10.3390/s21062222.

[41] Liu J., Dong X., Zhao H., and Tian Y. 'Predictive classifier for cardiovascular disease based on stacking model fusion'. *Processes*. 2022;10(4):749. https://doi.org/10.3390/pr10040749. Assessed 27 March 2022.

[42] Gupta C., Saha A., Reddy N.V., and Acharya U.D. 'Cardiac disease prediction using supervised machine learning techniques'. *Journal of Physics*. 2022;2161(1):012013. doi:10.1088/1742-6596/2161/1/012013.

[43] Salhi D.E., Tari A., and Kechadi M.T. 'Using machine learning for heart disease prediction'. In M.R. Senouci, M.E.Y. Boudaren, F. Sebbak, and M. Mataoui (eds.), *Advances in Computing Systems and Applications. CSA 2020. Lecture Notes in Networks and Systems, vol. 199*. Cham:Springer, 2021. https://doi.org/10.1007/978-3-030-69418-0_7. Assessed 27 March 2022.

[44] Jarrah Y.A., Asogbon Y.A., Samuel O.W., *et al.* 'A comparative analysis on the impact of linear and non-linear filtering techniques on EMG signal quality of transhumeral amputees'. In *IEEE International Workshop on Metrology for Industry 4.0 and IoT, MetroInd 4.0 and IoT 2021 – Proceedings*, 2021, pp. 604–608, 9488516.

[45] Sandrilla R. and Savitha D.M. 'Performance analysis of various ensemble feature selection'. *Design Engineering*. 2022; 1:228–247.

[46] Skin Disease dataset, www.https://archive.ics.uci.edu/ml/datasets/dermatology.

[47] Bol´on-Canedo V., S´anchez-Maro˜no N., and Alonso-Betanzo, A. 'Recent advances and emerging challenges of feature selection in the context of big data'. *Knowledge-Based Systems*. 2015;86:33–45. https://doi.org/10.1016/j.knosys.2015.05.014. Assessed 27 March 2022.

[48] Dzuida M.A. *Data Mining for Genomics and Proteomics Analysis of Gene and Protein Expression Data*. New York, NY: Wiley, 2010, p. 210.

[49] Mahajan S., Abhishek, and Singh, S. 'Review of feature selection approaches using gene expression data'. *Imperial Journal of Interdisciplinary Research*. 2016;2(3):356–364.

[50] Rim P. and Liu E. 'Optimizing the C4.5 decision trees algorithm using MSD-splitting'. *International Journal of Advanced Computer Science and Application*. 2020;11(10):41–47.

[51] Robnik-Sikonja M. and Kononenko I. 'Theoretical and empirical analysis of ReliefF and RReliefF'. *Machine Learning Journal*. 2003;53:23–69.

[52] Alelyani S. 'Stable bagging feature selection on medical data'. *Journal of Big Data*. 2021;8(1):1–18. https://doi.org/10.1186/s40537-020-00385-8. Assessed 20 February 2022.

[53] Liu H. and Yu L. 'Towards integrating feature selection algorithms for classification and clustering'. *IEEE Transactions on Knowledge and Data Engineering*. 2005;17(3):1–12.

[54] Zhang C. and Ma Y. *Ensemble Machine Learning: Methods and Applications*. London: Springer, 2012.

[55] Spencer T., Thabtah F., Abdelhamid N., and Thompson M. 'Exploring feature selection and classification methods for predicting heart disease'. *Digital Health*. 2020;6:1–10.

[56] Zhang J., Liang Q., Jiang R., and Li X. 'A feature analysis based identifying scheme using GBDT for DDoS with multiple attack vectors'. *Applied Science*. 2019;9(21):4633.

[57] Mazlin T.A.H.T., Sallehuddin R., and Zuriahati M.Y. 'Utilization of filter feature selection with support vector machine for tumour classification'. In *Joint Conference on Green Engineering Technology & Applied Computing. IOP Conference Series: Materials Science and Engineering*, Bangkok, Thailand, 2019, p. 551.

[58] Tama B.A. and Lim S. 'A comparative performance evaluation of classification algorithms for clinical decision support systems'. *Mathematics*. 2020;8:1814:1–25.

[59] Nagi S., Dhruba K., and Bhattacharyya D.K. 'Classification of microarray cancer data using ensemble approach'. *Network Model Analysis in Health Information and Bioinformatic*. 2013;2:159–173.

[60] Lu J., Song E., Ghoneim M., and Alrashoud, M. *Machine Learning for Assisting Cervical Cancer Diagnosis: An Ensemble Approach*. New York, NY: Springer, 2020, pp. 1–8.

[61] Ahishakiye E., Wario E., Nwangi W., and Taremwa D. 'Prediction of cervical cancer based on risk factors using ensemble learning'. In *IST-Africa 2020 Conference Proceedings*, 2013, pp. 1–12.

[62] Oded M. and Lio R. *Data Mining and Knowledge Discovery Handbook*, 2nd ed. London: Springer, 2010.

[63] Verma A.K., Pal S., and Kumar S. 'Prediction of skin disease using ensemble data mining techniques and feature selection method – a comparative study'. In *Applied Biochemistry and Biotechnology*, Springer Nature: London, 2019. Available from https://doi.org/10.1007/s12010-0-019-03093-z. Assessed 20 February 2022.

[64] Zhao Z., Morstatter F., Sharmer S., Alelyani S., Anand A., and Liu H. 'Advancing feature selection research'. In ASU Feature Selection Repository, 2010, pp. 1–28.

[65] Hoque N., Singh M., and Bhattacharyya D.K. 'EFS-MI: and ensemble feature selection method for classification'. In *Complex Intelligent System*. New York, NY: Springer, 2017. doi:10.1007/s40747-017-0060-x.

[66] Seijo-Pardo B., Porto-Diaz I., Bolon-Canedo V., and Alonso-Betanzos A. 'Ensemble feature selection: homogeneous and heterogeneous approaches'. *Knowledge-Based System*. 2017;118:124–139.

[67] Zhou Z.H. *Ensemble Methods: Foundation and Algorithms*. Cambridge, UK: Taylor & Francis Group, 2021.

[68] Michalska M. 'Overview of feature selection methods used in malignant diagnostics'. *IAPGOS*. 2021;1:32–35. http://doi.org/10.35784/iapgos.2455.

[69] Oreski D. and Novosel T. 'Comparison of feature selection techniques in knowledge discovery process'. *TREM Journal.* 2014;3(4):285–290.
[70] Holzinger A., Kieseberg P., Weippl E., and Tjoa A.M. 'Current advances, trends and challenges of machine learning and knowledge extraction: from machine learning to explainable AI'. In *CD-MAKE, LNCS,* vol. 11015, 2018, pp. 1–8. https://doi.org/10.1007/978-3-319-99740-7_1.

Chapter 8

Security-based explainable artificial intelligence (XAI) in healthcare system

Hüseyin Gürüler[1], Naveed Islam[2] and Alloud Din[2]

Abstract

Explainable Artificial Intelligence (XAI) is one of the most advanced research areas of Artificial Intelligence (AI). To explain the deep learning (DL) model is the main objective of XAI. It deals with artificial models which are understandable to humans, including the users, developers, policymakers, etc. XAI is very important in some critical domains like security, healthcare, etc. The purpose of XAI is only to provide a clear answer to the question of how the model made its decision. The explanation is very important before any system decision-making. As an example, if a system responds to a decision, it is necessary to have inside knowledge of the model about that decision. The decision can be positive or negative, but it is more important to know the decision based on characteristics. The decision of the model should be trusted when we know the internal structure of the DL model. Generally, DL models come under the black box models. So for security purposes, it is very necessary to explain a system internally for any decision-making. Security is very crucial in healthcare as well as in any other domain. The objective of this research is to provide a decision about security based on XAI which is a big challenge. We can improve security systems based on XAI for the next level. For medical/ healthcare security, when we recognize human action using transfer learning techniques, one pre-trained model is considered good for action and the same action is not good in terms of accuracy using another pre-trained model. This is called the black-box model problem, and it needs to know what is the internal mechanism of both models for the same action. Why one model considers good for action and why the same action is not very well using another model? Here need a model-specific approach of post-hoc interpretability to know the internal structure and characteristics of both models for the same action.

Keywords: Security in hospital; Abnormal action in healthcare; Security in healthcare; Smart healthcare; Smart medical security system; Health monitoring

[1]Department of Information Systems Engineering, Muğla Sıtkı Koçman University-Muğla, Turkey
[2]Department of Computer Science, Islamia College University, Pakistan

8.1 Introduction

Development in the science of Artificial Intelligence (AI) has led to its widespread use in a variety of fields, including finance, medical services, and security [1]. In this regard, one of the numerous study fields in which AI-based systems have achieved outstanding results or even outperform humans in computer vision, which belongs to machine learning (ML) algorithms to extract information from pictures [2]. For example, to recognize things. A neural network (NN) was able to outperform humans in identifying traffic signs early. The advancement of deep learning (DL) algorithms has been the backbone of numerous discoveries in this area. This is a famous field of ML that models the architecture of the human cerebral cortex and trains and applies multi-layer neural networks (NNs) using huge data [3]. DL is being studied more and more in the healthcare and security fields. It may be used for medical imaging as well as security. A DL model is used to identify abnormal activity in the healthcare system by surveillance [4]. Despite significant advances and achievements in this field, one issue with DL algorithms is their "black-box" nature. There is no intrinsically full knowledge of the underlying mechanisms of DL-based techniques such as NNs due to their great degree of complexity [5]. AI systems that suffer from this issue are frequently referred to as opaque. As a result, there is a closed relation between performance and explainability: as model performance improves, the explainability of these approaches reduces. Explainable Artificial Intelligence (XAI) approaches have been created to increase transparency, open the black box, and generate explanations for AI system decisions [6]. XAI aims to create transparent and explainable models with strong learning performance, or prediction accuracy, allowing human users to comprehend, fully trust, and control a new generation of artificially intelligent companions [1]. This study will concentrate on XAI and its possible effect on trust in healthcare security. Trust is undertaken across multiple disciplines, including philosophy, psychology, sociology, marketing, information systems (IS), and human–computer interface (HCI). Because AI is becoming more powerful and is increasingly being employed in crucial circumstances with potentially serious repercussions for people (e.g., auto driving, healthcare diagnosis), trust in such systems is becoming increasingly important. There are various notions and definitions of trust in the various lines of the trust literature review. We employ a paradigm to treat the trust as a formative second-order entity [1].

Our main goal is to develop a security system based on computer vision to detect abnormal actions from video frames (actions may be human fall, kick and punch, etc.). Different pre-trained models will be considered for the actions detection and recognition in healthcare for security purposes. The dataset obtained for this research is the "HMDB51 human action dataset" which consists of human 51 different human actions. The goal is then to develop explanations using XAI approaches of pre-trained models and use them to compare different actions in healthcare. We propose the following main question: using a pre-trained model, one model consider good for action and why the same action is not good using another model? Using XAI approaches, we will compare and explain the internal mechanism of the deep NNs for the same action [1]. Because DL algorithms are often black box models, there is no acceptable explanation for a given prediction, explaining them is challenging. This uncertainty impacts DL in healthcare [7] since a clinical practitioner has to know the rationale for a DL model's prediction. The issue of explainability in DL models has been addressed by a number of researchers [8]. Gradient-weighted Class Activation Mapping was established by [9] to reveal input areas that are significant for predictions (Grad-CAM). We can deduce where the ML model is concentrated when making a forecast and, as a result, why such values exist. Explainability is important in healthcare security because we need to be able to explain why a certain prediction for an input sample is correct [10]. Automated image analysis has seen breakthroughs because of DL. Previously, image analysis was mainly done with systems designed entirely by experts. For more details, a classifier for statistics that performs a job using handcrafted feature properties i.e. features. A system for image analysis may incorporate the classifier. Low-level picture qualities such as edges and corners, as well as high-level image aspects, contain on cancer's speculated border and high-interest points, were among the features. These characteristics are gained and learned through a deep NN (NN) in order to provide the optimal value or result after input data. A frame is passes through a model for an action as input, as a result the DL model may create the output result of the frame. Many nonlinear mixed interactions are frequently used to connect multiple layers of NNs [11]. It is hard to completely understand how the NN made its predictions even if all of these layers are investigated and their interactions are detailed. DL is sometimes referred regarded as a "black box" as a result of this. Concern has been expressed in a number of sectors and biased in some methods. Generally, this bias is hard to notice and may go unread. This has far-reaching implications, notably in medical and security applications. A request has been made for ideas on how it is possible to understand the black box clearly. The techniques which are followed as interpretable DL or XAI are mentioned in [12,13]. We will use the word XAI because these names are commonly used interchangeably. The US defense advanced research projects agency (DARPA) and the Association for Computing Machinery's Fairness, Accountability, and Transparency conferences are two notable XAI efforts. Decisions about healthcare security are often high-stakes affairs. Unexpectedly, some of the healthcare domain experts felt the need for explainability and they have raised concern about the black box of AI, which is the current state of the art in computer vision applications [14]. In addition, laws such as the General Data Protection Regulation of the European Union stipulate that a person has the right to information when a decision was made. Researchers are using the XAI techniques for healthcare security to get insight into their algorithms. The purpose of this study is to compile a comprehensive list of works that employ XAI in the medical computer vision domain. We narrowed our search to research that applied DL-based security-based XAI in healthcare systems [11].

8.1.1 XAI

In DL, we will present a quick introduction of XAI approaches for security in healthcare in this part. See also [12,13] for comprehensive XAI surveys. People can understand and explain how an AI system decides with XAI [7]. XAI is a set of approaches and tactics that allow humans to evaluate and trust the outcomes of ML algorithms. An AI model, its intended impact, and any biases are all referred to as XAI. It has to do with how AI-powered decision-making defines model correctness, fairness, transparency, and results. When it comes to putting AI models into production, XAI is crucial in terms of building trust and confidence. AI explainability also aids a company's approach to AI development in a responsible manner. There are many benefits to understanding how an AI-enabled system arrived at a particular decision. Explainability can assist developers in ensuring that the system is operating as intended, may be required to meet regulatory requirements, or may be essential in enabling individuals who will be impacted by the decision to contest or amend the result. It will be crucial in marketing [15], healthcare, industrial, security, and automobiles [10,16], according to recent research. The depth of the explanation will be divided into three categories: model-based vs. post-hoc, model-specific vs. model-agnostic, and global vs. local XAI techniques

8.1.2 Model-based explanation

Studied in and is defined as models that are simple to understand and consider good enough to fit a connection between the output and input data. Linear regression and SVM are examples of model-based explanations. In this context, traditional ML methods are typically applied. Ref. [13] offers instances of model-based explanations that require to have knowledge of the internal mechanism of the model, and human can also know how the model make a decision. That is sparsity, or that only allows some features which are very restricted and not enough to know the whole decision-making process. The models that impose sparsity are not the most but the least selection operator. As a consequence, the model's inner construct is explained by a subset of attributes that leads to output. Model-based explanation via mandated sparsity or simulatability is not possible because our study focuses on XAI techniques. Deep NNs have hundreds and millions of weights that are neither sparse nor appropriate for humans to explain the inside model whole mechanism. Refs. [11,13] discussed the techniques of feature engineering using model-based approach.

8.1.3 Post-hoc XAI

It is the process of analyzing a trained model, such as NN in DL, to gain an understanding of acquired relationships [11]. Apart from model-based explanation, the post-hoc explanation is that the latter caused the model to be explainable. After training the NN and then explaining the internal behavior of the black-box model. Post-hoc explanations can be found using approaches such as feature inspection, feature significance, and feature interaction [17].

8.1.4 Model-specific explanation

Only some types of models are suitable for model-specific explanation approaches. Such a method can make advantage of characteristics peculiar to a certain kind of NN. Finding a model-specific explanation limits our model choice and may be leave out models that could more closely match to the output. According to the research community, a model-specific explanation by definition is a model-based explanation. Despite certain post-hoc saliency mapping approaches being unique to a particular kind of CNN, they are not model-based explanation strategies [11,13].

8.1.5 Model-agnostic explanation

Regardless of the model used, model-agnostic explanations are fully dependent on the given data. The explanation of this approach is based on the input and output data of the model. By changing some input samples, the user can see the changes in the output of the same model [11]. As a consequence, it is easy to figure out what factors influence the model's output. By definition, post-hoc and model-agnostic explanation are very similar to each other.

8.1.6 Global explanation

The model's general relationships are described in a global explanation, also known as a dataset-level explanation. For example, at the dataset level, the global explanation may include feature significance ratings, showing how much characteristics contribute to the overall outcome [11]. The presentation of learned filters displays the extracted characteristics using a NNs they are very important for the process.

8.1.7 Local explanation

The term "local explanation" refers to the explanation of a single input. A single individual would be an input in the case of certain risks. Thus, a local explanation would explain why action is significant for security and risk for an event, but a global explanation would describe the relationship between an action and a threat to security throughout the whole dataset. A saliency map locating a useful region or features of action in a frame to explain which portion of the frame contributed the most to the classifier output for action is another example of a local explanation. This is a local explanation since it describes which element of the image causes the classifier to report abnormal action for security [11].

The following is how the paper is organized: the literature on AI and ML explanation, explanation of ML in healthcare, and explainable system research in human–computer interface (HCI), etc. is given first. Following that, we go into our study methodology, covering the transfer learning (TL) concepts and pre-trained networks we used, as well as XAI visual explanation approaches. The outcomes of our implemented NNs employing the TL techniques, as well as the generated explanations, are presented, followed by a discussion of the importance of XAI in terms of understandability, as well as consequences for study and experimentation. A conclusion is included at the end of the paper.

8.2 Literature review

This section discusses the current state of the explainable AI literature, with two main goals: one is to create, improve, and merge AI/ML algorithms, while the other is to identify and evaluate current infrastructure and models from a human perspective.

8.2.1 XAI and AI

The user experience has been significantly enhanced by ML and AI over the past 20 years by making computer systems more intelligent and secure. In fact, a significant number of ML and AI systems in use today have attained a level of relative autonomy, meaning they are capable of making decisions and carrying out activities without the need for human supervision. The fact that the current systems are black boxes and unable to explain why they made a certain choice is one of their biggest limitations [18]. As a result, customers who interact with these self-driving systems frequently struggle to trust and understand them, especially when making a decision. As a result, there has been a rise in interest in constructing XAI systems, in recent years. Good examples are XAI systems [19,20], classification systems [21,22], and activity recognition systems [18,23]. Based on our prior research [18], we focused our study on an explainable action recognition system for healthcare security. Many ML/AI communities have examined explainability and interpretability strategies in ML models and systems. For example, Ref. [24] provides a survey of various interpretability techniques, such as post-hoc explanations where the model is self-explanatory and designed to be interpreted globally or locally [25,26], and intrinsic explanations, where explanations are drawn from the model from local or global viewpoints, and model-specific/agnostic reasons [27]. Others have looked into an explanation by example, which involves a model presenting instances of important events from the training set for a specific input rather than attempting to explain the model's reasoning explicitly. For example, in the visual domain, Ref. [28] identified and investigated two types of example-based explanations, as well as their efficacy with humans: the first normative explanations present training examples to assist users to understand classifications, while the second comparative explanations display the most comparable cases which may be of different classes from the training set to the input [18].

Researchers have described distinct types of focus on what is being explained by various hypotheses. According to [29], transparency seeks to show how an AI system works, whereas post-hoc interpretability focuses on the whys in the AI system, providing a rationale for its results. Ref. [30] performs an in-depth analysis of a significant number of deep NN visualization papers and divides the literature into different categories based on how visualization may reveal different DNN properties. The following are the categories: Why visualize DL models? What data, features, etc. [31,32]. What data, features, etc. Article in ACM Trans. Comput.-Hum. Interact. Is it possible to visualize their relationships? When is visualization employed in DL? [33,34]. Who would benefit from and utilize DL visualization?

How to display data, features, and relationships? [34] and Where has DL visualization been used? (see, e.g., [67,68]). While these categories are intended to aid in the visualization of DL methodologies, the same problems apply to any AI system that may be interpreted and explained [18]. Our explainable system in this research adopts a similar strategy and concentrates on a crucial explainable system (explainable action recognition system), but it is distinct from the majority of the work in the ML community in that we employ explainable systems to assess and comprehend user behaviors as well as to compare pre-trained networks for further understandability. Then we will go through how to create ways for explaining to human users, as well as how visual explanations are feasible.

8.2.2 Explanation meaningfulness and veracity

Various types of explanations include textual, confidence scores, prediction accuracies [35], and saliency maps [27]. They can also deal with a variety of logic and model operating circumstances. Global explanations [36] seek to provide a high-level overview of how a model produces its output. Building models as global explanations has been the subject of several explainable systems aimed at data experts [18]. An interactive visual method for summarizing and illustrating DL models, as well as demonstrating how much each layer and attribute was used to make predictions. Global explanations are useful because they can reveal biases, help in the detection of model defects, and allow for hyperparameter change [18]. Global explanations, especially for complicated or DL models, have the disadvantage of being more difficult to achieve in practice [36]. In contrast to global explanation, local explanation attempts to justify particular outcomes based on specific instances of input. Users can obtain a deeper understanding of the model by engaging with it over time and examining multiple instances, but an individual local explanation seldom gives a comprehensive overview of how the model works. The instance-level explanation may be critical in improving the user's understanding of system output, depending on the task. Our goal was to look at a system and a scenario that did not require any unique data expertise. We employed post hoc explanations to visually compare and illustrate certain key characteristics of cases that were crucial to the conclusion. One drawback of local explanations is how quickly they might alter a user's perception of the system. According to [18], it is easy to lose trust in automation, but it is more difficult to rebuild trust after it has been lost. For that reason, poor or erroneous instance-level explanations might lead to a sudden loss of confidence. As a result, the quality of these explanations is extremely important. Researchers examine the quality of explanation from a variety of perspectives. Ref. [37] studied how the existence and fidelity of explanations i.e. how precisely the explanation represents the underlying model, as well as system accuracy, impact user confidence. Using two degrees of high and low fidelity explanations, they observed that system accuracy plays a vital role in establishing user confidence, but low fidelity explanations may erode trust. Other studies have looked into what is known as nonsensical explanations, or explanations that people do not understand. After completing three behavioral field investigations and determining that individuals adhere to an explanation when it is more instructive rather than meaningless. Mention in [18] recently developed a NN input reduction technique that reduced explanations while maintaining accuracy by removing extraneous material. The human assessment, on the other hand, demonstrated that these shortened explanations and summaries confound humans since they are meaningless, resulting in a decrease in task accuracy [18].

8.2.3 ML in healthcare

Some physiological movements are controlled by signals from some cognitive diseases [10]. For example, a stroke may result in a shift in movement. Several researchers have proposed utilizing wearable sensors to track users' behaviors, allowing for the recognition of various human physical functions [10,38]. Monitoring such activities can detect early warning signs of health problems. In this context, ML and DL technologies have made tremendous progress in healthcare security. While such technologies are unlikely to completely replace healthcare volunteers, they have the potential to reshape the sector, benefiting both patients and providers [10,39,40]. In addition to security in healthcare, ML and DL are crucial when it comes to ECG analysis [41–43]. Several methods for categorizing ECGs into arrhythmia categories have been presented [44]. DL was utilized to automate cardiac auscultation, which is the process of identifying aberrant heart rates. They described a time–frequency heat map representation-based deep convolutional NN (CNN)-based automated heart sound classification algorithm. Their CNN architecture is trained using a modified loss function, which directly improves the sensitivity-specificity trade-off. Ref. [45] presented a method for utilizing heart sounds to diagnose chronic heart failure. Traditional ML is used with end-to-end DL models in this method. While the DL model learns from expert features, the normal ML model learns from a spectro-temporal representation of the signal. Ref. [46] has developed an intelligent ECG classifier that uses rapid compression residual CNNs to give high-accuracy abnormality classification [10].

Although the study mentioned above appears promising, it may only be useful in the real world because it relies on centralized data collection methods. Users and data owners may get concerned about their privacy as a result of this. Typical centralized healthcare apps have limited applicability due to privacy concerns [47,48]. To address privacy problems in ML, researchers have been working on federated learning (FL) and TL. FL trains a ML model using a distributed architecture in which individual devices generate their own ML model using local data and a central global server aggregates all of the locally trained models before sending the aggregated model to all network nodes. FL has many applications in healthcare due to its privacy-preserving and efficient communication constraints. Refs. [49,50] addressed basic statistical, system, and privacy considerations, as well as the consequences and potentials of FL's application in healthcare. They demonstrate that training the model in a FL framework yields results comparable to those achieved in a traditional centralized learning environment. TL is the process of transferring knowledge from one trained model to another. The fundamental idea is to reduce disparities in distributions across various models. Instance re-weighting [51] and feature matching [52] are the two main approaches. Deep TL algorithms have shown a lot of success recently in a variety of domains. To address concerns about privacy and security. FedHealth, the first federated TL platform for wearable healthcare, was introduced by [53]. FedHealth uses FL to acquire data before using TL to create reasonably personalized models. FedHealth enables deep TL without requiring access to raw user data in the FL architecture [10]. To put it another way, as previously said, a lot of promising work has been done in the field of ML and DL healthcare. However, several of these efforts are prone to privacy problems. Fedhealth is a research project that uses FL and TL architecture to overcome privacy problems. Nonetheless, as previously mentioned, works like fedhealth have the limitations of explainability [10]. As a result, research is required to solve these issues. Also, security in healthcare is very important these days, the proposed research will look forward to security-based XAI in the healthcare system.

8.2.4 Intelligibility and explainable systems research in HCI

Researchers in the field of HCI are concentrating on how users engage with intelligent systems, and one important topic is the explanation. Researchers in HCI are particularly interested in the interaction between AI systems and users, and they have done a lot of work in this area. How a camera can monitor the patient and other person's abnormal actions? The strict notions of AI systems have been heavily criticized as being incompatible with human behavior patterns [54]. Context awareness, cognitive psychology, and software learnability are among the subjects covered by XAI in HCI. Context awareness is a technique for identifying user emotions and behaviors. With the introduction of mobile devices and sensors in the early 2000s, context awareness raised a lot of concerns [55,56]. People should be able to recognize what is being observed and what actions are being done in a particular frame. Users should be able to know "what they know, how they know it, and what they are going to do next" when using a context-aware system. Explainable AI requires simplified representations of the context to make for people to understand what is gained and what action will be taken by systems [54]. Theoretical explanations are the focus of cognitive psychology. Ref. [54] looked at cognitive explanations and discovered that they are closely linked to causality thinking. Furthermore, XAI focuses not only on human cognitive psychology but also on social context understanding. The capacity of software learnability to be learned is an essential aspect of usability. It focuses on how to utilize complicated software programs via demos or in-context videos [19] and assesses the system's ease of use [54]. Users require not only outcomes but also an account of their actions from systems. Furthermore, according to [54], research has been conducted on a customized interface that gives a visual or textual explanation for context-aware rules researchers also looked at interaction design methods and how to employ feedforward to assist users to predict system behavior [54]. A key development is how users understand and operate ML systems, which contributes to debuggable and understandable ML. In AI applications such as driverless cars, understandability and predictability are critical. Data visualization is a stream from the computational standpoint of HCI, which appears to be separated from what ML researchers perform in addition to algorithmic accountability, transparency, and fairness [54].

8.3 Methodology

8.3.1 Explainable video action recognition system

Our procedures were carried out with a video-based explainable action recognition system that we created ourselves. Because of the various real-world uses that it may have. For example, activity recognition is an excellent test bed for XAI research in fire detection [44], airport security [57], smart hospitals [58,59], and assisted living [59]. Our objective is to look at system understanding and efficacy in a non-specialist audience, i.e. people who have no prior familiarity with the subject or AI skills. That is why we decided to create a system for human abnormal action recognition for healthcare security [18]. Using a pre-trained architecture with different TL techniques, we used new trained layers on top of a deep uninterpretable layer, our system generates human-understandable explanations.

8.3.2 TL

In ML, the process of transferring previously acquired knowledge to a new task is known as TL or information transfer [60]. With tiny datasets, CNNs are prone to overfitting, hence TL with deep CNNs is a good option for training the model. Yet, increasing the size of the training data can prevent overfitting; however, giving a large amount of annotated data is time-consuming and costly. TL is advantageous in this context because it solves the problem by using a pre-trained deep representation as a source architecture for constructing the new model [61]. In this study, we compare pre-trained networks as a core architectural design to address the challenge of human action recognition using XAI. Generally, pre-trained models were trained on the ImageNet dataset and can use a 224 × 224 pixel RGB image as an input to classify a 224 × 224 pixel RGB image into the proper class. These networks, as shown in Figure 8.2, consist of multiple convolutional layers (conv1–conv5) as well as some fully interconnected layers (Fc1–Fc3). Due to the millions of parameters in this network design, learning all of them for a small training sample of a new task is difficult and time-consuming. As a consequence, we added several fully connected layers for training purposes and used the base framework as a feature extractor. This TL concept is called the freezing model. If the accuracy is increased by backpropagation, then we can call it fine-tune model approach. The proposed method is creative and fascinating to recognize human actions using DL models with a TL approach, freezing or fine-tuning the pre-trained model can easily improve the accuracy of human actions for security. The outcomes of the experiments confirm the proposed work's efficacy. Furthermore, rather than

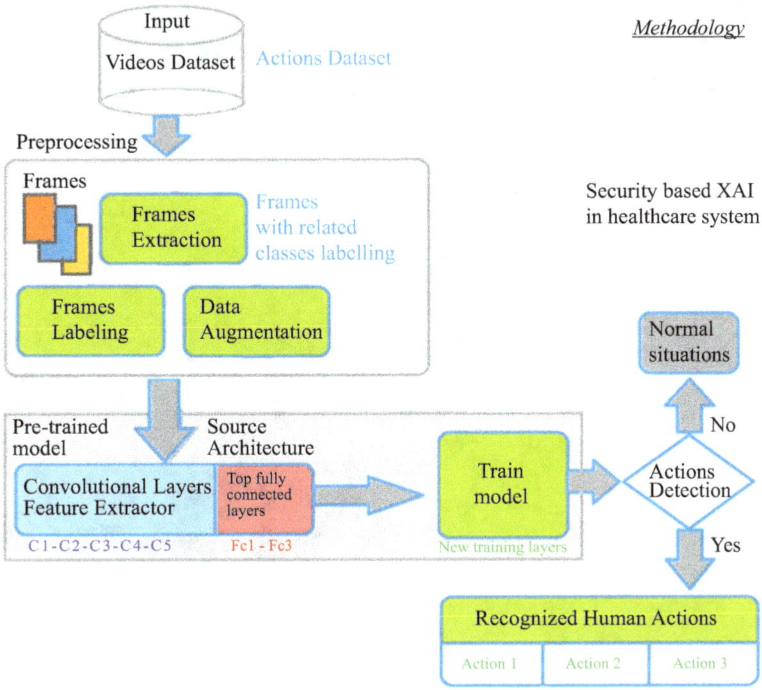

Figure 8.1 Methodology

building a new model from scratch, our studies indicate that using a TL pre-trained model enhances the classification and recognition system's accuracy. Figure 8.1 depicts a block diagram of the proposed methodology.

The followings make up the structure of the proposed approach for security-based XAI human action recognition.

1. Preprocessing module. 2. Model architecture. 3. Freeze model. 4. Fine-tune model. 5. The pre-trained model is followed by a new classifier.

8.3.3 Model architecture

After preprocessing module, the data are fed into the model to extract high-level features from video frames, the proposed method employs the pre-trained VGG19 [62] model. Figure 8.2 depicts the VGG19 structure diagram. The simplicity of this model and its demonstrated high descriptive potential for human action recognition [63] were the primary reasons for its selection. For image classification, VGG19 was trained on the ImageNet dataset. Its architecture is straightforward, with five convolutional and three fully connected layers. This model is used to extract features from the dataset we are working with. In this example, activating the first fully connected layer for classification generates a feature vector. It is worth

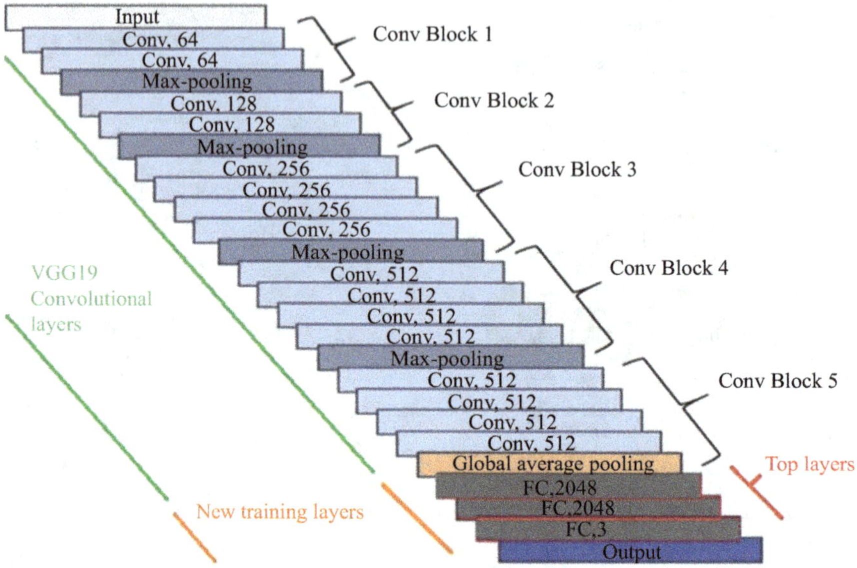

Figure 8.2 VGG19 internal structure

mentioning that some pre-trained models, such as GoogleNet [64] and ResNet [65], allow various types of low, medium, and high-level feature descriptors. Depending on the nature of the problem, different feature descriptors have advantages and disadvantages that must be evaluated. Figure 8.2 shows the proposed pre-trained VGG19 model architecture.

8.3.4 Freeze model

Pre-trained architecture is well defined and has five blocks of convolutional layers. The convolutional layers work as a feature extractor machine. In TL, there are two ways to train the target model: freeze the model and fine-tune the model. Using pre-trained models, the researchers only freeze the convolutional blocks for feature extraction. In the freeze model, the model does not update the weights of the neurons during backpropagation. In some cases, the freezing model will consider a very efficient approach to the target model. In the freeze model, the pre-trained model will use to rebuild with some changes. The top layers are removed from the pre-trained model that was trained on 1,000 classes, and new dense or fully connected layers are added to the trained new target model. The freezing model is a good strategy to train a new model for a similar problem. Most of the time, the accuracy increases with the freeze approach, and sometimes it is not applicable to increase accuracy. The proposed pre-trained model is VGG19 and freezes with new dense layers. In this paper, the model accuracy with other matrices is good as compared with other pre-trained models. The structure of the freezed model is shown in Figure 8.3.

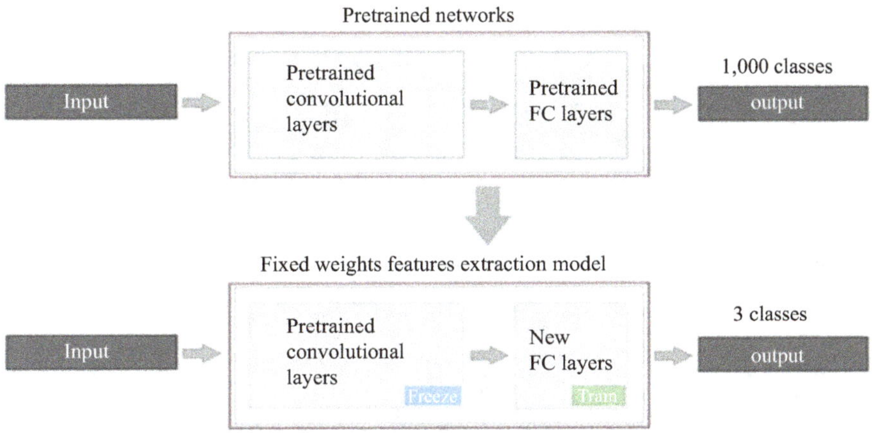

Figure 8.3 Freeze model architecture

8.3.5 *Fine-tune model*

As with like freeze model, the fine-tuned model can also work for a similar problem. In TL techniques, fine-tune method is popular for updating the weights of the neurons. In this technique, the model weights are updated during backpropagation. In this approach, the model will change the pre-trained model weights and can improve the accuracy of the problem. The pre-trained model consists of five convolutional layers for feature extraction, with some dense layers at the top of the model architecture. During fine-tune process, the model updates all the pre-trained convolutional block weights and trained target model with new fully connected layers. In fine-tune concept, the researchers can update some layers and can freeze some layers for better results. Using this method, many problems can be solved with high accuracy and are not applicable to all types of problems. It depends upon the nature of the problem. The researchers used both techniques of TL, but most of the problems can increase their accuracy through fine-tune approach. In this research, we have also used this approach for the security-based human action recognition problems. The model architecture is shown in Figure 8.4 and the experimental results are shown in the next section.

8.3.6 *Pre-trained model followed by a new classifier*

In TL, the pre-trained models are already existing and openly available. These pre-trained models are deep networks and are used for similar problems. When we reuse pre-knowledge for a new task, it is called knowledge adaptation. Therefore, many pre-trained architectures are available. The researcher uses these pre-trained models for similar problems. Many researchers use these pre-trained networks as a features' extraction machine, and some researchers freeze and fine-tune these models for classification problems. Adding new dense or fully connected layers to the top of the pre-trained model will work for the target task, but here some

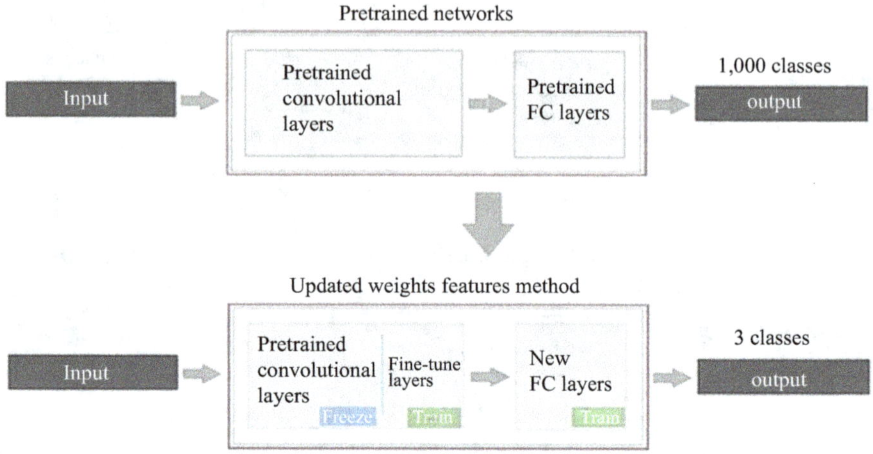

Figure 8.4 Fine-tune model architecture

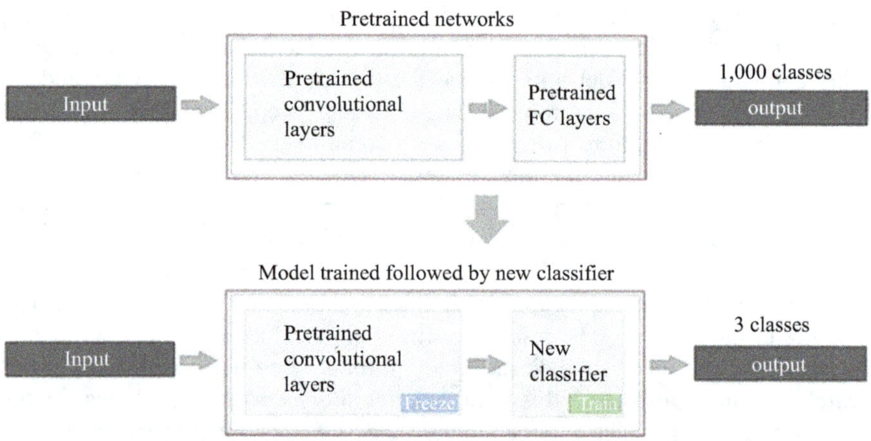

Figure 8.5 Model followed by a new classifier

researchers use the pre-trained model as just a descriptor to extract all key features from the target dataset. After features extraction using a pre-trained model, all the features are fed into a new ML classifier. In this technique, using deep representation followed by any classifier can result in better accuracy for specific action classification problems. Hence, all the mentioned techniques have their accuracy level, it depends upon the nature of the problems. The pre-trained convolutional blocks are followed by any classifier as shown in Figure 8.5, and the experimental result is shown in the next section.

8.3.7 Pre-trained CNNs implementation

Our objective is to train AI-based computer vision models to identify human abnormal actions through a camera: the models used for the purpose are ResNet50, VGG16, and VGG19. We will then use XAI to better understand and compare each model's detection (or "decision") technique to boost confidence. We used Keras to create these models and used the scikit-learn classification report to compute the metrics. We will go over the XAI techniques used in computer vision applications and how they differ from the approaches used in object recognition. We divide explanation techniques into two categories: visual and textual explanation.

8.3.7.1 Visual explanations

The most popular method of XAI in healthcare visual analysis is the visual explanation, also known as saliency mapping. Saliency maps reveal which aspects of an image are most significant for making a decision. The majority of saliency mapping techniques are based on backpropagation, although some are based on perturbation or multiple instance learning [11,66].

8.4 Experimental result

8.4.1 Human action dataset

There are many different actions performed by a human, but here in this research, we have only focused on some specific actions which are human fall, human punch, and human kick. The dataset which is used for these specific actions is HMDB51. This benchmark dataset is freely available on the Internet and consists of 51 different human actions. It is a large 3GB dataset. Finally, I used the same dataset for the proposed research, but I worked only on actions: human fall, punch, and kick. The actors in this dataset are random and have no restrictions. Both males and females are involved in the proposed actions. All the action consists of several videos without fixed time length. Video frames will be considered the input instance to the model after pre-processing. The dataset is very challenging, there is no restriction on fixed background, different genders with different dresses, and color intensity. One of the major challenges of the dataset is a moveable camera. There is no fixed camera to shoot each record on the same angle, but actions are generally recognized from any angle. The actions are recorded from different angles. Hence, for these challenges in the dataset, we proposed a security-based XAI in the healthcare system to get high accuracy with the explanation. The input samples of the dataset are overcome only for the purpose to train a model on a PC. Some of them as shown in Figure 8.6. Using the DL approach, the model needs huge dataset samples for better accuracy. The DL model generally does not work well with a small dataset. DL with a small dataset can cause an overfitting problem. Therefore, we proposed a TL technique to retrained pre-trained model with the said dataset, and no problem if the dataset is small. Besides these actions, I have also trained the same model for human fall detection. Human fall is also a very serious issue in the healthcare system. The proposed model will also recognize human falls.

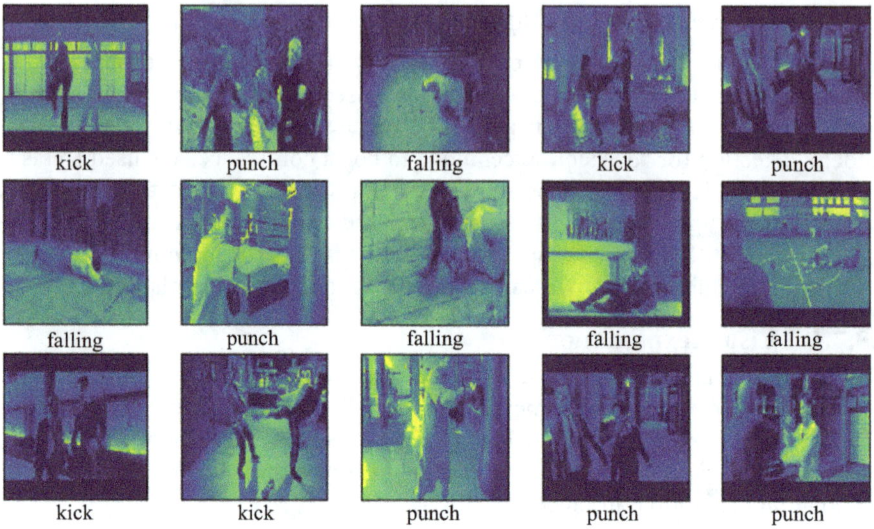

Figure 8.6 Dataset action samples

For these specific actions, I have followed the TL techniques in which the pre-trained model is used called VGG19. The model is freezed and trained with new dense layers. The pre-trained model works as a features extractor while the dense layers are fine-tuned. So from the experimental results, it is found that the model accuracy is average, approximately 75%, and val_accuracy is approximately 86%, which is considered good for real-world application because val_accuracy is more than the model accuracy. This model is also compared with other state-of-the-art pre-trained models called vgg16 and resnet50. See explanation in the next topics.

The experimental results of the dataset are examined and confirmed that the best approach for getting high accuracy is a freezing pre-trained model to recognize real-world human actions for healthcare security. Different approaches are applied to the dataset for getting high accuracy, but the freezing model approach confirmed that the accuracy is high as compared with other existing techniques. For experimental results, we used only three known pre-trained models (VGG16, VGG19, and ResNet50) as a base architecture for feature extraction. All the pre-trained model accuracy is good, but VGG19 improves the model accuracy as high as compared to other models. VGG19 model has been freezed for the new target model. The dataset, which is used to recognize human actions, is challenging, but the accuracy of the model using this dataset is increased, and good for the recognition task. The experimental results confirm that the accuracy of the model is high on the said dataset using the VGG19 pre-trained model architecture. For more explanation, the VGG16 and ResNet50 are also evaluated, first these models were freezed, fine-tuned, and then evaluated with a new classifier, but the accuracy of these models was not high as required for a recognition problem in terms of security. Finally, the VGG19 model improves the target model accuracy and

Table 8.1 Comparison of TL extraction approaches

Models	Freeze model	Fine-tune model	RF classifier	Model accuracy
VGG16	✓	×	×	**0.72**
VGG16	×	✓	×	0.32
VGG16	✓	×	✓	0.65
ResNet50	✓	×	×	**0.80**
RestNet50	×	✓	×	0.37
ResNet50	✓	×	✓	0.60
VGG19	✓	×	✓	0.68
VGG19	×	✓	×	0.39
VGG19	✓	×	×	**0.72**

Note: The bold values indicate the highest accuracy using three different approaches for one model. For example, VGG19 is a model and analyzed for three different approaches (freezed, finetune and model with RF classifier).

considers the source model for building the target model. All the experimental accuracy is shown in Table 8.1.

Table 8.1 compares different approaches.

8.4.2 ResNet50 visual explanations

ResNet-50 is as its name suggests 50-layer deep CNN. It can distinguish between 1,000 different object classes. As a result, the network has a large number of detailed feature representations for images. The ResNet50 model is used as a starting point in TL to train the target model for classification tasks. Here we will train the model for human abnormal action recognition in the healthcare system. In this research, first we used the ResNet50 model architecture for features extraction and we freeze the model, after evaluating this model, we confirmed that this model's accuracy was about 80%, which is high in terms of the human action video dataset. This model was evaluated on three different approaches, such as freezing the model, fine-tuning the model, and adding any new classifier after features extraction. All the approaches consider good and explainable using RestNet50 as shown in Table 8.1, we also used other pre-trained networks to explain further for better understandability. Using the freeze model, the confusion matrix of the actual and predicted labels is given in Figure 8.7.

The learning curve of training and validation accuracy and loss are shown in Figure 8.8.

8.4.3 VGG16 visual explanations

K. Simonyan and A. Zisserman proposed the VGG16 CNN and in ImageNet, a huge dataset with over 14 million images separated into 1,000 classes, the model achieves 92.7% test accuracy. This is a well-known model. Features are retrieved from the video footage using the VGG16 pre-trained model and then passed to a new classifier named RF. The accuracy of this deep representation with the RF classifier is not

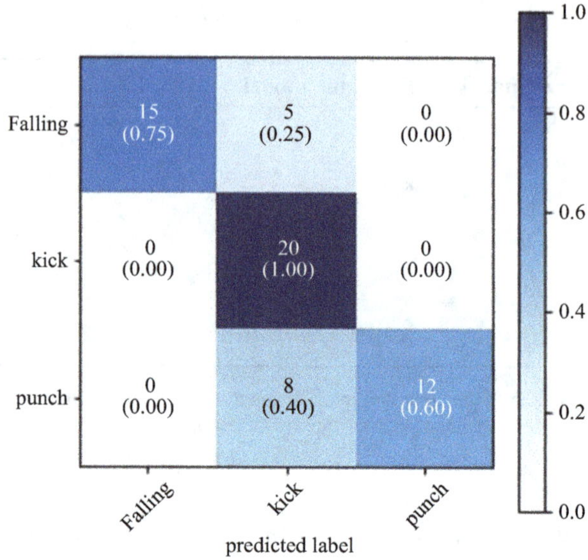

Figure 8.7   Freeze model confusion matrix

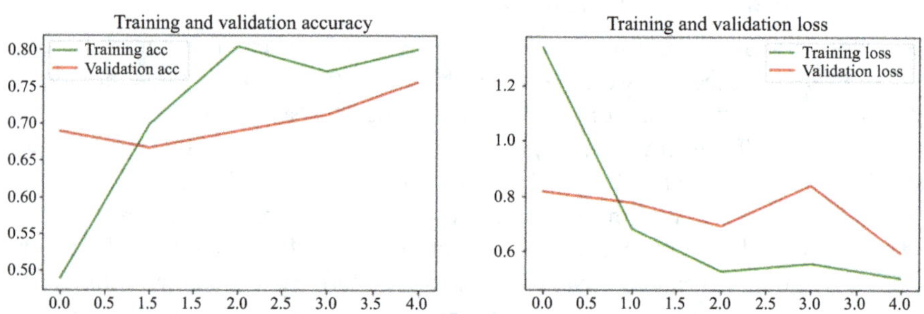

Figure 8.8   ResNet50 training and validation accuracy and loss

good because the traditional classifier is used with deep representation. So in TL, two different approaches were used to get high accuracy using the VGG16 model freezing model and fine-tuning the model. As compared to the ResNet50 pre-trained model, the VGG16 does not improve accuracy and the experiment confirmed that VGG16 is not performed well in this case. Using the VGG16 freeze model, the accuracy is about 72%, and the confusion matrix of the true label and predicted labels with learning curves is given in Figures 8.9 and 8.10.

8.4.4   VGG19 visual explanations

VGG19 is a 19-layer VGG variant that consists of 16 convolution layers, 3 dense or fully connected layers, 5 MaxPool layers, and 1 SoftMax layer. VGG is available in a number of different versions, including VGG11, VGG16, and others. VGG is a

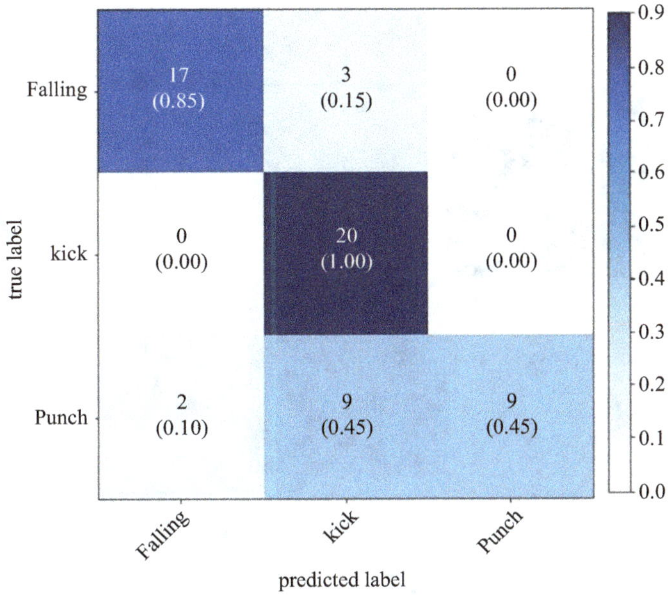

Figure 8.9 Confusion matrix VGG16

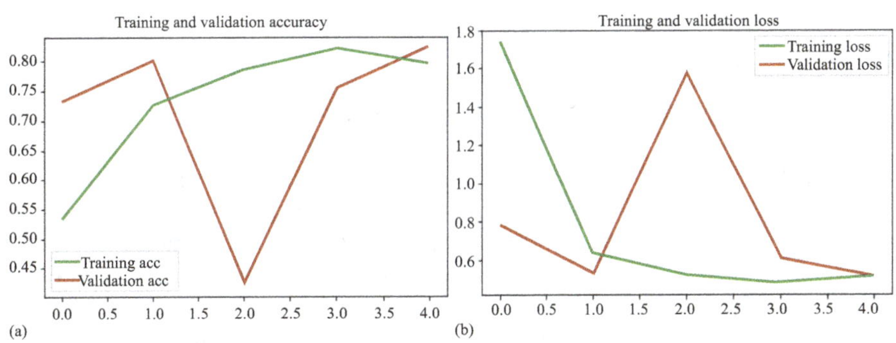

Figure 8.10 VGG16 training and validation accuracy and loss

successor to AlexNet, however, it was developed by a different Oxford group known as the Visual Geometry Group. It builds on and improves on some of its predecessors' concepts, and it employs deep convolutional neural layers to improve accuracy. VGG is a deep CNN that is used to solve classification problems in simple words. The Vgg19 model has 16 convolution layers and 3 fully connected layers. For the proposed system, the VGG19 model is freeze and the first top three layers are removed. For the target model, we add a new fully-connected layer for training purposes. VGG19 models are also evaluated on three different approaches for better results. The approaches are used for security-based action recognition

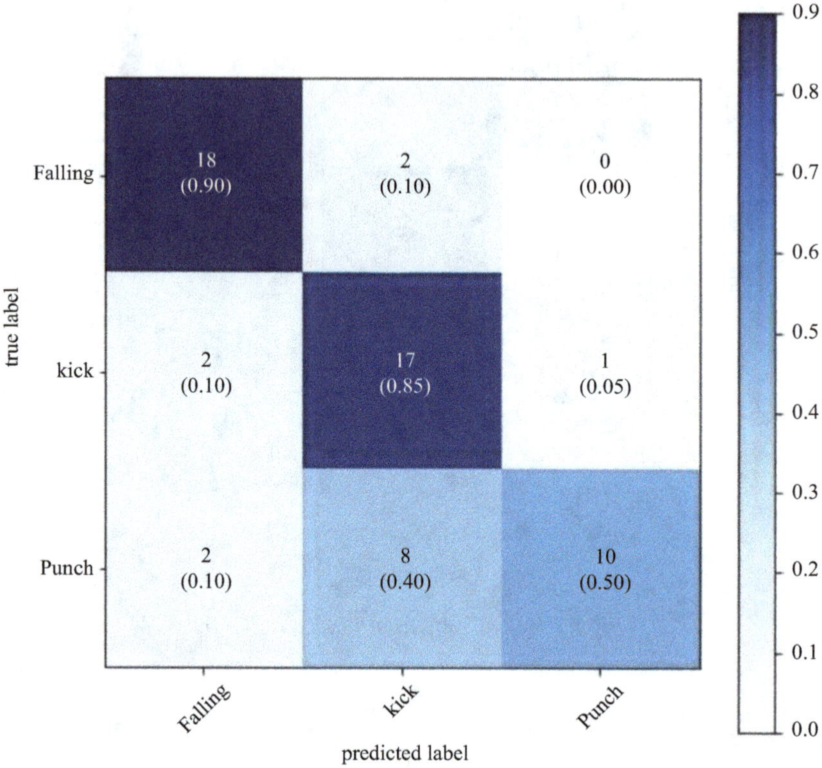

Figure 8.11 VGG19 freezed model confusion matrix

problems. The model is explained and evaluated using the freeze model, fine-tune model, and adding a new classifier after features extraction through convolutional layers. The above experiments confirmed that the VGG19 freezed model is the best in terms of accuracy. The model accuracy of the VGG19 freezed model is about 72% as same as VGG16, but this model is also considered best in terms of val_accuracy, which is about 86%. From all the matrices, it is understood that this model is best as compared to the previous different pre-trained models. Finally, the proposed pre-trained model for this research is VGG19 and the technique used for this research is the freezing model, which is a TL technique for security-based human action recognition in the healthcare system. For more explanation, the confusion matrix actual and predicted labels with the learning curve as shown in Figures 8.11 and 8.12.

For the learning curve of training and validation accuracy and loss, see Figure 8.12.

From experimental results, the learning curve for VGG19 is very smooth as compared with the other two models. It is also observed from the learning curves that other two models are not very smooth as VGG19. The performance of VGG19 is outstanding for training data as well as this model is also considered good.

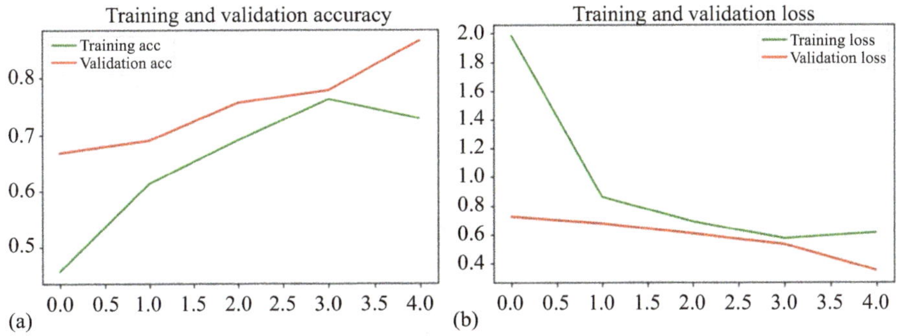

Figure 8.12 VGG19 training and validation accuracy and loss

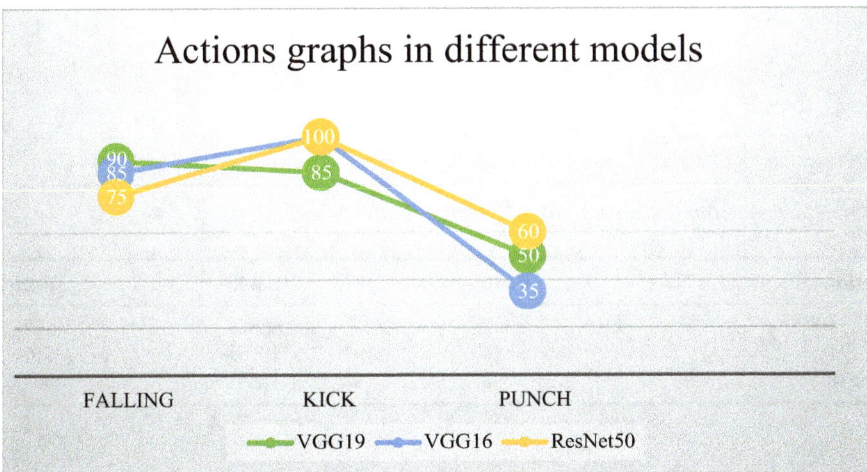

Figure 8.13 Action graphs

From the previous work, it is understood that the VGG19 model accuracy for the proposed actions has been evaluated. However, comparing the present model to other state-of-the-art pre-trained models in terms of each action, the accuracy is critical. As we know that the other pre-trained models are ResNet50 and VGG16 in this case, which are already trained on millions of training parameters with good accuracy. For more explanation of the proposed research, VGG16 and ResNet50 are also evaluated for the same proposed actions. The human actions accuracy in each model is shown in Figures 8.13 and 8.14, and Table 8.2.

It is also well observed that the VGG16 and ResNet50 models are good for the kick action, but the overall performance of these models shows us that these models are not as good as VGG19.

We compare the VGG19 with the other two models for more explanation. After the complete analysis of these three models, it is observed that the model

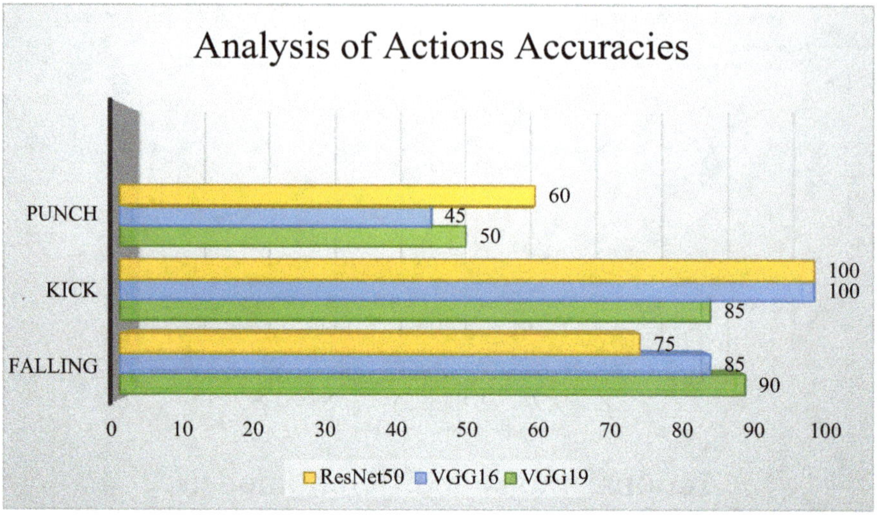

Figure 8.14 Action accuracies

Table 8.2 Actions accuracy on different models

Actions accuracies	Falling	Kick	Punch
VGG19	90%	85%	50%
VGG16	85%	100%	45%
ResNet50	75%	100%	60%

Table 8.3 Models accuracies

Model names	Model accuracy	Validation accuracy
VGG19	72%	86%
VGG16	72%	80%
ResNet50	80%	75%

accuracy of the RestNet50 is high as compared to the VGG19, while the model accuracy of the VGG16 is low, but the validation accuracy of all the models is not high than the VGG19 as shown in Table 8.3 and Figure 8.15.

8.4.5 Final discussion

After lots of experiments and details explanations, the VGG19 pre-trained model is considered the proposed model for security-based human abnormal action recognition. All the experiments confirm that the accuracy of the proposed model

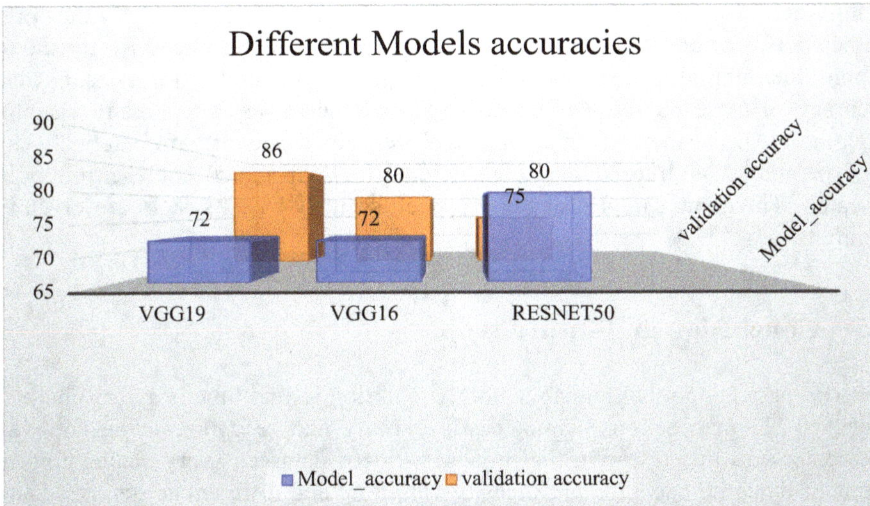

Figure 8.15 Model accuracies

Figure 8.16 Demo—real-time security-based abnormal action recognition

is high and also the accuracy of each action consider best. Initially, other pre-trained networks are also considered to retrain such as VGG16, and RestNet50 to explain the internal mechanism and performance of the models for users, but the accuracy of these models has not as high as VGG19, as it is already visually explained. All the methods applied to the pre-trained network are practical. The VGG19 model is freezed and considered the final model for the proposed research. The demo samples of the proposed security-based system are given in Figure 8.16.

8.5 Conclusion and future scope

Security-based abnormal human action recognition in real time is a very challenging task. Human actions cannot easily classify and recognize using complex datasets. So in this research, the proposed videos dataset is very challenging in terms of dynamic background, intensity variation, and different actors (male and female, young and old). For better results, it is followed that the proposed dataset as well as the techniques which are followed in this research considered best to overcome these key challenges, these challenges are generally a common problem in every complex video dataset. The researchers can easily implement the same dataset with new techniques for better performance. The proposed model is trained on a local system and then compared to other state-of-the-art pre-trained networks using XAI to visually describe the models' internal performance. In this research, VGG16, ResNet50, and VGG19 pre-trained models are explained visually in terms of performance, these pre-trained network architectures are deep NNs and all these models come under the black box of AI. Explaining the DL model is also the main objective of XAI. Hence, the complete analysis and matrices of all these pre-trained models using XAI confirmed that the VGG19 model is considered the best for security-based XAI in the healthcare system.

For future work, there are also some other challenges in the proposed research: the first challenge is the recognition of human abnormal action with new input samples. The second challenge is the model performance to recognize human abnormal action with different camera angles or locations. The mentioned challenges are achieved by this research, only the third challenge is still under observation for future work in the research community. The third challenge is, if a person touches another person's shoulder, head, or hair, then what will the model respond. This is a genuine problem for researchers in the field of AI specifically to recognize human abnormal actions.

Acknowledgment

The authors are thankful to the Higher Education Commission (HEC) of Pakistan and Turkish Higher Education in providing financial support for the proposed research project under the Pak-Turk Research Mobility Grant program.

References

[1] Meske, C. and E. Bunde. *Transparency and Trust in Human-AI-Interaction: The Role of Model-Agnostic Explanations in Computer Vision-Based Decision Support*. Cham: Springer International Publishing, 2020.

[2] Maedche, A., C. Legner, A. Benlian, *et al.* AI-based digital assistants. *Business & Information Systems Engineering*, 2019;61(4): 535–544.

[3] Lu, Y. Artificial intelligence: a survey on evolution, models, applications and future trends. *Journal of Management Analytics*, 2019;6(1): 1–29.

[4] Teso, S. and K. Kersting. Explanatory interactive machine learning, in *Proceedings of the 2019 AAAI/ACM Conference on AI, Ethics, and Society*, 2019, Honolulu, HI:Association for Computing Machinery, pp. 239–245.

[5] Zednik, C. Solving the black box problem: a normative framework for explainable artificial intelligence. *Philosophy & Technology*, 2021;34(2): 265–288.

[6] Gunning, D. and D. Aha. DARPA's explainable artificial intelligence (XAI) program. *AI Magazine*, 2019;40(2): 44–58.

[7] Gunning, D. and D.W. Aha. DARPA's explainable artificial intelligence program. *AI Magazine*, 2019;40(2): 44.

[8] Choo, J. and S. Liu. Visual analytics for explainable deep learning. *IEEE Computer Graphics and Applications*, 2018;38(4): 84–92.

[9] Mousavi, S., F. Afghah, and U.R. Acharya. HAN-ECG: an interpretable atrial fibrillation detection model using hierarchical attention networks. *Computers in Biology and Medicine*, 2020;127: 104057.

[10] Raza, A., K. Tran, L. Koehl, and S. Li. Designing ecg monitoring healthcare system with federated transfer learning and explainable AI. *Knowledge-Based Systems*, 2022;236: 107763.

[11] van der Velden, B.H., H. Kuijf, K. Gilhuijs, and M. Viergever. Explainable artificial intelligence (XAI) in deep learning-based medical image analysis. *Medical Image Analysis*, 2022: 102470.

[12] Adabi, A. and M. Berrada. Peeking inside the Black-Box: a survey on explainable artificial intelligence. *IEEE Access*, 2018;6: 52138–52160.

[13] Murdoch, W.J., C. Singh, K. Kumbier, R. Abbasi-Asl, and B. Yu. Definitions, methods, and applications in interpretable machine learning. *Proceedings of the National Academy of Sciences*, 2019;116(44): 22071–22080.

[14] Meijering, E. A bird's-eye view of deep learning in bioimage analysis. *Computational and Structural Biotechnology Journal*, 2020;18: 2312–2325.

[15] Yilmazer, R. and D. Birant. Shelf auditing based on image classification using semi-supervised deep learning to increase on-shelf availability in grocery stores. *Sensors*, 2021;21(2): 327.

[16] Došilović, F.K., M. Brčić, and N. Hlupić. Explainable artificial intelligence: a survey, in *2018 41st International Convention on Information and Communication Technology, Electronics and Microelectronics,* 2018. New York, NY: IEEE.

[17] Tsang, M., D. Cheng, and Y. Liu. Detecting Statistical Interactions from Neural Network Weights. arXiv preprint arXiv:1705.04977, 2017.

[18] Nourani, M., C. Roy, T. Rahman, *et al.* Don't Explain without Verifying Veracity: An Evaluation of Explainable AI with Video Activity Recognition. arXiv preprint arXiv:2005.02335, 2020.

[19] Cheng, Z., X. Chang, L. Zhu, R.C. Kanjirathinkal, and M. Kankanhalli. MMALFM: explainable recommendation by leveraging reviews and images. *ACM Transactions on Information Systems (*TOIS*)*, 2019;37(2): 1–28.

[20] Wang, X., X. He, F. Feng, and L. Nie. Tem: tree-enhanced embedding model for explainable recommendation, in *Proceedings of the 2018 World Wide Web Conference*, 2018.

[21] Alonso, J.M., A. Ramso-Soto, C. Castiello, and C. Mencar. Explainable AI beer style classifier, in *SICSA ReaLX*, 2018.

[22] Kim, Y. and J. Allan. Unsupervised explainable controversy detection from online news, in *European Conference on Information Retrieval,* 2019, New York, NY: Springer.

[23] Atzmueller, M., N. Hayat, M. Trojahn, and D. Kroll. Explicative human activity recognition using adaptive association rule-based classification, in *2018 IEEE International Conference on Future IoT Technologies (Future IoT)*, 2018, New York, NY: IEEE.

[24] Du, M., N. Liu, and X. Hu. Techniques for interpretable machine learning. *Communications of the ACM*, 2019;63(1): 68–77.

[25] Körber, M., L. Prasch, and K. Bengler. Why do I have to drive now? Post hoc explanations of takeover requests. *Human Factors*, 2018;60(3): 305–323.

[26] Laugel, T., M.J. Lesot, C. Marsala, X. Renard, and M. Detyniecki. The Dangers of Post-Hoc Interpretability: Unjustified Counterfactual Explanations. arXiv preprint arXiv:1907.09294, 2019.

[27] Ribeiro, M.T., S. Singh, and C. Guestrin. Anchors: high-precision model-agnostic explanations, in *Proceedings of the AAAI Conference on Artificial Intelligence*, 2018.

[28] Cai, C.J., J. Jongejan, and J. Holbrook. The effects of example-based explanations in a machine learning interface, in *Proceedings of the 24th International Conference on Intelligent User Interfaces*, 2019.

[29] Keane, M.T. and E.M. Kenny. How case-based reasoning explains neural networks: a theoretical analysis of XAI using post-hoc explanation-by-example from a survey of ANN-CBR twin-systems, in *International Conference on Case-Based Reasoning*, 2019, New York, NY: Springer.

[30] Hohman, F., M. Kahng, R. Pienta, and D.H. Chau, Visual analytics in deep learning: an interrogative survey for the next frontiers. *IEEE Transactions on Visualization and Computer Graphics*, 2018;25(8): 2674–2693.

[31] Lipton, Z.C. The mythos of model interpretability: in machine learning, the concept of interpretability is both important and slippery. *Queue*, 2018;16(3): 31–57.

[32] Montavon, G., W. Samek, and K.-R. Müller. Methods for interpreting and understanding deep neural networks. *Digital Signal Processing*, 2018;73: 1–15.

[33] Pezzotti, N., T. Hollt, J. van Gemert, *et al.* Deepeyes: progressive visual analytics for designing deep neural networks. *IEEE Transactions on Visualization and Computer Graphics*, 2017;24(1): 98–108.

[34] Kahng, M., P.Y. Andrews, A. Kalro, and D.H.P. Chau. A cti v is: visual exploration of industry-scale deep neural network models. *IEEE Transactions on Visualization and Computer Graphics*, 2017;24(1): 88–97.

[35] Schaffer, J., J. O'Donovan, J. Michaelis, *et al.* I can do better than your AI: expertise and explanations, in *Proceedings of the 24th International Conference on Intelligent User Interfaces*, 2019.

[36] Adadi, A. and M. Berrada. Peeking inside the black-box: a survey on explainable artificial intelligence (XAI). *IEEE Access*, 2018;6: 52138–52160.

[37] Papenmeier, A., G. Englebienne, and C. Seifert. How Model Accuracy and Explanation Fidelity Influence User Trust. arXiv preprint arXiv:1907.12652, 2019.

[38] Chen, H. and Y. Wang. Sschain: a full sharding protocol for public blockchain without data migration overhead. *Pervasive and Mobile Computing*, 2019;59: 101055.

[39] Miotto, R., R. Wang, S. Wang, *et al.* Deep learning for healthcare: review, opportunities and challenges. *Briefings in Bioinformatics*, 2018;19 (6): 1236–1246.

[40] Bhardwaj, R., A.R. Nambiar, and D. Dutta. A study of machine learning in healthcare, in *2017 IEEE 41st Annual Computer Software and Applications Conference (*COMPSAC*)*, 2017, New York, NY: IEEE.

[41] Sahoo, S., M. Dash, S. Behera, *et al.* Machine learning approach to detect cardiac arrhythmias in ECG signals: a survey. *Irbm*, 2020;41(4): 185–194.

[42] Liaqat, S., K. Dashtipour, A. Zahid, K. Assaleh, K. Arshad, and N. Ramzan. Detection of atrial fibrillation using a machine learning approach. *Information*, 2020;1(12): 549.

[43] Atal, D.K. and M. Singh. Arrhythmia classification with ECG signals based on the optimization-enabled deep convolutional neural network. *Computer Methods and Programs in Biomedicine*, 2020;196: 105607.

[44] Rubin, J., R. Abreu, A. Ganguli, S. Nelaturi, I. Matei, and K. Sricharan. Recognizing Abnormal Heart Sounds Using Deep Learning. arXiv preprint arXiv:1707.04642, 2017.

[45] Gjoreski, M., A. Gradišek, B. Budna, M. Gams, and G. Poglajen. Machine learning and end-to-end deep learning for the detection of chronic heart failure from heart sounds. *IEEE Access*, 2020;8: 20313–20324.

[46] Huang, J.-S., B.-Q. Chen, N.-Y. Zeng, X.-C. Cao, and Y. Li. Accurate classification of ECG arrhythmia using MOWPT enhanced fast compression deep learning networks. *Journal of Ambient Intelligence and Humanized Computing*, 2020: 1–18.

[47] Liu, B., M. Ding, S. Shaham, *et al.* When machine learning meets privacy: a survey and outlook. *ACM Computing Surveys (CSUR)*, 2021;54(2): 1–36.

[48] Waheed, N., X. He, M. Ikram, *et al.* Security and privacy in IoT using machine learning and blockchain: threats and countermeasures. *ACM Computing Surveys (CSUR)*, 2020;53(6): 1–37.

[49] Yang, Q., Y. Liu, T. Chen, and Y. Tong. Federated machine learning: concept and applications. *ACM Transactions on Intelligent Systems and Technology (TIST)*, 2019;10(2): 1–19.

[50] Xu, J., B.S. Glicksberg, C. Su, P. Walker, J. Bian, and F. Wang. Federated learning for healthcare informatics. *Journal of Healthcare Informatics Research*, 2021;5(1): 1–19.

[51] Huang, P., G. Wang, and S. Qin. Boosting for transfer learning from multiple data sources. *Pattern Recognition Letters*, 2012;33(5): 568–579.

[52] Qin, X., Y. Chen, J. Wang, and C. Yu. Cross-dataset activity recognition via adaptive spatial-temporal transfer learning. *Proceedings of the ACM on Interactive, Mobile, Wearable and Ubiquitous Technologies*, 2019;3(4): 1–25.

[53] Chen, Y., W. Lu, J. Wang, and X. Qin. Fedhealth: a federated transfer learning framework for wearable healthcare. *IEEE Intelligent Systems*, 2020;35(4): 83–93.

[54] Xie, Y., G. Gao, and X.A. Chen. Outlining the Design Space of Explainable Intelligent Systems for Medical Diagnosis. arXiv preprint arXiv:1902.06019, 2019.

[55] Dey, A.K., Understanding and using context. *Personal and Ubiquitous Computing*, 2001;5(1): 4–7.

[56] Schilit, B., N. Adams, and R. Want. Context-aware computing applications, in *1994 First Workshop on Mobile Computing Systems and Applications*, 1994, New York, NY: IEEE.

[57] Tripathi, R.K., A.S. Jalal, and S.C. Agrawal. Suspicious human activity recognition: a review. *Artificial Intelligence Review*, 2018;50(2): 283–339.

[58] Sánchez, D., M. Tentori, and J. Favela. Activity recognition for the smart hospital. *IEEE Intelligent Systems*, 2008;23(2): 50–57.

[59] Yeung, S., F. Rinaldo, J. Jopling, et al. A computer vision system for deep learning-based detection of patient mobilization activities in the ICU. *NPJ Digital Medicine*, 2019;2(1): 1–5.

[60] Aytar, Y. *Transfer Learning for Object Category Detection*, 2014, Oxford: Oxford University.

[61] Su, Y.-C., T.-H. Chiu, C.-Y. Yeh, H.-F. Huang, and W.H. Hsu. Transfer Learning for Video Recognition with Scarce Training Data for Deep Convolutional Neural Network. arXiv preprint arXiv:1409.4127, 2014.

[62] Krizhevsky, A., I. Sutskever, and G.E. Hinton. Imagenet classification with deep convolutional neural networks. *Advances in Neural Information Processing Systems*, 2012, 25.

[63] Jia, C., C. Jia, Y. Kong, Z. Ding, and Y.R. Fu. Latent tensor transfer learning for RGB-D action recognition, in *Proceedings of the 22nd ACM International Conference on Multimedia*, 2014.

[64] Szegedy, C., W. Liu, Y. Jia, *et al.* Going deeper with convolutions, in *Proceedings of the IEEE Conference on Computer Vision and Pattern Recognition*, 2015.

[65] He, K., X. Zhang, S. Ren, and J. Sun. Deep residual learning for image recognition, in *Proceedings of the IEEE Conference on Computer Vision and Pattern Recognition*, 2016.

[66] Vilone, G. and L. Longo. Classification of explainable artificial intelligence methods through their output formats. *Machine Learning and Knowledge Extraction*, 2021;3(3): 615–661.

[67] Robinson, C., F. Hohman, and B. Dilkina. A deep learning approach for population estimation from satellite imagery, in *Proceedings of the 1st ACM SIGSPATIAL Workshop on Geospatial Humanities (GeoHumanities'17)*. Association for Computing Machinery, New York, NY, USA, 2017, pp. 47–54.

[68] Zahavy, T., N. Ben-Zrihem, and S. Mannor. Graying the black box: Understanding dqns. *International Conference on Machine Learning*. PMLR, 2016.

Chapter 9

Explainable dimensionality reduction model with deep learning for diagnosing hypertensive retinopathy

Micheal Olaolu Arowolo[1,2], Hadassah Oluwadamilola Olumuyiwa[1], Ruth Omorinsola Adesina[1], Royal Afonime[1], Mobayonle Ayodeji Ajayi[1] and Paul Adeoye Omosebi[3]

Abstract

Artificial intelligence (AI) is a division of computer science that pacts with the formation and training of algorithms that attempt to mimic human intellect. Diabetic retinopathy is the major cause of eyesight loss worldwide. AI-based technologies have recently been employed to diagnose and assess diabetic retinopathy. Early identification allows for adequate therapy, preventing eyesight loss. Machine learning techniques can extract features from images and determine the existence of diabetic retinopathy. In computer-assisted medical image analysis for the identification of illnesses like hypertension, diabetes and diabetic nephropathy, and arteriosclerosis, automatic retinal picture segmentation is a crucial problem. The identification of retinal vessels allows for the primary discovery of diabetic retinopathy, the main cause of visual detachment. AI and dimensionality reduction techniques like linear discriminant analysis (LDA) with deep learning such as convolutional neural network (CNN), artificial neural network (ANN), and recurrent neural network is further recommended. Conventional identification of these retinal blood vessels is a time-consuming procedure that can be automated. In this study, the use machine learning algorithm LDA for the classification of the image with deep learning methods CNN, ANNs, and multi-layer perceptron for further classifications were used in the diagnosing of hypertensive retinopathy (HS). The data was first classified using LDA before being passed into CNN, ANN, and Resnet and results were obtained with the accuracy of 86.00%, 84.32%, and 43.29%, respectively, yet ANN required the shortest time to run, at 2.50 sec.

[1]Department of Computer Science, Landmark University, Nigeria
[2]Department of Electrical Engineering and Computer Science, University of Missouri Columbia, USA
[3]Department of Computer and Information Sciences, Trinity University Yaba, USA

Keywords: Hypertensive retinopathy; Artificial intelligence; Deep learning; CNN; ANN; ResNet; Dimensionality reduction; Diagnosis

9.1 Introduction

The layers of the retina that include axial structures, such as bipolar cells and photoreceptors, have less backscatter than the layers that contain nerve fibers and plexiform structures, both of which are organized in linear patterns. The inner retina (IR), which is comprised of the retinal nerve fiber layer, the inner plexiform layer, the inner ganglion cells, and the outer retina (OR), which is composed of the outer plexiform and photoreceptor layers, are both taken into consideration in this work. The IR and the outer retina (OR) are the two primary layers of the retina. Because it is possible to calculate the thickness of the entire retina by measuring the thickness of the layers of the retina, doing so will be a very effective method for identifying retinal disorders. They serve as the primary distinguishing qualities in therapeutic applications [1].

The ailment known as hypertension can manifest itself in several different ways within the human body. The damage to the retina of the eye that results from this condition is brought on by excessive blood pressure. Hypertensive retinopathy (HS) is the medical term for the damage that occurs to the retina as a result of high blood pressure. The retina is the primary component of the eye that is responsible for converting incoming stimuli into nerve signals and then transmitting those signals to the brain. The injury to the retina will eventually result in the loss of eyesight or possibly even blindness. Therefore, to alleviate this issue, several computerized systems that assist ophthalmologists in examining eye patients have been developed and implemented [2].

A systemic change in the arterial structure of the blood arteries in the retina is a symptom of HR, a disorder that is brought on by high blood pressure. HR is a disruption that can cause vision loss. Untreated high blood pressure symptoms are the leading cause of heart attacks in people. This accounts for the vast majority of all heart attacks. Cotton wool patches, retinal bleeding, and constriction of the arteriolar blood vessels are some of the symptoms of HR. Because of these variables, it is necessary to recognize the signs and symptoms of HR as soon as possible to more accurately target prevention and therapy. Deep learning strategies and Boltzmann machines are being utilized in the creation of a system that will aid in the initial diagnosis of the HR stage. To evaluate the classification of HR with deep neural networks (DNN) and Boltzmann machines, the proposed method uses retinal image parameters including the artery–vein diameter ratio (AVR) and changes in a location with the optic disk (OD). This is done to ensure that the classification of HR is accurate. We went with this approach because, according to the findings of previous studies, DNN models were superior to Boltzmann machines when it came to accurately recognize patterns in images. Boltzmann machines are typically used for learning neural networks since they entail rapid iteration. The creation of an original system for the primary discovery of the HR phase, as well as an evaluation of the effectiveness and accuracy of the projected methodologies, are the anticipated goals of this study. Additionally, the assessment of the efficacy and accuracy of the projected methodologies is also a goal [3].

There is a type of potentially fatal disease known as HR, which is instigated by high blood pressure in the retina. This condition can lead to impairment of the blood vessels in the eyes, damage to the nerves in the eyes, and even blindness in extreme cases. The quantity and dimensions of the data can be decreased using a technique that is known as "dimensional reduction," which is important for processing photos quickly and effectively [4].

Both principal component analysis (PCA) and the picture resize tool found on MATLAB were developed to achieve reduction rates of 96.875% and 99.032% of the original image, respectively. The backpropagation neural network (BNN) learning model that was proposed in this study has several distinguishing characteristics, including the number of layers, the number of nerve cells in the concealed layer, and the learning rate. There is a connection between the design of the BNN model, the model's performance, and its learning rate. According to the findings of the research, the learning model has an architecture consisting of 8 layers, 128 nerve cells in the input layer, 1 neuron in the output layer, and 258, 128, 64, 32, 8, and 2 neurons in each hidden layer. Its learning rate is 0.001%. The accuracy levels for learning can be found to be 88.46%, and the accuracy levels for testing can be found to be 86.36% accordingly [5].

Currently, there are hardly any electronic systems intended to identify HR. However, those systems primarily fixated on feature extraction utilizing approaches grounded on deep learning models (DLMs) or human intervention. DLMs have a hard time defining generic features for HR detection since custom features necessitate elaborate image processing techniques. Not only that, but even when employing deep-feature techniques the kind utilized by state-of-the-art HR diagnostics systems the classification accuracy still falls short. To deal with these issues, a HR-specific system (DenseHyper) was developed by adding a trained features layer (TF-L) and a dense feature transform (DFT) layer (DFT-L) to the deep residual learning (DRL) procedures. This is done to recognize the HR. The DenseHyper system is made up of a variety of multilayer dense architectures that integrate TF-L and convolutional neural networks (CNNs) to offer specialized features using DFT-L after learning features from a variety of lesions. These architectures are part of the DenseHyper system. An approach known as DFT, which is based on learning, was used in the development of the DenseHyper system to improve classification precision. Four online sources and one private source of data are gathered to evaluate the DenseHyper system and make comparisons to other systems. There were 4,270 photographs of the retina's fundus analyzed statistically to prove the DenseHyper system's efficacy. The area under the receiver operating curve (AUC), sensitivity (SE), specificity (SP), and accuracy (ACC) were used as performance indicators (AUC). When compared to contemporary research practices, the results were statistically significant. A 10-fold test of cross-validation resulted in an average standard error of 93%, standard performance of 95%, area under the curve (AUC) of 0.96, and accuracies of 95%. According to the experimental findings, the DenseHyper system can be put to use to make an accurate diagnosis of HR [6].

This study develops a model using an image analyzer, Recursive feature elimination, and deep learning methods CNN, artificial neural network (ANN), and recurrent neural network (RNN) in diagnosing HR from retinal images.

9.2 Overview and related works

9.2.1 Hypertension

The most important risk factor that may be modified to reduce all-cause mortality and morbidity around the world is systemic arterial hypertension, which is also connected with an improved likelihood of developing cardiovascular disease (CVD). Although proper management of hypertension can reduce the overall burden of disease and mortality, less than half of those who have the condition are aware that they have it, and many more are aware but are not treated or are managed incorrectly. This is even though treating hypertension correctly can lower the overall burden of disease and mortality [7]. Hypertension is a critical and pricey problem that affects the public's health. It is a substantial contributor to the development of CVD, although it can be altered. Based on randomized controlled trials, lowering blood pressure may lower the chance of having a stroke, diseases connected with coronary arteries, congestive heart failure, end-stage renal disease, peripheral vascular disease, and overall humanity. Even at a blood pressure of just 115/75 mm Hg, there is a persistent danger of getting certain hypertension-related issues. This risk increases with higher blood pressure readings. Despite the inherent dangers to one's health that are associated with hypertension that is not under control, the vast majority of individuals still receive insufficient therapy for their elevated blood pressure. In the United States, nearly one in three people have high blood pressure, and each year, two million new cases of the condition are identified. An additional 28% of people existing in the United States are affected by prehypertension, while 7% are completely ignorant that they even have hypertension. More than one billion individuals around the world are affected by hypertension, and it is anticipated that number will rise to 1.56 billion by the year 2025. Its death rates are the highest in the world, and the number of years of life lost due to disability is the second highest in the world. It has been demonstrated through randomized controlled trials that lowering one's blood pressure lowers the hazard of having coronary artery disease, stroke, peripheral vascular disease, congestive heart failure, end-stage renal disease, as well as death [8].

9.2.2 Machine learning

Using a method of data analytics known as "machine learning," computers may be instructed to learn new things from their experiences, just like persons and other animals do. Machine learning algorithms use various computational methods to "learn" information nonstop from the information. These algorithms do not rely on an equation as a model as traditional modeling techniques do. When there are more examples available for learning, the algorithms improve in terms of both their ability to adapt to their surroundings and their overall quality. Deep learning is a subfield of machine learning that emphasizes extremely complex problems [9].

Research in machine learning focuses on grasping and implementing "learning" processes, which are processes that use the information to improve performance on a given set of responsibilities. This is a subfield of computer science. It is generally measured to be a factor of artificial intelligence (AI). Algorithms that use machine learning construct a model by constructing it from samples of data, which are sometimes referred to as training data. This model is then used for predictions or judgments without specified instruction. Machine learning algorithms are put to use in a wide variety of application sectors, with speech recognition, computer vision, e-mail filtering, and medicine, where it is tough or infeasible to design outmoded algorithms that are proficient in doing the necessary tasks [10].

Although not all machine learning involves taking lessons from statistics, a subset of it does, and it is the part that focuses on how computers can make predictions. Computational statistics is closely tied to machine learning. The study of mathematical optimization provides several useful tools, theoretical frameworks, and application domains, all of which are of help to the focus of machine learning. Data mining which focuses on unsupervised learning for investigative information investigation is a ground of study that is analogous to cloud computing. Some bids of machine learning use data and neural networks in a manner that is very comparable to the way the neurons in a living brain communicate with one another. When it comes to finding solutions to challenges faced by businesses, one name for machine learning is predictive analytics [11].

Machine learning allows computer programs to complete tasks without having those duties specifically written into the program. Computers can learn specific tasks by analyzing the data that is accessible to them. For basic tasks that are delegated to computers, it is conceivable to construct algorithms that educate the device on how to carry out all of the steps that are necessary to handle the issue at hand; the computer does not need to learn anything to carry out these algorithms. It may be interesting for a human to physically construct the necessary algorithms to complete more complicated tasks. It may prove to be more successful in practice to aid the machine in building its algorithm than having a computer programmer specifically explain each necessary step [12].

In AI, machine learning refers to the process of teaching computers to carry out tasks for which there is no one, best answer. When many different responses could be given, one tactic that could be utilized is to acknowledge some of the correct answers as being correct. After that, the computer can utilize these examples as training data to perfect the algorithms that decide the correct responses. Both unsupervised learning and reinforcement learning are examples of approaches that are used in machine learning. Unsupervised learning can reveal patterns that were previously hidden or structures that were thought to be intrinsic to the input data. Reinforcement learning represents the other approach.

9.2.3 Related works

A method that uses the fundus image along with the input from a CNN to identify if there are indications [6]. The DRIVE image dataset is utilized to validate the proposed system, and the experimental outcomes indicate that the proposed method has an accuracy of 98.6%. The accuracy of the training increases proportionately with the iterations numbers used, so increasing the number of iterations results in a higher accuracy level. However, there is a need to progress the accuracy by including a blood vessel of the retina as an input feature of a CNN and by classifying not only two classes, which are regular and indications of HR, the classification ranking is based on four rankings. This will allow for a more accurate diagnosis of HR.

A strategy for determining AVR was proposed, in which first the vessels would be segmented by the use of the filtering approach, and then the ocular disk would be detected to regulate the Region of Interest [2]. A neural network is used to determine the AVR, which is then used to categorize the blood vessels as either arteries or veins. MATLAB R2014a.t is used in the execution of the work that was proposed. However, a comparison of these results with some earlier results reveals that the diagnostic accuracy of the neural network results for HR is slightly higher than that of the previous results.

BNNs were utilized as a method for accomplishing the task of locating the retinal fundus [13]. Before the identification process, pre-processing steps including the green channel, contrast limited adaptive histogram equalization (CLAHE), contextual elimination, morphologic close, thresholding, and linked component analysis were carried out. Feature extraction was performed with the help of zoning. The results of the study show that the approach proposed can detect the retinal fundus with an accuracy rate of 95% when using a maximum epoch of 1,500. This work benefited from solid research, but it did not include any examples of implementation.

An algorithm that initially extracts the blood vessels from the retinal image that has been preprocessed has been proposed [14]. To determine whether or not the pixels that have been detected belong to the blood vessel class, gray level and moment-based features are retrieved and analyzed. The change in intensity, along with color information, is utilized to determine whether the vessel in question is an artery or a vein. The arteriovenous (AV) ratio is measured using the vessel width estimate method and using this ratio, the several phases of HR can be diagnosed. Photographs of the retina were taken from the VICAVR record and were combined with images received from the Deepam eye facility in Chennai. Twenty-five of the photos were considered to be normal, while the remaining 76 were considered to be examples of HR.

However, in some extremely few instances, there is a concern with the disease being incorrectly identified as something else. It is possible to further increase the classification rate by introducing methods that efficiently detect the localized narrowing of retinal vessels as well as the right-angled crossing of those vessels.

Mean fractal dimension, tortuosity index, and arteriole-to-venule ratio can all be quantitatively evaluated simultaneously and without the need for an operator with the method that has been provided [15]. Both HR and cerebral autosomal central artery with subcortical infarctions and leukoencephalopathy, two conditions that are known to be associated with abnormalities in the retinal vasculature, were treated with this method (CADASIL). The findings substantiated the efficacy of our methodology in locating and estimating the severity of retinal vascular irregularities. Participants who had HR or CADASIL had altered aorto-venule ratios, tortuosity indices, and mean fractal dimensions as compared to controls of the same age and gender. Dependability among raters was quite high across all three criteria (intraclass correlation coefficient 85%). The technique provides an easy and extremely reproducible method for distinguishing between compulsive disorders that are categorized by abnormalities in the morphology of retinal vessels. These diseases are characterized by changes in the appearance of the vessels in the retina. Our method offers the advantages of simultaneity as well as independence from the operator. In addition, the integral of the warp as well as the number of maneuvering shifts that occur along the passage of the vessel are both reflected in TI as it is calculated by the Cioran program. It would appear that the number of directional changes is the most important component. This is since the integral's value, which is the AUC that represents the vessel pathway, did not offer statistically important findings.

A novel HR (HYPER-RETINO) framework has been established as part of this research. This framework assigns a grade to HR based on one of five levels [16]. The HYPER-RETINO system is built using HR-related lesions that have been pretrained. To create this HYPER-RETINO system, several steps had to be carried out, including preprocessing, the discovery of HR-related abrasions through semantic and instance-based segmentation, and the implementation of Dense Net planning to classify the stages of HR. In general, the HYPER-RETINO method identified five categories of HR by locating the limited regions in the input retinal fundus descriptions that corresponded to each grade. A 10-fold cross-validation test was performed on 1,400 images of HR, and the results showed an average of 90.5% sensitivity, 91.5% specificity, 92.6% accuracy, 91.7% precision, Matthew's correlation coefficient of 61%, 92% F1-score, and 0.915 area-under-the-curve. The results of the experiments show that the HYPER-RETINO approach can be used to accurately detect different phases of HR, proving its relevance in this regard. To the best of our knowledge, there have only been a select few frameworks given in the past that are based on deep learning (DL) models and are designed for the two stages of recognition (HR versus non-HR). These systems have been evaluated using relatively small datasets and did not involve any pre-processing processes. As a result of this, it is challenging to utilize them as a screening approach for the determination of the degree to which HR is present.

A process for the diagnosis of HR that makes use of PCA and BNNs [17]. The image of the retina was obtained from the STARE database, which divides its data into learning and testing in a proportion of seven to three. As a method for reducing the dimensions of fundus images, PCA has been successful in achieving a 99.9%

reduction in the amount of raw fundus image data, which reduces the amount of computing load required for neural network training. In this research, the back-propagation neural network, or BNN, was given as the major classification technique. This was accomplished by specifying the parameters of the algorithm, learning from the data, and then evaluating the data. Based on the BNN output result, the model was able to classify retinal images into one of two categories: typical retina and retina with high blood pressure. The outcome of the model that was proposed demonstrated that the testing accuracy can reach 86.36%. The learning approach that was proposed has an advantage over the other ways in that it can reduce the amount of image data by up to 99.9%, which enables it to use a very small amount of period and computation properties in comparison to those used by the other approaches. Because a clinical diagnosis is essential and has a substantial influence on the sequence of treatment that is administered to a patient, the accuracy of the model that has been proposed still needs to be improved.

Cross-sectional learning was carried proposed to investigate a process for gauging retinal vessel distances based on ethereal sphere ocular rationality imaging for the diagnosis of chronic HR [18] (SD-OCT), the artery-to-vein ratio (AVR), central retinal vein diameter (CRVD), and central retinal artery diameter were all measured. This was done to investigate a technique that could be used (CRAD). In this study, there were a total of 119 participants, each of whom had one of their 119 eyes examined. There were 63 subjects in the normotensive group, each of whom had one of their 63 eyes examined, and there were 56 subjects in the hypertensive group, each of whom had one of their 56 eyes examined. Both the CRAD ($t = 2.14$, $P = .04$) and the AVR ($t = 2.59$, $P = .01$) demonstrated a noteworthy gap among the two clusters. The cut-off points of 0.75 were determined with the use of the receiver operating characteristic (ROC) curve (AUC 0.786; 95% poise interval, 95% CI 0.70–0.87). According to multivariate logistic regression analysis, patients who had high systolic blood pressure (odds ratio OR 4.39; $P = .048$), patients who were male (OR 4.15; $P = .004$), and patients who smoked (OR 5.80; $P = .01$) were more likely to have an AVR of less than 0.75. According to the Bland–Altman plots, the two technicians who worked in the CRAD, CRVD, and AVR each produced measurements with a minor mean bias in their results. In conclusion, employing SD-OCT provides a method that is accurate, reproducible, and practical for measuring the sizes of the retinal blood vessels. It is useful in determining whether or not HR has progressed to the chronic stage. SD-OCT is utilized to quantitatively evaluate retinal hemorrhages, hard exudates, cotton-wool spots, and optic disc edema. SD-OCT will also be applied to differentiate the acute stages of HR from those of other illnesses. Standardizing categorization based on antihypertensive medication is something that we are going to give some thought to further examine the influence that diverse antihypertensive drug groupings have on retinal vascular variations as measured by SD-OCT.

It has been demonstrated that an automated approach for the early diagnosis of HR is beneficial for both ophthalmologists and patients [19]. An automatic approach for the identification of HR utilizing the AV ratio is presented in this research. The proposed system is composed of a brand-new method for classifying blood vessels as either arteries or veins through the utilization of a new feature vector and a hybrid classifier. In addition to that, this research proposes an innovative way to compute vessel width, which may be utilized while measuring AV ratio. Using the AV ratio that has previously been determined, the system determines whether or not the fundus image contains HR. In evaluating their method, two digital fundus image records were adopted, namely VICAVR and DRIVE, which are both open to the public. The findings from our experiments demonstrate that our proposed algorithm is sound. However, additional effort and investigation are required for this study.

Fundus images were reconstructed, and a model for classifying cases of HR was built using Restricted Boltzmann Machines (RBM). To add to this, the dataset known as Messidor was used for this study [20]. The results of the studies show that the model achieved a 99.05% accuracy rate in classifying images into one of nine categories according to the severity of HR. The goal of this study is to use RBM to create a classification scheme for HR. According to the experimental data, the model has a high rate of accuracy (99.05%) while reconstructing images, which shows that it can generalize image input into one of nine output classes well. But as it is still a picture, the output model must be combined with other layers like SoftMax to achieve the class label output. Building a Hypertension Retinopathy Classification Model is next on our list of things to investigate. We will use a hybrid of RBMs and CNNs to refine our classification results and produce a more reliable class label.

A new HR (Dense Hyper) system has been developed to recognize the HR based on a suggested TF-L and DFT-L to the DRL techniques. Dense Hyper is the name for this structure [21]. Several distinct multilayer dense architectures combine to form the Dense Hyper system. A CNN is combined with TF-L to generate these structures, which can then learn characteristics from a large dataset of lesions. Next, DFT-L is utilized to create domain-specific features. The Dense Hyper system's classification accuracy was enhanced by the use of a DFT method that was learned during development. The Dense Hyper system is evaluated and compared using information gathered from four different online and one offline sources. To further illustrate the efficacy of the Dense Hyper system, a statistical analysis of 4270 retinal fundus images is conducted utilizing sensitivity (SE), specificity (SP), accuracy (ACC), and area under the receiver operating curve (AUC) measures. The results gained were substantial, especially when compared to the results produced using the most modern and cutting-edge research methods. Average results for standard error (SE), predictive power (SP), and AUC (0.96) were obtained from a 10-fold cross-validation test (AUC). Evidence from these tests shows that the Dense Hyper system can be used to reliably diagnose HR. As a result, the Dense Hyper system for HR detection will benefit in the not-too-distant future from the addition of a larger dataset of retinography images that have been compiled from a wide range of sources. It is likely that, in addition to deep features, hand-crafted features will need to be incorporated into the model to increase its classification accuracy of the model.

A total of 4,000 fundus pictures [22], comprising both photos with and without fundus abnormalities, were employed in an investigation of continuous neural networks. This was done so that CNN could analyze fundus images for signs of macular degeneration and diagnose hypertension and arteriosclerosis in patients. To improve the performance of the convolutional neural network used in the deep learning structure, this article based its data preparation efforts on Turkey. As a part of training the DLM, these data sets were mixed with the local data sets. Additional data sets based on Turkey were also generated for this research as part of an effort to integrate the data globally, which can assist in standardizing the results and enhancing the accuracy. This equipment is used in the diagnosis of retinal vascular degeneration, such as fundus vascular disease and macular edema disease. The findings have been used to improve HR diagnosis and classification.

This basic understanding formed the basis for the application of the research. The author also discusses the constraints that are placed on the system. The requirement for sustained economic viability over the long term stands out as the most significant limitation among them. The fact that all of the categorized images were obtained from single imaging equipment is another restriction of the system that has been proposed. A considerable restricting effect is caused as a part of the extension of the system as a result of the fact that during the training of the system, different imaging equipment was not utilized.

The conventional method of enhancing retinal images, known as Contrast-Limited Adaptive Histogram Equalization (CLAHE), yields a result that is contingent on the user's selection of the clip limit (CL) and the number of sub-images (N) [23]. Adaptively Clipped-CLAHE (AC-CLAHE) and Fully Automated-CLAHE (FA-CLAHE) are the names given to the modified versions of CLAHE that were proposed by the study to eliminate the problems that are caused by these limiting variables. To improve the contrast between retinal landmarks and lesions on the retina, the proposed methods were evaluated and found to be successful. The new technology can be utilized directly in hospitals and at remote places as support to doctors for the screening of diabetic and HR. This allows for the inspection of the tiny details that are located on the retina. This issue is resolved by fully programmed auto-CLAHE, in contrast to AC-CLAHE, which still retains its subjectivity as a result of the choice of a fixed number of sub-pictures. It is important to evaluate the suggested adaptive CLAHE methods in terms of quantitating quality measures such as entropy, global contrast, and absolute mean brightness error in the context of the future work scope for this project.

To test for HR, an algorithm has been developed for the identification of AV nicking in fundus pictures [24]. Within the scope of this investigation, the input fundus images have been pre-processed by employing green channel extraction and the histogram technique to achieve improved contrast. Utilizing the butylated hydroxytoluene (BHT) approach, blood vessels have been isolated from the images that have been pre-processed. After that, the center line was taken from the blood vessels that had been segmented. Through the utilization of the crossing number method, we were able to identify places of crossover in the center-line extracted image. The computation of the distance between the crossover and the termination point is what's used to accomplish the removal of spurious. According to the value of intensity that is lowest, veins have become distinct. Finally, the thickness of the vessel was measured in the AV crossing places. The normal AV crossing has an average thickness of 11, while AV nicking has an average thickness of 16. Therefore, the thickness of the vessels in a normal AV crossing at the crossover area is significantly less than the thickness of the vessels in an AV nicking. The vessel thickness is what allows for the detection of normal AV crossings as well as AV nicking in retinal fundus images. In the future, there will be a greater emphasis on using visuals.

An effective and speedy method for removing the retinal vasculature has been presented in this paper [25]. To complete the segmentation process, the RGB color space is broken down into three planes, and in the first stage, only the green plane is analyzed and processed. Increasing contrast can be accomplished by first applying a sigmoid function, and then excluding the backdrop from the image. In the final step, hysteresis thresholding and morphological processing are used to extract vessels to enhance the fine features in the image that was produced. Producing promising accuracy and other metrics is one way to minimize the impact of a tradeoff between the accuracy of segmentation and the amount of time used by the segmentation algorithm. The algorithm is validated and assessed using the images of the retinal fundus contained within databases such as STARE and DRIVE. The created findings were assessed and compared to other state-of-the-art work, and they showed to be superior and exceeded the other work in the majority of instances. The proposed method does have one drawback, and that is the fact that during the hysteresis thresholding stage, changing the lower threshold value, which can be anywhere from 1 to 12 in value, can alter the results. Our work will be able to be expanded in the future by the automatic selection of lower thresholds that produce the greatest results.

This chapter demonstrates how to use a fundus image of the retina and a fractal analysis method to detect HR at an early stage. The fractal analysis relies on the fractal dimension and lacunarity as independent variables, while the ensemble Random Forest and the k-fold cross-validation serve as classification and validation strategies, respectively [26]. Accuracy, positive predictive value (PPV), negative predictive value (NPV), sensitivity, specificity, and AUC are all metrics used to evaluate a test's efficacy (AUC). The test's 10-fold cross-validation statistics showed an accuracy of 88.0%, a positive predictive value of 84.05, a negative predictive value of 92.0%, a sensitivity of 91.3%, a specificity of 85.19%, and an AUC of 88.25%. The best results can be achieved with lacunarity while using a box size of 22. The results of the study suggest that the AUC generated by utilizing fractal analytic approaches for HR early diagnosis is in the good to excellent range.

In this study, we improved and segmented the retinal vasculature, which is required for quantifying measurements of the AV ratio based on its physical properties including width and length (AVR). The NEI and the NIH have funded this study [27]. To enhance retinal fundus images, the tophat transform is used, and iterative thresh-holding is used to segment the blood vessels. Fifty digital fundus photos from the public MESSIDOR dataset are used to test the efficacy of the proposed technique. HR can be measured quantitatively through the detection of a narrowing of the retinal artery–vein ratio compared to normal images. When compared to regular pictures, this ratio serves as a useful yardstick. Patients with HR on the MESSIDOR dataset showed an AV ratio of 0.203–0.495 after applying the proposed methodology. The usual range for this ratio was discovered to be between 0.62 and 0.73. This kind of dataset was not done in the past. The stages of HR can be estimated by measuring the AV ratio. It is feasible that in the following investigations, an automatic classification of blood vessels will be applied. To determine the clinical usefulness and reliability of our approach, it is possible to conduct evaluations on a broad group of photos collected from people whose AVR ranges are substantially more diverse.

Developed methods for identifying significant retinal vessels and dissecting them into arteries and veins for A/V ratio determination have been implemented [28]. The photos utilized in this study were from the DRIVE database. For the sake of developing and evaluating vessel detection algorithms, we have compiled a database with 20 unique examples. The database's reference standard for vascular segmentation served as a starting point, and from there, the principal arteries and veins were selected manually for the superior and inferior temporal regions, respectively. The black top-hat transformation and the double-ring filter are used to identify blood vessels in the retina. The A/V ratio was calculated with a focus on the large vessels that ran from the optic disk to the temporal regions. These blood vessels were chosen out of the ones that were taken out. Image features were extracted from blood vessel segments located between a quarter and one optic disc diameter from the disc periphery. Linear discriminant analysis (LDA) was used to classify the target segments in the training examples as either arteries or veins. Then, the selected parameters were used on the target segments in the test instances. In the 20 test cases, the arteries and veins were correctly classified in 30 out of 40 pairs, which is a 75% success rate. The outcome can be utilized in the computerized process of calculating the AV ratio. The vessel segmentation is an area that needs to be worked on further in upcoming research.

A probabilistic neural network (PNN) for the diagnosis of HR using images of the retinal fundus [29]. In the proposed method, the image processing and feature extraction techniques employed before the identification procedure are called box counting and invariant moments, respectively. According to the findings of the trial, the method that was suggested was successful in identifying HR with an accuracy that reached one hundred percent. It has been suggested that future studies should make use of an adequate image processing method so that the object segmentation process, in particular the identification of retinal vascular structures, can be improved. In addition, the utilization of additional training data is recommended to improve the precision of the testing data. In addition, a wide variety of methods for the extraction of features can be put into practice. The technique can be evaluated alongside deep learning in forthcoming studies by utilizing the same types of research information.

Among the most significant blood arteries in the retinal picture, from which arterial and venous blood can be extracted mechanically. Using the graphical vascular tree derived from retinal scans, this article proposes an automated approach for artery and vein categorization [30]. The proposed method distinguishes the graphical retinal network by classifying each graphical node as an endpoint, an intersection point, or a distinct point node. Each visual connection is tagged as either an artery or a vein. The final step in the process involves not only structural features but also intensity-based characteristics to properly categorize arteries and veins. The publicly available DRIVE database has been used to validate this method's findings. An automated retinal image analysis package will be studied further. This will provide several metrics that can aid in the early diagnosis of systemic diseases like diabetes, hypertension, and vascular disorders, including AVR, branching angles, vessel tortuosity, and fractal dimensions.

9.3 Materials and methods

This study proposes to develop a model using an image analyzer, recursive feature elimination, and deep learning methods CNN, ANN, and RNN in diagnosing HR from retinal images.

9.3.1 Data description

The dataset adopted for this study was obtained from the Kaggle repository (https://www.kaggle.com/code/meenavyas/retinopathy-detection). Kaggle is a platform for developers to get data and framework which is crowdsourced. This dataset was discovered by searching for "retina images" on the Kaggle platform. This dataset of retina images will be analyzed using CNN, ANN, and RNN to accomplish the evaluation performance metrics; processing time, false positive rate, negative predictive value, false discovery rate, precision, false negative rate, accuracy, sensitivity, Matthews correlation coefficient, specificity, F1 score, and average recall [31]. Table 9.1 shows the dataset and features.

The dataset contains images of hypertensive samples which entails healthy, proliferate, moderate, and severe. The proposed model uses a convolutional neural network, the dataset comprises retina descriptions, and they are trained using ANN, CNN, and Resnet models to predict the model with a better performance result.

9.3.2 Data preprocessing:

This technique is adopted for changing and cleaning data. The process of cleaning and changing data is done to make the dataset suitable for analysis. Feeding an unprocessed dataset into a system reduces the accuracy of the results. The method of preprocessing adopted in this system of the neural network is LDA.

Table 9.1 Number of samples and features of the retina image dataset

Samples	Features
1,500	4

9.3.2.1 LDA

For feature extraction and dimension reduction, a well-known method is LDA. It has been extensively employed in a variety of applications, including face recognition, image retrieval, categorization of microarray data, and more. To achieve maximum discrimination, data are projected onto a lower-dimensional vector space with the proportion of the between-class distance to the within-class distance exploited. Applying the Eigen decomposition to the scatter matrices will make it simple to calculate the best projection (transformation) [32].

9.3.2.2 Convolutional neural network

Convolution involves a method of scheming the combination of two functions combined while one indicator is reversed. Let's take the two functions as x and y they are passed over each other, it states the sum of their overlay. This system is used for the combination of two components. With signal o, it rotates 180° on the horizontal axis. And then flipped across x and y while multiplying and maintaining the values [33].

9.3.2.3 ANN

They are a type of non-parametric prediction tool that mimics the linked nature of the biological nervous system by using a network of artificial neurons. It is possible to employ a data processing unit based on a connection technique for different kinds of pattern categorization and recognition. This neural network is used to perform complex functions. It can be used in classification, identification, pattern recognition, speech, control systems, visions, and control systems and it is used to solve difficult problems for computers or human beings [34]. Algorithm 9.1 shows ANN pseudocode.

Algorithm 9.1 Artificial neural network source [35]

```
process ANN (Input, Neurons, Reiteration)
generate input record
Record with the conceivable mutable groupings → Input
for input =1 to end the input do
for neurons =1 to 20 do
 Train ANN
ANN- Loading ← save maximum test R^2
end for
end for
ANN- Loading← save top forecasting ANN dependent on inputs
end for
return ANN- Loading
end procedure
```

9.3.2.4 ResNet

He *et al.* put forth ResNet in 2015 [36]. ResNet is made to optimize the network layer because hierarchical networks have a lot of redundant components. The completion of the identity mapping and ensuring that the identity layer's input and output are identical are the two goals of ResNet. The person through training, the network's layer is automatically determined. ResNet altered multiple layers of a block of the old network remain.

9.3.2.5 Research tool

The Python implementation for this study is created in a Jupyter Notebook environment. To complete this research, a 16 GB RAM and a 64-bit system, Intel® core^{TM}, 2.60 GHz processor was used.

9.4 Results and discussions

On the Jupyter and Google Collaborator platform, LDA to pre-process the data. In this part, we provide the findings from our investigation using the proposed model. This necessitated separating the information into a training set and a test set. This research applies a HR model using deep learning algorithms such as CNN, ResNet, and ANN.

9.4.1 Importing the dataset

To run the code in Jupiter, the dataset has to be imported into the environment. Figure 9.2 shows the imported loaded dataset.

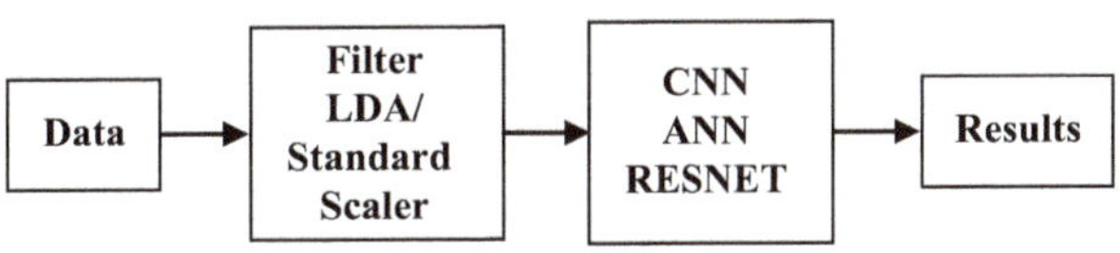

Figure 9.1 Proposed model

```
[4]: #===============importing the dataset from the directory=======================#
Healthy = os.listdir("D:\kaggle\input\diabetic-retinopathy-dataset\Healthy")
Mild = os.listdir('D:\kaggle\input\diabetic-retinopathy-dataset\Mild DR')
Moderate = os.listdir('D:\kaggle\input\diabetic-retinopathy-dataset\Moderate DR')
Proliferate = os.listdir('D:\kaggle\input\diabetic-retinopathy-dataset\Proliferate DR')
Severe = os.listdir('D:\kaggle\input\diabetic-retinopathy-dataset\Severe DR')
```

Figure 9.2 Imported dataset

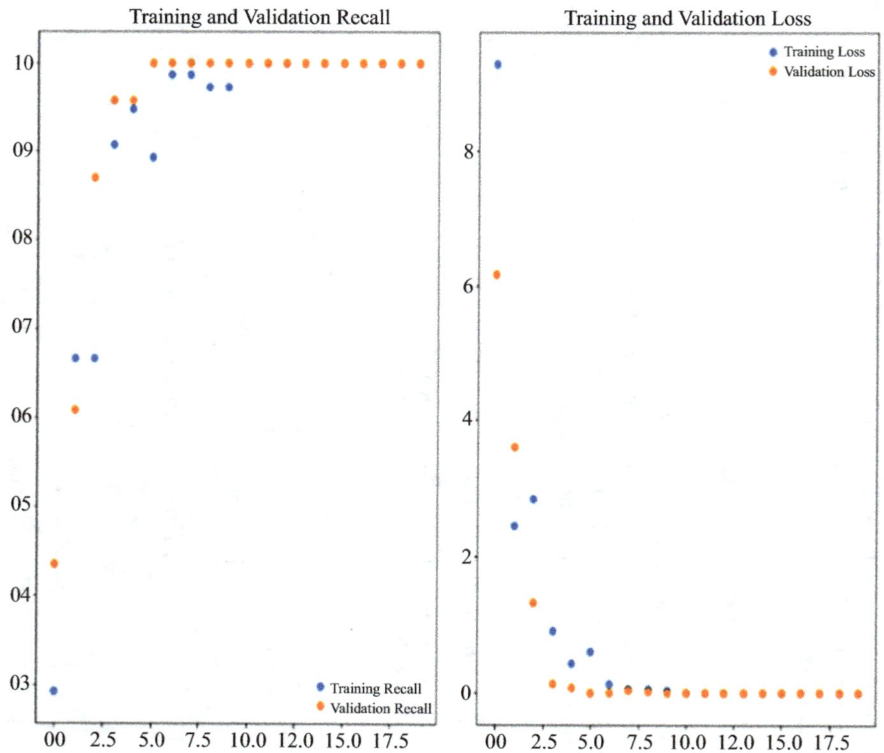

Figure 9.3 Scatter plot for ANN model

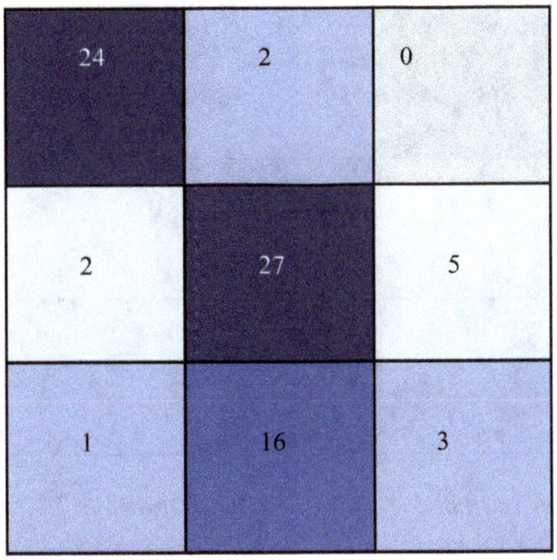

Figure 9.4 Confusing matrix for ANN model with LDA (TP=54, FP=26, TN=134, FN=26)

9.4.2 Resizing and converting the images to array

When the images have been imported into the environment, they come in different shapes and sizes, so the next step was to resize all the images into a unique shape to be able to work with it and then put all the images in an array.

9.4.3 Data splitting

Splitting data into train and test sets is common in machine learning. For the respective algorithm, the data was separated into training and testing subsets. To fit the model and conduct testing, the training set was used as a basis for evaluation.

When the data has been split into the train and test data, the data generator tool was used to perform data augmentation on the images.

9.4.4 Pre-processing the data with LDA

To achieve better accuracy, the dataset was pre-processed with the LDA, and the needed data was retained to achieve better accuracy and performance.

9.4.5 Training the ANN model with and without LDA

When the image has been pre-processed and divided into the train and test, the training data is therefore passed into the ANN model for training.

9.4.6 Plotting the scattered plot and confusion matrix for the ANN model with and without LDA

After the model has been trained, it is custom for us to plot the scattered plot and confusion matrix to enable us to visually evaluate the performance of the model. The scattered plot and confusion matrix is shown in Figures 9.3–9.9.

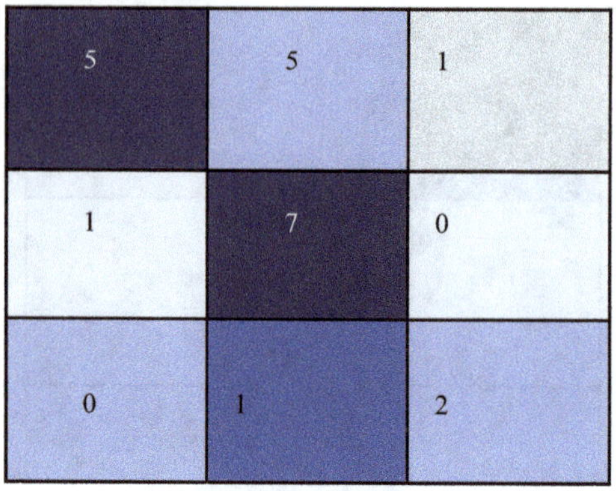

Figure 9.5 *Confusion matrix for ANN model without LDA (TP=11, FP=8, TN=36, FN=15)*

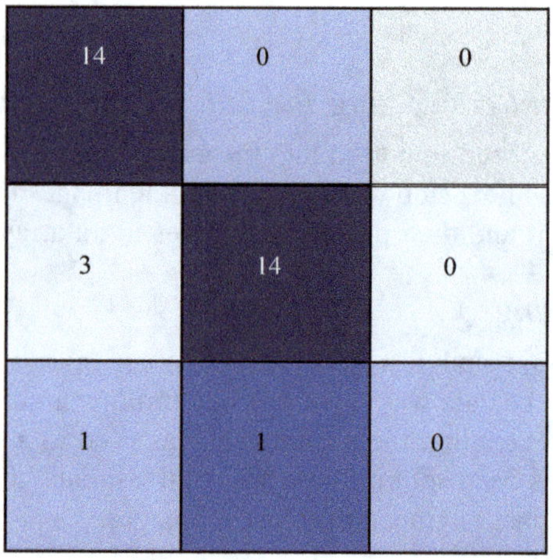

Figure 9.6 *Confusion matrix for CNN model with LDA (TP=28, FP=5, TN=61, FN=5)*

When the image has been pre-processed and split into the train and test, the training data is therefore passed into the CNN model for training.

When the image has been pre-processed and split into the train and test, the training data is therefore passed into the Resnet model for training.

In this study, LDA is used for the classification of the image with deep learning methods CNN, ANNs, and ResNet for further classifications. Evaluation measures such as sensitivity, specificity, precision, accuracy, and F1 score were used to assess the confusion matrices obtained. The performance measures of each of the classifiers with and without LDA are shown in Table 9.2.

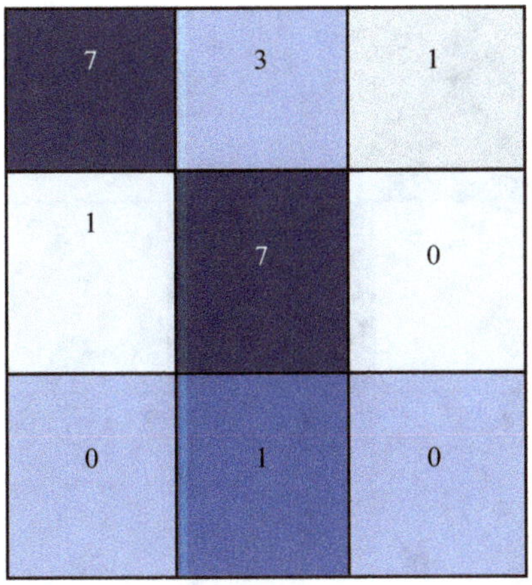

Figure 9.7 Confusion matrix for CNN model without LDA (TP=14, FP=6, TN=34, FN=6)

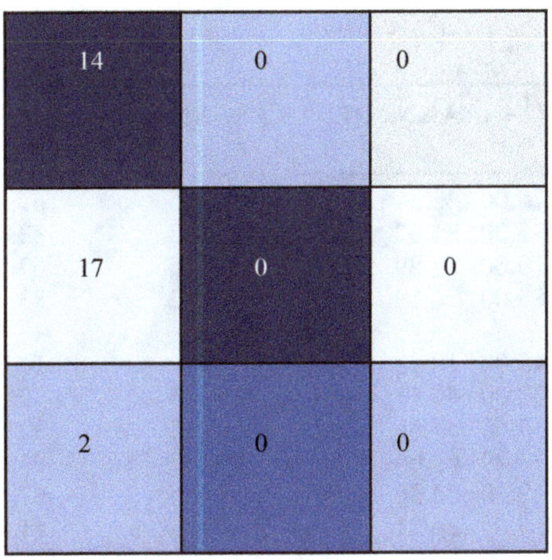

Figure 9.8 Confusion matrix for ResNet model with LDA (TP=14, FP=24, TN=52, FN=24)

9.4.7 Comparison with previous works

Several trials were carried out in this study, and the results are shown in Table 9.2. CNN outperformed the other models with 86.42% accuracy; however, ANN took the shortest time to execute at 2.50 sec and Resnet took the longest amount of time and gave the lowest accuracy of 46.66%.

Table 9.3 displays the comparison of the results gained from various works.

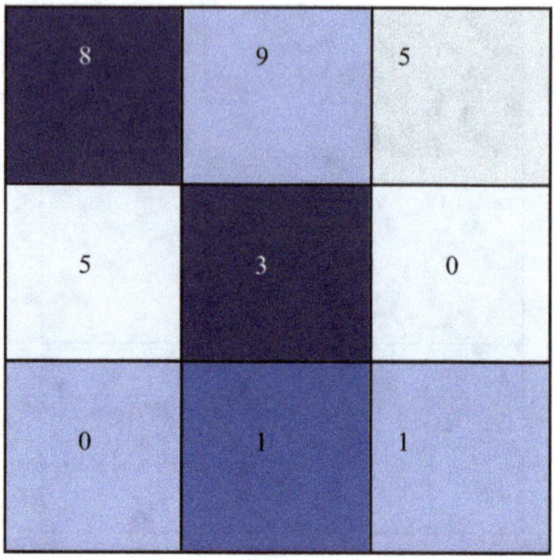

Figure 9.9 Confusion matrix for ResNet model without LDA (TP=12, FP=20, TN=44, FN=20)

Table 9.2 Performance measures

Performance measures (%)	CNN	ANN	RESNET	CNN + LDA	ANN + LDA	RESNET + LDA
Sensitivity	70.00	42.31	37.50	84.85	67.50	36.84
Specificity	85.00	81.82	68.75	92.42	83.75	68.42
Precision	70.00	57.89	37.50	84.85	67.50	36.84
Negative predictive value	85.00	70.59	68.75	92.42	83.75	68.42
False positive rate	15.00	18.18	31.25	7.58	16.25	31.58
False discovery rate	30.00	42.11	62.50	15.15	32.50	63.16
False negative rate	30.00	57.69	62.50	15.15	32.50	63.16
Accuracy	73.29	63.64	36.43	86.42	84.20	46.66
F1 score	49.30	64.30	37.50	84.85	67.50	36.84
Matthews correlation coefficient	55.00	26.21	6.25	77.27	51.25	5.26

Table 9.3 Comparison of the findings

Authors	Techniques	Results
[37]	PCA and BNN	96.875% and 86.36%
[38]	AVR with neural network	93.9%
[39]	Dense hyper system	95%
[40]	CNN	98.6 %
[41]	DSF-Net, DSA-Net	97.26% and 97.25%
[20]	RBM	99.05%

The purpose of this study was to create a HR detection model from a publicly available dataset utilizing DMLs such as CNN, ANN, and RESNET with and without LDA. The training set (80%) was used to find the optimal combinations of variables to generate a successful predictive model, while the testing set (20%) was utilized to offer an overall analysis of the model fit on the training dataset. Using LDA as a preprocessor before feeding the data into CNN, ANN, and RESNET yielded the best accuracy of 86%; other evaluation outcomes were also acquired and displayed in the findings.

9.5 Conclusions

When it comes to the human body, hypertension is a condition that can take many forms. Due to excessive blood pressure, the retina of the eye might be damaged. HR is a term used to describe retinal damage brought on by high blood pressure. The retina is the primary structure in the eye responsible for converting visual information into nerve signals for transmission to the brain. Retinal injury can cause permanent vision impairment or blindness. So many automated devices have been developed to aid ophthalmologists in their examinations.

This study developed a model using an image analyzer, LDA, and deep learning methods CNN, ANN, and ResNET in diagnosing HR from retinal images and to analyze and evaluate the performance of the developed model. The dataset used adopted available at https://www.kaggle.com/code/meenavyas/retinopathy-detection. In summary, the study involved getting the dataset from the Kaggle repository and passing it through the LDA to pre-process the data furthermore the unprocessed data was passed raw to the algorithms as well as the data that has been pre-processed through the LDA algorithm.

The testing set (20%) was used to offer an objective assessment of the final model fit on the training dataset, while the training set (80%) was used to discover the best combinations of variables that will produce a successful predictive model. The data was first passed through LDA before being passed into CNN, ANN, and ResNET and results were obtained with the accuracy of 86%, 84%, and 48%, respectively which was later compared to the results obtained in the related works.

This study aids in detecting HR. The study was based on a Kaggle dataset. The model was first created using deep learning techniques like CNN, ANN, and ResNET, and then processed via LDA. Several trials were carried out in this study, and the results are shown in the table. CNN outperformed the other models with 86% accuracy; however, ANN took the shortest time to execute at 2.50 sec. In conclusion, this study can be adopted for predicting HR.

This study recommends that for further investigation, more data should be used for training the model as it will enable the algorithm to train better and gain better accuracy. It is important to remember that the goal of developing a system for diagnosing HR is to enhance its accuracy. To increase the model's precision, future studies should broaden the model's field of application. Also, future researchers could investigate using a larger dataset to improve the model's accuracy. This study would also suggest that alternative methods, such as graphical neural networks be used to improve the results.

References

[1] Willermain F, Libert S, Motulsky E, *et al.* Origins and consequences of hyperosmolar stress in retinal pigmented epithelial cells. *Front Physiol.* 2014;5:199. Available from: http://journal.frontiersin.org/article/10.3389/fphys.2014.00199/abstract

[2] Faheem MR and Mui-zzud-Din. Diagnosing hypertensive retinopathy through retinal images. *Biomed Res Ther.* 2015;2(10):25. Available from: http://www.globalsciencejournals.com/article/10.7603/s40730-015-0025-x

[3] Triwijoyo BK and Pradipto YD. Detection of hypertension retinopathy using deep learning and Boltzmann machines. *J Phys Conf Ser.* 2017;801: 012039. Available from: https://iopscience.iop.org/article/10.1088/1742-6596/801/1/012039

[4] Bandara AMRR and Giragama PWGRMPB. A retinal image enhancement technique for blood vessel segmentation algorithm. In: *2017 IEEE International Conference on Industrial and Information Systems (ICIIS).* New York, NY: IEEE, 2017, pp. 1–5. Available from: http://ieeexplore.ieee.org/document/8300426/

[5] Mohamad-Saleh J and Hoyle B. Improved neural network performance using principal component analysis on Matlab. *J Comput Internet Manag.* 2008;16 (2):1–8. Available from: http://ijcim.th.org/past_editions/2008V16N2/P1-IJCIM16n2-May-Aug-2008-Improved Neural Network Performance-IJCIM-p1-8.pdf

[6] Triwijoyo BK, Budiharto W, and Abdurachman E. The classification of hypertensive retinopathy using convolutional neural network. *Procedia Comput Sci.* 2017;116:166–73. Available from: https://linkinghub.elsevier.com/retrieve/pii/S1877050917321166

[7] Oparil S, Acelajado MC, Bakris GL, *et al.* Hypertension. *Nat Rev Dis Prim.* 2018;4(1):18014. Available from: http://www.nature.com/articles/nrdp201814

[8] Carey RM, Muntner P, Bosworth HB, and Whelton PK. Prevention and control of hypertension. *J Am Coll Cardiol.* 2018;72(11):1278–93. Available from: https://linkinghub.elsevier.com/retrieve/pii/S0735109718354676

[9] Sarker IH. Machine learning: algorithms, real-world applications and research directions. *SN Comput Sci.* 2021;2(3):160. Available from: https://link.springer.com/10.1007/s42979-021-00592-x

[10] Algren M, Fisher W, and Landis AE. Machine learning in life cycle assessment. In: *Data Science Applied to Sustainability Analysis*. New York, NY: Elsevier, 2021, pp. 167–90. Available from: https://linkinghub.elsevier.com/retrieve/pii/B9780128179765000097

[11] Dada EG, Bassi JS, Chiroma H, Abdulhamid SM, Adetunmbi AO, and Ajibuwa OE. Machine learning for email spam filtering: review, approaches and open research problems. *Heliyon*. 2019;5(6):e01802. Available from: https://linkinghub.elsevier.com/retrieve/pii/S2405844018353404

[12] Marinescu R, Seceleanu C, Le Guen H, and Pettersson P. A Research Overview of Tool-Supported Model-based Testing of Requirements-based Designs, *Advances in Computers*, 2015, pp. 89–140. Available from: https://doi.org/10.1016/bs.adcom.2015.03.003

[13] Syahputra MF, Amalia C, Rahmat RF, *et al.* Hypertensive retinopathy identification through retinal fundus image using backpropagation neural network. *J Phys Conf Ser*. 2018;978: 012106. Available from: https://iopscience.iop.org/article/10.1088/1742-6596/978/1/012106

[14] Narasimhan K, Neha VC, and Vijayarekha K. Hypertensive retinopathy diagnosis from fundus images by estimation of Avr. *Procedia Eng*. 2012;38:980–93. Available from: https://linkinghub.elsevier.com/retrieve/pii/S1877705812020371

[15] Cavallari M, Stamile C, Umeton R, Calimeri F, and Orzi F. Novel method for automated analysis of retinal images: results in subjects with hypertensive retinopathy and CADASIL. *Biomed Res Int*. 2015;2015:1–10. Available from: http://www.hindawi.com/journals/bmri/2015/752957/

[16] Abbas Q, Qureshi I, and Ibrahim MEA. An automatic detection and classification system of five stages for hypertensive retinopathy using semantic and instance segmentation in DenseNet architecture. *Sensors*. 2021;21(20):6936. Available from: https://www.mdpi.com/1424-8220/21/20/6936

[17] Arasy R and Basari. Detection of hypertensive retinopathy using principal component analysis (PCA) and backpropagation neural network methods, 2019, pp. 040002. Available from: http://aip.scitation.org/doi/abs/10.1063/1.5096735

[18] Feng X, Wang H, Kong Y, *et al.* Diagnosis of chronic stage of hypertensive retinopathy based on spectral domain optical coherence tomography. *J Clin Hypertens*. 2020;22(7):1247–52. Available from: https://onlinelibrary.wiley.com/doi/10.1111/jch.13935

[19] Khitran S, Akram MU, Usman A, and Yasin U. Automated system for the detection of hypertensive retinopathy. In: *2014 4th International Conference on Image Processing Theory, Tools and Applications* (*IPTA*). New York, NY: IEEE; 2014, ppp. 1–6. Available from: http://ieeexplore.ieee.org/document/7001984/

[20] Triwijoyo BK, Sabarguna BS, Budiharto W, and Abdurachman E. Restricted Boltzmann machines for fundus image reconstruction and classification of hypertension retinopathy. *J Comput Sci*. 2021;17(2):156–66.

[21] Abbas Q and Ibrahim MEA. DenseHyper: an automatic recognition system for detection of hypertensive retinopathy using dense features transform and deep-residual learning. *Multimed Tools Appl*. 2020;79(41–42):31595–623. Available from: https://link.springer.com/10.1007/s11042-020-09630-x

[22] Şüyun SB, Taşdemir Ş, Biliş S, and Milea A. Using a deep learning system that classifies hypertensive retinopathy based on the fundus images of patients of wide age. *Trait du Signal*. 2021;38(1):207–13. Available from: http://www.iieta.org/journals/ts/paper/10.18280/ts.380122

[23] Patil BP. Retinal fundus image enhancement using adaptive CLAHE methods Retinal fundus image enhancement using adaptive CLAHE methods. *J Seybold Rep*. 2020;15(9):3476–84.

[24] Devi S. Detection of arteriovenous nicking in retinal fundus images for screening hypertensive retinopathy. *Int Res J Eng Technol*. 2020;7(7): 4347–4350

[25] Khan RA and Minallah N. An improved blood vessel extraction approach from retinal fundus images using digital image processing. *Proc Pakistan Acad Sci Part A*. 2017;54(2):135–44.

[26] Wiharto W, Suryani E, and Kipti MY. Assessment of early hypertensive retinopathy using fractal analysis of retinal fundus image. *TELKOMNIKA (Telecommunication Comput Electron Control)*. 2018;16(1):445. Available from: http://telkomnika.uad.ac.id/index.php/TELKOMNIKA/article/view/6188

[27] Rani A and Mittal D. Measurement of arterio-venous ratio for detection of hypertensive retinopathy through digital color fundus images. *J Biomed EngMed Imaging*. 2015;2(5). Available from: http://scholarpublishing.org/index.php/JBEMi/article/view/1577

[28] Muramatsu C, Hatanaka Y, Iwase T, Hara T, and Fujita H. Automated detection and classification of major retinal vessels for determination of diameter ratio of arteries and veins. Proc. SPIE 7624, Medical Imaging 2010: Computer-Aided Diagnosis, 76240J. Available from: https://doi.org/10.1117/12.843898

[29] Dai G, He W, Xu L, *et al.* Exploring the effect of hypertension on retinal microvasculature using deep learning on East Asian population. *PLoS One*. 2020;15(3):e0230111. Available from: https://dx.plos.org/10.1371/journal.pone.0230111

[30] Dipak M and Aditi S. An automatic approach to segment retinal blood vessels and its separation into arteries/Veins. *Proceedings of the International Conference on Data Engineering and Communication Technology*, 2016, pp. 191–199. Available from: https://doi.org/10.1007/978-981-10-1675-2_21

[31] Afolayan JO, Adebiyi MO, Arowolo MO, Chakraborty C, and Adebiyi AA. Breast cancer detection using particle swarm optimization and decision tree machine learning technique. In: *Intelligent Healthcare*. Singapore: Springer Nature Singapore; 2022, pp. 61–83. Available from: https://link.springer.com/10.1007/978-981-16-8150-9_4

[32] Wan H, Guo G, Wang H, and Wei X. A new linear discriminant analysis method to address the over-reducing problem. PReMI 2015. *Lecture Notes in Computer Science*, 2015, pp. 65–72. Available from: https://doi.org/10.1007/978-3-319-19941-2_7

[33] Raitoharju J. Convolutional neural networks. In: *Deep Learning for Robot Perception and Cognition*. New York, NY: Elsevier; 2022,

pp. 35–69. Available from: https://linkinghub.elsevier.com/retrieve/pii/B9780323857871000087

[34] Jeswal SK and Chakraverty S. Fuzzy eigenvalue problems of structural dynamics using ANN. In: *New Paradigms in Computational Modeling and Its Applications*. New York, NY: Elsevier; 2021, pp. 145–61. Available from: https://linkinghub.elsevier.com/retrieve/pii/B9780128221334000104

[35] Balduin S, Tröschel M, and Lehnhoff S. Towards domain-specific surrogate models for smart grid co-simulation. *Energy Informatics*. 2019;2(S1):27. Available from: https://energyinformatics.springeropen.com/articles/10.1186/s42162-019-0082-2

[36] Chen Z and Xiu D. On generalized residual network for deep learning of unknown dynamical systems. *J Comput Phys*. 2021;438: 110362. Available from: https://linkinghub.elsevier.com/retrieve/pii/S0021999121002576

[37] Arasy R and Basari. Detection of hypertensive retinopathy using principal component analysis (PCA) and backpropagation neural network methods. *AIP Conf Proc*. 2019;2092(April):040002.

[38] Faheem MR and Mui-zzud-Din. Diagnosing hypertensive retinopathy through retinal images. *Biomed Res Ther*. 2015;2(10):385–8.

[39] Abbas Q and Ibrahim MEA. DenseHyper: an automatic recognition system for detection of hypertensive retinopathy using dense features transform and deep-residual learning. *Multimed Tools Appl*. 2020;79(41–42):31595–623.

[40] Triwijoyo BK, Budiharto W, and Abdurachman E. The classification of hypertensive retinopathy using convolutional neural network. *Procedia Comput Sci*. 2017;116:166–73.

[41] Arsalan M, Haider A, Choi J, and Park KR. Diabetic and hypertensive retinopathy screening in fundus images using artificially intelligent shallow architectures. *J Pers Med*. 2022;12(1):7.

Chapter 10

Understanding cancer patients with diagnostically influential factors using high-dimensional data embedding

Ameer Sohail Syed[1], Hajderanj Laureta[1], Kun Guo[2] and Daqing Chen[1]

Abstract

Analysing breast cancer data is a long-established research topic from both medical diagnosis and data modeling perspectives. Enormous predictive models have been employed in modeling breast cancer data, e.g., predicting a patient's survival rate given certain medical circumstances and a patient's demographics. However, these predictive models tend to take a black-box approach to the modeling and therefore can hardly provide any explainable results to be applied for diagnostic purposes, in particular, if neural network-based models are utilised. On the other hand, identifying diagnostically influential factors with exploratory descriptive models has been proven difficult due to the high dimensionality of breast cancer data under consideration. For instance, the breast cancer data provided by SEER, The Surveillance, Epidemiology, and End Results Program, typically has more than 100 dimensions of numeric and categorical data types and could expend to about some 1,000 dimensions for analysis if orthogonal (one-hot) encoding is applied. Hence, effectively interpreting and understanding high-dimensional data becomes crucial in modelling cancer data, and it is because of this that dimensionality reduction algorithms and manifold learning algorithms have been studied intensively and many relevant algorithms are available, with each having pros and cons of its own. In this chapter, a comparative study is presented aiming at providing visualized, explainable insights in breast cancer survival rate analysis and identifying critical influential factors that strongly determine the likelihood of a patient's survival. Two dimensionality reduction algorithms are considered in this study for comparison purpose: one is a typical and popular *t*-distributed stochastic neighbor embedding (*t*-SNE) algorithm and another is a relevant new same degree distribution (SDD) algorithm. The relevant experiments have demonstrated that based

[1]Division of Computer Science and Informatics, School of Engineering, London South Bank University, UK
[2]AVIC Xi'an Aeronautics Computing Technique Research Institute, China

on the same embedding performance assessment metrics, the SDD algorithm can achieve much better data embedding results which could be impossible or difficult if *t*-SNE is used. Furthermore, using the reliable embedding results from SDD, meaningful and explainable factors have been identified that reflect crucially the similarities of the patients who have survived and the diversities of the patients who, unfortunately, have died. Clusters of patients who survived are clearly recognizable in a two-dimensional embedding space, whereas the embedded points of patients who died are significantly scattered in the space. The entire package of the codes used for the analysis is available for replication.

Keywords: Breast cancer survivability; *t*-SNE; Same degree distribution algorithm; Dimensionality reduction; Visualization; Data embedding; classification

10.1 Introduction

Cancer is one of the most deadly diseases and the number of patients diagnosed with cancer has been increasing year by year, although slowly. Because of its mysterious nature and depending on the way it spreads out, most people by the time they have diagnosed with cancer, it is almost in the last stages and the disease could have completely invaded body organs. As such, unfortunately, many patients have passed away [1]. The success stories of cancer survivors are occasional and the cure for the treatment is very far from the human-beings [2].

There are over 200 different types of cancer but only 13 of them are predominant [3]. Cancer often called as malignant neoplasm happens due to alterations in genetics of somatic cells, while some of the alterations do happen due to hereditary origins. Several biological agents, such as viruses, bacteria, and parasites, can exacerbate a carcinogenic process in humans in addition to the different chemical and physical (ionising radiation, UV light) carcinogens. A number of studies have been conducted to describe cancer cells. Hanahan and Weinberg provided the most significant characterization of cancer cells [4]. Many possible risk factors, including nutrition, lifestyle, smoking, drinking, viral infection, and others, have been identified with the goal of reducing the occurrence of cancer. Among them, the strongest link between smoking and lung cancer was discovered [2]. According to estimates, 20% of all breast cancer cases globally are caused by preventable risk factors like obesity, inactivity, and alcohol consumption. By encouraging a healthy lifestyle, the spread of the disease might be reduced [5].

There are many screening tests as well as the clinical tests that patients need to go through for their effective treatment and we are very sure one of the risk factors is genetics but it is not the genetics all the time. The clinical findings' reports include more than 60–70 findings and doctors do have a very big challenge as to get an understanding about the disease and how is it going to behave like. An increase in clinical trials and screening tests over the past few decades have reduced the mortality rate but still there is lot undiscovered and new risk factors are being added to the list National Cancer Institute (NCI) every year [6].

Although analysing breast cancer data has been a long-established research topic, models applied are mainly predicative models, e.g., predicting the likelihood of a patient's survival given certain medical circumstances and a patient's demographics. However, these predictive models usually take a black-box approach to the modelling and therefore can hardly provide any explainable results to be applied for diagnostic purposes, in particular, if neural network-based models are utilised. On the other hand, identifying diagnostically influential factors with descriptive models has been proven difficult due to the high dimensionality of breast cancer data under consideration.

This chapter presents a case study of using dimensionality reduction algorithms to create explainable insights in breast cancer survivability analysis and to identify critical influential factors that strongly determine the likelihood of a patient's survival. Two dimensionality reduction algorithms are used in this study for comparison's purpose:

- The typical t-distributed stochastic neighbor embedding (*t*-SNE) [7].
- The same degree distribution (SDD) algorithm [8].

The breast cancer data from The Surveillance, Epidemiology, and End Results Program (SEER, https://seer.cancer.gov/data/) has been used in this study. The data typically has more than 100 dimensions of numeric and categorical types and can become about some 1,000 dimensions for analysis if one-hot (orthogonal) encoding applied.

It has been demonstrated in the study that based on the same embedding performance metrics, the SDD algorithm can achieve much better data embedding results which would be impossible or difficult if *t*-SNE is used. Furthermore, using the reliable embedding results from SDD, meaningful and explainable factors have been identified that reflect crucially the similarities of the patients who have survived and the diversities of the patients who, unfortunately, have died. Clusters of patients who survived are clearly recognizable in a two-dimensional embedding space, whereas the embedded points of patients who died are significantly scattered in the space. To the best knowledge of the authors', this study is the first one of this kind.

The remaining of the chapter is organised as follows. Section 10.2 provides a literature review of the relevant work with conclusions. Section 10.3 focuses on the state-of-the-art of the unsupervised dimensionality reduction algorithms, in particular, the SDD algorithm. The methodology of this study is discussed in Section 10.4, including the analytical procedure, the data used, the performance metrics applied for dimensionality reduction, and data pre-processing. In Section 10.5, the experiments of the study are discussed and the relevant environment settings and codes are explained. The results of the experiments are examined with essential findings and insights provided, and finally in Section 10.7 end with conclusion and future works.

10.2 Literature review

The research so far till the date has always given lot of emphasis on building predictive models such as classification models. A large variety of datasets from

Table 10.1 Classification models accuracy on SEER dataset [9]

Year	Main author	Model	Accuracy	Sensitivity	Specificity
2020	Ming-Huan Zhang	ANN	76.8%	77.7%	76.4%
2017	Shi-Chao Xin	ANN	73.18%	59.64%	83.54%
2017	Zhi-Gang Huang	C5.0	87.26%	75.22%	92.98%

different regions and people diagnosed with different types of cancer are available and lot of papers in the recent years have published the built models with different accuracies. The problem here is that there is not a really good understanding about the features or risk factors that the doctors can really learn from this. There is always going to be a better model with classification and it is also possible to build a own classifier and tune it with regards to the data but there is very little progress in the way the data is made to speak by itself as there is an ocean of information hidden inside it. There is a large gap at the pace at which the research is being done to build a better classification model or prediction model than really understanding about the data and giving some insights with the use of Artificial Intelligence. In short, the study on SEER data or any large data had given less emphasis on unsupervised learning and search for something unknown from the vast data available.

For example, the latest paper submitted at the international conference on Orange technologies used the SEER data set to build classification models [9]. The above models were used and there is quite a good trade-off between sensitivity and specificity following each year work. It is more like a prediction system that based on the rules and hyper parameters tuned based on the data fed built the model (Table 10.1).

A recent paper published at the Iranian Biomedical Engineering conference used the PCA techniques on Wisconsin Diagnostic dataset with 569 records. The same work was done as the earlier but the accuracies are different as the data is different and the model as well.

Table 10.2 from the above paper shows few techniques that reduce the data into two-dimensional space but the question arises that does the data presented in two dimensions after reduction actually represent its originality because every time a dimensionality reduction technique is applied on a data there is some data loss in different forms and the data we get is not the exact same as that in high-dimensional space. Indeed this is one of the good models but there have been better dimensionality reduction algorithms than PCA and they outperform PCA in many aspects and one such method is *t*-SNE and we had used our own algorithm which did perform quite better than *t*-SNE during the simulation of this data.

t-SNE is one of the best dimensionality reduction techniques at present and there are various modified versions of it and tuned for speed convergence to work around with large datasets. *t*-SNE solves the crowding problem that many other dimensionality reduction algorithms face and it can locate the structure and usually does whenever other dimensionality-reduction methods have been unable to.

Table 10.2 Various feature selections and there accuracies [13]

Feature selection method	Number of features	Accuracy	Sensitivity	Specificity	F1-score	Kappa
PCA	8	97.2%±2.1%	98.3%	95.3%	97.8%	0.93
	4	96.3%±1.8%	97.5%	94.3%	97.1%	0.92
	2	94.2%±3.0%	96.9%	89.6%	95.4%	0.87
Factor analysis	8	97.2%±2.0%	99.2%	93.9%	97.8%	0.94
	4	95.4%±3.8%	97.5%	92.0%	96.4%	0.90
	2	91.6%±5.6%	95%	85.8%	93.4%	0.82
Incremental PCA	8	96.7%±2.0%	98.3%	93.9%	97.4%	0.93
	4	96.1%±1.5%	97.2%	94.3%	96.9%	0.92
	2	94.6%±3.0%	96.9%	90.6%	95.7%	0.88
ICA	8	82.8%±4.0%	99.7%	54.3%	87.9%	0.59
	4	72.6%±4.9%	100%	26.4%	82.1%	0.31
	2	66.3%±1.7%	100%	9.30%	78.8%	0.12
SVD	8	97.2%±2.1%	98.3%	95.3%	97.8%	0.93
	4	96.3%±1.8%	97.5%	94.3%	97.1%	0.92
	2	94.2%±3.0%	96.9%	89.6%	95.4%	0.87
Kernel PCA	8	94.0%±2.4%	97.8%	87.7%	95.4%	0.87
	4	92.8%±3.5%	99.4%	81.6%	94.5%	0.84
	2	84.7%±4.5%	97.2%	63.7%	88.9%	0.65
Feature importance	8	97.4%±2.4%	98.6%	95.3%	97.9%	0.94
	4	97.0%±2.2%	98.3%	94.8%	97.6%	0.94
	2	94.5%±2.3%	96.6%	91.0%	95.7%	0.88

Even the stochastic neighbouring has the crowding problem i.e., the reduced two-dimensional space. A huge number of applications and research have been done on various kinds of data whether it is image compression, microwaves filtering. Dimensionality reduction is used basically to visualize high-dimensional data in two-dimensional space. The catch here is to check the similarity between the patients who have passed away and also look into patients who survived then look upon the attributes where they are more similar in for further research or make conclusions from it.

These models are definitely a good prediction models but there is really a little information that a doctor or researchers can learn something new about the features or the risk factors, as an increase in understanding of risk factors means the deeper the understanding of the environment of cancer disease, it may give a new direction to work on and build the medicines so we can narrow down the disease based on some factors and develop better medicines accordingly so we can further improve the survival rate.

In conclusion, from the previous works, it is very clear that the research has emphasised a lot on predictive modelling, there was no work done on descriptive modelling and there was no change in shift of the aim behind researches were carried out. It was more like tunnel vision where lots of flavours of predictive modelling techniques were developed.

10.3 Dimensionality reduction methods

This section introduces various unsupervised dimensionality reduction algorithms, in particular, the SDD algorithm [10]. Generally speaking, there are two main approaches for reducing the dimensionality: one is the Projection and another is the manifold learning.

10.3.1 Projection

Generally, not all the dimensions are spread in all the dimensions; many tend to be very close to each other. Let us look at the figure below. If we project these points perpendicularly in an imaginary plane downwards by connecting them with lines, then we have just reduced the data from 3D to 2D. But projection technique is not always the best; many times, points tend to overlap over one another and this does not carry any significance. The picture on the left shows the 3D representation of the data and the picture on the right shows the actual data that we would want to visualise.

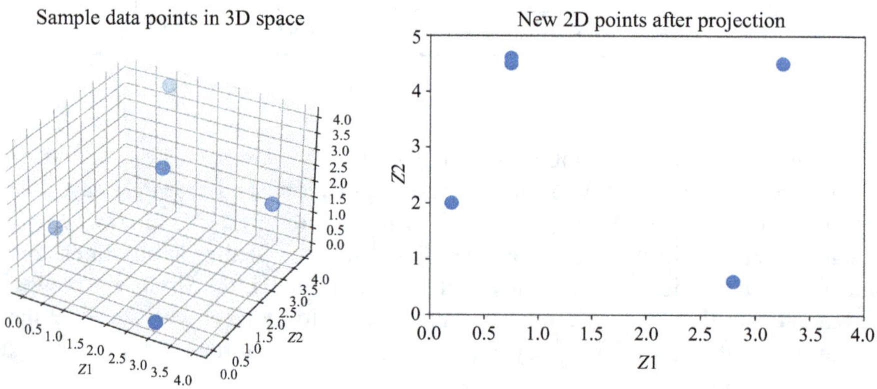

10.3.2 Manifold learning

The below Swiss roll dataset is an example of manifold learning to put it in simple, manifold learning is any 2D image that can be twisted or bent in high-dimensional space. It is one of the popular approaches for dimensionality reduction of non-linear data.

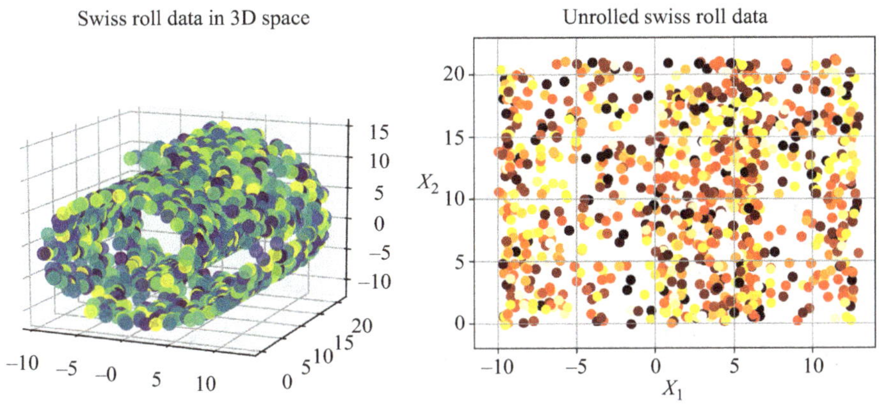

10.3.3 PCA

One of the most often used procedures for reducing the number of linear dimensions is principal component analysis (PCA). The data is transformed using a projection-based technique that projects the data onto a set of perpendicular axes. When data from a higher-dimensional space is mapped to data into a lower-dimensional space, PCA requires that the variance or spread of the data in the lower-dimensional space be as little as possible. Despite the entire efficacy that PCA offers, it might be challenging to comprehend the principal components when there are a lot of variables. PCA works best when variables are related to one another linearly. PCA is also sensitive to significant outliers. An oldest technique that has received much investigation is PCA. Numerous modifications of basic PCA exist that address its drawbacks, including kernel PCA, and incremental PCA.

10.3.4 t-SNE

t-SNE is a new method invented by Maaten *et al.* in 2008 [7]. It can capture the local data structure of the high dimension and reveal the global structure such as the presence of clusters on some degrees. *t*-SNE is a non-linear dimension reduction approach that computes the conditional probability using Gaussian distribution.

10.3.5 SDD

The SDD [10] is a new non-linear dimensional reduction approach that captures both the global and the local data structures. SDD determines which degree of distribution can best capture data structure. Large distances tend to cause very little effect on degree distributions while it is contradictory on the small distances. Kullback-Leibler is the loss function used in SDD to approximate the degree-distribution in the low-dimensional space with the degree-distribution in the high-dimensional space:

$$C_1 = \sum\nolimits_{i \neq j} p_{deg_m} log\left(\frac{(p_{deg_m})_{ij}}{(q_{deg_m})_{ij}}\right) \tag{10.1}$$

where deg_m is the degree of degree-distribution m, $m = 1{:}\ n$. SDD intends to minimize the cost function C_1 as (10.2):

$$loss_1 = \min(C_1) \tag{10.2}$$

where

$$\left(p_{deg_m}\right)_{ij} = \frac{\left(1 + dis\left(x_i, x_j\right)\right)^{-deg_m}}{\sum_{k \neq l} \left(1 + dis(x_k, x_l)\right)^{-deg_m}} \tag{10.3}$$

$$\left(q_{deg_m}\right)_{ij} = \frac{\left(1 + dis\left(y_i, y_j\right)\right)^{-deg_m}}{\sum_{k \neq l} \left(1 + dis(y_k, y_l)\right)^{-deg_m}} \tag{10.4}$$

and $dis(\cdot, \cdot)$ represents the Euclidean distance.

Algorithm. SDD

Require: Input $X \in R^{N \times D}$, number of iterations H, learning rate η, momentum α, number of degree-distributions n, degree $degm$, and initial low-dimensional data $Y^0 = y_1, y_2, \ldots, y_N \in N\left(0, 10^{-4} I\right)$.

Step 1: Compute similarities in high-dimensional space p_{deg_m} using (10.3).
Step 2: Compute the similarities in high-dimensional space q_{deg_m} using (10.4).
Step 3: Compute the gradient $\frac{\partial C_1}{\partial y_i}$, where C_1 is defined as (10.1).
Step 4: Minimize the objective function (10.2) using gradient descent optimisation algorithm:

$$Y^h = Y^{h-1} + \eta \frac{\partial C_1}{\partial y_i} + \alpha\left(Y^{h-1} - Y^{h-2}\right)$$

Output: Low-dimensional space representation $Y_{bestdeg_m}$.

10.4 Methodology

High-dimensional data visualisation is a critical step in studying high-dimensional data. As a result, lowering dimensionality is an important stage in data analytics. The ideal dimensionality reduction strategy for visualisation is one that preserves both the global and local data structures.

Data structure can be effectively captured by SDD with a significant portion of small and medium distances. In contrast, it performs worse in datasets with a high proportion of big distances because there are so many samples in the degree-low distribution. In data where short and medium distances predominate, SDD outperforms benchmark approaches for structure capture including *t*-SNE and Isomap [10]. The one degree distribution used in the studies generated the best low-dimensional data representation in terms of retaining structure. We have demonstrated that one degree distributions can effectively capture data structure.

10.4.1 Procedure

The data preprocessing steps include converting the original data to one hot-encoded columns, dealing with the missing data. The dimensionality reduction and clustering is performed in the below experiment with *t*-SNE and SDD algorithms, then the patient ID is mapped to the original database into excel to find the patterns. The steps involved in data preprocessing are summarised in the Figure 10.1 below.

10.4.2 Data used

The raw data used is the SEER breast cancer which has 798k records with 133 variables and predominantly of categorical type.

Table 10.4 gives all the variables contained in the data. Refer to the original document, available at https://seer.cancer.gov/data-software/documentation/seer-stat/nov2019/TextData.FileDescription.pdf.

10.4.3 Performance assessment metrics

There are several performance assessment metrics as discussed below.

1. Kendall's Tau (τ): A non-parametric correlation coefficient calculated using the rankings of x and y is Kendall's τ and it is distribution-free. Second, it does not assume that x and y be linearly related, making it usable to both discrete random and continuous variables. Kendall's Tau produces value range between 0 and 1 where 0 means no relationship and 1 for perfect relationship, in some instances, can also produce a negative value, the negative sign has no significance and can be discarded. There are various versions of Tau formula.
2. Sheppard diagram: A Shepard diagram examines the spacing between your data points before and after you change them. For data reduction methods like PCA, MDS, or *t*-SNE, Shepard diagrams might be employed.
3. Spearman's Rho (ρ): A non-parametric test called Spearman's Rho is used to assess how strongly two variables are related; a value of $r = 1$ denotes a perfect

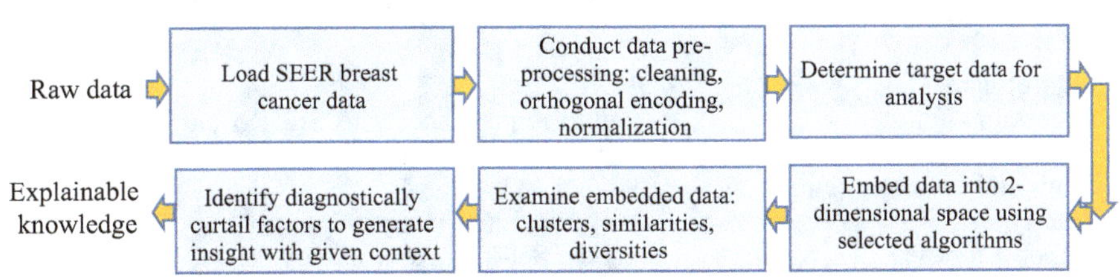

Figure 10.1 Block diagram of proposed approach

positive correlation, while a value of $r = -1$ denotes a perfect negative correlation.
4. Co-ranking matrix: It is a tool used to assess the performance of dimensionality reduction.

Multiple metrics from the above can be chosen depending upon the task needed to be assessed. Kendall's Tau has been used throughout the experiments as the performance metric to assess the quality of dimensionality reduction.

10.5 Experiments

Table 10.3 provides the information about the dataset and the working environment setup.

In the experiments, we are interested in these following questions:

1. Is there any similarity between patients who survived and how do they look like in 2D space?
2. What is the difference in risk factors between survived patients and patients who have died?
3. Compare unsupervised *t*-SNE and unsupervised SDD and find which algorithm best replicates the high-dimensional data into low dimensional space.

Libraries overview:
Pandas is used to create and manage tables, while numpy for numerical calculations. Scipy is a scientific python library that consists of statistics and other suite built in it to measure and use packages like Kendall's tau for comparison. The Sk-Learn Library consists of all the algorithms that we need, basically we will make use of *t*-SNE and our own algorithm which will be using as a class. The visuals are going to be generated with the help of matplotlib and seaborn, seaborn library offers more flexibility and gives us the ability to make changes and we prefer for customization. And visuals are very important as part of this experiment. It is highly recommended to create an environment in python using Anaconda and then

Table 10.3 Tools used and the environment setting

Tools used	Python 3.9.1, Tableau, and SAS Miner
OS	Windows 10
Dataset name	SEER
Access link	How to request access to SEER Data – SEER datasets (cancer.gov)
Rows and columns count	798,625,133
Data types	Categorical and integers
Python Libraries	Pandas, Sci-kit Learn, numpy, scipy, matplotlib, seaborn

start using it. A tip is to use multicore tsne instead of the regular tsne or the cuda library if nvidia graphics card is present to enable faster computations.

We have to first load the entire dataset into python, so we performed the following steps in order. The important steps in the data preprocessing are the treatment of missing values and handling outliers.

Few commonly used preprocessing functions of pandas

1. Dataframe.shape – to check for the number of rows and columns.
2. Dataframe.info() – to check for the data types and get a glance of null values.
3. Dataframe.isnull().sum(axis=1) – to check for missing values along the column side.
4. Dataframe.dropna(axis=1) – to drop missing values.
5. There is no direct function to detect outliers, prefer using scatter plots or box plot and remove outliers accordingly.
6. One hot encoding – from sklearn.preprocessing import OneHotEncoder then create an instance of it and one hot encode the categorical data.

Note that the following approach is applicable to any dataset and if the objectives are same for the comparison of data points purposes.

The SEER data set was pre-processed and it consisted of 291,761 records and 63 categorical features. The features were one hot encoded and the resulting data had 160,000 records and 947 features. For the demonstration in regards to the scope of the chapter, we have conducted the following experiments as in Table 10.4 [11].

For the two experiments that were conducted, the highest difference found was low with 42% when the experiment was conducted with 500 samples. The confidence of difference similarity grew in the case of 1,000 samples experiment to 75%. This might be due to fact that the algorithm needed more data to generalize well. This experiment can be repeated with more samples all the way on the entire dataset but training the entire dataset requires very large computational power and it is hard to do it with a single machine. As the subset of 1,000 samples included the previous 500 samples as well we investigated in the 1,000 samples plot and looked into the older patients from the previous experiment and we found that patients who appeared together in the former experiment, they appear still together in the latter experiment as well.

Table 10.4 The data originality percentages of t-SNE and SDD

Algorithm (unsupervised)	**Number of samples**	**Data originality after reduction (%)**
t-SNE	500	58.8
SDD	500	72.9
t-SNE	1,000	56.1
SDD	1,000	72.9

The code for the SDD can be found in the references below please copy paste the code as the algorithm has yet to be submitted to the sklearn [11]. The use of ray library is optional it has been used here just to speed up the loading process. The magic functions are used as needed. As we will be using randomly 500 and 1,000 records for the experiment to pick the samples, we will be using random library. The seed value has been set to 10; we will make use of indexing to pick the records. First we will be generating random numbers and use those numbers to pick the records with the indexes. So each time, the simulation is run, it will generate the same exact results and gives the same samples output if the seed value is not set, then each time the simulation runs, the results will be different because the randomly selected records will be different. Here in this simulation below, I have generated 5,000 random numbers and then picked equal proportions of patients who have survived and passed away. The experiment procedure is the same for the 1,000 patients as well but the limit will get changed to 1,000 in the next run.

We have combined both the data frames dead and alive for the dimensionality reduction. The best value of the list with the highest correlation will tell us the best k value where the data represents the best high-dimensional data. Once we find the best k value, then we will be running the SDD again with the best degree of freedom found at. Then we can move into plotting once we have the reduced 2D data. We have created a new column where we have mapped the data and then created a unique label to each patient from the index we picked up the records initially, this list of id will be the key to compare patients and look them into the 2D space on the map.

A total of 5 clusters were visible but the density was quite less meaning that it is quite noisy but this algorithm adds support to fact that the patients who have survived are indeed quite similar with each other. We have noted down the points of the clusters when we investigated the yellow points that were close together. We did find that many factors were similar but the results were not strong and that did not give any significance as there was very little that we could make use of from it.

First, when we look at the picture of SDD implementation, we can see that the yellow points in each cluster are very near to each other like a group while the blue points are spread out like they are quite present everywhere. A total of five clusters with high density were visible. The distance separation between the clusters was quite significant. There was very little disorder in points among the patients who survived while in the case of patients who died, there was large noise (lot of disorder between points). When we further investigated the yellow points and mapped them back to the original data, we found some similarity among them in many factors.

A good visual tool to display data that is grouped is Tree map. The Tree map below shows the people who had survived had the following factors in common with percentage showing the highest similarity they were common together in, the larger the size of rectangle that means they are largely common together.

We looked at the patients who have died and were still close together to those who survived and we did find through their records that they were indeed much similar like the factors were same but they had some risk factors as opposite to that of patients survived and that might be the reason why they have passed away.

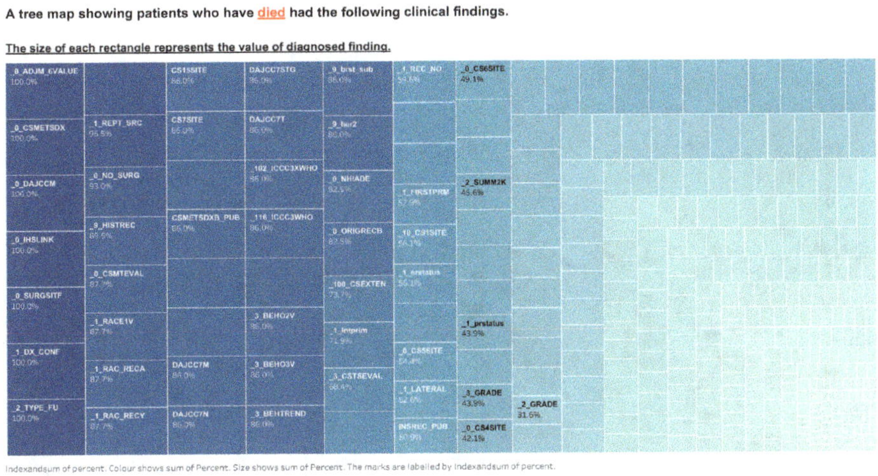

(a)

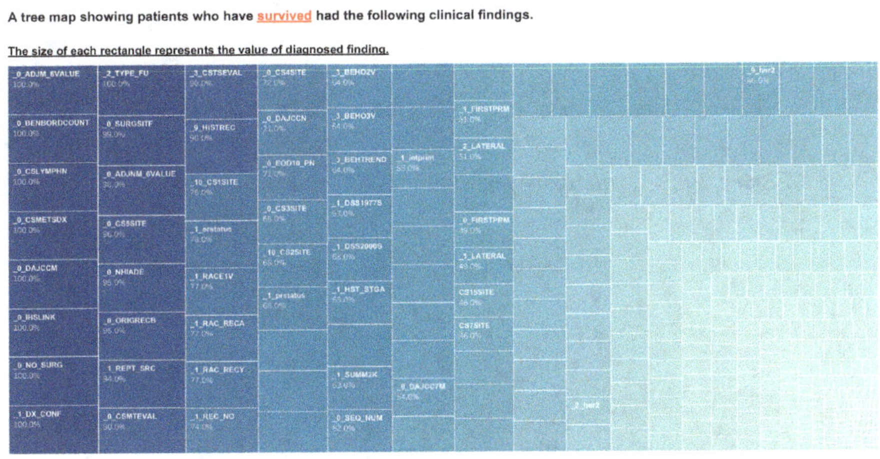

(b)

One strange thing to notice is that there are risk factors like adjusted AJCC value (_0_ADJM_6VALUE) which shows 100% in both patients who have died and passed away. What exactly has happened then? Well these factors really do not tell us anything significant because the dataset has majority of positive value for these factors and no matter what the record is the _0_ADJM_6VALUE is always going to be positive either patient has passed away or survived. The next step would be to store the similarity of both survived and passed away and check for factors where there is a large difference.

The tree maps did give us some information but it is still not much clear what exactly are the factors, whereas the risk factors were found after the difference and

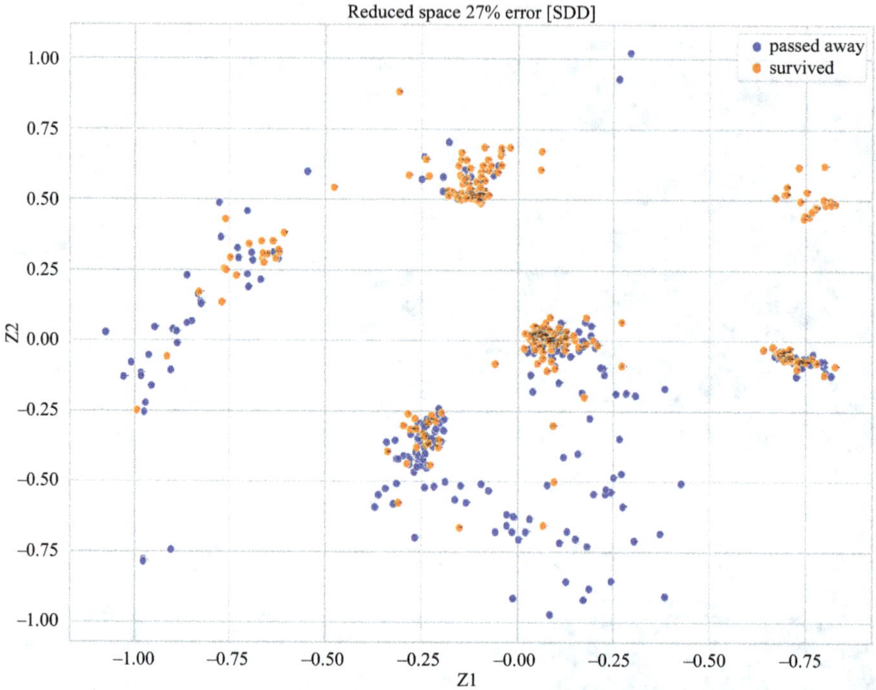

Figure 10.2 Patients representing in 2D space by using SDD

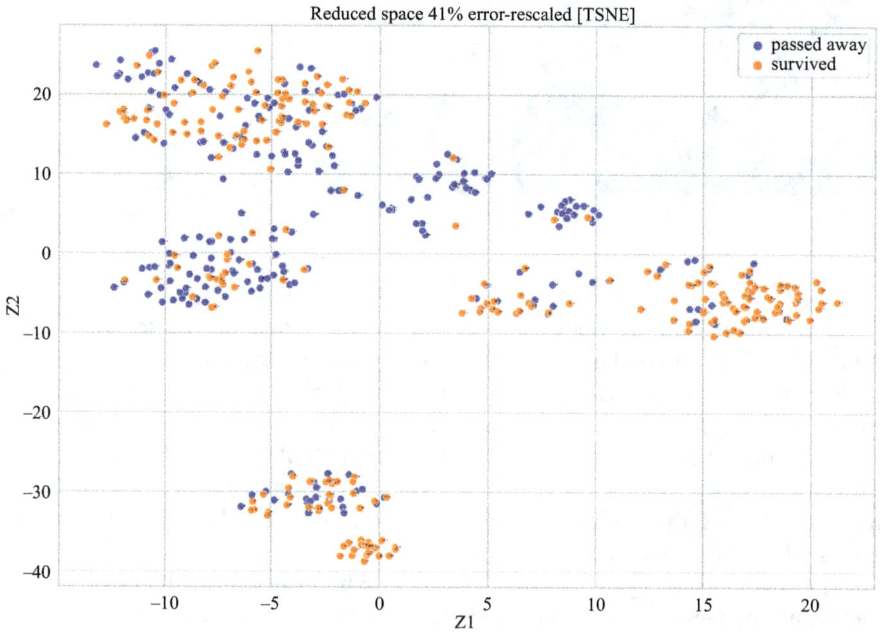

Figure 10.3 Patients representing in 2D space with t-SNE

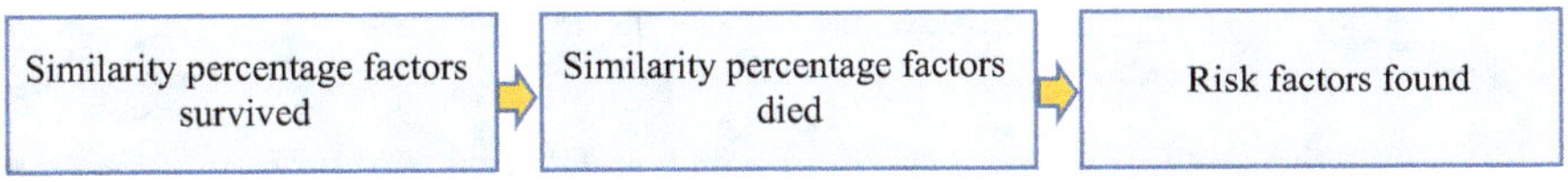

Figure 10.4 Block diagram illustrating how risk factors were found

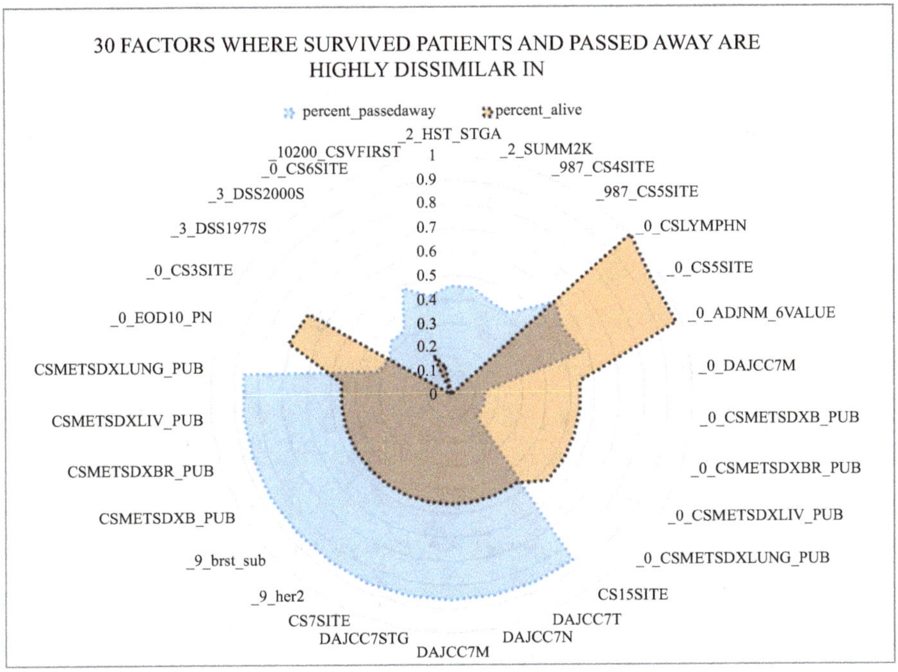

Figure 10.5 The pink region shows the distinct attributes where patients survived exhibit while the blue region belongs attributes of patients who passed away (sample size = 500)

when presented using the radar chart. The area for the patients died is quite large the factors are vast but when we look at the pink area cross section, the points are very sharp and clear that the cancer site 3 and site 5 are among the patients who survived and none of the patients had regional lymph nodes (Figure 10.5).

To summarise, the cluster points that were seemed together in SDD implementation were also seen together in the *t*-SNE cluster points but due to error rate being quite high *t*-SNE was not able to group all the patients together but did capture some similarity. Again, the factors contributing to these similarities were checked between SDD and *t*-SNE and factors found in *t*-SNE were present in the SDD as well but the SDD algorithm gave us little more information.

The next experiment consisted of 1,000 samples which included the previous 500 samples from the previous experiment. When we look into the graph, we notice

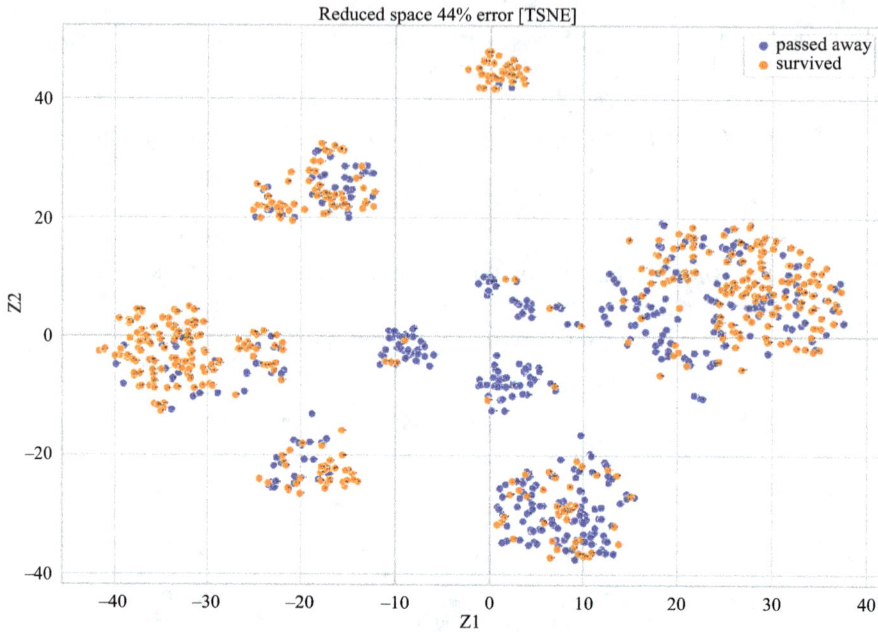

Figure 10.6 Patients representing in 2D space by using t-*SNE*

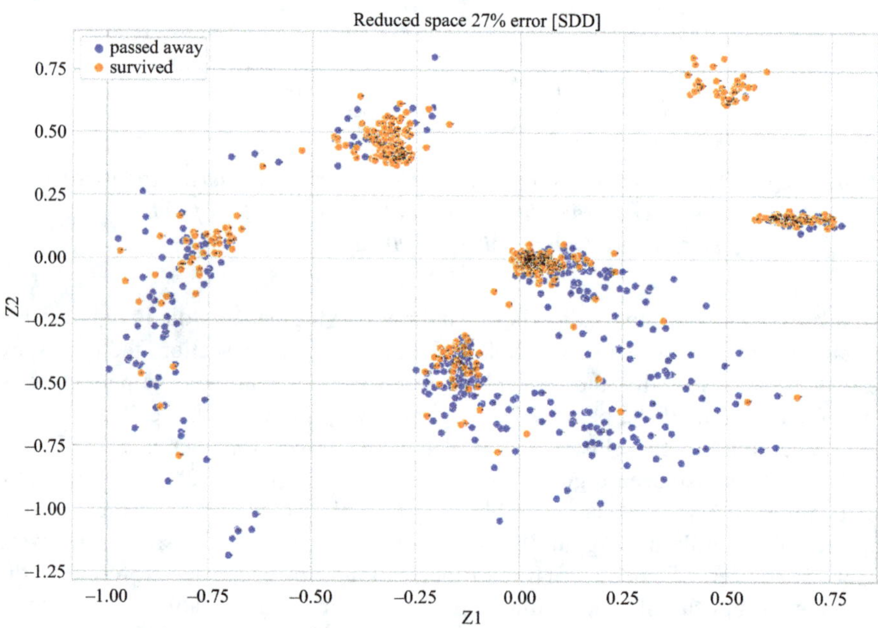

Figure 10.7 Patients representing in 2D space by using SDD

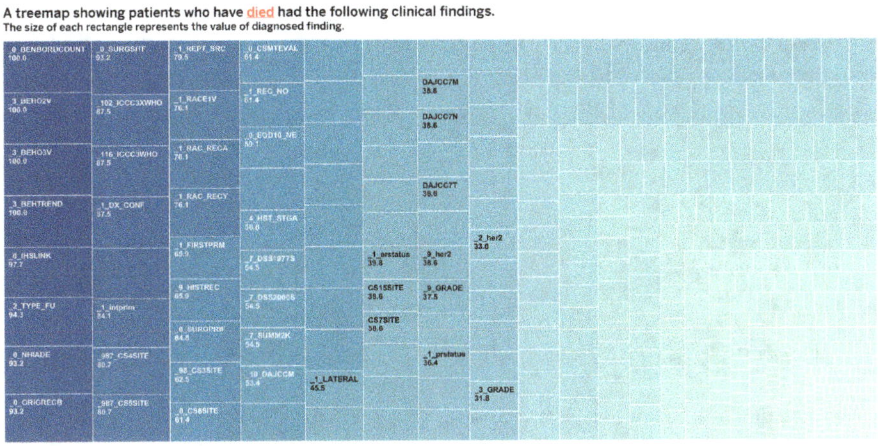

(a)

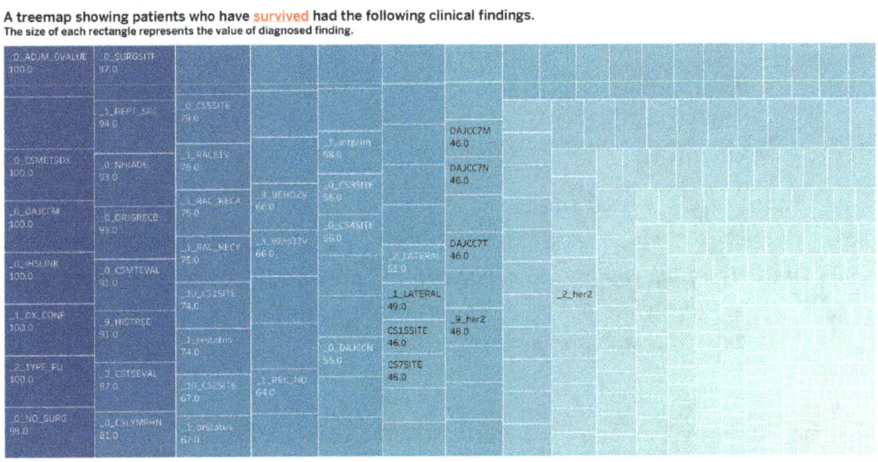

(b)

that *t*-SNE has formed clusters well, indeed with less representation of originality with 44% error, *t*-SNE still was able to create groups of patients together but the results with SDD were quite more informative. The SDD algorithm was able to generalize very well when compared with *t*-SNE. The samples that were close together in the previous experiment were still together in this experiment as well.

A total of 6 clusters were visible; the density was decent it means that there was very little noise among the data reduced, this experiment still points and adds support to answer our question that survived patients are indeed quite similar with each other and that is the reason why they always appear very near to each other in the reduced space while it is not the case with the patients who have passed away. But there are few clusters of blue points as well and upon inspection, these points

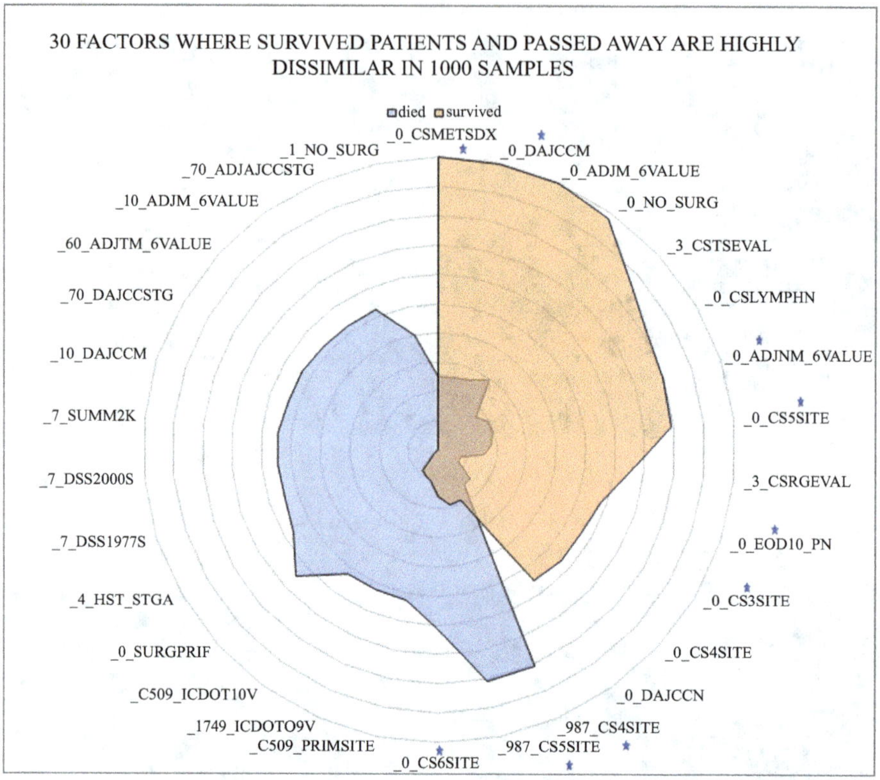

Figure 10.8 The pink region shows the distinct attributes where patients survived exhibit while the blue region belongs attributes of patients who passed away (sample size = 1,000)

had some similarities among them but vastly the dominant tightly packed clusters are of yellow points that are those of who survived. We have noted down the points of the clusters when we investigated the yellow points that were close together; we did find many factors that were similar in but the results were not strong it means that the similarity percentage was less which is not bad but we will not be much confident about it and that did not give any significance as there was very little that we could make use of from it.

The SDD simulation produced clusters as well and the survived patient clusters are very tightly packed. We have randomly picked up 100 samples from the graph above and studied the survived patients and we found high similarity percentage when compared with the previous experiment. This might be due to an increase in the number of records and the algorithm was able to generalise well. The tree map for the two categories of labels is being shown below; the experiment followed the same procedure like the previous one.

10.6 Discussion of results

'All happy families are alike, but every unhappy family is unhappy in its own way' (Leo Tolstoy and Anna Karenina, 1878) [12]. The Anna Karenina principle meant that happy families share a common set of attributes for the reason behind their happiness whilst in the case of unhappy families, it can be any attribute which causes unhappiness in their life. We found this concept to be aligned and fits in the scenario of the cancer patients as well. We saw in both the experiments that the yellow points (survived) appeared always close together while blue points (died) appeared quite scattered and the points on the graph appeared quite noisy and had their own factors. The clinical findings among the patients who survived were quite similar and many of the patients shared some common characteristics. While it was not the same in the case of patients who passed away. From the 947 one hot encoded factors, we narrowed down the list to 30 factors based on the similarity threshold. This narrowed down list will help to study only the few findings more elaborately and lot of time will be saved and burden on the researchers will be decreased as well and we can speed up the process and focus on working with important clinical risk factors instead of looking at all the factors.

In the following discussion, several influential variables will be focused and examined closely based on the experiments results. Table 10.5 shows each of such variable with their meaning and the code used after one-hot encoding for a quick look-up.

Remark 1: It has been found that these five distinct tumour sites (_0_CS3SITE, _0_CS5SITE, _0_CS6SITE, _987_CS4SITE, _987_CS5SITE) were predominant in both the experiments. The first two sites are predominant among patients who survived while the remaining three are predominant among patients who have died. These sites might be added to new risk factors list and a strong emphasis and further research can be conducted to find out what is special about this site that is

Table 10.5 Influential variables with their original meaning

Attribute code	Meaning
_0_CSMETSDX	No distant metastasis
0 NO_SURG	Surgery performed
_0_CSLYMPN	No involvement of Lymph nodes
_0_CS5SITE	Test not done
_3_CSRGEVAL	Microscopic examination of regional node
_0_EOD10_PN	All nodes examined are negative
_0_CS3SITE	Tumour site
_0_CS4SITE	Regional lymph nodes negative
_987_CS4SITE	Not applicable CS Lymph nodes not coded
_987_CS5SITE	Tumour site
_0_CS6SITE	Test not done
_0_SURGPRIF	No systematic therapy/surgical procedures
_4_HST_STGA	Distant neoplasm
_70_ DAJCCSTG	Stage of cancer

why survived patients at those two specific sites are able to make out and this experiment also asks for why majority of people who died who had tumour at the three other sites have less mortality.

Among the survived patients, 80% patients had tumours at site _0_CS5SITE while the patients who passed away had less than 0.2% tumours at this site. This indicates that if a patient is diagnosed with cancer and has tumour at this sites, then there is a very good chance that the treatment might be successful and they might survive. Fifty-six per cent of survived patients had tumours at site _0_CS3SITE while the patients who passed away had less than 0.1% tumours at this site. The patients had low mortality rate at tumour location of _0_CS6SITE, 61% of patients died off while less than 2% of survived patients had tumours at this site. The _987_CS5SITE,_987_CS4SITE attributes had 80% of patients' similarity who passed away while less than 2% of patients who survived had tumours at this site. By comparing the results of tumours against a new patient, we can expect how the treatment of the patient would go like and predict the chances of mortality and further study can be done to find more about the location sites so we can gain deeper understanding of the tumour sites.

Remark 2: With regard to _0_NO_SURG, it has been found that among the patients who have survived 98% of them had a surgery after diagnosis while the patients who died never underwent a surgery after being diagnosed. The mortality rate does even depend on this factor from the following result. Among the patients who have died, 39.7% had one surgery in their life time after diagnosis. It is highly recommended to follow the clinical advice and get surgery done if it is suggested.

Remark 3: Lymph nodes, _0_CSLYMPHN, have shape of beans, they are widely found across the body and specially at the lymphatic pathways where they filter the lymph before it goes into the blood. Among the patients who have survived, 81% of them did not have any lymph node involvement with the disease while they were diagnosed while patients who died had more than one lymph node involved with the disease at their time of diagnosis. This is another risk factor which has significant contribution for predicting the chances of mortality.

Remark 4: Age does not really say much, from the results of the data, there is no pattern to catch up as there are few cases where age is quite high but still patients were able to survive and patients with age less than 40 still pass away. Even race does not give any significance and has any importance with relation to the disease no matter what the race, it has no impact in high or low mortality.

Remark 5: The historic stage A, 4_HST_STGA, showed us that patients who died had excessive growth in tissues and the growth had spread out to the other organs of the body as well. This indicates that the extent up to which the neoplasm has spread across the body has very strong impact if the cancer cells have started to invade other organs, the mortality gets very high and among the patients who died, 56.8% of them had gone through this stage and they passed away.

Remark 6: In relation to variable _0_SURGPRIF, while the patients if they ever had a tumour in life and who never got it destroyed or undergone tissue removal are more prone to die and this was the case here observed in our experiment as 64.7% of the patients who died never got there first tissue removed or destroyed. This really does stress on the importance of screening and negligence in treatment can cause major mishappening. Proper screening and immediate treatment play a key role. Once the patient finds a tissue or abnormal growth they may need to get rid of the primary tumour as soon as possible as there is very high chance the cells will start to divide and spread to regional parts of the organs first and then start invading other distant organs as well and this may become fatal.

Remark 7: In relation to _0_CSMETDX, all the patients who survived never had metastasis; metastasis is the movement of cancer cells first from where they were formed to nearby parts of the organs or the body. The movement can be through lymph nodes or through the blood or any other means. While among the patients who had died, the results are contradicting this means that metastatic cells do play a part in determining the survival chances of the patient. If the cells are not metastatic, then there is a 75% chance that the patients can survive because there are 25% cases of patients who died despite they have not had metastatic cells but still they passed away.

Remark 8: This finding is related to _70_DAJCCSTG and supports the fact that at the advanced stages of cancer, people cannot make it and pass away among the patients who have died, 53.4% of them were at the fourth stage of the cancer while 22.7% patients data from the died category about the stage is unknown. While we look into the patients who have survived 34% of the patients were at the stage 0 and 31% at stage 1 and 21% at stage 2 of the disease, this is the reason why patients among who survived were able to fight the cancer. Once the disease progresses into the advanced stages, the mortality rate tends to get high.

Remark 9: Examined with the variable _0_EOD10_PN, 58% of the patients who had survived when they got examined during the diagnosis had their regional lymph nodes tested negative, well the number is not large but on the contrary basis, surprisingly 59% of the people who had died never got there nodes examined. We exactly do not know was there difference in treatment as why these patients lymph nodes were not examined by the pathologists as this could also be a trigger or point to look at as a risk factor. If in case the patients were examined and the treatment would have started, may be the patients could have survived.

To summarise the points mentioned above, while the patients who survived showed us something in similar that we discussed above, we narrowed down the scope of the investigation into the findings to 30 from 947 variables or findings and then we found some uniqueness in tumour sites in patients who survived and passed away. We found that lymph nodes play a huge role as they spread quite across the body and we also found that if the cancer cells start spreading to the nearby regions or organs, then the chances of survival gets very low. Excessive growth in tissues can also be a warning sign and proved to be fatal while patients who survived had less growth of tissues in nearby part of region of body. Metastasis is also an important factor once the cancer cells start to move into blood stream or get in movement through lymph nodes, the case might get complicated and it increases the complexity of survival. Screening is very important, we also saw people who survived were at initial stages of the cancer while they were diagnosed, the people who were at advanced or later stages could not make it. If necessary, it is very important to get the tissue removed or the tumour to be removed as in the cases where people who survived got their tissue or tumours destroyed initially while it was not the case with the patients who passed away. And lots of surgeries happening are also a bad sign; this might be an indication that the cells or the disease has started invading other parts and eventually patients who got multiple surgeries after diagnosis could not make it. Well the age and race had no effect on the disease.

This experiment was successfully completed with our SDD algorithm. This algorithm proved very helpful as we were able to derive a good number of findings from with the help of it.

10.7 Concluding remarks and future work

This study presents a comparative case study demonstrating how to use high dimensionality reduction algorithms like SDD to effectively embed data in a 2D embedding space to create insight into the data. Diagonally influential factors have been identified in relation to the likelihood of a breast cancer's survival. The procedure applied is valid and efficient. It has been shown that SDD is much more effective in dimensionality reduction than *t*-SNE, and therefore, it deserves further research effort in the future.

It is our intention to extend SDD to multiple degrees distribution in order to obtain better data embedding performance. In addition, conducting dimensionality reduction with data from our domain, such as lip-reading and micro facial expression.

References

[1] P. Bhardwaj, Y. Kumar, and G. Bhandari, "AI-enabled computational techniques for cancer diagnosis," in *2021 IEEE 8th Uttar Pradesh Section International Conference on Electrical, Electronics and Computer Engineering* (*UPCON*), 2021, pp. 1–7, doi:10.1109/UPCON52273.2021.9667624.

[2] M. Bahrami and M. Vali. "Wise feature selection for breast cancer detection from a clinical dataset," in *2021 28th National and 6th International Iranian Conference on Biomedical Engineering (ICBME)*, 2021, pp. 160–164, doi:10.1109/ICBME54433.2021.9750287.

[3] Cancer – NHS (www.nhs.uk).

[4] Y.A. Fouad and C. Aanei. "Revisiting the hallmarks of cancer," *Am J Cancer Res*. 2017;7(5):1016–1036.

[5] N. Harbeck, F. Penault-Llorca, J. Cortés, *et al.* "Breast cancer," Nat Rev Dis Primers. 2019;5:66, doi:10.1038/s41572-019-0111-2.

[6] R.J. Santen, N.F. Boyd, R.T. Chlebowski, *et al.* "Critical assessment of new risk factors for breast cancer: considerations for development of an improved risk prediction model," *Endocr Relat Cancer*. 2007;14(2):169–168.

[7] L. van der Maaten and G. Hinton. "Visualizing data using t-SNE," *J Mach Learn Res*. 2008;9:2579–2605.

[8] L. Hajderanj, D. Chen, and I. Weheliye. "The impact of supervised manifold learning on structure preserving and classification error: a theoretical study," *IEEE Access*. 2021;9:43909–43922, doi:10.1109/ACCESS.2021.3066259.

[9] X.-Y. Zhong, Z.-F. Cai; S.-J. Zhou, *et al.*, "A data mining experiment on SEER database using artificial neural network," in *2020 8th International Conference on Orange Technology* (*ICOT*), 2020, pp. 1–4, doi:10.1109/ICOT51877.2020.9468764.

[10] L. Hajderanj, D. Chen, E. Grisan, and S. Dudley. "Single- and multi-distribution dimensionality reduction approaches for a better data structure capturing," *IEEE Access*. 2020;8:207141–207155, doi:10.1109/ACCESS.2020.3038460.

[11] A.S. Syed, L. Hajderanj, D. Chen. 2022. See https://github.com/syedameersohail/understanding-cancer-patients.git.

[12] L. Bornmann and W. Marx. The Anna Karenina principle: a concept for the explanation of success in science, 2011.

[13] S. Hussein, P. Kandel, C. W. Bolan, M. B. Wallace, and U. Bagci, "Lung and pancreatic tumor characterization in the deep learning era: novel supervised and unsupervised learning approaches," *I*EEE Trans Med Imaging, 2019; 38(8):1777–1787, doi:10.1109/TMI.2019.2894349.

[14] D. Li, W. Yang, Y. Zhang, *et al.*, "Comprehensive analysis of pulmonary adenocarcinoma in situ (AIS) revealed new insights into lung cancer progression," in *2017 IEEE International Conference on Bioinformatics and Biomedicine* (*BIBM*), 2017, pp. 792–797, doi:10.1109/BIBM.2017.8217756.

Appendix

```
from time import time
import numpy as np
import pandas as pd
from scipy import linalg
from scipy.spatial.distance import pdist
from scipy.spatial.distance import squareform
from scipy.sparse import csr_matrix, issparse
from sklearn.neighbors import NearestNeighbors
from sklearn.base import BaseEstimator
from sklearn.utils import check_random_state
from sklearn.utils._openmp_helpers import _openmp_effective_n_threads
from sklearn.utils.validation import check_non_negative
from sklearn.utils.validation import _deprecate_positional_args
from sklearn.decomposition import PCA
from sklearn.metrics.pairwise import pairwise_distances
from warnings import simplefilter
simplefilter(action='ignore', category=FutureWarning)
from MulticoreTSNE import MulticoreTSNE as TSNE
from scipy import stat
MACHINE_EPSILON = np.finfo(np.double).eps
```

The use of ray library is optional. It has been used here just to speed up the loading process. The magic functions are used as needed. As we will be using randomly 500 and 1,000 records for the experiment to pick the samples, we will be using random library. The seed value has been set to 10. We will make use of indexing to pick the records. First, we will be generating random numbers and use those numbers to pick the records with the indexes. So each time the simulation is run, it will generate the same exact results and give the same samples output. If the seed value is not set, then each time the simulation runs, the results will be different because the randomly selected records will be different. Here in this simulation below, I have generated

```
•[3]: import ray
      ray.init(ignore_reinit_error=True)

      2022-07-04 09:56:50,912 INFO services.py:1477 -- View the Ray da

 [3]: RayContext(dashboard_url='127.0.0.1:8265', python_version='3.9.1
      7.0.0.1', 'raylet_ip_address': '127.0.0.1', 'redis_address': Non
      1:8265', 'session_dir': 'C:\\Users\\home\\AppData\\Local\\Temp\\
      '127.0.0.1:64984', 'node_id': '1085cdf4010621d40e9da54d262bdc26a

 [4]: @ray.remote(num_cpus=8)
      def read_files(x):
          x=pd.read_csv(x,nrows=160000,low_memory=False)
          return x

 [6]: %%time
      df=ray.get(read_files.remote('original947.csv'))

      CPU times: total: 688 ms
      Wall time: 52.9 s

 [9]: import random
      random.seed(10)
      random_num=random.sample(range(len(df)),5000)

[11]: random_num[:5]

[11]: [149789, 8541, 112430, 126500, 151543]

[12]: sample=df.iloc[random_num]

[13]: dead=sample[sample['Survive']==0].reset_index().loc[:499,:]
      alive=sample[sample['Survive']==1].reset_index().loc[:499,:]
      DF=pd.concat([dead,alive],axis=0)
```

5,000 random numbers and then picked equal proportions of patients who have survived and passed away. The experiment procedure is the same for the 1,000 patients as well but the limit will get changed to 1,000 in the next run.

Combine both the data frames dead and alive for the dimensionality reduction. We will be using unsupervised techniques so we will be dropping the target column and store the target in the variable *y* which will be used later for the purpose of plotting. It is always a good habit to store the original dataframe and use the copy of dataframe for the simulation purposes.

```
[54]: y=DF[['Survive']]
      DF=DF.drop(['Survive'],axis=1)
      X=DF.loc[:,'_1_MAR_STAT':].copy()
      X.shape

[54]: (500, 946)

[55]: X.columns=range(X.shape[1])
      X
```

[55]:

	0	1	2	3	4	5	6	7	8	9	...	936	937	938	939	940	941	942	943	944	945
0	0.0	0.0	0.0	0.0	1.0	0.0	0.0	1.0	0.0	0.0	...	0.0	0.0	0.0	0.0	0.0	0.0	0.0	1.0	0.0	0.0
1	0.0	0.0	0.0	0.0	1.0	0.0	0.0	1.0	0.0	0.0	...	0.0	0.0	0.0	0.0	0.0	0.0	0.0	1.0	0.0	0.0
2	1.0	0.0	0.0	0.0	0.0	0.0	0.0	1.0	0.0	0.0	...	0.0	0.0	0.0	0.0	0.0	1.0	0.0	0.0	0.0	0.0
3	0.0	0.0	0.0	0.0	1.0	0.0	0.0	1.0	0.0	0.0	...	0.0	0.0	0.0	0.0	0.0	0.0	0.0	0.0	1.0	0.0
4	0.0	0.0	0.0	0.0	1.0	0.0	0.0	1.0	0.0	0.0	...	0.0	0.0	0.0	0.0	1.0	0.0	0.0	0.0	0.0	0.0
...	...	...	...	...	...	...	...	...	...	...	...	...	...	...	...	...	...	...	...	...	...
245	0.0	0.0	0.0	1.0	0.0	0.0	0.0	1.0	0.0	0.0	...	0.0	0.0	0.0	0.0	0.0	0.0	1.0	0.0	0.0	0.0
246	0.0	1.0	0.0	0.0	0.0	0.0	0.0	1.0	0.0	0.0	...	0.0	0.0	0.0	0.0	0.0	0.0	1.0	0.0	0.0	0.0
247	1.0	0.0	0.0	0.0	0.0	0.0	0.0	1.0	0.0	0.0	...	0.0	0.0	0.0	0.0	0.0	0.0	1.0	0.0	0.0	0.0
248	1.0	0.0	0.0	0.0	0.0	0.0	0.0	1.0	0.0	0.0	...	0.0	0.0	0.0	0.0	0.0	1.0	0.0	0.0	0.0	0.0
249	0.0	1.0	0.0	0.0	0.0	0.0	0.0	0.0	0.0	0.0	...	0.0	0.0	1.0	0.0	0.0	0.0	0.0	0.0	0.0	0.0

500 rows × 946 columns

The magic function time returns the cell runtime and then we are calculating the spatial distance in *X* matrix and then running a for loop over the SDD class and then reducing the dimensionality and storing all the correlation values in a list. The best value of the list with the highest correlation will tell us the best *k* value where the data represents the best high-dimensional data.

```
•[63]: %%time

       A=scipy.spatial.distance.pdist(X, metric='euclidean')
       kendSDD=[]

       for k in range(1,15):
           print('loop started:',k)
           embedding = SDD()
           X_SDD = embedding._fit(X,degrees_of_freedom=k)
           B=scipy.spatial.distance.pdist(X_SDD, metric='euclidean')
           kendSDD.append(scipy.stats.kendalltau(A, B))
```

Once we find the best *k* value, then we will be running the SDD again with the best degree of freedom found at. Then we can move into plotting once we have the reduced 2D data. We have created a new column where we have mapped the data and then created a unique label to each patient from the index we picked up the records initially. This list of id will be the key to compare patients and look them into the 2D space on the map.

The same applies for *t*-SNE. We are running *t*-SNE with a step size of 5 and appending the values of correlation to a list.

```
[87]: embedding=SDD()
      X_SDD = embedding._fit(X,degrees_of_freedom=(kendSDD.index(max(kendSDD))+1))
      #X_SDD = embedding._fit(X,degrees_of_freedom=(4))

      C:\Users\home\AppData\Local\Temp\ipykernel_17304\3264856897.py:42: DeprecationWar
      tself. Doing this will not modify any behavior and is safe. If you specifically w
      Deprecated in NumPy 1.20; for more details and guidance: https://numpy.org/devdoc
        error = np.finfo(np.float).max
      C:\Users\home\AppData\Local\Temp\ipykernel_17304\3264856897.py:43: DeprecationWar
      tself. Doing this will not modify any behavior and is safe. If you specifically w
      Deprecated in NumPy 1.20; for more details and guidance: https://numpy.org/devdoc
        best_error = np.finfo(np.float).max

[88]: y.shape

[88]: (500, 2)

[89]: mapping={0:'passed away',1:'survived'}
      y['mapped']=y['Survive'].map(mapping)

[90]: indexes=DF[['index']].values.flatten()

[91]: import seaborn as sns
      import matplotlib.pyplot as plt

      li=['p-'+str(i) for i in indexes]
```

We are using the figure size of 11,7 and have applied some customization to make the visuals more appealing. We have exported the pictures with 1,080 dpi. We have the ability to set the font size and use hue and palette as required. The same applies for TSNE. We are running the TSNE with a step size of 5.

```
[93]: sns.set(rc={'figure.figsize':(11.7,8.27)})
      palette = sns.color_palette("bright", 2)
      sns.set_style('whitegrid')

      fig, ax = plt.subplots()
      ax=sns.scatterplot(X_SDD[:,0], X_SDD[:,1], hue=np.array(y['mapped']).flatten(), legend='full',palette=palette)
      for i, txt in enumerate(li):
          ax.annotate(txt, (X_SDD[i]),fontsize=0.07)
      resolution_value = 1080
      plt.xlabel('Z1')
      plt.ylabel('Z2')
      plt.title('Reduced space 39% error [SDD]')
      plt.savefig("pic500_60.9confid_SDD_1.png", format="png", dpi=resolution_value)
```

```
[95]: A=scipy.spatial.distance.pdist(X, metric='euclidean')
      kendTSNE=[]

      for k in range(1,500,5):
          embedding = TSNE(n_jobs=8, perplexity=k,n_components=2)
          X_tsne = embedding.fit_transform(X)
          B=scipy.spatial.distance.pdist(X_tsne, metric='euclidean')
          kendTSNE.append(scipy.stats.kendalltau(A, B))

      CPU times: total: 1h 18min 42s
      Wall time: 13min 28s

[96]: kendTSNE

[96]: [KendalltauResult(correlation=0.18725543594261398, pvalue=0.0),
       KendalltauResult(correlation=0.20597671266942566, pvalue=0.0),
       KendalltauResult(correlation=0.24511085115745077, pvalue=0.0),
       KendalltauResult(correlation=0.2892615040544596, pvalue=0.0),
       KendalltauResult(correlation=0.2841508818351283, pvalue=0.0),
       KendalltauResult(correlation=0.3146066084700592, pvalue=0.0),
       KendalltauResult(correlation=0.31266616888264437, pvalue=0.0),
       KendalltauResult(correlation=0.3135130064319658, pvalue=0.0),
       KendalltauResult(correlation=0.3463750987286461, pvalue=0.0),
       KendalltauResult(correlation=0.3340427262096364, pvalue=0.0),
       KendalltauResult(correlation=0.3534142773619266, pvalue=0.0),
       KendalltauResult(correlation=0.4138628011838061, pvalue=0.0),
       KendalltauResult(correlation=0.4063928936265193, pvalue=0.0),
       KendalltauResult(correlation=0.4537817119436237, pvalue=0.0),
       KendalltauResult(correlation=0.5014969238545498, pvalue=0.0),
       KendalltauResult(correlation=0.41895370912595764, pvalue=0.0),
       KendalltauResult(correlation=0.4923861042587252, pvalue=0.0),
       KendalltauResult(correlation=0.4421009001636018, pvalue=0.0),
       KendalltauResult(correlation=0.4583046325648521, pvalue=0.0),
       KendalltauResult(correlation=0.41346910778170165, pvalue=0.0),
       KendalltauResult(correlation=0.5368122999811045, pvalue=0.0),

0  5  Python 3 (ipykernel) | Idle
```

Chapter 11

Explainable neural networks in diabetes mellitus prediction

Solomon Chiekezi Nwaneri[1], Chika Yinka-Banjo[2], Ugochi Chinomso Uregbulam[1], Oluwakemi Ololade Odukoya[3] and Agbotiname Lucky Imoize[4,5]

Abstract

Artificial Intelligence (AI) has been widely applied in healthcare for several purposes, especially in disease prediction enabling physicians to more accurately diagnose patients' conditions. Results generated by traditional AI models are difficult to justify due to the opaqueness of the models. Thus, making it difficult for physicians to trust the results and use them in real-life practice. Recent advancements in explainable AI (XAI) have made the results more reliable, making it possible for physicians to embrace AI in clinical practice. Explainable deep neural network (xDNN) is a machine learning technique that can enhance diabetes mellitus disease prediction and explain the results. This chapter focuses on using explainable neural networks in diabetes mellitus prediction. It provides valuable insights on key steps and techniques for diabetes mellitus prediction using explainable neural networks (xNNs). In particular, the sequence for implementing the model using R programming software was discussed. In order to demonstrate the implementation of xNNs in diabetes mellitus prediction, the Pima Indian diabetes mellitus datasets were used. The model was assessed based on accuracy, sensitivity, specificity, precision, recall, and F1 score. Additionally, the chapter discussed the different methods of implementing explainability in XAI's and provided a clear illustration using the variable importance tool in R. The results revealed the effect of each variable on the overall model. We found that the

[1]Department of Biomedical Engineering, Faculty of Engineering, University of Lagos, Nigeria
[2]Department of Computer Science, Faculty of Science, University of Lagos, Nigeria
[3]Department of Community Health and Primary Care, College of Medicine, University of Lagos, Nigeria
[4]Department of Electrical and Electronics Engineering, Faculty of Engineering, University of Lagos, Nigeria
[5]Department of Electrical Engineering and Information Technology, Institute of Digital Communication, Ruhr University, Germany

variable importance varies with the network architecture. Overall, diabetes pedigree functions are the least important predictor of diabetes mellitus in the model.

Keywords: Artificial Intelligence; Diabetes mellitus; Explainable neural networks

11.1 Introduction

The recent advances in Artificial Intelligence (AI) have reasonably transformed healthcare systems, particularly the process of clinical decision-making. In clinical practice, decision-making is vital as it is a major determinant of the effective diagnosis and treatment of patients. In the pre-AI era, this was very difficult and required a great deal of experience garnered from many years of clinical practice. Computers can now be programmed to reason like humans to perform previously difficult or even impossible tasks for traditional machines, given the large volumes of data they can handle [1]. Despite these outstanding solutions, there is a general reluctance to embrace the use of AI systems in real-life situations. Genuine efforts have been made to improve user acceptability and transparency. The need to address these challenges necessitated innovations in the field of explainable AI (XAI). The concept of XAI is novel as it comes with new features that enable it to generate explainable results [2]. Explainability or interpretability is generally defined as the degree to which a human can understand the cause of a decision [3]. With XAI's, users can now understand and trust the results created by machine learning algorithms [4].

Explainability has become necessary in providing convincing evidence to experts and regulatory bodies that the results generated by AI models are reliable and justifiable. Unlike the inner workings of conventional AI systems considered complex and difficult to understand, XAI's are usually more transparent to users with no black boxes, making it easier for clinicians to trust their results [5]. One example of XAI is the explainable deep neural network (xDNN) which provides a transparent means of classifying complex networks from a wide range of inputs. This classification is essential because it creates a means to solve the problem of lack of explainability in traditional DNNs [6]. Traditional DNNs consist of several neurons with the capacity to mimic the human brain [7]. When fed with input data, DNNs have an internal opaque architecture that trains the input data, thus, producing results with reasonably high levels of clarity. The inability of the results to be explained due to black boxes in traditional DNNs often affects the levels of confidence an average user has on the results. With xDNNs, the understanding of the user regarding reasons for specific results from a deep neural network (DNN) training model is guaranteed. Various approaches have been used in explainable deep learning models to enhance decision-making [8]. One approach uses explainable neural symbolic learning (X-NeSyL), combining deep learning and domain expert knowledge [9]. X-NeSyL comprises three key stages summarized as:

1. Processing of symbolic representation.
2. Classification of an object by its detected parts is usually done using the explainable part-based classifying network architecture (EXPLANet).

3. Ensuring that model aligns its output with symbolic explanations using tools such as Shapley Additive explanation (SHAP) [10].

The healthcare sector has benefitted immensely from innovations in AI, especially in enhancing clinical decision-making. Applications of AI in healthcare are well reported [11,12]. Clinical decisions are usually difficult to make, even with the most experienced physicians. With clinical decision support systems, clinicians make better diagnostic decisions by combining clinical knowledge with patient information [13,14], expressing the need to go beyond XAI and focus on causability to attain a level of explainable medicine. One of the several diseases that xDNNs can be used to predict is diabetes mellitus.

Diabetes mellitus is a metabolic disease characterized by high plasma glucose. There are generally three types of diabetes mellitus: Type 1, Type 2, and gestational diabetes. While in Type 1 diabetes mellitus, the pancreas is unable to produce insulin, Type 2 diabetes mellitus is associated with impaired glucose regulation caused by insulin resistance and dysfunctional pancreatic beta-cells [15,16]. Gestational diabetes mellitus (GDM) is associated with pregnancy and is the most common medical complication in pregnant women [17,18]. Type 2 diabetes mellitus is pervasive not just in developed countries but in developing countries. The global prevalence and incidence of diabetes mellitus are quite problematic, particularly with the high level of ignorance about the disease in many communities [19]. According to [20], 537 million adults between 20 and 79 suffer from DM. The most common risk factors for diabetes mellitus include genetics, obesity, advanced age, hypertension, tobacco use and diet [17,21,22]. Prolonged uncontrolled diabetes mellitus is usually associated with several medical complications such as cardiovascular diseases, kidney failure, stroke, amputation, blindness and, in some cases, death [23]. Early identification and detection of DM, followed by glycemic control, are crucial to preventing complications [24].

Hence, researchers are working on predicting the chances of people developing diabetes mellitus before its onset using AI [23,25]. However, the models are not explainable. This chapter focuses on the application of xNNs in predicting diabetes mellitus. This would help to improve the diagnosis of the disease by providing a pathway for readers to understand the methods and strategies for using xNNs to predict the disease. Moreover, since the model reveals the level of importance of each predictor, it will help readers to understand the impact of each predictor. Furthermore, the concept of DNNs was exhaustively discussed.

11.2 Related work

AI applications in disease prediction have been well reported [25–30]. Predictive models are vital tools for the pragmatic management of diabetes mellitus [19]. The most common AI models widely used include NNs, [27], logistic regression [31,32], decision tree [33], support vector machines [34], and random forest [35]. Many AI algorithms were applied singly or combined with other algorithms [31]. In some studies, ensemble models were developed by combining two or more AI models to obtain better results than the individual models [36]. In [19], the authors applied improved K-means clustering and logistic regression models for diabetes mellitus prediction. Larabi-Marie-Sainte *et al.* [37] did a comparative analysis of the following machine learning classifiers; RepTree, Lazy, Rules, Functions, and Bayes, which were tested on the Pima Indian Diabetes Dataset and obtained the highest classification accuracy of 74.48%.

Deep learning has become popular, especially with the rapid advancements in the speed and functionality of modern computers. Indeed, deep learning is seen to be computationally intensive [38]. The minimum requirements for any computer running a deep learning program include graphics processing units (GPUs), large memory ($\geq$ 8 GB), and high computational speed. Often, DL is preferred to traditional machine learning techniques because of its better performance [39]. Consequently, there is a growing interest by researchers in medical applications of DL such as liver cancer diagnosis [40,41], breast cancer [42], detection of acute intracranial hemorrhage (ICH) to support expert radiologists [43], prediction of mental disorders [44], and diabetic retinopathy [45]. In some cases, deep learning algorithms have performed better than human experts in medical image analysis [46]. The prediction of diabetes mellitus using various deep learning algorithms has continued to generate research interest. Improved performance in artificial neural network models trained using an artificial backpropagation scaled conjugate gradient neural network (ABP-SCGNN) algorithm to predict diabetes mellitus has been demonstrated by [47]. The persistent memory of recurrent neural networks (RNNs) has been utilized in deep learning algorithms because it can overcome the vanishing gradient problem. The vanishing gradient problem is generally characterized by a drastic reduction in the value of the gradient of multilayer networks to abysmally small values that are too difficult to train. In Rhee *et al.* [48], RNN long short-term memory network performed better than conventional networks.

Jang *et al.* [49] proposed explainable diabetic retinopathy (ExplainDR) classification model, leveraging neural-symbolic learning. Linden *et al.* [50] developed an explainable multimodal neural network architecture for predicting the time-dependent risk of six common comorbidities of epilepsy patients based on administrative claims data. In [51], the authors compared the performance of DNNs, extremely gradient boosting (XGBoost), and random forest in predicting Type 2 and found DNN to outperform the other algorithms.

Despite notable advancements in the application of DNNs in medical science, there are limited data sources for DNNs and challenges of knowledge representation [46]. The major limitation of machine learning models is the difficulty in interpreting complex models [52]. Chetoui and Akhloufi [45] proposed an explainable deep learning algorithm for detecting diabetic retinopathy (DR) on retinal fundus images. Diabetic retinopathy (DR) is a significant complication of diabetes mellitus, which causes damage to the retina. Using neural symbolic learning methods, the authors achieved a high-level neural representation. Results showed higher classification rates using this deep learning neural network prediction method than other methods. El Rashidy *et al.* [18] proposed a medically intuitive and cost-effective solution that focused on early predicting gestational diabetes mellitus by implementing deep learning algorithms. In another study, Mpreno-Sanchez [53] proposed explainable classification models for chronic kidney disease using ensemble tree classifiers. By using two different explainability approaches, the study yielded the explainability of the results. To the best of our knowledge, using XDNNs is an emerging subject that needs further investigation.

11.3 Methodology

11.3.1 Key implementation requirements and strategies for xDNNs

The implementation of DNN and NN models is generally achieved using various techniques and strategies. The software tool and programming languages used for the implementation are of critical importance. The most common programming languages used for DNN implementation are Python, R programming, C++, Java, and MATLAB. There are also many commercially available DNNs, such as Deepy and NVIDIA Deep Learning software. The user is at liberty to implement the xDNN in any of the aforementioned programs. The R programming language was chosen in this chapter to implement the xDNN model. The use of R programming has several benefits. It is an open-source programming language which is very easy to use.

Before the implementation of xDNNs, the choice of the type of learning is paramount. The three learning techniques for DNNs consist of three types:

1. Supervised DNN—A learning technique that provides a target or set of examples for the network to learn from Ref. [54].
2. Unsupervised DNN—This learning technique has no target values as learning is without the support of a teacher [55].
3. Hybrid DNN—A learning technique that combines both the features of both supervised and unsupervised learning.

For illustration purposes in this chapter, the supervised DNN was used to implement the xDNN. In order to achieve this, target values are included while training the DNN to help the model learn by examples. In this case, the target values are assigned a value of 1 for diabetic patients and 0 for non-diabetic patients. The model is trained using machine learning learnable parameters and hyperparameters. An important hyperparameter is the learning rate α, which determines the rate at which the other parameters are updated. The learning rate ranges between 0 and 1.

Another important strategy for training an xDNN is to optimize the loss function. Optimization is an important phenomenon which is common in any discipline. When a DNN is optimized, the motive is to train the network to minimize the difference in values between the target output and the desired output. Thus, optimization of DNNs is necessary to minimize the loss function and enhance the accuracy of the classification algorithm. The variety of optimization algorithms in

DNN demonstrates the need to improve the performance of the training algorithm. The loss function is generally determined by key evaluation metrics such as mean squared error (MSE), mean absolute error (MAE), binary cross entropy, and many others based on the criteria set by the investigator. Eq. (11.1) shows the MSE:

$$MSE = \frac{1}{N}\sum_{i=1}^{N}(T - A)^2 \tag{11.1}$$

where N is the number of observations; T is the target values; A is the actual values.

The traditional optimization algorithms frequently used for DNNs are Gradient Descent and Stochastic Gradient Descent. These optimization algorithms are widely used in healthcare for disease prediction. For instance, Nawaz *et al.* [56] proposed an intelligent cardiovascular disease prediction model empowered with Gradient Descent Optimization. The gradient descent algorithm in DNNs and conventional neural networks should be convex and differentiable. Computation of the gradient descent algorithm is an iterative process as shown in Eq. (11.2):

$$p_{n+1} = p_n - \alpha \nabla \mathrm{f}(p_n) \tag{11.2}$$

where p_n is the initial point; p_{n+1} is the Next *point*; α is the learning rate.

Other examples include Adaptive moment estimation (Adam) optimizer, Adagrad, Stochastic Gradient Descent, Root Mean Square Propagation (RMSProp), and Stochastic Gradient Descent with Momentum. The Adam optimizer was used in this study of the different optimization algorithms. The Adam optimizer was proposed in [57] as an optimization algorithm. Having been developed from the AdaGrad and the RMSP algorithm, the Adam optimizer combines the strengths of both algorithms, which makes it very suitable for implementing xDNNs [58,59]. To derive the Adam optimization algorithm, the aggregate gradient (m_t) is calculated at time t and the sum of squares of past gradients considering the exponential decay rates β_1, β_2, learning rate α, as indicated in Eqs. (11.3) and (11.4):

$$m_t = \beta_1 m_{t-1} + (1 - \beta_1) \cdot g_t \tag{11.3}$$

The exponential moving average is updated as:

$$v_t = \beta_2 v_{t-1} + (1 - \beta_2) \cdot g_t^2 \tag{11.4}$$

However, the results are usually biased. Therefore, to enhance the quality of the optimization results, a bias-corrected m_t and v_t are added in (11.5) and (11.6), respectively:

$$\widehat{m}_t = \frac{m_t}{1 - \beta_1^t} \tag{11.5}$$

$$\widehat{v}_t = \frac{v_t}{1 - \beta_2^t} \tag{11.6}$$

The Adam optimizer is iteratively computed by updating the general equation as:

$$w_{t+1} = w_t - \widehat{m}_t\left(\frac{\alpha}{\sqrt{\widehat{v}_t} + \varepsilon}\right) \tag{11.7}$$

11.3.2 DNN architecture

The architecture for DNN has been broadly classified into unsupervised pretrained networks, convolutional neural networks (CNNs), recurrent neural networks, and recursive neural networks [60]. The CNN architecture is widely utilized in computer image analysis. The architecture of CNNs is similar to the organization of the visual cortex in the brain [61]. The CNN is based on the mathematical theory of convolution. DNNs generally consist of several layers, including multiple hidden layers.

11.3.3 Activation function

Activation functions provide the driving force for DNNs as they are designed to boost the power of the network, adding nonlinearity to the network. Activation functions may be classified as linear or non-linear. Common activation functions include sigmoid, linear, rectified linear unit (ReLU), Tanh, and Smish activation functions [44,62]. While binary step, linear activation functions, sigmoid, Tanh, ReLU, and its variants are the most prominent non-linear activation functions. The choice of activation function is critical to the efficiency of DNN. In recent times, most DNNs have been designed using ReLU activation functions. The ReLU activation function is a very efficient activation function developed to solve the vanishing gradient problem. The ReLU activation function is given by:

$$f(x) = \max(0, x) \tag{11.8}$$

11.3.4 Procedures for xDNN model implementation

The xDNN model was implemented in R software (version 4.2.1). Figure 11.1 displays a flowchart for the xDNN model implementation.

All the libraries required for the xDNN model implementation were installed. The libraries include keras, tensorflow, mlbench, dplR, corrplot, and shapr. Afterwards, the diabetes datasets were loaded. The dataset used in this study is the PIMA Indian diabetes dataset obtained from the University of California Irvine (UCI) machine learning repository [64]. The dataset consists of 768 data samples of Indian women living in Arizona. Table 11.1 shows the basic characteristics of PIMA Indian datasets. It has nine attributes (9) and one (1) output. The input features include plasma glucose concentration, body mass index, age, diabetes pedigree function, triceps skin food thickness, diastolic blood pressure, 2-Hour Serum Insulin, and number of times pregnant. The status of the subject, whether diabetic or not diabetic, is the only output feature. There are 268 instances of diabetic patients and 500 cases of non-diabetic patients. All the data are numeric values

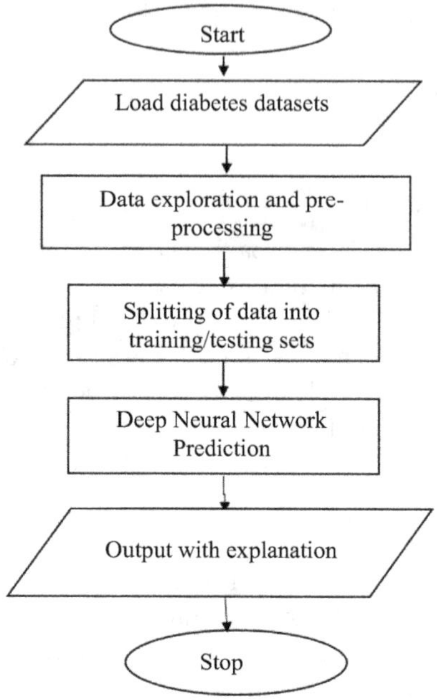

Figure 11.1 Flow chart for the xDNN model

comprising integer and floating-point values. Table 11.2 describes the attributes and data type representation in python.

Data exploration is performed after loading the datasets. This is necessary to determine if multicollinearity exists among the variables. Multicollinearity was determined using correlation. Figure 11.2 displays the correlation between the variables.

Data exploration is followed by data preprocessing. This involves cleaning and normalization of data. Data cleaning involves the removal of outliers and fixing the issues of missing values. There are diverse approaches to addressing missing values, including deletion or replacement of missing values with the mean value of each variable. Data normalization was done using the min–max normalization. For each variable *d*, the normalized value *dn* is calculated as:

$$dn = \left(\frac{d - \min(d)}{\max(d) - \min(d)}\right) \tag{11.9}$$

Implementing xDNNs comes with challenges, such as overfitting and under-fitting the models. These are addressed by splitting data into training and testing sets. Another technique is the process of k-fold cross validation. In recent studies on the application of xDNNs in healthcare, k-fold cross validation was used [65–68]. There are several deep learning libraries for implementing xDNN,

Table 11.1 Review of related studies

Authors/references	Problem type	Remarks
Wu *et al.* [19]	Diabetes mellitus	Improved K-means clustering and logistic regression models were developed for diabetes mellitus prediction with no provision for explainability
Larabi-Marie-Sainte *et al.* [37]	Diabetes mellitus	Reviewed trends in diabetes mellitus prediction and carried out a case study involving unpopular machine learning classifiers. No provision was made for the explainability of the machine learning classifiers
Rhee *et al.* [48]	Type 2 diabetes mellitus	A deep learning system for the prediction of diabetes mellitus was developed with no provision for explainability
Jang *et al.* [49]	Diabetic retinopathy	The authors developed a diabetic retinopathy classification model by proposing a human-readable symbolic representation
Moreno-Sanchez [53]	Chronic kidney disease	Developed an explainable classification model for chronic kidney disease using various ensemble trees
El Rashidy *et al.* [18]	Gestational diabetes mellitus	Developed a framework for the continuous monitoring of gestational diabetes mellitus. The study combined data finding methodology and xDNNs
Sadeghi *et al.* [51]	Type 2 diabetes mellitus	The authors developed diabetes mellitus risk models in the presence of class imbalance using DNNs, extremely gradient boosting (XGBoost) and random forest

Table 11.2 Characteristics of Pima Indian Dataset [64]

S. no.	Attributes	Descriptions and attribute values	Data type in Python
1	Number of times pregnant (NTP)	Numerical values	Integer value
2	Plasma glucose concentration (PGC)	Numerical values	Integer value
3	Diastolic blood pressure (DBP)	Numerical values (mm Hg)	Integer value
4	Triceps skin food thickness (TSFT)	Numerical values in mm	Integer value
5	2-Hour Serum Insulin	Numerical values in (muU/ml)	Integer value
6	Body Mass Index (BMI)	Numerical value in $(kg/m)^2$	Float values
7	Diabetes pedigree function (DPF)	Numerical value	Float values
8	Age	Numerical values	Integer value
9	Diagnosis of Type 2 diabetes disease	Yes=1 No=0	Integer value

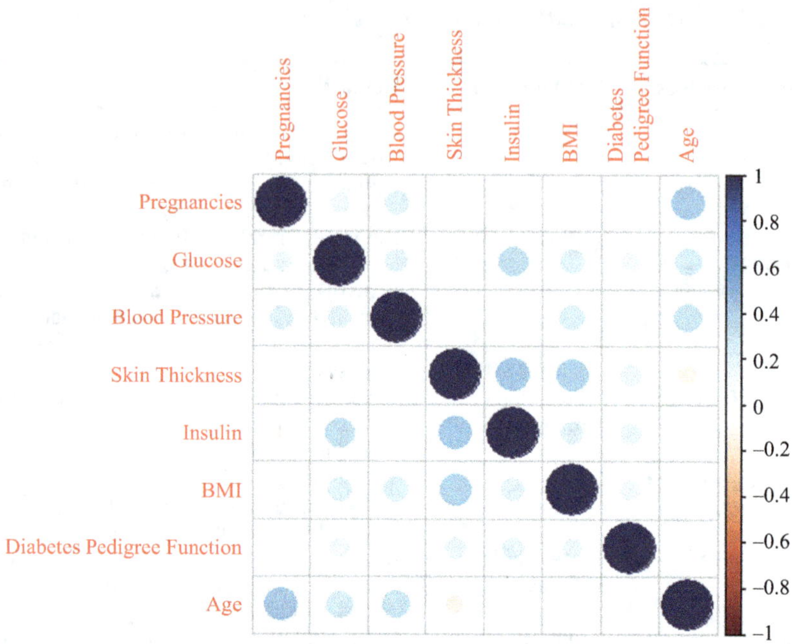

Figure 11.2 Correlation between the variables

Table 11.3 Results: comparison of optimization algorithms

Optimizer	Loss	MAE
Adam	0.206	0.415
rmsprop	0.173	0.335

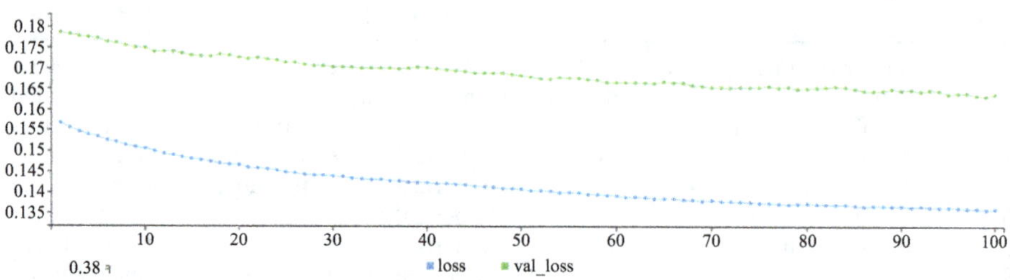

Figure 11.3 Time versus of values of the loss function of the model using Adams Optimizer

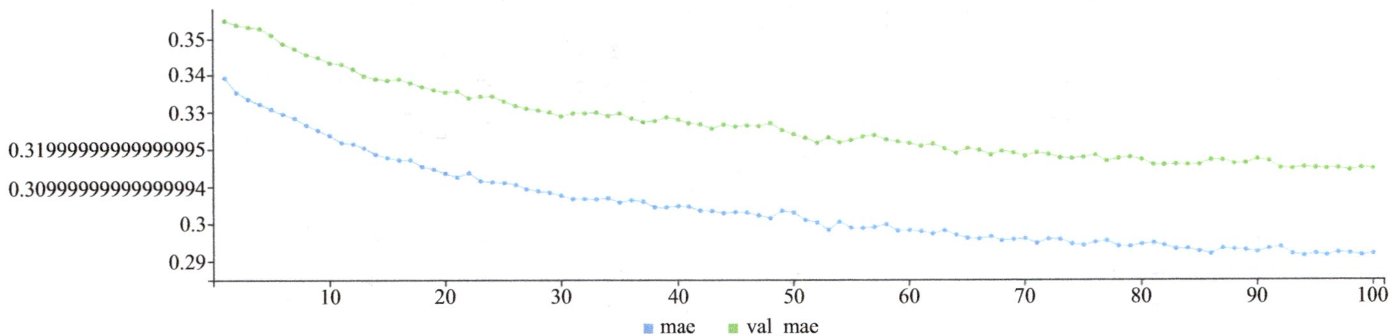

Figure 11.4 Time versus values of the MAE model using Adams Optimizer

Table 11.4 Confusion matrix for the DNN model

	0	1
0	129	41
1	16	43

Table 11.5 Performance metrics for DNN in R

Performance metrics	**Values (%)**
Accuracy	75.11
Sensitivity	51.19
Specificity	88.97
Precision	72.88
Recall	51.19
F1 score	60.14

such as keras, tensorflow, deepnet, deepr, H_2O, and darch. In some healthcare-related studies, these tools have been shown to be immensely beneficial [69].

11.3.5 Model parameters and hyper-parameters

A sequential deep learning model consisted of eight neurons in the input layer, one hidden layer, and an output layer. The model was implemented using a number of parameters and hyper-parameters. These include 100 epochs, a batch size of 32, and Relu activation function.

11.3.6 Evaluation and explainability metrics

The model evaluation metrics determined are accuracy, sensitivity, specificity, precision, recall, and F1-measures from the true positive (TP), the true negative (TN), the false positive (FP), and the false negative (FN) as shown in Eqs. (11.10)–(11.15), respectively:

$$\text{Accuracy} = \frac{\text{TP} + \text{TN}}{\text{TP} + \text{TN} + \text{FP} + \text{FN}} \tag{11.10}$$

$$\text{Sensitivity} = \frac{\text{TP}}{\text{TP} + \text{FN}} \tag{11.11}$$

$$\text{Specificity} = \frac{\text{TN}}{\text{TN} + \text{FP}} \tag{11.12}$$

$$\text{Precision} = \frac{\text{TP}}{\text{TP} + \text{FP}} \tag{11.13}$$

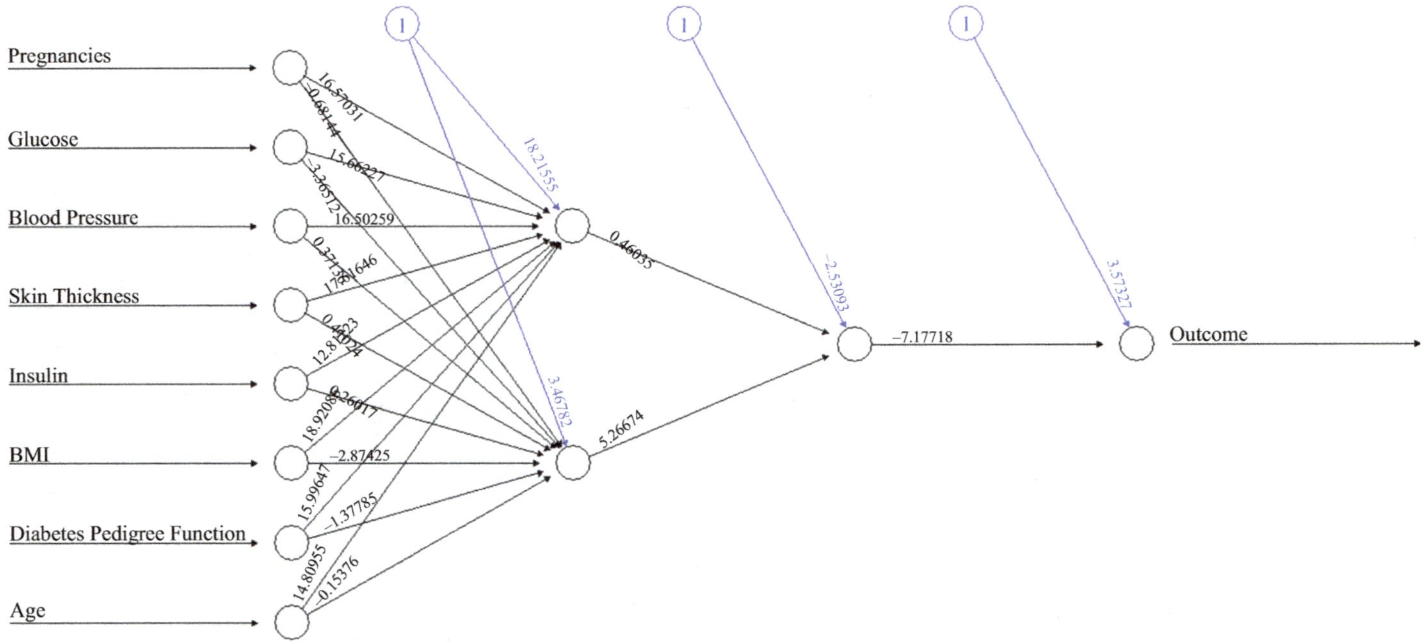

Error: 43.18755 Steps: 2842

Figure 11.5 Network architecture for 8–2–1 neural network model

Table 11.6 Results of the 8–2–1 neural network model

Parameters	Values
Error	38.35
Reached.threshold	0.010
steps	23,765
Intercept.to.1layhid1	9.04
Pregnancies.to.1layhid1	−1.05
Glucose.to.1layhid1	−6.36
BloodPressure.to.1layhid1	−3.13
SkinThickness.to.1layhid1	4.29
Insulin.to.1layhid1	−4.16
BMI.to.1layhid1	−5.64
DiabetesPedigreeFunction.to.1layhid1	−3.39
Age.to.1layhid1.	1.10
Intercept.to.1layhid2	−8.38
Pregnancies.to.1layhid2	3.01
Glucose.to.1layhid2	24.8
BloodPressure.to.1layhid2	−26.66
SkinThickness.to.1layhid2	36.17
Insulin.to.1layhid2	−38.2
BMI.to.1layhid2	3.22
DiabetesPedigreeFunction.to.1layhid2	4.37
Age.to.1layhid2	52.94
Intercept.to.2layhid1	−0.73
1layhid1.to.2layhid1	−8.43
1layhid2.to.2layhid1	8.80
Intercept.to.Outcome	−4.28
2layhid1.to.Outcome	6.47

$$\text{Recall} = \frac{\text{TP}}{\text{TP} + \text{FN}} \tag{11.14}$$

$$F1\ Score = 2 \times \frac{Precision \times Recall}{Precision + Recall} \tag{11.15}$$

Model explainability is generally achieved with different approaches and libraries. One of the approaches is the determination of variable importance. This would help to know the impact of each variable on the model. Another technique is the use of common explainability tools like the Shapley additive explanations (SHAP), and the Local Interpretable Model Agnostic Explainer (LIME). LIME has been used for a number of medical applications [70–72]. Magesh *et al.* [70] utilized LIME to provide explainability from a machine learning model designed to detect early Parkinson disease by from DatSCAN images.

11.4 Results and discussion

11.4.1 Results for DNN models

A comparison between the optimization algorithms in Table 11.3 revealed that Rmsprop performed better than Adam optimizer and was chosen for the model. The confusion matrix for the model is shown in Table 11.4. The performance metrics for DNN in R are presented in Table 11.5.

11.4.2 Results for neural network models

A neural network model was implemented using 8–2–1 architecture consisting of eight neurons in the input layer, two neurons in the hidden layer, and one neuron in the output layer, as shown in Figure 11.5. Results of the model are shown in Table 11.6.

Explainability is demonstrated by the graphical display of the variable importance of the various diabetes mellitus predictors as shown in Figure 11.6 with the level of importance on the *x*-axis and the predictors on the *y*-axis. Glucose is the most important predictor of diabetes mellitus, followed by BMI and pregnancy. Skin thickness is the least important predictor of diabetes mellitus in the model.

There is a considerable difference in the contribution of each predictor in the 8–2–1 and the 8–2–2–1 network architecture. In Figure 11.7, the age was the most important predictor, followed by glucose and skin thickness.

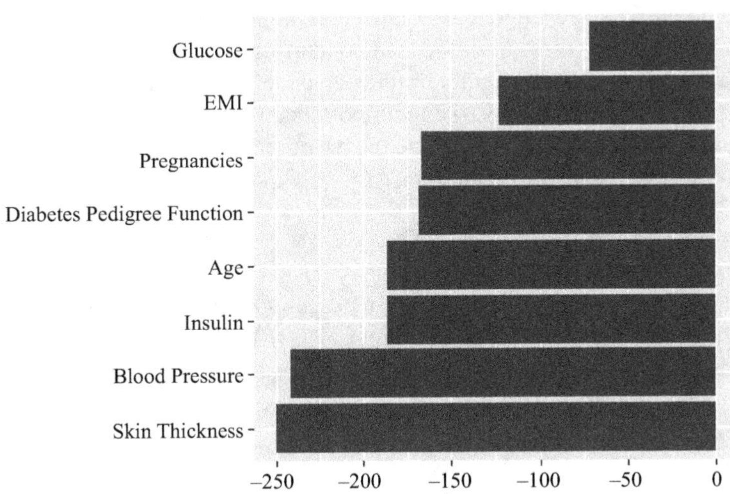

Figure 11.6 Variable Importance of features for 8–2–1 architecture

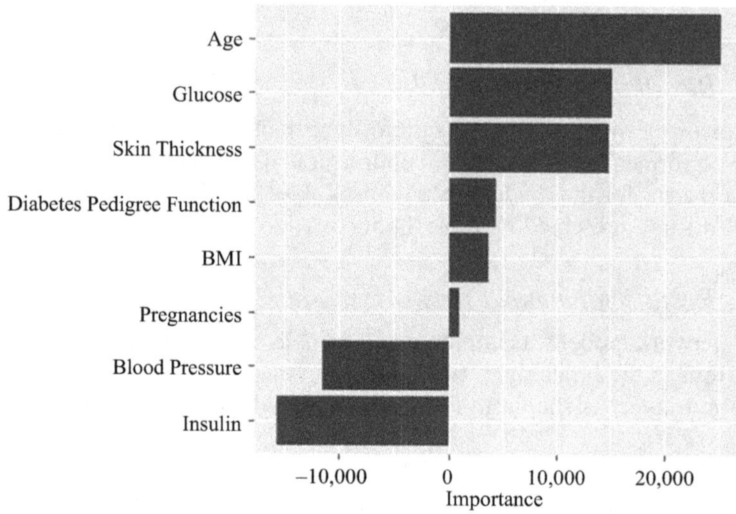

Figure 11.7 Variable importance of features for 8–2–2–1 architecture

11.5 Conclusion and future scope

Explainability in deep learning models has become necessary in recent times, particularly in healthcare systems where the reliability of results from deep learning models is important. In this chapter, we have discussed explainable neural networks in diabetes prediction. A general definition of explainability and the benefits of explainable deep learning algorithms were provided. The literature was exhaustively examined for related works to highlight the recent trends in research on the topic and identify knowledge gaps. Different approaches and tools for implementing explainability were discussed. In conclusion, xDNNs have been shown to reveal the impact of the different variables considered as risk factors for diabetes mellitus. We found that the variable importance varies with different network architectures. Future work will examine the proposed neural network-based models in XAI for accurate disease prediction.

Acknowledgment

The work of Agbotiname Lucky Imoize is supported in part by the Nigerian Petroleum Technology Development Fund (PTDF) and in part by the German Academic Exchange Service (DAAD) through the Nigerian-German Postgraduate Program under Grant 57473408.

References

[1] Miles J. and Walker A. 'The potential application of artificial intelligence in transport'. *IEE Proceedings – Intelligent. Transport Systems*, 2006;153:183–198.

[2] Janssen F.M., Aben K.K.H., Heesterman B.L., Voorham Q.J.M., Seegers P. A., and Moncada-Torres A. 'Using explainable machine learning to explore the impact of synoptic reporting on prostate cancer'. *Algorithms*, 2022; 15(2):49, doi:10.3390/a15020049. ISSN 1999-4893

[3] Miller T. 'Explanation in artificial intelligence: Insights from the social sciences'. *Artificial Intelligence*, 2019;267:1–38.

[4] Samek W., Wiegand T., and Muller K. 'Explainable artificial intelligence: understanding, visualizing and interpreting deep learning models'. *ITU Journal: ICT Discoveries*, 2018;1(1):39–41.

[5] Kubben P., Dumontier M., and Dekker A. *Fundamentals of Clinical Data Science*. Cham: Springer International Publishing, 2021.

[6] Rudin C. 'Stop explaining black box machine learning models for high stakes decisions and use interpretable models instead'. *Nature Machine Intelligence*, 2019;1(5), 206–215.

[7] Gupta R., Kewalramani M.A., and Goel A. 'Prediction of concrete strength using neural-expert system'. *Journal of Materials in Civil Engineering*, 2006;18(3):462–466.

[8] Kamath U. and Liu J. 'Explainable deep learning'. In: *Explainable Artificial Intelligence: An Introduction to Interpretable Machine Learning*. Cham: Springer, 2021, https://doi.org/10.1007/978-3-030-83356-5_6

[9] Díaz-Rodríguez N., Lamas A., Sanchez J., *et al.* 'EXplainable Neural-Symbolic Learning (X-NeSyL) methodology to fuse deep learning representations with expert knowledge graphs: the MonuMAI cultural heritage use case'. *Information Fusion*, 2022;79:58–83.

[10] Scott M. and Lundberg Su-In Lee A. 'Unified approach to interpreting model predictions'. In: *Proceedings of the International Conference on Neural Information Processing Systems*, 2017, pp. 4765–4774.

[11] Paul D., Sanap G., Shenoy S., Kalyane D., Kalia K., and Tekade R.K. 'Artificial intelligence in drug discovery and development'. *Drug Discovery Today*, 2021;26(1):80–93.

[12] Vaishya R., Javaid M., Khan I.H., and Haleem A. Artificial Intelligence (AI) applications for COVID-19 pandemic. *Diabetes & Metabolic Syndrome: Clinical Research & Reviews*, 2020;14(4):337–339.

[13] Sutton R.T., Pincock D., Baumgart D.C., *et al.* An overview of clinical decision support systems: benefits, risks, and strategies for success. *npj Digital Medicine,* 2020;3:17, https://doi.org/10.1038/s41746-020-0221-y

[14] Holzinger A., Langs G., Denk H., Zatloukal K., and Müller H. Causability and explainability of artificial intelligence in medicine. *Wiley Interdisciplinary Review Data Mining and Knowledge Discovery*, 2019;9(4): e1312, doi:10.1002/widm.1312

[15] Blair M. 'Diabetes mellitus review'. *Urologic Nursing*, 2016;36(1):27–36.

[16] Li, W., Huang E., and Gao S. 'Type 1 diabetes mellitus and cognitive impairments: a systematic review'. *Journal of Alzheimers Disorders*, 2017;57(1):29–36, doi: 10.3233/JAD-161250

[17] McIntyre H.D., Catalano, P., Zhang C., Desoye G., Mathiesen, E.R., and Damm, P. 'Gestational diabetes mellitus'. *Nature Reviews Disease Primers*, 2019;5(1):1–19.
[18] El Rashidy N., El Sayed N.E., El-Ghamry A., and Talaat, F.M. 'Utilizing fog computing and explainable deep learning techniques for gestational diabetes prediction'. *Research Square*, 2022. doi:10.21203/rs.3.rs-1098270/v1.
[19] Wu H., Yang S., Huang, Z., He J., and Wang X. 'Type 2 diabetes mellitus prediction model based on data mining'. *Informatics in Medicine Unlocked*, 2018;10:100–107.
[20] International Diabetes Federation. *Diabetes Atlas*, 10th ed. Brussels: International Diabetes Federation, 2021.
[21] Kaul K., Tarr, J.M., Ahmad, S.I., Kohner, E.M., and Chibber, R. Introduction to diabetes mellitus. *Diabetes*, 2013;771:1–11.
[22] Alam, U., Asghar, O., Azmi S., and Malik R.A. 'General aspects of diabetes mellitus'. *Handbook of Clinical Neurology*, 2014;126:211–222.
[23] Daanouni O., Cherradi, B., and Tmiri A. 'Type 2 diabetes mellitus prediction model based on machine learning approach'. In: *The Proceedings of the Third International Conference on Smart City Applications*. Cham: Springer, 2019, pp. 454–469.
[24] Odukoya O., Nwaneri S., Odeniyi I., *et al.* 'Development and comparison of three data models for predicting diabetes mellitus using risk factors in a Nigerian population'. *Health Informatics Research*, 2022;28(1):58–67.
[25] Hassan A.S., Malaserene, I., and Leema, A.A. 'Diabetes mellitus prediction using classification techniques'. *International Journal of Innovative Technology and Exploring Engineering*, 2020;9(5):2080–2084.
[26] Chang, C.L. and Hsu M.Y. The study that applies artificial intelligence and logistic regression for assistance in differential diagnostic of pancreatic cancer. *Expert System Application*, 2009;36:10663–10672, https://doi.org/10.1016/j.eswa.2009.02.046
[27] Nwaneri S.C., Nwoye E.O,. Irurhe N.K., and Babatunde A.M. 'Application of artificial neural networks in breast cancer classification: a comparative study'. *University of Lagos, Journal of Basic Medical Sciences,* 2014; 2(1):32–38.
[28] De Ramón-Fernández A., Fernández D.R., and Prieto Sánchez M.T. 'A decision support system for predicting the treatment of ectopic pregnancies'. *International Journal of Medical Informatics*, 2019;129:198–204, https://doi.org/10.1016/j.ijmedinf.2019.06.002
[29] Lu J., Song E, Ghoneim A, and Alrashoud M. 'Machine learning for assisting cervical cancer diagnosis: an ensemble approach'. *Future Generation Computer Systems*, 2020;106:199–205.
[30] Lei Y., Yin M., Yu M., *et al.* 'Artificial intelligence in medical imaging of the breast'. *Frontiers in Oncology*, 2021;11:600557, https://doi.org/10.3389/fonc.2021.600557
[31] Fisher M.A. and Taylor G.W. A prediction model for chronic kidney disease includes periodontal disease. *Journal of Periodontology,* 2009;80(1):16–23.

[32] Dwivedi, A.K. Analysis of computational intelligence techniques for diabetes mellitus prediction. *Neural Computing and Applications*, 2018; 30(12):3837–3845.

[33] Verma A.K., Chakraborty M., and Biswas S.K. 'Breast cancer management system using decision tree and neural network'. *SN Computer Science*, 2021;2:234, https://doi.org/10.1007/s42979-021-00644-2.

[34] Yu W, Liu T., Valdez R., Gwinn M., and Khoury M.J. Application of support vector machine modeling for prediction of common diseases: the case of diabetes and pre-diabetes. *BMC Medical Informatics and Decision Making*, 2010;10:16, doi:10.1186/1472-6947-10-16

[35] Macaulay B.O., Aribisala B.S., Akande S.A., Akinnuwesi B.A., and Olabanjo O.A. 'Breast cancer risk prediction in African women using Random Forest Classifier'. *Cancer Treatment and Research Communications*, 2021;28:100396.

[36] Kumari S., Kumar D., and Mittal M. 'An ensemble approach for classification and prediction of diabetes mellitus using soft voting classifier'. *International Journal of Cognitive Computing in Engineering*, 2021;2:40–46, https://doi.org/10.1016/j.ijcce.2021.01.001

[37] Larabi-Marie-Sainte S., Aburahmah L., Almohaini R., and Saba T. 'Current techniques for diabetes prediction: review and case study'. *Applied Sciences*, 2019;9(21):4604, https://doi.org/10.3390/app9214604

[38] Goel A., Tung C., Lu Y.H., and Thiruvathukal G.K. 'A survey of methods for low-power deep learning and computer vision'. In: *2020 IEEE 6th World Forum on Internet of Things (WF-IoT)*, 2020; pp. 1–6, doi: 10.1109/WF-IoT48130.2020.9221198.

[39] Lee H., Eun Y., Hwang J.Y., and Eun L.Y. 'Explainable deep learning algorithm for distinguishing incomplete Kawasaki disease by coronary artery lesions on echocardiographic imaging'. *Computer Methods and Programs in Biomedicine*, 2022;223, https://doi.org/10.1016/j.cmpb.2022.106970

[40] Wu K., Chen X., and Ding M. 'Deep learning-based classification of focal liver lesions with contrast-enhanced ultrasound'. *Optik*, 2014;125(15):405.

[41] Amin J., Anjum M.A., Sharif M., Kadry S., Nadeem A., and Ahmad S.F. 'Liver tumor localization based on YOLOv3 and 3D-semantic segmentation using deep neural networks'. *Diagnostics,* 2022;12:823, https://doi.org/10.3390/diagnostics12040823

[42] Lee H., Park J., and Hwang J.Y. 'Channel attention module with multiscale grid average pooling for breast cancer segmentation in an ultrasound image'. *IEEE Transactions on Ultrasonics, Ferroelectrics. Frequency Control,* 2020;67(7):1344–1353.

[43] Lee H., Yune S., Mansouri M., *et al.* 'An explainable deep-learning algorithm for the detection of acute intracranial haemorrhage from small datasets'. *Nature Biomedical Engineering*, 2019;3(3):173–182.

[44] Wang X., Ren H., Wang A., and Smish A. 'Novel activation function for deep learning methods'. *Electronics*, 2022;11(4):540, https://doi.org/10.3390/electronics11040540

[45] Chetoui M., and Akhloufi M.A. 'Explainable diabetic retinopathy using EfficientNET'. In: *2020 42nd Annual International Conference of the IEEE Engineering in Medicine & Biology Society* (*EMBC*), 2020, pp. 1966–1969.

[46] Singh A., Sengupta S., and Lakshminarayanan V. 'Explainable deep learning models in medical image analyses. *Journal of Imaging*, 2020;6(6):52.

[47] Bukhari M.M., Alkhamees B.F., Hussain S., Gumaei A., Assiri A., and Ullah S.S. 'An improved artificial neural network model for effective diabetes prediction'. *Complexity*, 2021:1–10, https://doi.org/10.1155/2021/5525271

[48] Rhee S.Y., Sung J.M., Kim S, Cho I.J., Lee S.E., and Chang H.J. 'Development and validation of a deep learning based diabetes prediction system using a nationwide population-based cohort'. *Diabetes & Metabolism Journal*, 2021;45(4):515–525, doi: 10.4093/dmj.2020.0081

[49] Jang S., Girard M.J.A., and Thiery A.H. 'Explainable diabetic retinopathy classification based on neural-symbolic learning'. NeSy, 2021:104–114. Available from: http://ceur-ws.org/Vol-2986/paper8.pdF. Accessed 1 July 2022.

[50] Linden T., De Jong J., Lu C., Kiri V., Haeffs K., and Fröhlich H. 'An explainable multimodal neural network architecture for predicting epilepsy comorbidities based on administrative claims data'. *Frontiers in Artificial Intelligence*, 2021;4:610197, https://doi.org/10.3389/frai.2021.610197

[51] Sadeghi S., Khalili D., Ramezankhani A., Mansournia M.A., and Parsaeian M. 'Diabetes mellitus risk prediction in the presence of class imbalance using flexible machine learning methods'. *BMC Medical Informatics and Decision Making*, 2022;22:36. Available from: https://bmcmedinformdecismak.biomedcentral.com/articles/10.1186/s12911-022-01775-z#citeas. Accessed 23 July 2022.

[52] Khedkar S., Subramanian V., Shinde G., and Gandhi P. 'Explainable AI in healthcare'. In: Presented at the 2nd International Conference on Advances in Science & Technology; Bahir Dar, Ethiopia, 2019.

[53] Mpreno-Sanchez P.A. 'An explainable classification model for chronic kidney disease', 2021. Available from: https://arxiv.org/abs/2105.10368v1. Accessed 26 July 2022.

[54] Masolo, C. Supervised, Unsupervised and Deep Learning, 2017. Available from https://towardsdatascience.com/supervised-unsupervised-and-deep-learning-aa61a0e5471c. Accessed 11 July 2022.

[55] Nwaneri S.C. and Anyaeche C.O. 'Application of deep learning in disease prediction'. In *Advancing Industrial Engineering in Nigeria Through Teaching, Research and Innovation, A Book of Reading*. Ibadan: Department of Industrial Engineering, University of Ibadan, 2020, pp. 195–219.

[56] Nawaz M.S., Shoaib B., and Ashraf M.A. 'Intelligent cardiovascular disease prediction empowered with gradient descent optimization. *Heliyon*, 2021;7 (5):e06948. https://doi.org/10.1016/j.heliyon.2021.e06948

[57] Kingma D. and Ba J. 'Adam: a method for stochastic optimization'. In: *Conference Paper at the 3rd International Conference for Learning Representations*, San Diego, 2015. Available from: https://arxiv.org/pdf/1412.6980.pdf. Accessed 11 July 2022.

[58] Duchi J., Hazan E., and Singer Y. Adaptive subgradient methods for online learning and stochastic optimization. *The Journal of Machine Learning Research*, 2011;12:2121–2159.

[59] Tieleman T. and Hinton G. Lecture 6.5 – RMSProp, COURSERA: Neural Networks for Machine Learning. Technical Report, University of Toronto, 2012. Available from: https://www.cs.toronto.edu/~tijmen/csc321/slides/lecture_slides_lec6.pdf (accessed 26 July 2022).

[60] Patterson J. and Gibson A. *Deep Learning*. Sebastopol, CA: O'Reilly Media Inc., 2022.

[61] Alzubaidi L., Zhang J., Humaidi A.J., *et al.* 'Review of deep learning: concepts, CNN architectures, challenges, applications, future directions'. *Journal of Big Data*, 2021;8:53, https://doi.org/10.1186/s40537-021-00444-8

[62] Zhu M., Min W., Wang Q., Zou S., and Chen X. 'PFLU and FPFLU: two novel non-monotonic activation functions in convolutional neural networks'. *Neurocomputing,* 2021;429:110–117.

[63] Eckle K. and Scmidt- Hieber J. 'A comparison of deep networks with ReLU activation function and linear spline-type methods'. *Neural Networks*, 2019;110:232–242.

[64] Dua D. and Graff C. UCI Machine Learning Repository, 2017. Available from: http://archive.ics.uci.edu/ml. Accessed 11 July 2022.

[65] Casiraghi E., Malchiodi D., Trucco G., Frasca M., Cappolletti L., and Fontana T. 'Explainable machine learning for early assessment of COVID-19 risk prediction in emergency departments'. *IEEE Access*, 2020;8:196299–196325. doi:10.1109/ACCESS.2020.3034032

[66] Yan S., Ramazanian T., Sagheb G., *et al.* 'DeepTKAClassifier: brand classification of total knee arthroplasty implants using explainable deep convolutional neural networks'. In: *Advances in Visual Computing. ISVC 2020. Lecture Notes in Computer Science*, vol. 12510. Cham: Springer, 2020, https://doi.org/10.1007/978-3-030-64559-5_12

[67] Sourab S.Y., Shuvo H.R., Hasan R., and Masruf T. 'Diagnosis of COVID-19 from chest X-ray images using convolutional neural networking with K-fold cross validation'. In: *2021 IEEE International Power and Renewable Energy Conference* (*IPRECON*), 2021, pp. 1–5, doi: 10.1109/IPRECON52453.2021.9640744

[68] Dasari C.M. and Bhukya R. 'Explainable deep neural networks for novel viral genome prediction'. *Application Intelligence,* 2022;52:3002–3017. https://doi.org/10.1007/s10489-021-02572-3

[69] Ashraf M., Ahmad S.M., Ganai N.A., Shah R.A., Zaman M. and Khan S.A. Prediction of cardiovascular disease through cutting-edge deep learning technologies: an empirical study based on TENSORFLOW, PYTORCH and KERAS. In: Gupta, D., Khanna, A., Bhattacharyya, S., Hassanien, A.E., Anand, S., and Jaiswal, A. (eds.), *International Conference on Innovative Computing and Communications. Advances in Intelligent Systems and Computing*. Singapore: Springer, 2020, p. 1165. https://doi.org/10.1007/978-981-15-5113-0_18

[70] Magesh P.R., Myloth R.D., and Tom R.J. An explainable machine learning model for early detection of Parkinson's disease using LIME on DaTSCAN imagery. *Computers in Biology and Medicine*, 2020;126:104041. ISSN 0010-4825, https://doi.org/10.1016/j.compbiomed.2020.104041

[71] Gabbay F., Bar-Lev S., Montano O., and Hadad N. 'A LIME-based explainable machine learning model for predicting the severity level of COVID-19 diagnosed patients'. *Applied Science*, 2021;11(21):10417. https://doi.org/10.3390/app112110417

[72] Kumar R.L., Wang Y., Poongodi T., and Imoize A.L. (eds.), *Internet of Things, Artificial Intelligence and Blockchain Technology*. 1st ed. Switzerland AG: Springer Nature, 2021.

Chapter 12

A KNN and ANN model for predicting heart diseases

Sulaiman Olaniyi Abdulsalam[1], *Micheal Olaolu Arowolo*[2,3], *Enobong Chidera Udofot*[2], *Ayodeji Matthew Sanni*[1], *Damilola David Popoola*[1] and *Marion Olubunmi Adebiyi*[2,4,5]

Abstract

The heart is the single most important organ in the human body. Patients, professions, and medical systems are all bearing the brunt of heart failure's devastating effects on contemporary society. Since cardiac arrest may well be demonstrated as a better understanding or conceivably go unobserved, particularly in the vast population of clients that have other cardiovascular disorders, the true prevalence of heart failure is likely to be underestimated, accounting for only 1–4% of all hospitalized patients as test procedures in developed nations. A person with heart failure has a heart that is unable to circulate sufficient blood through the body, but the term "heart failure" does not explain why this happens. The clinical picture is confusing since there are several possible causes of heart problems, many of which are diseases in and of themselves. Many cases of heart failure can be avoided if the underlying medical conditions that cause them are identified and treated promptly. The study and prediction of cardiac conditions must be precise because numerous diseases have been connected to the cardiovascular system. The resolution of this problem requires intensive online research on the relevant topic. Since incorrect illness prognoses are a leading cause of death among heart patients, learning more about effective prediction algorithms is crucial.

This research utilizes K-nearest neighbor (KNN) and artificial neural network (ANN) to assess cardiovascular diseases using data collected from Kaggle. The highest accuracy (96%) was achieved by ANN trained with the standard scalar. Medical experts, specialists, and academics can all benefit greatly from this study. Based on the results of this study, cardiologists will be able to make more knowledgeable decisions about the inhibition, analysis, and handling of heart disease.

[1]Department of Computer Science, Kwara State University, Nigeria
[2]Department of Computer Science, Landmark University, Nigeria
[3]Department of Electrical Engineering and Computer Science, University of Missouri Columbia, USA
[4]SDG 3 (Good Health and Well-being), Landmark University, Nigeria
[5]Covenant Applied Informatics and Communications-Africa Centre of Excellence (CApIC-ACE), Covenant University, Nigeria

Keywords: Heart; Cardio; Disease; Machine learning; Prediction; KNN; ANN; CNN

12.1 Introduction

As the largest cause of death worldwide, cardiovascular disease (CVD) is a critical health concern no matter its form. CVD is not just one disease but a collection of disorders that affect the cardiovascular system (the heart and blood vessels). Diseases affecting the cardiovascular system and the central nervous system rank the highest in prevalence. The majority of persons who will acquire a kind of CVD may already have symptoms by the age of 35, according to a famous cardiologist [1]. These diseases typically afflict persons in the future (with prevalence drastically cumulative around the 30–44 age range).

When plaque shapes up in the veins and blocks plasma movement to the heart, this is called a heart attack. Excessive cholesterol, smoking, high blood pressure, extreme alcohol consumption, high blood sugar, insufficient physical activity, and a hypertensive heart are all risk factors for developing CVD [2].

Death from cardiovascular causes has surpassed all others in recent decades, both in high-income nations and in poorer ones. The mortality rate can be lowered through the early diagnosis of heart disorders and the constant monitoring of patients by medical professionals. Unfortunately, due to the additional intelligence, time, and expertise required, 24-hour medical consultation and the precise diagnosis of heart disorders are not currently options for patients [3].

To improve the accuracy with which heart illnesses may be predicted, this research makes use of machine learning methods. Essential features are selected using a standard scaler technique.

Since cardiac issues have such a high death rate, many lives can be spared if they are detected and treated promptly by utilizing the tools developed for heart ailment forecast. Data mining and machine learning are tools for sifting through mountains of information and distilling it into usable insights. The term "data mining" is used to describe the process of extracting non-trivial meaning from datasets by uncovering latent, previously unknown, and potentially useful information. Massive amounts of healthcare data are collected by the healthcare business every year, and these data can be mined for insights using a variety of data mining methods. This information can subsequently be put to use in a diversity of fields, in the medical, marketing, and financial sectors. In the medical field, treating cardiovascular illness is a crucial but challenging endeavor that calls for haste, efficiency, and the right kind of automation. Heart disease prediction systems have been examined, and it has been shown that varying amounts of medical characteristics and risk variables have been applied using various data mining approaches.

Removing valuable data from huge amounts of information is essential in the healthcare sector. Predicting outcomes and obtaining a deeper understanding of medical data are two areas where data mining and machine learning are rapidly emerging as crucial professions. The World Health Organization (WHO) intellects that 17 million persons yearly lose their lives due to CVD. Accurate predictions of cardiovascular illnesses can be made utilizing data mining techniques. The prediction can aid doctors in making more informed diagnoses by providing answers to difficult questions about heart disease [4].

Medical experts, specialists, and researchers can learn a great deal from this study. Because of this research, cardiologists will be able to make more knowledgeable results on the inhibition, analysis, and handling of heart disease. Using deep learning strategies including ANN, CNN, and KNN, this research aims to detect and forecast heart problems. To do this, researchers compiled data from a wide range of sources and applied predictive analytics to cut down on cases of heart disease. This research creates a standardized scalar approach to data pre-processing using ANN, CNN, and KNN for predicting cardiac disorders.

12.2 Overview of the literature

12.2.1 Heart diseases

Coronary artery disease (CAD) is a complex process that narrows the coronary arteries, leading to heart disease (HD). The three most frequent kinds of IHD are angina, myocardial infarction (MI), and heart failure (HF), all of which are connected but have distinct clinical manifestations.

Approximately 31% of all fatalities worldwide in 2012 were attributed to CVDs, conferring to the WHO. It is projected that 7.4 million of these demises were brought on by IHD. The incidence of IHD was identified as the main cause of mortality worldwide in 2012, accounting for 13.2% of all deaths. In 2012, IHD was the main cause of mortality in both middle- and high-income nations, accounting for around 45% of demises in middle-income nations and 37% in high-income republics, respectively. Heart disease is a foremost source of death and disability worldwide. HD exerts a gigantic economic drain on patients, in addition to reducing their health-related quality of life (HRQOL) and life expectancy. Costs associated with HD patients' medical care were $851 million in 2004, placing them third most expensive out of seven major disorders. "A state of total psychological, physical, and societal welfare, and not only the absenteeism of ailment," as stated by the WHO. Changes in the incidence and severity of diseases should be included when gauging health and the effects of health treatment. Improved HRQOL can be used as a proxy for overall well-being in an evaluation.

Many different conditions affecting the heart are collectively referred to as "heart disease." The most prevalent kind of heart ailment in the United States is CAD, which limits the blood supply to the heart. Heart Disease and Stroke Figures in 2021 appraise a statement from the American Heart Association, states that a decrease in blood movement can lead to a heart attack [5].

Someone in the United States suffers a heart attack every 40 sec on an average. The average rate of death in the United States due to heart disease is higher than one person every 60 sec. There are an estimated 720,000 annual new heart attacks and 335,000 annual recurrences in the United States, according to the study. On an average, a man will have his first heart attack at the age of 65.6 years old, while a woman will be 72.0 years old [6].

A human heart will beat 2.5 billion times during its lifespan, distributing millions of gallons of blood to all parts of the body. Oxygen, fuel, hormones, and other substances, along with a wide variety of cells, are carried by this constant flow. It also helps get rid of metabolic waste. When the heart stops beating, life support systems collapse almost instantly. Incredibly, the heart can stay beating for so long and for so many people, given how hard it works all the time. However, it can falter for a variety of reasons [7], including a poor diet and lack of exercise, smoking, illness, and unfortunate genes.

The development of atherosclerosis is a serious health concern. Atherosclerosis is the hardening and narrowing of arteries due to a buildup of cholesterol-rich plaque. The coronary blood vessel, which delivers plasma to the heart, and other arteries in the body can be impeded by these plaque pockets. When a portion of the panel breaks off, it can cause a heart attack or a stroke [8].

A stroke occurs when a clot in a cerebral artery grazes off body fluid supply to the brain. Heart disease occurs when the heart is incapable to pump sufficient blood to the body's tissues and organs. Initial symptoms can include a rapid heart rate, difficulty breathing, discomfort in the chest, abrupt confusion, nausea, swollen feet, and a cold sweat. Cardiovascular ailment is a broad term relating to the variability of situations disturbing the heart and blood containers, and it is quite common as individuals' age. However, it is not inevitable. When adopted at an early age, a healthy lifestyle greatly reduces the probability of emerging vascular ailment. High blood pressure and high cholesterol are both perilous features of CVD, but they can be prevented with healthy lifestyle choices and medication. Drugs, treatments, and devices can help keep a damaged heart functioning. Those who suffer from heart disease have an abnormality involving the cardiovascular system (such as coronary heart disease, arrhythmia, or heart-valve defect).

A heart valve ailment happens when the heart's valves do not function normally. Naturally, occurring heart valves are delicate and smooth structures. They control how the heart's blood is pumped from one chamber to the next. Furthermore, they stop blood from re-entering the heart from the atria.

CAD, strokes, temporary ischemic attacks (TIAs), peripheral artery ailment, and aortal ailment are the key types of cardiac ailment.

12.2.2 Machine learning

Algorithms developed for use on computers that can learn from their own experience and incorporate new information into their operation are the focus of machine learning research. They classify it under AI. The field of machine learning focuses on the study of algorithms that may be programmed to learn and improve on their user data and previous examples. They classify it under AI. In data science, it refers to a technique that uses programming to construct analytical models automatically. It is a subfield of AI that seeks to automate as much of the learning, pattern recognition, and decision-making processes as possible. Applications, where it would be impractical or impossible to build traditional statistics which is focused on making predictions using computers, include e-mail filtering and computer vision, two areas where machine learning methods are widely employed. Results for huge datasets can be predicted using machine learning methods. Prediction systems can benefit greatly from the application of machine learning, an AI method.

Supervised, unsupervised, and reinforcement learnings are types of machine learning algorithms.

Supervised learning: Supervised learning requires labeled or known data for training. Learning is supervised or geared toward productive action because the data is already known. Once the data is prepared, the machine learning method helps train the model. Large amounts of data are needed to adequately grasp the patterns and enable effective prediction of test data. By comparing the training outputs to the actual ones and making adjustments to the algorithms based on the discrepancies, the method can be refined [9].

Supervised machine learning is the evolving algorithms that can use the information provided by an outside source to form general hypotheses and then predict new data. A compact model of the circulation of class labels in terms of predictor attributes is the goal of supervised learning. The resulting classifier is applied to testing instances where predictor feature values are identified but class label values are unidentified [10].

An algorithm is used to discover the association among a set of input (x) and output (Y) variables (Y=f) (X)

The objective is to get close enough to the mapping function so that new data may be used to reliably predict the output variables (Y) (X). Supervised learning is the procedure of teaching an algorithm how to learn from a given dataset; in this case, the training data set. As the algorithm iteratively processes the training data, it makes predictions that are then corrected by the instructor, who knows the right answers. The learning process concludes when the algorithm performs to the desired standard. Once the model has been trained on identified data, new results can be generated by feeding in unknown input [11].

Today, the most popular supervised learning algorithms include random forest, polynomial regression, K-nearest neighbors (kNNs), linear regression, decision tree, Naive Bayes, and logistic regression [12].

In classification, a supervised learning strategy is used to assign labels to new observations based on the labels assigned to the training data. The software learns from a collection of observations and then uses that knowledge to categorize new observations into one of many groups. Predicting a class label is an example of supervised learning. The detection of spam in electronic messages is a shining illustration of the success of categorization machine learning. There are primarily different types of categorization, and these are [13]; linear classification methods include, for instance, logistic regression and support vector machine (SVM).

Naive Bayes, decision tree, and random forest classification are all forms of non-linear classification [14].

Mathematical methods known as regression allow data scientists to make predictions about a continuous outcome (y) given the values of one or more predictor variables (x). The reason why linear regression is so popular is that it can be easily implemented in the context of prediction and forecasting. Predicting a number label is an example of supervised learning. Linear, logistic, polynomial, support vector, decision tree, random forest, ridge, and Lasso Regressions are just a few of the many forms of regression that can be performed [15].

In unsupervised learning, the data used for training purposes has never been seen by a human being and is therefore unlabeled. The term "unsupervised" refers to the inability to guide the algorithm's input with the benefit of prior knowledge. The machine learning algorithm is fed this information to train a model. For the machine learning method to work, the model must be trained using the information that will eventually be used as input [16]. A trained model is one that actively searches for patterns and then takes action based on what it finds.

In contrast to supervised methods, unsupervised machine learning can process data that has not been labeled. Since no human effort is needed to make the dataset machine-readable, significantly larger datasets can be processed by the program. Relational data modeling is a type of challenge wherein a model is used to either define or extract relationships within a dataset. Due to the lack of labels, it is unable to build anything transparent. Without any guidance from humans, the program can abstractly perceive relationships between data pieces. Partial least squares, singular value decomposition, Fuzzy means, K-means clustering, hierarchical clustering, a priori, and principal component analysis are currently utilized as unsupervised learning techniques [17].

How humans learn from data in their daily lives is a major inspiration for the field of reinforcement learning. It has a learning mechanism based on trial and error that can adapt to new situations. Outputs that are deemed unfavorable are "punished," while those that are deemed desirable are bolstered. Through the application of the psychological concept of conditioning, reinforcement learning places the algorithm in a working setting with an interpreter and a reward system. Every time the algorithm completes an iteration, the result is sent to an interpreter who determines whether or not the result was useful. In this class of challenges, an agent is placed in its natural environment and tasked with figuring out how to best interact with it. Positive and negative reinforcement are two distinct types of reinforcement learning [18].

CNN: An example of an artificial neural network (ANN) is a CNN. To accomplish both generative and descriptive tasks, they rely on deep learning, making them a subfield of artificial intelligence (AI) [19].

CNN is a type of deep learning model used to evaluate data with a grid form, such as photographs. With inspiration from the visual brain of animals, CNN is programmed to robotically and adaptively learn longitudinal orders of qualities, beginning with basic features and progressing to more complex ones. The three categories of layers (or "structure blocks") that make up a typical CNN are convolution, pooling, and fully linked [20].

ANN: Synonymous with the word "neural network," an ANN is a mathematical model that is biologically inspired and consists of a collection of artificial neurons that are connected. The architecture has three layers: an input layer, a transition layer(s), and an output layer(s). For computation, it employs a connectionist model.

This method is a highly advanced analytical technique, capable of modeling exceedingly complex non-linear functions. Multilayer perceptron (MLP) is a famous ANN architecture (MLP). The input layer, the output layer, and the hidden layer are its three components. It takes data from the output layer and feeds it into the input layer. As a result, we may tell it to add as many hidden layers as we like to the model. The MLP is widely recognized as a highly effective function estimate for classification and forecasting issues. Given the right parameters, MLP is capable of efficiently learning non-linear functions of arbitrary complexity and precision. Nonlinear neurons (perceptrons) are the building blocks of the MLP, which consists of many layers of neurons coupled via a feed-forward architecture.

Interesting patterns and new insights can be mined from massive datasets using data mining techniques. For the resolution of investigation and decision-making, medical data mining techniques have been increasingly applied. In poor and middle-income countries, CVD was responsible for almost 80% of deaths. Big, unstructured data sets on heart disease are generated by the healthcare industry every year.

12.2.3 Related work

A CNNs model was proposed as part of the Cardio-Help system [21] to predict CVD in persons. For early HF prediction, the suggested method uses CNN and is concerned with temporal data modeling. The heart disease dataset was developed, and the results were encouraging when compared to cutting-edge approaches. Experiment results achieve better results in terms of evaluation metrics. There is a 97% success rate with the proposed method.

It was suggested [22] that researchers compile and report on results from analyses of different categorizations of learning models used for predicting cardiovascular disease. The review focuses on three main areas: CVD classification approaches. Performance measures, datasets, and tools used to report accuracy, prediction, and category of these approaches are also gathered and stated.

An innovative deep learning model using 1D CNN for classification between healthy and non-healthy individuals with a balanced dataset was suggested [23], which aims to overcome the restrictions of conventional machine learning methods. Multiple scientific parameters are utilized to determine a patient's risk profile, which is useful for making an early prognosis. To prevent overfitting, the proposed model employs several regularization procedures. The proposed model obtains over 97% training accuracy and over 96% test accuracy on the dataset. The effectiveness of the proposed model is demonstrated through in-depth comparisons to other machine learning techniques utilizing multiple performance metrics.

Prognosis prediction using recurrent neural networks (PPRNNs) [24] suggest the use of deep recurrent neural networks (RNNs) to predict high-risk prognosis from patient diagnostic histories, like that of language models. To learn from patient diagnostic code sequences and predict the existence of high-risk illnesses, the proposed PP-RNN uses several RNNs. Our findings also imply that our proposed technique has the potential to improve upon previous efforts.

It looked into several different methods for silent heart attack predicting [25], including machine learning algorithms and deep neural networks, but none of them yielded satisfactory results. Deep learning techniques, in particular RNN, are proposed as part of a heart attack prediction architecture for determining the severity of a patient's cardiovascular illness. Due to the findings of this study, the author has decided to implement RNN and GRU to enhance the system's ability to detect silent heart attacks and provide the user with timely notifications. This method has increased the accuracy of heart attack prediction to 92%, making it a reliable tool for predicting even silent heart attacks.

Most of the article [26] is devoted to determining which types of patients are more likely to acquire heart disease based on a variety of factors. To anticipate whether patients are diagnosed with heart diseases based on the patient's medical history. Using several machine learning procedures, including logistic regression and KNN, for classifying and predicting cardiac patients. A very useful method was applied to regulate how the model may be utilized to progress the accuracy of heart attack prediction in individuals. The proposed model's strength was seen in its ability to predict symptoms of heart disease in an individual, utilizing KNN and logistic regression, with respectable accuracy after comparing with other classifiers.

The diagnosis of CVD by the use of data mining techniques was presented [27], with only 14 parameters such as gender, age, BMI, down sloping, sugar, and fat being used as examples.

Data from the UCI machine learning for heart disease were analyzed and compared using a variety of machine and deep learning methods. The accuracy target was met with 94.2% [28].

Strong pre-processing models, such as feature selection and clustering with DBSCAN, were recommended in addition to HDPM via SMOTE and ENN algorithm [29]. The XGBoost algorithm was also utilized for cardiac illness prognosis. In some cases, the deployed model can achieve an accuracy of 99.4%.

The use of a deep neural network in a self-operating diagnostic model for the detection of cardiac problem disease [30]. The classification of healthy and unwell persons is greatly improved by the use of machine learning methods. A method for predicting a patient's risk profile based on aspects of their clinical data has been created in this study. The proposed model is built utilizing both deep neural networks and the 2-statistical model. Lack of a proper fit or an excessive one is no longer concern. The training and test data model performed exceptionally well. Using DNN and ANN, analyze the performance of a model that can accurately predict if a person has heart disease.

Based on a trained recurrent fuzzy neural networks (RFNNs), a genetic approach for cardiac illness diagnosis was proposed [31]. This research proposes utilizing genetically based trained RFNNs for cardiac disease diagnosis. The Cleveland Heart Disease dataset from the University of California, Irvine (UCI) is used for this investigation. Only 45 are used for the actual testing, while the remaining 252 are used for training. The outcomes presented an accuracy of the testing set was 98% with some other metrics that can be used to evaluate a test's efficacy. Upon inspection, it was decided that the outcomes were adequate.

A high-quality heart disease prediction model that uses the unsupervised K-means clustering method to identify outliers in healthcare data was proposed [32]. Most existing methods for anomaly detection focus on building profiles of typical occurrences. But these models need a large enough sample of typical people to be convincing. Our proposed model uses the Silhouette method to determine the best possible value for K. As a next step, it intends to identify outliers that are more than a predetermined distance from their respective clusters. Five of the most common categorization methods KNN, random forest, SVM, Naive Bayes, and logistic regression were adopted. The effectiveness of the proposed strategy is proved on a benchmark dataset for CVD.

Using a genetic algorithm for feature selection, researchers at the University of Toronto were able to create a heart disease prediction model based on 303 samples from the Cleveland dataset. Through this process, they were able to collect seven features that were then utilized to train models for heart prediction with four different machine learning strategies: SVM, MLP, jackknife (J48), and kernel (KNN). The framework was further assessed by comparing its findings to those of models built using the traditional feature selection techniques and by employing a 10-fold cross-validation procedure. When the SVM was used in conjunction with the genetic algorithm, it was shown to have an accuracy of 88.34%, up from 83.76% when using the original dataset alone.

An evaluation and categorization of machine learning procedures for the analysis of cardiac disease were proposed [33], using four different datasets from diverse locations. Several different classification methods, including KNN, decision tree NB, J48, SVM, JRip, AB, decision tree, and stochastic gradient descent classifiers, were applied to establish the classification and prediction of cases of heart disease. Results showed that using various classification algorithms for heart disease classification improves accuracy, with KNN (N=1), JRip, and Decision tree J48, all achieving above 98% accuracy in their respective classifications. Extraction of features using the classifiers subset evaluator enhanced the performance of KNN (N=1) and decision table classifiers. The feature selection technique for CVD prediction improved when only 4 out of 13 features were included.

There was a suggestion [34] to compare the efficacy of six data mining tools such as Weka, Orange, Knime, RapidMiner, MATLAB, and Scikit-Learn to classify heart disease with the Cleveland data comprising 297 annotations and 13 features. In total, there are 164 healthy individuals and 139 individuals with CVD in the dataset used for this study. Using these three performance indicators, a comparison of the accuracy, sensitivity, and specificity of the procedures available in each instrument was evaluated. Based on the results, it was clear that Matlab's Artificial Neural Network model was the most efficient approach. After generating a Matlab Receiver Operating Characteristic Curve, we made many suggestions for the most appropriate tool to utilize according to the customers' level of experience with data mining.

Changing mislaid values with the mean through pre-processing was presented as a method to determine cardiac disease presence or absence [35] using Naïve Bayes, SVM (linear and radial basis function), and KNN classifiers. A better preprocessing was used to increase the precision of predictions about cardiovascular illness. It helps medical professionals establish if a patient has cardiovascular disease and, if so, what stage of the condition they may be in. To prove the dependability of the outcomes, several machine learning algorithms were tested utilizing accuracy, precision, f1-score, and recall as performance measures. This research analyzed machine learning algorithms based on a variety of performance metrics to improve their accuracy. In the preprocessing phase, where data is missing, the average is substituted. The results show how well the mean work as a stand-in for missing variables. Using SVMs and a linear kernel, 86.8% accuracy in scoring was attained.

In the medical field, data science is used to foresee cardiac issues [36]. Numerous studies have been done on the topic, but more research is needed to improve forecasting precision. This study examines numerous heart disease datasets to analyze experiments and show how accuracy can be improved. Algorithms and techniques for selecting features are then discussed. The use of the Rapid Miner instrument allows for the use of the decision tree, logistic regression, SVM, Naive Bayes, and random forest algorithms for feature selection and enhancement. Accuracy for several models such as the decision tree, random forest, logistic regression, Naive Bayes, and the logistic regression support vector machine is 82.22%, 82.56%, 84.17%, and 84.85%, respectively. Naive Bayes and logistic regression (SVM) are two methods utilized in this paper, and the results show that the accuracy of both methods has improved.

12.3 Materials and methods

Feature extraction and classification are two of the main methods used in this study's dataset. In this research, we load data from an open source, process it with a feature extraction method (standard scalar), which extracts the necessary data from the dataset, and then classify it using deep learning algorithms (ANNs and Kernel Naive Bayes). Finally, the results are compared using ANN and KNN algorithms using performance evaluation. Figure 12.1 shows the proposed workflow.

Dataset was downloaded from the Kaggle repository (https://www.kaggle.com/johnsmith88/heart-disease-dataset/version/2) for use in the study. Developers and other technical specialists can use Kaggle, an online data science platform that hosts a wide range of crowdsourced datasets and frameworks, to collaborate and advance their work. Using the search term "heart diseases data set," we were able to locate the dataset.

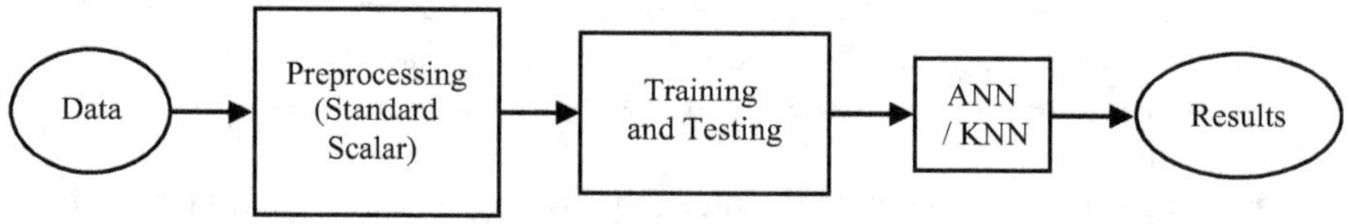

Figure 12.1 Workflow

The Cleveland, Hungary, Switzerland, and Long Beach V databases make up this 1988 bundle of information. It has 76 properties, including the anticipated attribute, but only 14 have been used in any of the published trials. If the patient has a heart illness, that fact will be included in the "target" field. The value ranges from 0 (no disease) to 1 (severe disease). Attributes include age, sex, chest pain type (4 values), resting blood pressure, serum cholesterol in mg/dl, fasting blood sugar > 120 mg/dl, resting electrocardiographic results (values 0, 1, 2), maximum heart rate achieved exercise-induced angina, old peak = ST depression induced by exercise relative to rest, the slope of the peak exercise ST segment, the number of major vessels (0–3) colored by fluoroscopy, thal: 0 = normal. Recently, real patient information was scrubbed from the database and replaced with fictitious data.

The existence of the cardiac disease in the patient is what the "target" field is tracking. It takes on the form of an integer, with 0 denoting the absence of disease and 1 indicating the presence of illness. I plan to employ random forest, naive Bayes, and SVMs to evaluate this dataset of people with heart disease.

12.3.1 Standard scalar

To standardize the range of functionality of the input dataset, many machine learning models execute a standard scalar as a preprocessing step [37]. It standardizes values by converting variance to a fixed value of 1.0 and disregarding the mean. To calculate the standard score, z, for training sample x, we divide x by the mean, u, of the training samples, which is zero if mean=False, and by the standard deviation, s, which is one if std=False.

Each feature undergoes its centering and scaling by estimating the appropriate statistics on the samples in the training set. When data is converted, the mean and standard deviation are retained for analysis.

Many machine learning estimators need that a dataset is standardized; otherwise, they may not perform as expected if the underlying features do not roughly conform to the assumptions of normal distribution (e.g. Gaussian with 0 mean and unit variance).

The RBF kernel of SVMs and the L1 and L2 regularizers of LMs are just illustrations of components used in the optimal solution of a learning algorithm that assume all features are symmetrical around 0 and have the same variance. For example, if one feature's variance is hundreds of times bigger than the rest, it could overwhelm the objective function and prevent the estimator from learning as predicted from the other features. To preserve the sparsity structure of sparse CSR or CSC matrices, this scalar can be applied to them with the with mean=False argument.

12.3.2 ANNs

Computable algorithms known as ANNs have been shown to mimic the learning and problem-solving abilities of the human brain. The idea was to model the functioning of biological systems made up of "neurons." The inspiration for ANNs comes from the structure and function of the central nervous systems of animals. The ability to learn new things and recognize patterns are two of its many talents [25].

For the sake of visualization, an ANN is best depicted as a directed graph with artificial neurons serving as the nodes and the weights. Directional edges with weights represent the connection between neuron outputs and neuron inputs. The ANN takes in data from the outside world in the form of a vector representing a pattern or an image. For each set of *n* inputs, a corresponding mathematical notation $x(n)$ is written [38].

Then, each input is multiplied by its weight (these weights are the details utilized by the ANNs to solve a specific problem). Commonly, the strength of the connections between neurons within the ANN is described by these weights. A compilation of all the weights used in the calculation is stored in the computer system [39].

A set of transfer functions, or the activation function, is applied to some input to produce a specific output. While there are a wide variety of activation functions, most can be placed into two main categories: linear and non-linear sets of operations. Some of the most common activation functions are the binary, linear, and tan hyperbolic sigmoidal sets.

12.3.3 K-nearest neighbor

K-nearest neighbor (kNN) is an abbreviation for this. This algorithm is used in supervised machine learning. Statements of classification and regression problems can both be solved using the procedure. With K, we signify the number of neighbors of a new unknown variable that needs to be predicted or categorized [40].

The kNN method is widely used since it does not rely on the data having a normal distribution or being homoscedastic. As a form of supervised machine learning, kNN can be used for classification and modeling in numerous forest mapping applications. It is also not too difficult to put together. To forecast the value of an unknown pixel, we use the weighted observations of k plots established from the training sample that is furthest from the projected pixel in the feature space [41,42].

12.3.4 Performance metrics

Researching this topic requires a computer with at least 8 GB of RAM and a 2.30 GHz processor, running Microsoft Windows 10 (64-bit edition) or a later version. For this study, we also used data from https://www.kaggle.com/johnsmith88/heart-disease-dataset/version/2.

Accuracy: A model's accuracy is measured by how well it predicts future values for a target variable given a set of historical values for those variables (the "training data"). The accuracy of a classification algorithm is typically used as a benchmark for its performance. Accuracy=((TP+TN))/((FP+FN)).

The degree of specificity is defined as the percentage of actual negatives that matched the projection of a negative outcome. Which is true (or in the latter case). more predicted false negatives A consequence of this is the potential for unintended "positives," or false positives. False positive rate is a word for this phenomenon. Specificity=TN/((FP+TN).

Specifically, precision is the fraction of correct predictions as a percentage of all correct guesses (i.e., the number of true positives plus the number of false positives). Precision=TP/((TP+FP)).

Sensitivity is a measure of how effectively a model can predict the real positive results of the model for each available category. To put it another way, specificity measures how well a model can predict the true negative of each available category. We can calculate the sensitivity as TP/(TP+FN) [43].

12.4 Results and discussions

Discussion of the outcomes of the models' implementations and evaluations form the bulk of this chapter. The evaluation findings justify the study's completion because they show that the aims and objectives were met. The data was preprocessed using a regular scalar technique. In this part, we provide the findings from our investigation using the proposed model. By running it through a typical scalar preprocessor, we were able to normalize the data. This necessitated separating the information into a training set and a test set. In this study, we use deep learning techniques, including KNN and neural networks, to simulate the effects of cardiac disease.

For pandas to be able to read the dataset, it must be imported into the environment. Df = PD.read CSV("/content/heart.csv") is the code to use for accessing the data. The complete dataset is displayed in Figure 12.2.

Information is used to describe datasets in the software. The describe() function can be used to determine the mean, median, and standard deviation of a data frame's numerical values. In this data set, we estimated the number of occurrences, the average, the standard deviation, and the minimum percentile (25th, 50th, 75th). In machine learning, it is common practice to separate data into train and test sets.

```
[5] data.head(10)
```

	Age	Sex	Chest pain type	BP	Cholesterol	FBS over 120	EKG results	Max HR	Exercise angina	ST depression	Slope of ST	Number of vessels fluro	Thallium	Heart Disease
0	70	1	4	130	322	0	2	109	0	2.4	2	3	3	Presence
1	67	0	3	115	564	0	2	160	0	1.6	2	0	7	Absence
2	57	1	2	124	261	0	0	141	0	0.3	1	0	7	Presence
3	64	1	4	128	263	0	0	105	1	0.2	2	1	7	Absence
4	74	0	2	120	269	0	2	121	1	0.2	1	1	3	Absence
5	65	1	4	120	177	0	0	140	0	0.4	1	0	7	Absence
6	56	1	3	130	256	1	2	142	1	0.6	2	1	6	Presence
7	59	1	4	110	239	0	2	142	1	1.2	2	1	7	Presence
8	60	1	4	140	293	0	2	170	0	1.2	2	2	7	Presence
9	63	0	4	150	407	0	2	154	0	4.0	2	3	7	Presence

Figure 12.2 Dataset

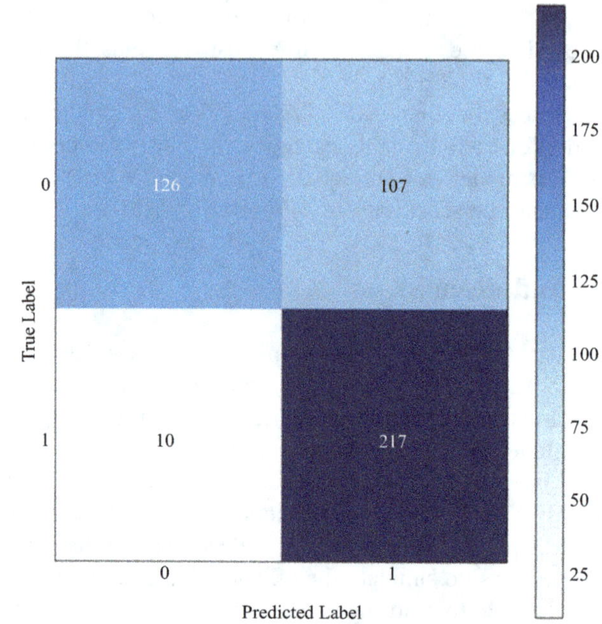

Figure 12.3 Confusion matrix for KNN model without standard scaler (TP=126; TN=217; FP=107; FN=10)

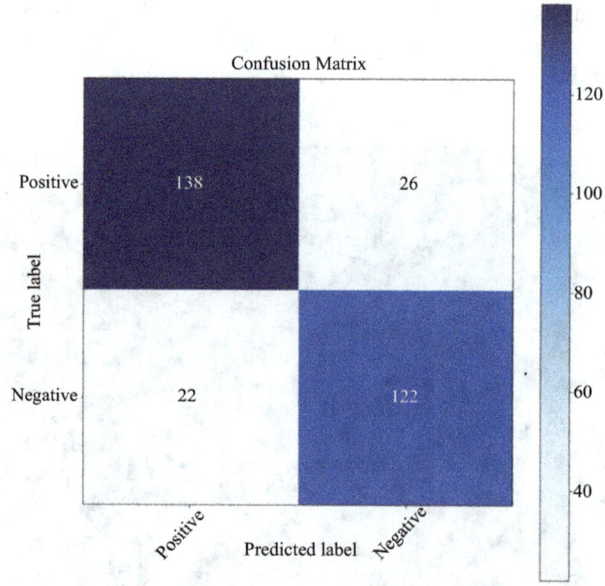

Figure 12.4 Confusion matrix for KNN with standard scaler model (TP=138; TN=122; FP=26; FN=22)

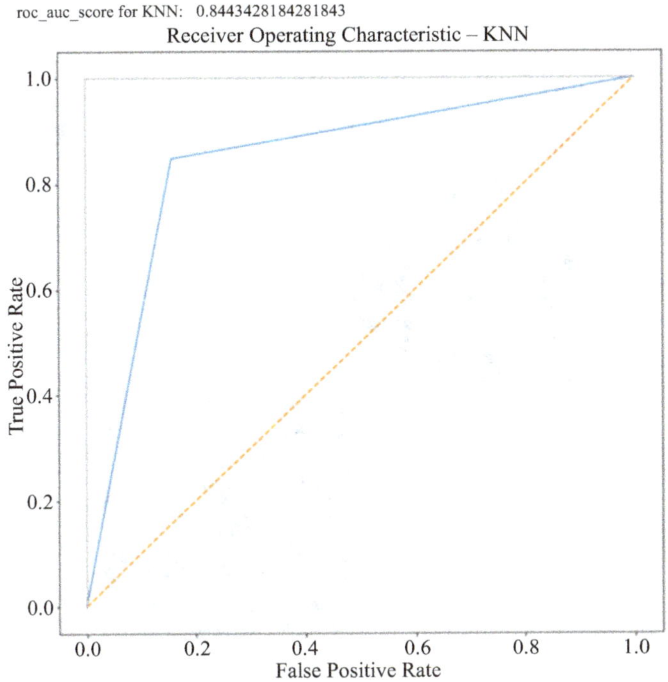

Figure 12.5 ROC curve for the KNN model

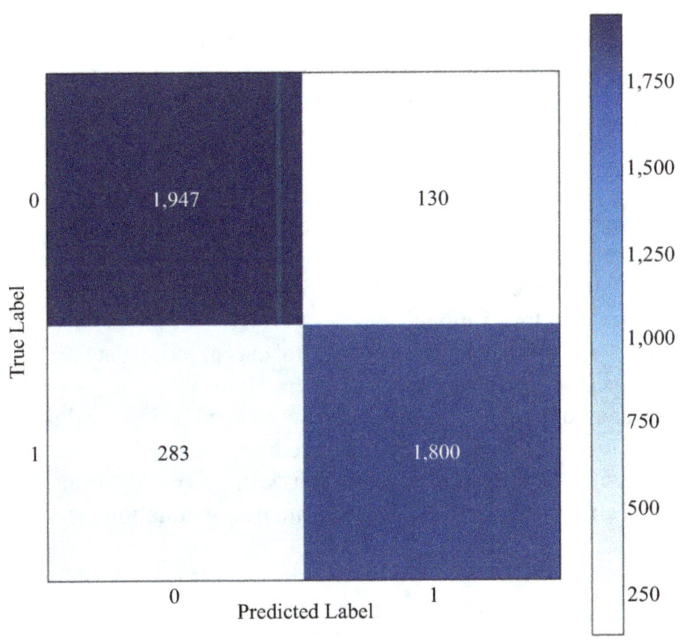

Figure 12.6 Confusion matrix for ANN model without standard scaler (TP=157; TN=140; FP=4; FN=7)

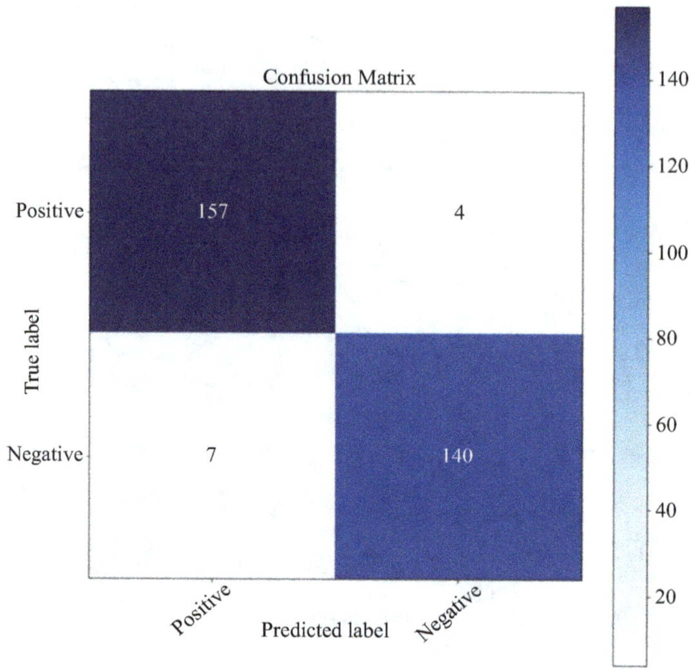

Figure 12.7 Confusion matrix for the ANN with standard scalar model (TP=157; TN=140; FP=4; FN=7)

It was necessary to split the data into training and testing sets for each method. Model fitting and testing were both performed using the training set as the foundation for comparison.

After data is partitioned into a training set and a testing set, the former is used to teach the latter. If you want to visually assess the efficacy of a classification model, plot the confusion matrix. Here you can see the KNN model's confusion matrix, roc curve, and accuracy. Figures 12.3–12.8 show the confusion matrix and ROC curve of the developed model.

KNN and ANN use a common scalar to classify the data in this analysis. Standard scaler is also given into the classifiers along with the data.

The resultant confusion matrices were evaluated using evaluation methods such as sensitivity, specificity, precision, accuracy, and F1 score. Metrics comparing each classifier's performance with and without the addition of supplementary features. Table 12.1 shows the performance evaluation of the developed models.

12.4.1 Comparison with previous work

This study included multiple experiments, the outcomes of which are displayed in Table 12.2. With an accuracy of 96%, ANN easily beat out the other models.

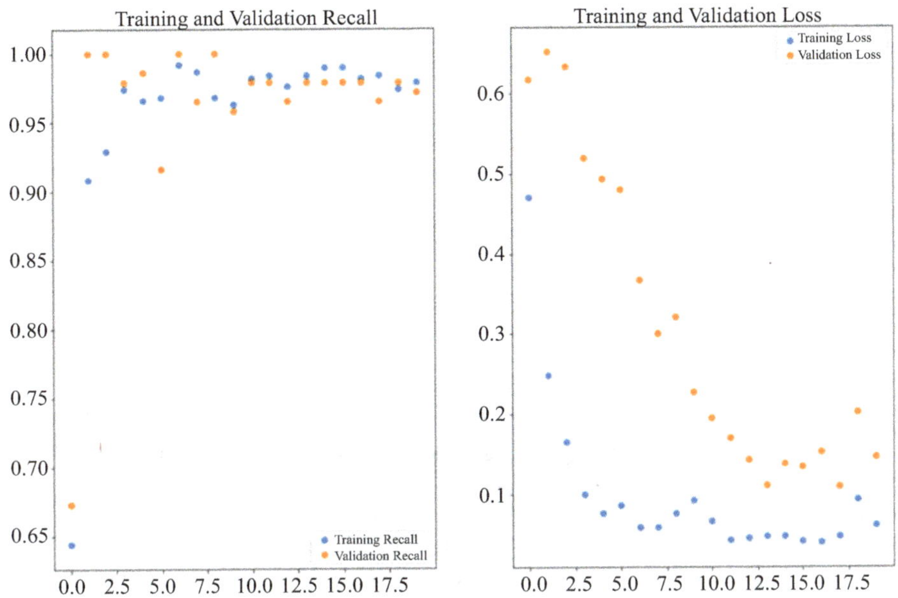

Figure 12.8 Scattered plot for the ANN model

Table 12.1 Performance metrics

Performance measures	**KNN**	**ANN**	**ANN+SC**	**KNN+SC**
Sensitivity	86.25	87.31	95.73	92.65
Specificity	82.43	93.26	97.22	66.98
Precision	84.15	93.74	97.52	54.08
Accuracy	84.42	90.07	96.43	74.57
F1 score	85.19	90.41	96.62	68.29

Table 12.2 Comparison of the findings

Reference	**Technique**	**Results**
[44]	SVM	83.7%
[25]	RNN	92%
[32]	SVM	88%
[36]	Random forest	82%

Results from different works are compared in Table 12.2.

Data mining models SC with ANN were shown to be the most accurate predictors of cardiovascular illnesses, as shown in Table 12.1. Existing works such as Nashif were used to inform the development of a heart disease prediction system that achieved 95.8% accuracy in experiments. Standard scalar was used to find the missing values in the dataset, and those missing values were then replaced with the most suitable filter values. The results of previous research are shown in Table 12.2 and compared to the built model, demonstrating that a neural network has better accuracy than other data mining techniques. Heart disease is fatal because it can cause heart attacks and other potentially fatal complications. The suggested approach aims to reduce this by identifying the most effective model across the two datasets where gold-standard methods for cardiac disease prognosis have been tested.

12.5 Conclusions

Machine learning techniques such as KNN and ANN were used to diagnose heart conditions in this research. If you had like access to the dataset used for this study, you may get it here: https://www.kaggle.com/johnsmith88/heart-disease-dataset/version/2. With heart disease being the field's "goal," it is clear that this patient is suffering from a cardiac condition. If the value is 0, no disease is present; if it is 1, the disease is present. After running the data through a normal scalar, we fed it into KNN and ANN, and we got accuracy rates of 84%, 96%, and 88%, respectively, which we compared to those of comparable research.

The heart is an essential organ. Heart conditions necessitate higher precision and accuracy in diagnosis and analysis. More study is needed to determine whether or not they can be detected in real-time. Using a dataset of heart disorders, this work presents a reliable and early prediction of these conditions. Several machine learning algorithms are needed to implement the proposed methodology. Kaggle data is used in this endeavor. The model was constructed using an advanced learning algorithm, like ANN or KNN, and fed data using the standard scalar algorithm. The purpose of this study is to provide a summary of the methods and results used.

It is crucial to stress that the goal of the methods used in creating a model for predicting cardiac disease is to enhance it. The scope of this model might be expanded by using a more comprehensive dataset, which is why I propose doing just that in future research. Among these algorithms, the ANN algorithm has the highest accuracy, hence it should be used.

References

[1] Chen Z and Xiu D. On generalized residual network for deep learning of unknown dynamical systems. *J Comput Phys*. 2021;438: 110362. Available from: https://linkinghub.elsevier.com/retrieve/pii/S0021999121002576

[2] Buttar HS, Li T, and Ravi N. Prevention of cardiovascular diseases: role of exercise, dietary interventions, obesity and smoking cessation. *Exp Clin Cardiol*. 2005;10(4):229–49. Available from: http://www.ncbi.nlm.nih.gov/pubmed/19641674

[3] Amini M, Zayeri F, and Salehi M. Trend analysis of cardiovascular disease mortality, incidence, and mortality-to-incidence ratio: results from global burden of disease study 2017. *BMC Public Health*. 2021;21(1):401. Available from: https://bmcpublichealth.biomedcentral.com/articles/10.1186/s12889-021-10429-0

[4] Shafique U, Majeed F, Qaiser H, and Mustafa IU. Data mining in healthcare for heart diseases. *Int J Innov Appl Stud*. 2016;10(January):1312.

[5] Sakellarios A, Correia J, Kyriakidis S, *et al.* A cloud-based platform for the non-invasive management of coronary artery disease. *Enterp Inf Syst*. 2020;14(8):1102–23. Available from: https://www.tandfonline.com/doi/full/10.1080/17517575.2020.1746975

[6] Adegoke O, Awolola NA, and Ajuluchukwu JN. Prevalence and pattern of cardiovascular-related causes of out-of-hospital deaths in Lagos, Nigeria. *Afr Health Sci*. 2018;18(4):942. Available from: https://www.ajol.info/index.php/ahs/article/view/180231

[7] O'Callaghan J. How does the heart work? Wikimedia Commons. 2012. Available from: http://www.howitworksdaily.com/whats-inside/inside.

[8] Dvorkin N, Clark MW, and Hamkalo BA. Ultrastructural localization of nucleic acid sequences in *Saccharomyces cerevisiae* nucleoli. *Chromosoma*. 1991;100(8):519–23. Available from: http://www.ncbi.nlm.nih.gov/pubmed/1764970

[9] Nartey OT, Yang G, Wu J, and Asare SK. Semi-supervised learning for fine-grained classification with self-training. *IEEE Access*. 2020;8:2109–21. Available from: https://ieeexplore.ieee.org/document/8943213/

[10] Kopper AE and Apelian D. Predicting quality of castings via supervised learning method. *Int J Met*. 2022;16(1):93–105. Available from: https://link.springer.com/10.1007/s40962-021-00606-7

[11] Sakhrawi Z, Sellami A, and Bouassida N. Software enhancement effort prediction using machine-learning techniques: a systematic mapping study. *SN Comput Sci*. 2021;2(6):468. Available from: https://link.springer.com/10.1007/s42979-021-00872-6

[12] Osisanwo FY, Akinsola JET, Awodele O, Hinmikaiye JO, Olakanmi O, and Akinjobi J. Supervised machine learning algorithms: classification and comparison. *Int J Comput Trends Technol*. 2017;48(3):128–38. Available from: http://www.ijcttjournal.org/archives/ijctt-v48p126

[13] Arumugam K, Swathi Y, Sanchez DT, *et al.* Towards applicability of machine learning techniques in agriculture and energy sector. *Mater Today Proc*. 2022;51:2260–3. Available from: https://linkinghub.elsevier.com/retrieve/pii/S2214785321074186

[14] Pranckevičius T and Marcinkevičius V. Comparison of naive Bayes, random forest, decision tree, support vector machines, and logistic regression classifiers for text reviews classification. *Balt J Mod Comput*. 2017;5(2). Available from: https://doi.org/10.22364/bjmc.2017.5.2.05

[15] Weinberg GH and Schumaker JA. Diagnosis of heart disease using genetic algorithm based trained recurrent fuzzy neural networks. In: *Stat an Intuitive Approach*. Baltimore, MD: Brooks, 2009, pp. 227–245.
[16] Bengio Y. Deep learning of representations: Looking forward. In *Statistical Language and Speech Processing*, Lecture Notes in Computer Science, Springer, Berlin, Heidelberg, 2013, pp. 1–37. Available from: https://doi.org/10.1007/978-3-642-39593-2_1
[17] Adadi A. A survey on data-efficient algorithms in big data era. *J Big Data*. 2021;8(1):24. Available from: https://journalofbigdata.springeropen.com/articles/10.1186/s40537-021-00419-9
[18] Deo RC. Machine learning in medicine. *Circulation*. 2015;132(20):1920–30. Available from: https://www.ahajournals.org/doi/10.1161/CIRCULATIONAHA.115.001593
[19] Yamashita R, Nishio M, Do RKG, and Togashi K. Convolutional neural networks: an overview and application in radiology. *Insights Imaging*. 2018;9(4):611–29. Available from: https://insightsimaging.springeropen.com/articles/10.1007/s13244-018-0639-9
[20] Lindsay GW. Convolutional neural networks as a model of the visual system: past, present, and future. *J Cogn Neurosci*. 2021;33(10):2017–31.
[21] Mehmood A, Iqbal M, Mehmood Z, *et al.* Prediction of heart disease using deep convolutional neural networks. *Arab J Sci Eng*. 2021;46(4):3409–22. Available from: http://link.springer.com/10.1007/s13369-020-05105-1
[22] Swathy M and Saruladha K. A comparative study of classification and prediction of cardio-vascular diseases (CVD) using machine learning and deep learning techniques. *ICT Express*. 2022;8(1):109–16. Available from: https://linkinghub.elsevier.com/retrieve/pii/S2405959521001119
[23] Hussain S, Nanda SK, Barigidad S, Akhtar S, Suaib M, and Ray NK. Novel deep learning architecture for predicting heart disease using CNN. In: *2021 19th OITS International Conference on Information Technology* (OCIT). New York, NY: IEEE, 2021, pp. 353–357. Available from: https://ieeexplore.ieee.org/document/9719326/
[24] Pillai NSR and Bee KK. Prediction of heart disease using Rnn algorithm. *Int Res J Eng Technol*. 2019;06(03):4452–8.
[25] Pandey AK, Basandrai AK, Basandrai D, *et al.* Field-relevant new sources of resistance to anthracnose caused by *Colletotrichum truncatum* in a mungbean mini-core collection. *Plant Dis*. 2021;105(7):2001–10. Available from: https://apsjournals.apsnet.org/doi/10.1094/PDIS-12-20-2722-RE
[26] Jindal H, Agrawal S, Khera R, Jain R, and Nagrath P. Heart disease prediction using machine learning algorithms. *IOP Conf Ser Mater Sci Eng*. 2021;1022(1):012072. Available from: https://iopscience.iop.org/article/10.1088/1757-899X/1022/1/012072
[27] Dhara Mehta NV. Comparative analysis of data mining classification techniques for heart disease prediction. *Int Res J Eng Technol*. 2018;5(12):206–10.

[28] Bharti R, Khamparia A, Shabaz M, Dhiman G, Pande S, and Singh P. Prediction of heart disease using a combination of machine learning and deep learning. *Comput Intell Neurosci*. 2021;2021:1–11. Available from: https://www.hindawi.com/journals/cin/2021/8387680/

[29] Alfian G, Syafrudin M, Fitriyani NL, *et al.* Deep neural network for predicting diabetic retinopathy from risk factors. *Mathematics*. 2020;8(9):1620. Available from: https://www.mdpi.com/2227-7390/8/9/1620

[30] Ramprakash P, Sarumathi R, Mowriya R, and Nithyavishnupriya S. Heart disease prediction using deep neural network. In: *2020 International Conference on Inventive Computation Technologies* (ICICT). New York, NY: IEEE, 2020, pp. 666–70. Available from: https://ieeexplore.ieee.org/document/9112443/

[31] Uyar K and Ilhan A. Diagnosis of heart disease using genetic algorithm based trained recurrent fuzzy neural networks. *Procedia Comput Sci*. 2017;120:588–93.

[32] Ripan RC, Sarker IH, Hasan Furhad M, Musfique Anwar M, and Hoque MM. An effective heart disease prediction model based on machine learning techniques. *Adv Intell Syst Comput*. 2021;1375 AIST:280–8.

[33] Almustafa KM. Prediction of heart disease and classifiers' sensitivity analysis. *BMC Bioinform*. 2020;21(1):278. Available from: https://bmcbioinformatics.biomedcentral.com/articles/10.1186/s12859-020-03626-y

[34] Tougui I, Jilbab A, and El Mhamdi J. Heart disease classification using data mining tools and machine learning techniques. *Health Technol (Berl)*. 2020;10(5):1137–44. Available from: https://link.springer.com/10.1007/s12553-020-00438-1

[35] Louridi N, Amar M, and Ouahidi B.El. Identification of cardiovascular diseases using machine learning. In: *2019 7th Mediterranean Congress of Telecommunications* (*CMT*). New York, NY: IEEE, 2019, pp. 1–6. Available from: https://ieeexplore.ieee.org/document/8931411/

[36] Bashir S, Khan ZS, Hassan Khan F, Anjum A, and Bashir K. Improving heart disease prediction using feature selection approaches. In: *2019 16th International Bhurban Conference on Applied Sciences and Technology* (*IBCAST*). New York, NY: IEEE, 2019, pp. 619–23. Available from: https://ieeexplore.ieee.org/document/8667106/

[37] Singh D and Singh B. Investigating the impact of data normalization on classification performance. *Appl Soft Comput*. 2020;97:105524. Available from: https://linkinghub.elsevier.com/retrieve/pii/S1568494619302947

[38] Montesinos López OA, Montesinos López A, and Crossa J. Fundamentals of artificial neural networks and deep learning. In: *Multivariate Statistical Machine Learning Methods for Genomic Prediction*. Cham: Springer International Publishing, 2022, pp. 379–425. Available from: https://link.springer.com/10.1007/978-3-030-89010-0_10

[39] Al-Hroob A, Imam AT, and Al-Heisa R. The use of artificial neural networks for extracting actions and actors from requirements document. *Inf*

Softw Technol. 2018;101:1–15. Available from: https://linkinghub.elsevier.com/retrieve/pii/S0950584918300752

[40] Garcia-Carretero R, Vigil-Medina L, Mora-Jimenez I, Soguero-Ruiz C, Barquero-Perez O, and Ramos-Lopez J. Use of a K-nearest neighbors model to predict the development of type 2 diabetes within 2 years in an obese, hypertensive population. *Med Biol Eng Comput*. 2020;58(5):991–1002. Available from: http://link.springer.com/10.1007/s11517-020-02132-w

[41] McRoberts RE, Næsset E, and Gobakken T. Optimizing the k-nearest neighbors technique for estimating forest aboveground biomass using airborne laser scanning data. *Remote Sens Environ*. 2015;163:13–22. Available from: https://linkinghub.elsevier.com/retrieve/pii/S0034425715000851

[42] Arowolo MO, Adebiyi MO, Adebiyi AA, and Olugbara O. Optimized hybrid investigative based dimensionality reduction methods for malaria vector using KNN classifier. *J Big Data*. 2021;8(1):29. Available from: https://journalofbigdata.springeropen.com/articles/10.1186/s40537-021-00415-z

[43] Afolayan JO, Adebiyi MO, Arowolo MO, Chakraborty C, and Adebiyi AA. Breast cancer detection using particle swarm optimization and decision tree machine learning technique. In: *Intelligent Healthcare*. Singapore: Springer Nature Singapore, 2022, pp. 61–83. Available from: https://link.springer.com/10.1007/978-981-16-8150-9_4

[44] Gokulnath CB and Shantharajah SP. An optimized feature selection based on genetic approach and support vector machine for heart disease. *Cluster Comput*. 2019;22(S6):14777–87. Available from: http://link.springer.com/10.1007/s10586-018-2416-4

[45] Bashir S, Qamar U, Khan FH, and Javed MY. MV5: a clinical decision support framework for heart disease prediction using majority vote based classifier ensemble. *Arab J Sci Eng*. 2014;39(11):7771–83. Available from: http://link.springer.com/10.1007/s13369-014-1315-0

[46] Tiwari V, Garg B, and Sharma UP. Significant impact of improved machine learning algorithm in the processes of large data sets. *Int J Sci Res Comput Sci Eng Inf Technol*. 2020;458–67. Available from: https://doi.org/10.32628/cseit206133

[47] Aldahiri A, Alrashed B, and Hussain W. Trends in using IoT with machine learning in health prediction system. *Forecasting*. 2021;3(1):181–206. Available from: https://www.mdpi.com/2571-9394/3/1/12

Chapter 13

Artificial Intelligence-enabled Internet of Medical Things for COVID-19 pandemic data management

Agbotiname Lucky Imoize[1,2], Peter Anuoluwapo Gbadega[3], Hope Ikoghene Obakhena[4,5], Daisy Osarugue Irabor[6], K.V.N. Kavitha[7] and Chinmay Chakraborty[8]

Abstract

The dreaded coronavirus (COVID-19) disease traceable to Severe Acute Respiratory Syndrome Coronavirus 2 (SARS-CoV2) has killed thousands of people worldwide, and the World Health Organization (WHO) has proclaimed the viral respiratory disease a human pandemic. The adverse flare of COVID-19 and its variants has triggered collaborative research interests across all disciplines, especially in medicine and healthcare delivery. Complex healthcare data collected from patients via sensors and devices are transmitted to the cloud for analysis and sharing. However, it is pretty difficult to achieve rapid and intelligent decisions on the processed information due to the heterogeneity and complexity of the data. Artificial intelligence (AI) has recently appeared as a promising paradigm to address this issue. The introduction of AI to the Internet of Medical Things (IoMT) births the era of AI of Medical Things (AIoMT). The AIoMT enables the

[1]Department of Electrical and Electronics Engineering, Faculty of Engineering, University of Lagos, Nigeria
[2]Department of Electrical Engineering and Information Technology, Institute of Digital Communication, Ruhr University, Germany
[3]Discipline of Electrical, Electronic & Computer Engineering, School of Engineering, Howard College Campus, University of KwaZulu-Natal, South Africa
[4]Department of Electrical and Electronics Engineering, Faculty of Engineering and Technology, Ambrose Alli University, Nigeria
[5]Department of Electrical and Electronics Engineering, Faculty of Engineering, University of Benin, Nigeria
[6]Department of Electrical and Computer Engineering, Tandon School of Engineering, New York University, USA
[7]Department of Communication Engineering, School of Electronics Engineering, Vellore Institute of Technology, India
[8]Department of Electronics and Communication Engineering, Birla Institute of Technology, India

autonomous operation of sensors and devices to provide a favourable and secure environmental landscape to healthcare personnel and patients. AIoMT finds successful applications in natural language processing (NLP), speech recognition, and computer vision. In the current emergency, medical-related records comprising blood pressure, heart rate, oxygen level, temperature, and more are collected to examine the medical conditions of patients. However, the power usage of the low-power sensor nodes employed for data transmission to the remote data centres poses significant limitations. Currently, sensitive medical information is transmitted over open wireless channels, which are highly susceptible to malicious attacks, posing a significant security risk. An insightful privacy-aware energy-efficient architecture using AIoMT for COVID-19 pandemic data handling is presented in this chapter. The goal is to secure sensitive medical records of patients and other stakeholders in the healthcare domain. Additionally, this chapter presents an elaborate discussion on improving energy efficiency and minimizing the communication cost to improve healthcare information security. Finally, the chapter highlights the open research issues and possible lines of future research in AIoMT.

Keywords: Artificial Intelligence; Medical information; Internet of Medical Things; AIoMT; COVID-19 pandemic data management; Energy-efficient devices; Wireless sensor nodes; Security and privacy schemes

13.1 Introduction

The new pandemic disease SARSCoV-2, currently known as COVID-19, has infected the whole world. The virus originated in Wuhan, China, in late December 2019. Since then, it has emerged as the world's fastest-spreading contagious epidemic, posing a new threat to global public health. A COVID-19 epidemic has shown and highlighted the present organizations' shortcomings. The COVID-19 disease is unparalleled, having affected the lives of millions of people worldwide and crippled the economy. World Health Organization (WHO) has released detailed technical guidelines to all nations on diagnosing and managing patients based on the middle east respiratory syndrome (MERS) and severe acute respiratory syndrome (SARS) illness experiences. The WHO and its international collaborators have cooperated to speed up the establishment of critical health procedures and devices. The committee concluded that new COVID-19 diagnostics, treatments, and vaccinations are necessary to maintain healthy systems. This epidemic has created many research challenges and possibilities that our society can and must address to prepare for the current and future crises [1]. Medical practitioners and researchers are now searching for innovative tools to test for and limit the outburst of the unpredictable pandemic in this global health crisis [2]. As noted by [3], it is critical for healthcare providers and public health officials to monitor viral infections to ensure appropriate patient isolation and real-time containment measures. The COVID-19 epidemic has caused devastation, and a rapid remedy for the disease will be a therapeutic medication with a history of use in patients to address the present pandemic.

Hence, the pharmaceutical sector is seeking innovative, cutting-edge technology to detect, manage, and limit the outbreak of COVID-19 disease [4]. The distinctive combination of IoMT and other techniques has been widely deployed in several countries to combat COVID-19, protect the front-line employees, boost efficacy by minimizing the detrimental impact of the pandemic on human lives, and lower death tolls. IoMT technologies are becoming more diversified and common and have appeared as key enablers for preventing, forecasting, and monitoring new virulent illnesses such as COVID-19. Cloud-based remote health tests, AI, wearable health-monitoring devices, and wireless body area networks (WBANs) are all used in IoMT as a health-monitoring structure to offer real-time supervision [5,6]. It is interesting to note that significant advancements ranging from technology and applications to security have been made, amplified by the quick and broad deployment of IoMT throughout the world. An enormous count of ongoing studies demonstrates that integrating security measures with technology can lead to the adoption of secure IoMT applications. Furthermore, when novel IoMT frameworks mix with Blockchain, Big Data, and AI, more feasible options become available [7]. Using IoMT functional components such as analytics, data collecting, and storage, transmission, it is beneficial to create timely detection mechanisms to restrict the spread of virulent illnesses such as the COVID-19 pandemic.

Sophisticated computational research such as AI and Internet-of-Things (IoT) promise technological enablers to address key clinical concerns related to COVID-19 in this situation [8,9]. Interestingly, the recent advancements in AI via the distinctive interplay of machine learning (ML) algorithms (reinforcement learning and extreme learning machine) and deep learning (DL) models, blockchain technology, integration of IoT in wireless communication networks, big-data analytics, cloud computing, and Industry 4.0 can yield lasting solutions to combat the dreaded pandemic. In addition, these technologies can aid the diagnosis, treatment of diseases, and prevention of their spread. These interconnected technologies may help with real-time data collection from people in distant areas utilizing decision making based on big-data analytics and AI IoT; interpreting, processing, predicting; data backup employing cloud computing; and secure data networks utilizing blockchain technology [10,11]. AI, Big Data, and IoT are three connected study disciplines that significantly influence the design and implementation of better-customized healthcare infrastructures. Most importantly, the IoT has revolutionized healthcare procedures and spawned a sustainable evolution known as the IoMT.

Hitherto, the excruciating outburst and divergence of medical data had presented a major bottleneck to data access, information processing, security and privacy in the IoMT. Fortunately, with the widespread deployment of tetherless connectivity, ultra-high reliability, and medical informatization based on 5G-and-beyond networks, the IoMT has attracted fast-growing interest in maximizing the accuracy and productivity of electronic equipment in the healthcare industry. By integrating the existing healthcare services and medical resources, researchers are helping to create an efficient digital healthcare system [12,13]. Essentially, the

IoMT is a practical embodiment of the medical industry's IoT technology and the heart of digital medical transformation. More so, when hybridized with network communication devices, mobile terminals, and other accessories, the IoT technologies (IoMT) such as sensor technology, positioning technology, and radio frequency identification (RFID) infrastructure can be adapted in the medical field to achieve high-quality medical staff–patient–medical interaction. Thus, spurring automation, intelligence and digitalization of the healthcare environment [14]. The IoMT also finds application in virtually all elements of the healthcare industry, including but not limited to remote monitoring, signs monitoring, identity identification, and waste and equipment monitoring. The ability to process, evaluate, and make rapid decisions on gathered medical data in real-time is critical since it directly affects the health and well-being of patients. Furthermore, because medical data concerns patients' privacy, it is paramount to protect sensitive data and patient privacy data. The IoMT technologies are shown to actively fill this gap by ensuring the remote availability of the patients' real-time physiological data (blood pressure, glucose level, heart rate, body temperature, oxygen level, ECG, etc.) alongside the psychological data (expression, speech, etc.) to healthcare providers [11].

Besides, the unprecedented demand for intelligent frameworks necessitates a sophisticated data processing mechanism, which stimulates the deployment of AI-enabled Internet of Medical Things (IoMT) in the healthcare field. AI is undoubtedly one such parallel technology that can aid in the fight against this pandemic through numerous approaches, namely notification, population screening, medical assistance, and infection control directions. Moreover, as revealed in Ref. [15], the state-of-the-art can potentially enhance planning, medication discovery, therapy and reported results for the COVID-19 patient, being an evidence-based medical tool.

AI-enabled technology also finds practical application in field design, where learning-prediction models dimensioned to perform quick virtual screening are designed to show consistent outcomes effectively. Drugs that can potentially treat rare diseases like COVID-19 can be quickly screened using AI-based technologies. This technology can prove helpful in the COVID-19 scenario via medical aid, thanks to technical developments in AI combined with greater processing capacity. AI can swiftly discover medicines to combat new diseases like COVID-19 using a drug discovery approach. AI will have numerous helpful uses in various industries, ranging from agriculture, banking, medical procedures, and military operations, after it has been wholly developed within electronic systems, mainly by minimizing human activity in vitally difficult cases [16]. Depicted in Figure 13.1 is a diagrammatic illustration of IoMT in a practical environment. Advanced diagnostic techniques and devices are adapted to collect patient vitals transmitted over the Internet to the IoMT applications where advanced regimens are performed. The information is forwarded to a medical centre for health practitioners to act on. Finally, a timely response is forwarded to the patient in concern.

Therefore, it is critical to exhaustively characterize smart healthcare systems based on IoT to effectively combat the dreaded COVID-19 outbreak in the era of sophisticated digital technologies. This paper presents a comprehensive discussion on the deployment of AI in conducting remote screening and monitoring for

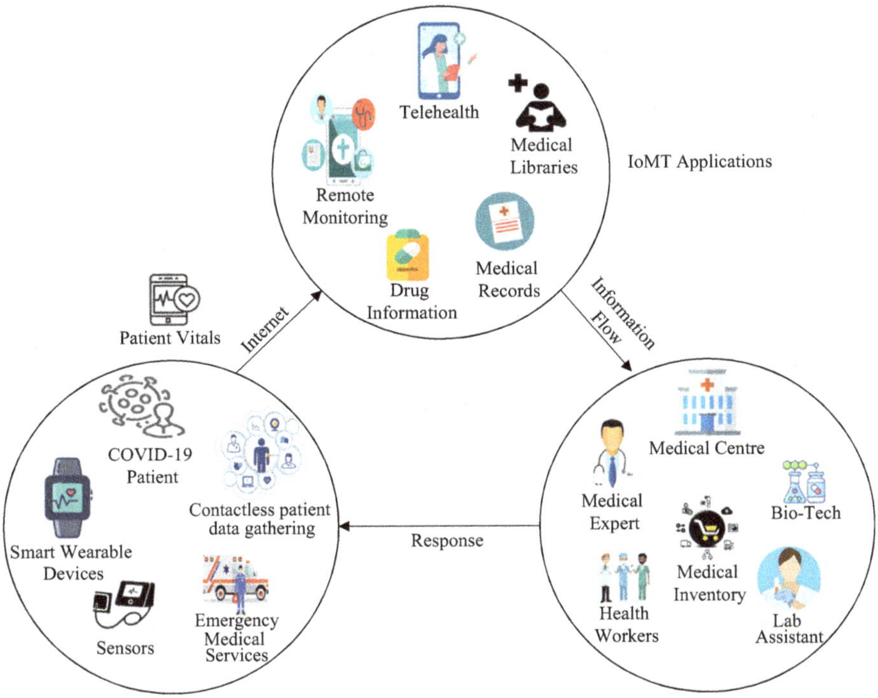

Figure 13.1 A typical illustration of an IoMT framework

COVID-19 symptoms. Further, novel applications of AI-enabled IoMT to address the global issues in this domain are considered. The possibility of adapting ML and image/signal processing approaches for monitoring various vital indicators, including cough, oxygen saturation, blood pressure, heart and respiratory rates, cough, and using basic cameras and no specialist equipment is demonstrated. It is envisaged that patients would be considerably more conscious of their vital signs due to this model, resulting in a higher overall quality of life.

13.2 Related work

The research on IoMT architecture has recently garnered considerable attention from academia and industry. Ref. [4] proposed the cognitive IoMT to mitigate the widespread problem of the COVID-19 virus. The proposed model depicts an innovative application of the cognitive radio (CR)-based IoT dimensioned to serve the healthcare industry. Oniani *et al.* [17] extensively surveyed various applications of IoMT in the medical domain. Specifically, an in-depth analysis of different devices and techniques employed in therapeutic and diverse areas to collect clinical data, conduct analysis, and diagnosis is presented. In addition, a brief introduction to the application of AI methodologies in medical IoT is presented. Guo *et al.* [18]

characterized the performance of an insightful AI-based semantic IoT (AI-SIoT) hybrid service infrastructure to support intelligent services by incorporating heterogeneous IoT devices. Also, real-time application scenarios alongside the opportunities and challenges associated with the proposed model were discussed.

Mohanty *et al.* [19] provided a comprehensive survey of research trends in AI-based drug repurposing to combat the virulent pandemic. The superiority of the AI approach in enhancing drug discovery is validated. The survey also revealed that, based on a handful of the old medicines in patients, they might be easily used to treat COVID-19 patients if they proved to be efficacious against SARS-CoV-2. Ref. [16] considered the distinctive combination of IoT and big data in the medical domain. It addressed a variety of technologies that can lower total expenses for chronic disease prevention or management. Some technologies investigated include those that continually monitor health indicators, capture real-time health data, and auto-administer treatments. The work [20] describes a privacy-aware, energy-efficient framework for securing a patient's medical information. The primary goal of this article is to reduce communication costs to increase security features and energy efficiency against unauthorized access. Joyia *et al.* [12] highlighted the researchers' contributions of IoT in the healthcare sector, its application, and future difficulties of IoT in the context of medical services in healthcare. Ref. [21] proposed integrating AI into IoT to realize a quicker, greener, smarter, and safer model. The AIoT architecture was briefly introduced with regard to fog computing, cloud computing, and edge computing.

Furthermore, current trends and recent advances in the state-of-the-art were highlighted in the context of seeing, learning, reasoning, and acting. Some intriguing AIoT applications that have the potential to change our environment dramatically were also summarized. Finally, the difficulties of AIoT were discussed, as well as some possible research possibilities. Finally, Sun *et al.* [13] critically analyzed the beneficial interplay between IoMT, AI technologies, cloud computing, and edge computing. The paper provided a detailed evaluation of the rapid processing and analysis of big medical data and the deployment of high-quality medical resources while working within the limits of the current medical-related equipment and medical environment.

13.3 IoMT for COVID-19 pandemic data management

The synergic deployment of IoT architectures in the healthcare industry has been advocated as a highly promising technology to enable the digital administration of medical information [22]. Thus, allowing healthcare personnel to focus on patients rather than documenting and organizing a huge amount of tedious medical data, allowing them to deliver better medical services. With an attendant rise in the number of casualties associated with the virulent COVID-19 pandemic, the medical field has exploited novel technologies and frameworks to combat the pandemic. Emerging technologies have the potential to provide a cure for the global problem. The IoMT, a practical incarnation of the IoT, can help with disease detection,

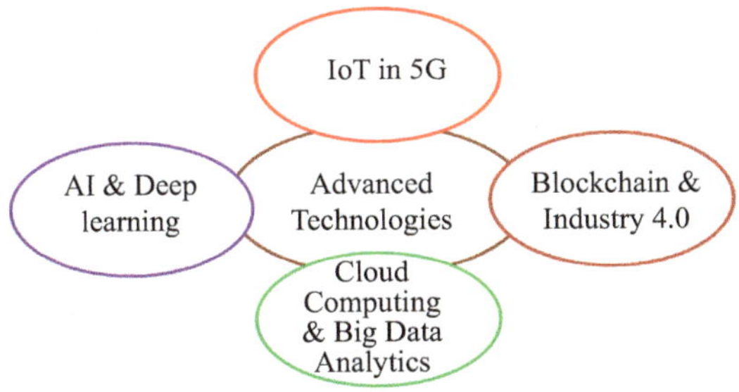

Figure 13.2 Advanced technologies to combat the COVID-19 pandemic

monitoring, contact tracking, and control [23]. IoMT analyses data in the healthcare sector by combining medical equipment and digital applications. It is becoming known as mobile health to combat health crises. Smart healthcare has combined sophisticated technology and new applications to create a potential COVID-19 solution that offers treatment modalities for monitoring, screening, tracking, and controlling disease transmission [24]. IoMT is well associated with providing reliable online medical services according to the medical emergency. IoMT can also be extended to provide intelligent medical platforms that connect or track patients at various locations and relay the information to the telecare database. Besides, the overwhelming outbreak of the dreaded virus worldwide makes it difficult to control the situation without access to real-time information. Figure 13.2 presents a pictorial illustration of advanced technologies dimensioned to provide long-term solutions to the COVID-19 pandemic in this digital age of advanced technology.

13.3.1 Architecture of IoMT

The ever-evolving nature and the recent successes in innovations, science, and technology have transformed the landscape of the medical environment dramatically. Hence, this led to the design and development of smart medical devices and the innovation of novel medical procedures. Furthermore, due to the advances in communication technology, it is worth mentioning that various medical services have been transformed into virtual systems and applications that may be accessed from a distance. As a springboard to improved public life, quality healthcare, and multifold gains, the IoT and its integration into the medical field have appeared as a preferred candidate. Thus, researchers and companies advocate IoMT applications as a highly promising technology to realize cheaper and more accessible healthcare [5,25].

As alluded to earlier, the architecture of the IoT has been carefully examined in academia, and the industry and certain components can be applied in developing IoMT. The IoMT is a condensed version of IoT technology used in the health sector. The three tiers of an IoT application architecture, namely the perception layer, network layer, and transmission layer, are characterized in [13]. The description of the concept of the architecture of IoMT is shown in Figure 13.3.

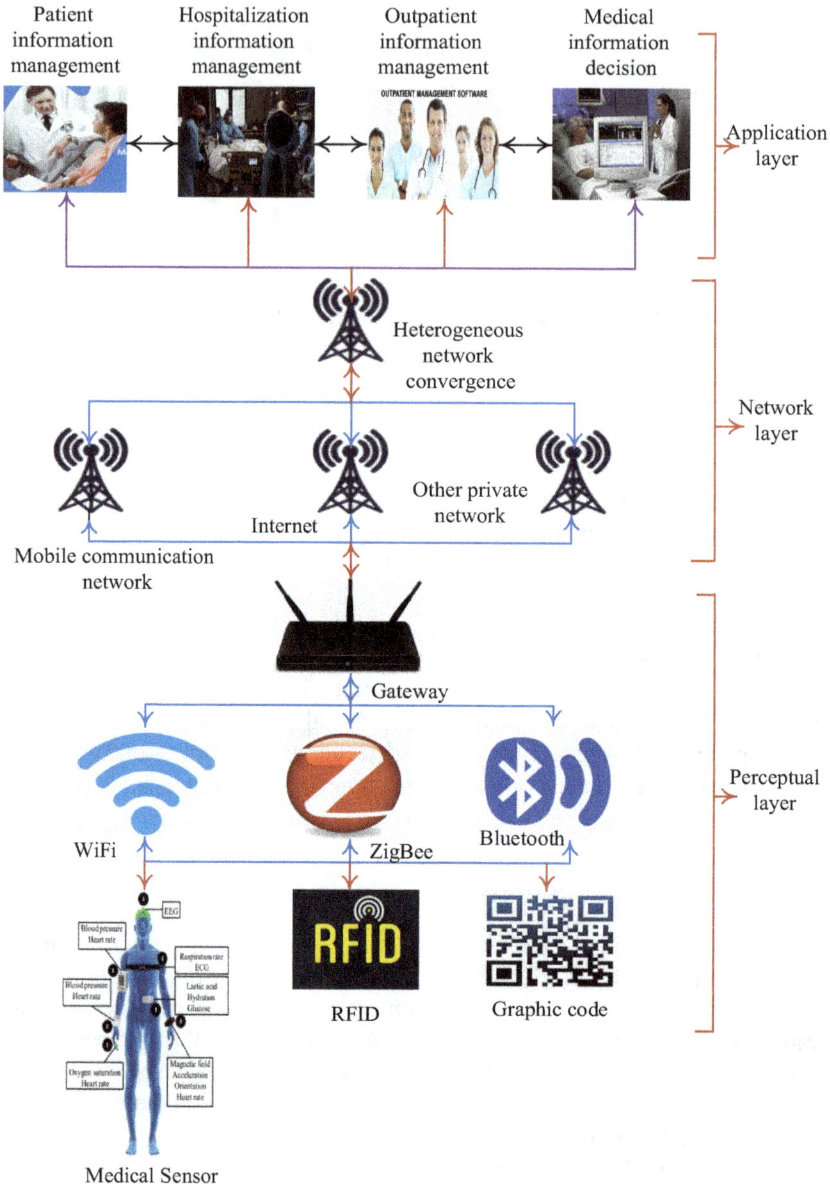

Figure 13.3 The generic architecture of the IoMT

The various health and governing agencies set testing policies at the local level. Although this is not a generally followed method, it is a good example of a policy that might be anticipated to be followed. Many countries, in particular, have battled with poor testing rates [26], which presents a prime challenge in testing a

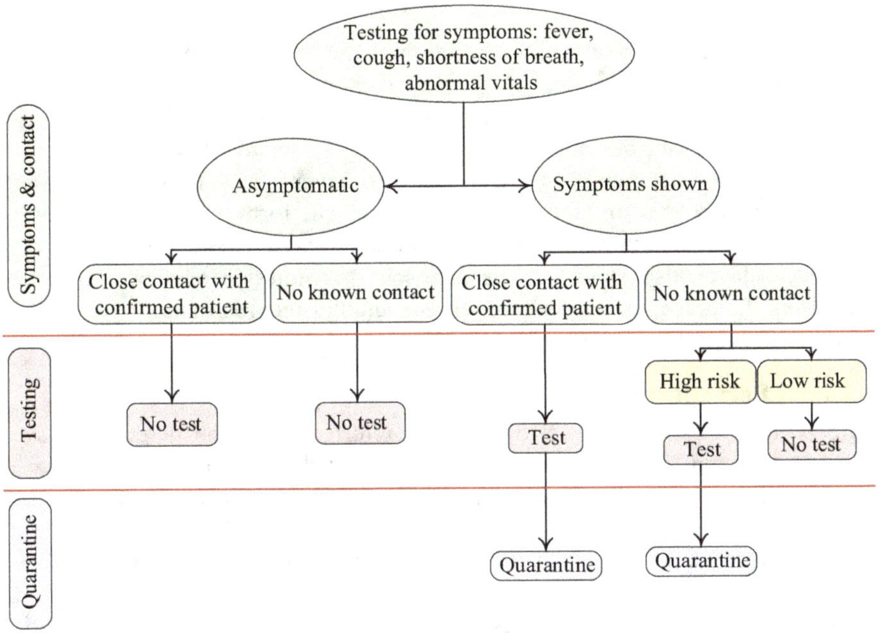

Figure 13.4 COVID-19 testing flowchart

substantial number of individuals. As a result, the testing flowchart delineated in Figure 13.4 has gained widespread adoption in several countries to prioritize testing for high-risk populations (those over 60 years old or with underlying medical conditions such as immune system disorders, diabetes, lung or heart disease, as defined by the WHO). In many situations, there has been a failure to comply with quarantine, burdening the testing system with additional individuals to track down and test. In the current situation, "flattening the curve" is critical to guarantee that the number of cases remains within the capability of healthcare institutions [27].

13.3.1.1 Application layer

The application layer, the topmost layer of the IoMT architecture, is driven by cloud computing platforms and provides customers with customized services such as data analysis and storage. Basically, a cloud computing platform is deployed to process and analyze data gathered from sensors, which is forwarded over the network to end devices [21]. The application layer is composed of two levels, namely medical information application and medical information decision-making application.

Medical data management applications include inpatient treatment data management, patient and outpatient data management, medical equipment and material information management, and so on. Patient, diagnosis and therapy, disease, pharmaceutical information analysis, and so on are examples of medical information decision-making applications [13].

13.3.1.2 Network layer

The network layer is central to the operation of the IoMT framework. It is essentially an integration of different networks such as the Internet, LANs, and various devices such as gateways, routers, and hubs that are connected via innovative technologies including but not limited to 5G mobile networks, Wi-Fi, Bluetooth, and LTE. Like the application layer, the network layer can be decoupled into two sublayers: network transmission and service layer. The former is at the heart of the IoMT network and is analogous to the human brain and nerve centre. The network transmission layer also facilitates instantaneous transmission of data information obtained by the perception layer in a reliable and barrier-free manner via wireless networks or insightful technologies [13,21].

13.3.1.3 Perceptual layer

The bottom layer of the IoMT framework, referred to as the perception layer, comprises different devices, actuators, and sensors that gather data and forward them to higher levels. The major emphasis and difficulty in the practical implementation of the IoMT framework stem from the perceptual layer, subdivided into the data acquisition and access layers. The former is actualized using various kinds of signal acquisition equipment and medical perception equipment to acquire the data information of people and things and perform the perception and identification of nodes in the IoMT. This layer adopts signal acquisition techniques, including image recognition technology, graphic code, and RFID technology. Conversely, the latter creates a connection between the data collected by the data acquisition sublayer and the network layer via several access techniques and limited-distance data transmission mechanisms, namely Bluetooth, Wi-Fi, and ZigBee [13,21].

13.3.2 Applications of the IoMT in COVID-19 data management

As previously stated, decades of smart healthcare research have resulted in a steadily increasing use of IoMT. Rather than relying on traditional treatment methods, this technique allows for substantial advancements in COVID-19 management. This problem, covering daily new cases, is being diagnosed, monitored, tracked, and controlled in real-time [11]. Regarding COVID-19 emergency preparedness, IoMT is used extensively in providing patients with online medical services, adequate medical care, and self-testing at any location and quarantine facility. As demonstrated in Figure 13.5, it may also be employed to create a well-established medical platform for administering datasets valuable for healthcare and government activities.

As illustrated in Figure 13.5, the IoMT enables intelligent medical treatment and effective administration of people and things, resulting in improved quality of health and lower medical costs. In the IoMT architecture, the level of security provided by the identity recognition system is critical. The rapidly emerging field of mobile medicine and telemedicine in correspondence with smartphones [28], tablet computers, laptops, and other devices [29] can also be deployed in cases

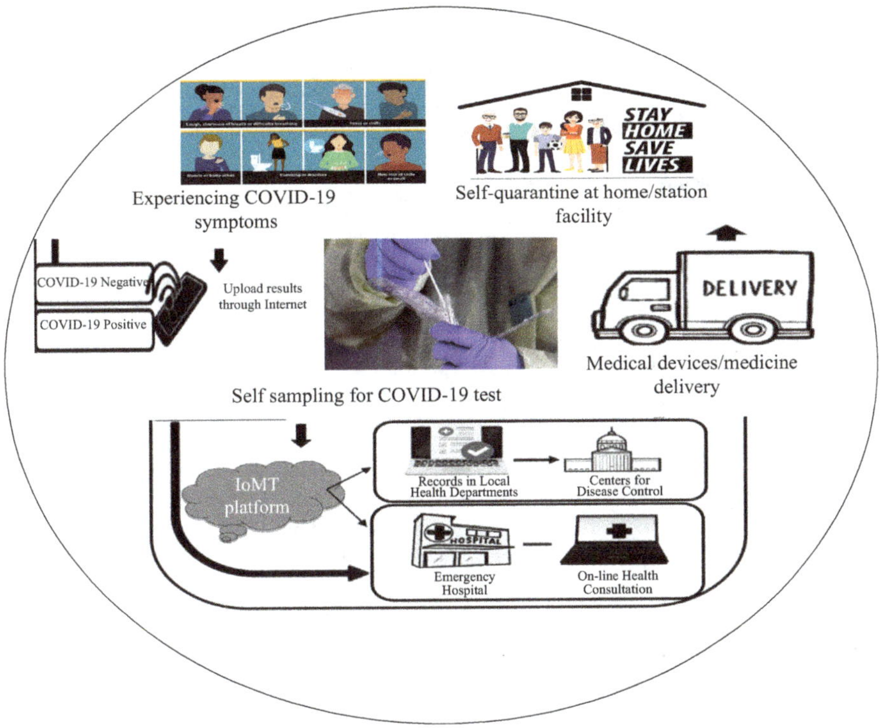

Figure 13.5 COVID-19 suppression using IoMT

where patients do not have time to visit doctors or are unable to do so. The state-of-the-art has also provided significant improvements in the medical field regarding work efficiency, hospital business processes, and resource utilization. More specifically, the various applications of IoMT in the health sectors are but not limited as follows: prevention and control, reducing the workload of the medical industry, screening and surveillance, contact tracing and clustering, rapid diagnosis, remote monitoring of the patient, real-time tracking, etc. The descriptions of the mentioned applications of IoMT in the context of medical services are well explained in [4]. In addition, a pictorial representation of various applications of IoMT for combatting the deadly COVID-19 disease is presented in Figure 13.6.

13.3.2.1 Screening and surveillance

In many countries, the COVID-19 epidemic has wreaked havoc on the healthcare system. Healthcare workers are overworked and in danger of infectious diseases from COVID-19 patients. It is difficult to screen and monitor the health status of an enormous count of sick or vulnerable individuals. To achieve this objective, the use of appropriate systems for real-time remote patient monitoring. The continuous advancements in ML and DL, which depicts the fundamental AI technologies, have increased the capability of imaging methods and may now be utilized to remotely

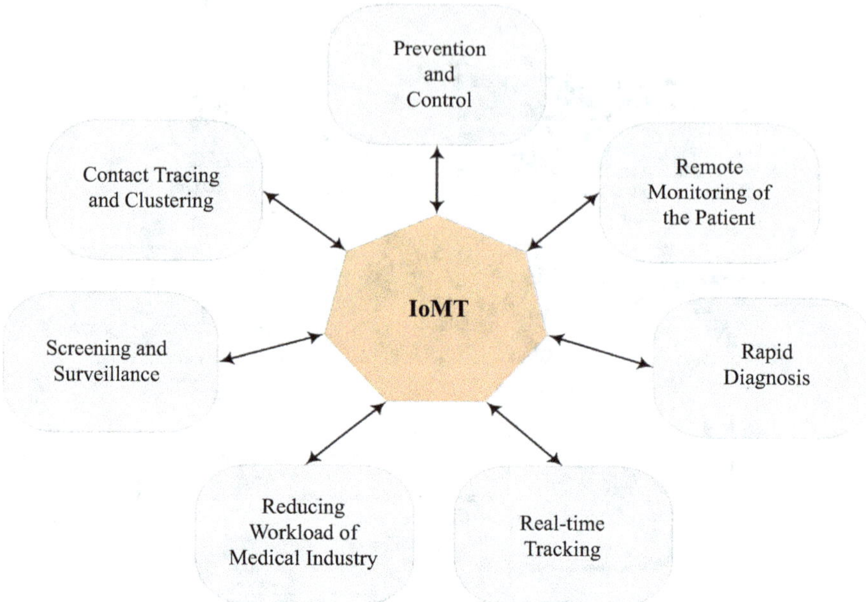

Figure 13.6 Various applications of IoMT for combatting the deadly COVID-19 disease

execute numerous activities which could not be conducted without the physical presence of a healthcare expert. IoMT can obtain thermal imaging-based face recognition data at sensitive entrance points such as hotels, railway stations, airports, and other locations for screening and surveillance reasons [15,30]. Automating surveillance of probable and confirmed cases may aid in controlling infection transmission [11,31].

13.3.2.2 Contact tracing and clustering

Contact tracing is one possible approach to minimizing the outbreak of virulent disease. However, the technique is time-consuming and ineffective considering the large number of people living in a locality. The process can be simplified if the location history of the patient tested positive is obtainable in a database that healthcare authorities can easily access. Therefore, the distinctive interconnection of healthcare and medical units through IoT is critical to collating the number of positive instances in real-time by location. The government may gain access to this information and issue alerts for health checks in the impacted region, all of which can be done quickly using an AI framework. Public authorities can also use zone clustering to enact different social distancing and lockdown laws and regulations [31,32].

13.3.2.3 Real-time tracking

This technology allows for an overview of global updates on COVID-19 cases, highlighting key data such as the current cases in various localities, the number of cured patients, and fatalities. Hence, AI may be used to model disease severity and predicts disease activity, allowing health authorities and policymakers to make better decisions and be better prepared for control [20]. More so, each person connected to the IoMT network can access health care preventative measures, treatment process updates, and government initiatives [11,30].

13.3.2.4 Prevention and control

Individual preparedness and prompt action by healthcare and governmental authorities can aid in restricting the overwhelming outbreak of the pandemic. The IoMT allows one to be aware of positive cases in the neighbourhood and to remain attentive by utilizing certain applications [3]. The disease transmission is further limited by geographical clustering areas mentioned above under contact tracing and clustering [31].

13.3.2.5 Remote monitoring of the patient

Doctors and healthcare professionals are particularly vulnerable to the COVID-19 pandemic because it is extremely infectious. The IoMT enables medical personnel to perform remote monitoring of a patient using fingertip medical data such as glucose, pulse rate, blood pressure, temperature, electromyography (EMG), electroencephalogram (EEG), electrocardiogram (ECG), heart rate, and breathing rate. Wearable IoT sensors can collect clinical parameters data [33,34]. Due to the wide Internet connectivity of the COVID-19 facilities, real-time medical data communication is feasible, saving cost, effort, and time. Particularly, the use of IoMT is beneficial for the aged or individuals with numerous illnesses [11].

13.3.2.6 Rapid diagnosis

Many countries have mandated that travelers and probable patients be held in quarantine even if they display no clinical symptoms, ensuring rapid identification of critical cases. IoMT can yield multiple gains through specific network applications by enabling quick diagnosis for migrants with travel history. In these scenarios, AI-enabled visual sensors can be effectively deployed to analyze results obtained remotely in a control room/service centre from computed tomography (CT) scan or X-ray, allowing them to diagnose and confirm cases in less time. This also allows for contact-free and early viral identification [32].

13.4 Reducing the workload of the medical industry

The scarcity of medical personnel in practically all nations of the world is alarming, thus increasing the workload on the available healthcare workers. IoMT plays a huge role in filling this gap by aiding with diagnosing, monitoring, and treating patients. As mentioned in the remote monitoring of patients, the IoMT allows for remote disease monitoring, which decreases effort even further. AI integrated with IoT sensory data, modelling and forecasting of the infection are also aided by AI [35,36]. Hospitals can also collaborate with blockchain firms to provide fast telemedicine consultations and medicines delivered to patients' homes.

13.4.1 Applications of AI-enabled IoMT

The scourge of the COVID-19 pandemic has overwhelmed the healthcare system of many countries, and the shortage of healthcare workers exacerbates its impact. In some regions, the healthcare workers are currently overstretched and at a significantly higher risk of contracting the dreaded virus. It is difficult to screen and monitor the health of a huge number of people who are vulnerable or sick. While expert medical care and possibly admittance to the hospital are required for individuals with severe symptoms, providing care at home is a successful method for individuals with mild symptoms and individuals who have been isolated and are at risk of infection. Achieving this objective is predicated on deploying appropriate systems and telehealth technologies for remote patient monitoring. The continuous advancements in ML and DL have increased the capability of imaging methods, allowing them to perform various activities that could not be conducted without the physical presence of a healthcare expert. The basic idea underpinning the introduction of IoMT is to maximize the quality of the healthcare industry via intelligent data collection and storage from medical applications and devices. An excess supply of sensors, including but not limited to blood pressure sensors, gyroscope sensors, electromyogram/electroencephalogram sensors, visual sensors, blood oxygen saturation sensors, carbon dioxide sensors, humidity sensors, accelerometer sensors, temperature sensors, and respiration sensors are adapted to monitor the symptoms of a patient in real-time [37]. These medical equipment keep track of their patient's health status, which is later forwarded to a therapist through cloud data mechanisms. The primary issue for IoMT is regulating clinical applications that create a huge volume of medical data from linked devices [38,39]. This motivated the AI-empowered IoMT to tackle the challenges in the health sector. Recent AI advancements have considerably improved human lives, and further sophisticated AI-enhanced models are currently being developed to assist humans in managing and overcoming this virulent bottleneck. ML, expert systems, computer vision, and NLP are some of the most important AI technologies [40,41]. ML is the most important AI technique among them [42,43]. The superiority of the ML technique over artificial diagnosis is demonstrated in terms of effective disease forecasting and diagnosis, minimization of severe diseases associated with artificial diagnosis, and high accuracy and efficiency. An expert system is an AI system that includes image recognition and vast knowledge and expertise in certain medical disciplines. With inspiration, pertinence, transparency, and adaptability, it can substantially optimize the diagnostic and treatment process of a patient through exact simulation of medical activity and the decision-making process of healthcare personnel. Implementing AI technology in medical care results in technological innovation and a shift in medical service delivery [44]. Figure 13.7 shows the various applications of AI-based IoMT in the health sector.

13.4.2 Applications of AI-enabled IoMT for drug repurposing

Drug repositioning is faced with numerous challenges, some of which include the establishment of a specific medication-disease connection and diagnosing.

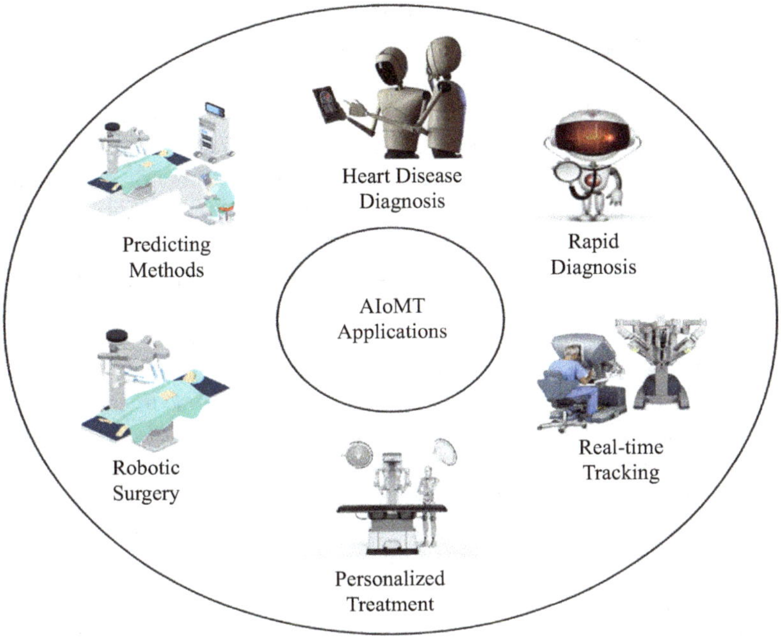

Figure 13.7 Various applications of AIoMT in the health sector

A plethora of techniques such as experimental biological approaches, computational approaches (such as AI), and hybrid techniques have been developed to solve this challenge. Among the proposed approaches, the potential of implementing AI-based models for drug repurposing in a practical environment is quite high [45]. On the other hand, researchers have compared the SARS virus of 2003 and the current COVID-19 virus and found some similar clinical manifestations and symptoms. Thus, from the available data and clinical procedures on SARS, insightful AI-based models and medication structures can be designed and developed to manage the pandemic [46]. AI and ML can help with this process by quickly identifying effective medicines against COVID-19, removing any barriers between scores of repurposed pharmaceuticals, clinical testing, and final drug approval. Drug repurposing can be further optimized by exploiting DL approaches in the age of big data. Drug repurposing based on AI is a less expensive, quicker, and more effective method that can reduce clinical trial failures. Without going through the first trials and toxicity assessments, the repurposed medication might go straight to the advanced phase of trials. Though AI-assisted drug repurposing is still in its early stages, it appears to be a viable strategy for creating possible COVID-19 curative medicines. AI-powered medication repurposing might be useful in the COVID-19 scenario, thanks to technical improvements in AI combined with greater processing capacity.

13.5 Privacy-aware energy-efficient framework using AIoMT for COVID-19

The primary goal of this framework is to regulate data publication during the diagnostic and therapeutic process. In this case, external users are restricted from altering the storage options, such as changing privacy settings or publishing data. On the other hand, authorized healthcare personnel have the authority to set aside sensitive data to reveal the infection rate progressively. This framework manages medical data to affect data sensitivity by categorizing the type of multimedia elements. Furthermore, significant energy savings between medical sensors can be achieved, resulting in improved security of transmitted medical data [9]. The technology also guarantees a dependable, effective, and trustworthy method of monitoring patients' physical activity, from the mobile sink to medical centres. Medical data may be monitored on a patient's body regularly to manage data transfer to medical experts using intelligent technology. However, certain malevolent nodes may attack the network architecture, rendering essential features like integrity, user authentication, and data privacy unsatisfactory [20,47].

13.6 Open research issues

The concept of AI-enabled IoMT for healthcare applications is still in its infancy [48]. While advantageous perspectives and significant progress have been realized recently, significant bottlenecks and unresolved issues are mitigating the practical deployment of the state-of-the-art. This section provides a holistic overview of open research issues and exciting research trends for future work. The lines of research requiring further investigation are highlighted as follows.

13.6.1 Security and privacy

IoMT-based solutions are expected to optimize medical procedures, improve people's health and convenience and ultimately minimize the financial implication for the healthcare industry. However, it is unfortunate that involved users and stakeholders in modern IoMT settings are less aware of the security threats and vulnerability to ransomware and other attacks. Information security control alongside the conventional neat, physical walls deployed for security is insufficient against the underlying vulnerabilities [49,50]. The security challenge is further heightened by transmitting sensitive data over the insecure internet and adopting interconnected heterogeneous multimodal systems for e-health applications. Therefore, it is of paramount importance to protect the privacy and security of medical data against malicious nodes and traffic. Specifically, core security and privacy controls are required to safeguard cloud-connected databases and IoMT frameworks from malicious attacks, thus, enhancing data privacy and integrity [51,52].

13.6.2 Energy efficiency

The design of energy-efficient devices while ensuring low latency communication and timely response in chronic disease treatment/emergencies is critical for realizing

IoMT in a practical environment [53,54]. IoMT applications are notable for high speed and immediate delivery, thus, increasing the energy consumption and communication overhead between biosensor nodes. The biosensor nodes, whose primary objective is to collect healthcare data through various mechanisms such as mobile devices, wearable bands, and implanted surgical devices and transmit the patient's information to remote centres, are quite limited concerning transmission power and battery power [55,56]. Therefore, addressing these energy-related challenges is of paramount importance. More specifically, developing energy-efficient devices, energy management strategies, renewable energy resources, and energy-related trade-offs in healthcare networks and facilities while ensuring immediate response and sustainability are interesting areas worth investigating [57,58].

13.6.3 Integration of emotion-aware abilities

Designing and developing emotion-aware recognition frameworks is an interesting approach to ensure healthy living and offer emotional solicitude during the dreaded epidemic. Moreover, the emotional problems related to the outbreak are envisaged to linger during the post-covid era for the elderly, young children, infants and mentally ill persons. Confronting these emotional problems depends on intelligent remote health monitoring, data gathering, information supervision and personalized therapeutic solutions. Therefore, building insightful emotion-aware detection modules is critical to effectively dealing with the challenge of emotional solicitude. However, numerous challenges ranging from the available machine power factors, multiple models, the types of big data, the way of the signal acquisition, the environment, accuracy of the database, and pattern (images, voice, or video) persist. Integrating emotion-aware abilities into IoMT systems is also a novel difficulty and an interesting area worth investigating.

13.6.4 Interoperability

Lately, the rapid proliferation of IoMT technologies is overwhelming and is envisioned to increase at a breakneck pace in the next few years. Due to the heterogeneity of IoMT solutions (data semantics, data structure, communication protocols, network interfaces), interoperability is a prime challenge. It is critical to exchange data without restrictions through a dynamic and connected interoperability architecture. Thus, exploring the heterogeneity of various IoMT components and incorporating IoMT-based frameworks in an interoperable environment is an area worth investigating. Particularly, dynamic and homogenous models are required to merge these sophisticated digital architectures.

13.6.5 AI in IoMT

AI-powered medical technologies are proliferating and transforming the medical industry's various features. AI-based solutions can be effectively deployed to automatically capture patient information, provide advanced diagnostics and tailored regimens, support intelligent-decision making, and predict future conditions with quick delivery time using supervised or unsupervised learning. Although the integration of AI in IoMT makes a compelling case to provide precision medicine models, the research is still at a very early stage. Moreover, adopting ML and natural language processing (NLP) in the medical domain has numerous challenges and is an exciting area worth considering.

13.6.6 Ethical issues

In addition to collecting health-related data in a surveillance environment, conveniently transmitting the same, and providing immediate response during emergencies, different ethical issues must be addressed before successfully deploying IoMT in a real-time environment. These ethical issues may range from the private use of information, accessibility, property rights, and user awareness of attack risks to the integrity of information. International legal bodies are responsible for developing dynamic modifiable policy rules to enforce accountability and specialized control of IoMT applications [59,60]. Overall, all stakeholders in the healthcare ecosystem should adopt ethical policies and regulations to safeguard massive devices linked to all AIoMT platforms to ensure the security and privacy of sensitive user data across all open communication channels [59–68].

13.7 Conclusion

The resultant impact of the COVID-19 pandemic on all segments of the human race is particularly disruptive and detrimental. The unprecedented loss of human lives worldwide, the significant constraint of economic activities and deepened poverty rate have accentuated the development of insightful technologies. AI-based models and ML frameworks have appeared promising technological enablers to manage virulent diseases and transition to better solutions. This paper highlights the key benefits and application areas of AIoMT in suppressing the COVID-19 pandemic and providing improved healthcare. AIoMT is a potential technique for rapid diagnosis and improved therapy and management while preventing the virus from spreading to others and has found significant applications in these trying times. Many organizations and government parastatals are adopting this technique to address the current healthcare concerns. Ultimately, this chapter brings to limelight, the need for innovative research to address the critical pandemic plaguing the world with the COVID-19 virus. Hospitals will have fewer qualified personnel than necessary as the number of COVID-19 patients grows. As a result, any new procedures must be automated and need minimal engagement from medical experts, other than periodic monitoring. In such scenarios, an intelligent centralized system monitors all vital signs. As used in surveillance networks, a practical video summary might help give crucial insights into future work. Once completely implemented, these approaches will benefit vulnerable individuals in a pandemic situation like COVID-19 since they can detect symptoms peculiar to the disease at an early stage. Due to the enormous benefits outlined above, this field of research is envisioned to find even higher applications in the post-COVID era and is an exciting area worth investigating.

Acknowledgment

The work of Agbotiname Lucky Imoize is supported in part by the Nigerian Petroleum Technology Development Fund (PTDF) and in part by the German

Academic Exchange Service (DAAD) through the Nigerian-German Postgraduate Program under Grant 57473408.

References

[1] M. Gupta, M. Abdelsalam, and S. Mittal, "Enabling and enforcing social distancing measures using smart city and its infrastructures: a COVID-19 use case," *arXiv preprint arXiv*, 2004. *09246*, 2020.

[2] Z. Allam and D. S. Jones, "On the coronavirus (COVID-19) outbreak and the smart city network: universal data sharing standards coupled with artificial intelligence (AI) to benefit urban health monitoring and management," in *Healthcare*, vol. 8, p. 46, 2020.

[3] A. Haleem, M. Javaid, and R. Vaishya, "Effects of COVID-19 pandemic in daily life," *Current Medicine Research and Practice*, vol. 10, p. 78, 2020.

[4] S. Swayamsiddha and C. Mohanty, "Application of cognitive Internet of medical things for COVID-19 pandemic," *Diabetes & Metabolic Syndrome: Clinical Research & Reviews*, vol. 14, pp. 911–915, 2020.

[5] E. Christaki, "New technologies in predicting, preventing and controlling emerging infectious diseases," *Virulence*, vol. 6, pp. 558–565, 2015.

[6] A. Alabdulatif, I. Khalil, A. R. M. Forkan, and M. Atiquzzaman, "Real-time secure health surveillance for smarter health communities," *IEEE Communications Magazine*, vol. 57, pp. 122–129, 2018.

[7] A. H. M. Aman, W. H. Hassan, S. Sameen, Z. S. Attarbashi, M. Alizadeh, and L. A. Latiff, "IoMT amid COVID-19 pandemic: application, architecture, technology, and security," *Journal of Network and Computer Applications*, p. 102886, 2020.

[8] R. P. Singh, M. Javaid, A. Haleem, and R. Suman, "Internet of things (IoT) applications to fight against COVID-19 pandemic," *Diabetes & Metabolic Syndrome: Clinical Research & Reviews*, vol. 14, pp. 521–524, 2020.

[9] R. Vaishya, M. Javaid, I. H. Khan, and A. Haleem, "Artificial Intelligence (AI) applications for COVID-19 pandemic," *Diabetes & Metabolic Syndrome: Clinical Research & Reviews*, vol. 14, pp. 337–339, 2020.

[10] M. Javaid, A. Haleem, R. Vaishya, S. Bahl, R. Suman, and A. Vaish, "Industry 4.0 technologies and their applications in fighting COVID-19 pandemic," *Diabetes & Metabolic Syndrome: Clinical Research & Reviews*, vol. 14, pp. 419–422, 2020.

[11] T. Yang, M. Gentile, C.-F. Shen, and C.-M. Cheng, *Combining Point-of-Care Diagnostics and Internet of Medical Things (IoMT) to Combat the COVID-19 Pandemic.* Basel: Multidisciplinary Digital Publishing Institute, 2020.

[12] G. J. Joyia, R. M. Liaqat, A. Farooq, and S. Rehman, "Internet of medical things (IoMT): applications, benefits and future challenges in healthcare domain," *Journal of Communications*, vol. 12, pp. 240–247, 2017.

[13] L. Sun, X. Jiang, H. Ren, and Y. Guo, "Edge-cloud computing and artificial intelligence in internet of medical things: architecture, technology and application," *IEEE Access*, vol. 8, pp. 101079–101092, 2020.

[14] R. C. Shit, S. Sharma, D. Puthal, and A. Y. Zomaya, "Location of Things (LoT): a review and taxonomy of sensors localization in IoT infrastructure," *IEEE Communications Surveys & Tutorials*, vol. 20, pp. 2028–2061, 2018.

[15] D. S. W. Ting, L. Carin, V. Dzau, and T. Y. Wong, "Digital technology and COVID-19," *Nature Medicine*, vol. 26, pp. 459–461, 2020.

[16] D. V. Dimitrov, "Medical Internet of Things and big data in healthcare," *Healthcare Informatics Research*, vol. 22, pp. 156–163, 2016.

[17] S. Oniani, G. Marques, S. Barnovi, I. M. Pires, and A. K. Bhoi, "Artificial intelligence for internet of things and enhanced medical systems," in *Bio-inspired Neurocomputing*, New York, NY: Springer, 2021, pp. 43–59.

[18] K. Guo, Y. Lu, H. Gao, and R. Cao, "Artificial intelligence-based semantic internet of things in a user-centric smart city," *Sensors*, vol. 18, pp. 1341, 2018.

[19] S. Mohanty, M. H. A. Rashid, M. Mridul, C. Mohanty, and S. Swayamsiddha, "Application of artificial intelligence in COVID-19 drug repurposing," *Diabetes & Metabolic Syndrome: Clinical Research & Reviews*, vol. 14, p. 5, 2020.

[20] F. Al-Turjman and B. D. Deebak, "Privacy-aware energy-efficient framework using the internet of medical things for COVID-19," *IEEE Internet of Things Magazine*, vol. 3, pp. 64–68, 2020.

[21] J. Zhang and D. Tao, "Empowering things with intelligence: a survey of the progress, challenges, and opportunities in artificial intelligence of things," *IEEE Internet of Things Journal*, vol. 8, pp. 7789–7817, 2020.

[22] Z. Ning, P. Dong, X. Wang, *et al.*, "Mobile edge computing enabled 5G health monitoring for Internet of medical things: a decentralized game theoretic approach," *IEEE Journal on Selected Areas in Communications*, vol. 39, pp. 463–478, 2020.

[23] L. Bai, D. Yang, X. Wang, *et al.*, "Chinese experts' consensus on the Internet of Things-aided diagnosis and treatment of coronavirus disease 2019 (COVID-19)," *Clinical eHealth*, vol. 3, pp. 7–15, 2020.

[24] R. P. Singh, M. Javaid, A. Haleem, R. Vaishya, and S. Ali, "Internet of Medical Things (IoMT) for orthopaedic in COVID-19 pandemic: roles, challenges, and applications," *Journal of Clinical Orthopaedics and Trauma*, vol. 11, pp. 713–717, 2020.

[25] A. Ghimire, S. Thapa, A. K. Jha, A. Kumar, A. Kumar, and S. Adhikari, "AI and IoT solutions for tackling COVID-19 pandemic," in *2020 4th International Conference on Electronics, Communication and Aerospace Technology (ICECA)*, 2020, pp. 1083–1092.

[26] Y.-Y. Tsou, Y.-A. Lee, C.-T. Hsu, and S.-H. Chang, "Siamese-rPPG network: remote photoplethysmography signal estimation from face videos," in *Proceedings of the 35th Annual ACM Symposium on Applied Computing*, 2020, pp. 2066–2073.

[27] H. Rohmetra, N. Raghunath, P. Narang, V. Chamola, M. Guizani, and N. R. Lakkaniga, "AI-enabled remote monitoring of vital signs for

COVID-19: methods, prospects and challenges," in *Computing*, New York, NY: Springer, 2021, pp. 1–27.

[28] Z. Ning, F. Xia, X. Hu, Z. Chen, and M. S. Obaidat, "Social-oriented adaptive transmission in opportunistic Internet of smartphones," *IEEE Transactions on Industrial Informatics*, vol. 13, pp. 810–820, 2016.

[29] X. Huang, Y. Li, J. Chen, *et al.*, "Smartphone-based blood lipid data acquisition for cardiovascular disease management in internet of medical things," *IEEE Access*, vol. 7, pp. 75276–75283, 2019.

[30] R. Vaishya, A. Haleem, A. Vaish, and M. Javaid, "Emerging technologies to combat the COVID-19 pandemic," *Journal of Clinical and Experimental Hepatology*, vol. 10, pp. 409–411, 2020.

[31] A. S. S. Rao and J. A. Vazquez, "Identification of COVID-19 can be quicker through artificial intelligence framework using a mobile phone–based survey when cities and towns are under quarantine," *Infection Control & Hospital Epidemiology*, vol. 41, pp. 826–830, 2020.

[32] F. Shi, J. Wang, J. Shi, *et al.*, "Review of artificial intelligence techniques in imaging data acquisition, segmentation, and diagnosis for COVID-19," *IEEE Reviews in Biomedical Engineering*, vol. 14, pp. 4–15, 2020.

[33] X.-B. Pan, "Application of personal-oriented digital technology in preventing transmission of COVID-19, China," *Irish Journal of Medical Science (1971-)*, vol. 189, pp. 1145–1146, 2020.

[34] S. K. Sood and I. Mahajan, "Wearable IoT sensor based healthcare system for identifying and controlling chikungunya virus," *Computers in Industry*, vol. 91, pp. 33–44, 2017.

[35] F. Ibrahim, T. H. G. Thio, T. Faisal, and M. Neuman, "The application of biomedical engineering techniques to the diagnosis and management of tropical diseases: a review," *Sensors*, vol. 15, pp. 6947–6995, 2015.

[36] R. M. Elavarasan and R. Pugazhendhi, "Restructured society and environment: a review on potential technological strategies to control the COVID-19 pandemic," *Science of The Total Environment*, vol. 725, pp. 138858, 2020.

[37] G. Manogaran, N. Chilamkurti, and C.-H. Hsu, "Emerging trends, issues, and challenges in Internet of medical things and wireless networks," *Personal and Ubiquitous Computing*, vol. 22, pp. 879–882, 2018.

[38] P. Kaur, N. Sharma, A. Singh, and B. Gill, "CI-DPF: a cloud IoT based framework for diabetes prediction," in *2018 IEEE 9th Annual Information Technology, Electronics and Mobile Communication Conference (IEMCON)*, 2018, pp. 654–660.

[39] I. Dankwa-Mullan, M. Rivo, M. Sepulveda, Y. Park, J. Snowdon, and K. Rhee, "Transforming diabetes care through artificial intelligence: the future is here," *Population Health Management*, vol. 22, pp. 229–242, 2019.

[40] K. Zhang, S. Leng, X. Peng, L. Pan, S. Maharjan, and Y. Zhang, "Artificial intelligence inspired transmission scheduling in cognitive vehicular communications and networks," *IEEE Internet of Things Journal*, vol. 6, pp. 1987–1997, 2018.

[41] W. Qi, B. Landfeldt, Q. Song, L. Guo, and A. Jamalipour, "Traffic differentiated clustering routing in DSRC and C-V2X hybrid vehicular networks," *IEEE Transactions on Vehicular Technology*, vol. 69, pp. 7723–7734, 2020.

[42] W. Wang, J. Chen, J. Wang, J. Chen, and Z. Gong, "Geography-aware inductive matrix completion for personalized point-of-interest recommendation in smart cities," *IEEE Internet of Things Journal*, vol. 7, pp. 4361–4370, 2019.

[43] W. Wang, J. Chen, J. Wang, J. Chen, J. Liu, and Z. Gong, "Trust-enhanced collaborative filtering for personalized point of interests recommendation," *IEEE Transactions on Industrial Informatics*, vol. 16, pp. 6124–6132, 2019.

[44] O. Iliashenko, Z. Bikkulova, and A. Dubgorn, "Opportunities and challenges of artificial intelligence in healthcare," in *E3S Web of Conferences*, 2019, p. 02028.

[45] R. C. Mohs and N. H. Greig, "Drug discovery and development: role of basic biological research," *Alzheimer's & Dementia: Translational Research & Clinical Interventions*, vol. 3, pp. 651–657, 2017.

[46] H. S. Gns, G. Saraswathy, M. Murahari, and M. Krishnamurthy, "An update on drug repurposing: re-written saga of the drug's fate," *Biomedicine & Pharmacotherapy*, vol. 110, pp. 700–716, 2019.

[47] J. Mu, X. Liu, and X. Yi, "Simplified energy-balanced alternative-aware routing algorithm for wireless body area networks," *IEEE Access*, vol. 7, pp. 108295–108303, 2019.

[48] R. L. Kumar, Y. Wang, T. Poongodi, and A. L. Imoize, (eds.), *Internet of Things, Artificial Intelligence and Blockchain Technology*. 1st ed. Switzerland AG: Springer Nature, 2021.

[49] C. Meshram, R. W. Ibrahim, S. G. Meshram, A. L. Imoize, S. S. Jamal, and S. K. Barve, "An efficient remote user authentication with key agreement procedure based on convolution-Chebyshev chaotic maps using biometric," *Journal of Supercomputing*, vol. 78, pp. 1–23, 2022. Available: https://doi.org/10.1007/s11227-021-04280-8.

[50] C. Meshram, M. S. Obaidat, K.-F. Hsiao, A. L. Imoize, and A. Meshram, "An effective fair off-line electronic cash protocol using extended chaotic maps with anonymity revoking trustee," in *2021 International Conference On Communication, Computing, Cybersecurity, and Informatics*, pp. 1–5, 2021, doi:10.1109/ccci52664.2021.9583217.

[51] V. O. Etta, A. Sari, A. L. Imoize, P. K. Shukla, and M. Alhassan, "Assessment and test-case study of Wi-Fi security through the wardriving technique," *Mobile Information Systems*, vol. 2022, pp. 7936236, 2022, doi:10.1155/2022/7936236.

[52] C. Meshram, A. L. Imoize, S. S. Jamal, A. Aljaedi, and A. R. Alharbi, "SBOOSP for massive devices in 5G WSNs using conformable chaotic maps," *Computers, Materials \& Continua*, vol. 71, no. 3. pp. 4591–4608, 2022, doi:10.32604/cmc.2022.022642.

[53] S. Adetona, L. Ahemba, and A. L. Imoize, "Design and implementation of a low cost experimental testbed for wireless sensor networks," *Nigerian*

Journal of Technology, vol. 37, no. 1, pp. 226–232, 2018, doi:10.4314/njt.v37i1.30.

[54] A. L. Imoize, O. A. Ajibola, T. R. Oyedare, J. O. Ogbebor, and S. O. Ajose, "Development of an energy-efficient wireless sensor network model for perimeter surveillance," *International Journal of Electrical Engineering Applied Sciences*, vol. 4, no. 1, 2021.

[55] A. L. Imoize and O. O. Orodeji, "Development of a low-latency wireless telemetry system for monitoring patients heart rates," *International Journal of Electrical Engineering Applied Sciences*, vol. 3, no. 2, pp. 61–73, 2020.

[56] A. L. Imoize and A. E. Babajide, "Development of an infrared-based sensor for finger movement detection," *Journal of Biomedical Engineering and Medical Imaging*, vol. 6, no. 4, pp. 29–44, 2019, doi:10.14738/jbemi.64.7639.

[57] O. Alamu, A. Gbenga-Ilori, M. Adelabu, A. Imoize, and O. Ladipo, "Energy efficiency techniques in ultra-dense wireless heterogeneous networks: an overview and outlook," *Engineering Science and Technology, an International Journal*, vol. 23, pp. 1308–1326, 2020, doi:10.1016/j.jestch.2020.05.001.

[58] A. L. Imoize, O. A. Ajibola, T. R. Oyedare, J. O. Ogbebor, and S. O. Ajose, "Development of an energy-efficient wireless sensor network model for perimeter surveillance," vol. 4, no. 1, pp. 1–17, 2021.

[59] A. L. Imoize, , S. C. Mekiliuwa, and I. M. B. Omiogbemi, "Recent trends on the application of cost-effective economics principles to software engineering development," *International Journal of Information Security and Software Engineering*, vol. 6, no.1, pp. 39–49, 2020.

[60] A. L. Imoize, S. C. Mekiliuwa, I. M. B. Omiogbemi, and D. O. Omofonma, "Ethical issues and policies in software engineering," *International Journal of Information Security and Software Engineering*, vol. 6, no. 1, pp. 6–17, 2020.

[61] C. Meshram, A. L. Imoize, A. Aljaedi, A. R. Alharbi, S. S. Jamal, and S. K. Barve, "A provably secure IBE transformation model for PKC using conformable Chebyshev chaotic maps under human-centered IoT environments," *Sensors*, vol. 21, no. 21, pp. 7227, 2021, doi:10.3390/s21217227.

[62] C. Meshram, A. L. Imoize, A. Aljaedi, A. R. Alharbi, S. S. Jamal, and S. K. Barve, "An efficient electronic cash system based on certificateless group signcryption scheme using conformable chaotic maps," *Sensors*, vol. 21, no. 21, pp. 7039, 2021.

[63] C. Meshram, A. L. Imoize, S. S. Jamal, P. Tambare, A. R. Alharbi, and I. Hussain, "An efficient three-factor authenticated key agreement technique using FCM under HC-IoT architectures," *Computers, Materials \& Continua*, vol. 72, no. 1. pp. 1373–1389, 2022, doi:10.32604/cmc.2022.024996.

[64] C. Meshram, A. L. Imoize, A. Elhassouny, A. Aljaedi, A. R. Alharbi, and S. S. Jamal, "IBOOST: a lightweight provably secure identity-based online/offline signature technique based on FCM for massive devices in 5G wireless sensor networks," *IEEE Access*, vol. 9, pp. 131336–131347, 2021, doi:10.1109/ACCESS.2021.3114287.

[65] C. Meshram, R. W. Ibrahim, S. G. Meshram, S. S. Jamal, and A. L. Imoize, "An efficient authentication with key agreement procedure using Mittag–Leffler–Chebyshev summation chaotic map under the multi-server architecture," *Journal of Supercomputing*, vol. 78, no. 4, pp. 4938–4959, 2022. Available: https://doi.org/10.1007/s11227-021-04039-1.

[66] S. O. Adetona, L. Ahemba, and A. L. Imoize, "A Cluster Head Assisted Routing (CHAR) scheme for improved energy load balancing in wireless sensor networks," *Journal of Engineering Technology*, vol. 9, no. 2, pp. 77–95, 2018.

[67] J. O. Ogbebor, A. L. Imoize, and A. A.-A. Atayero, "Energy efficient design techniques in next-generation wireless communication networks: emerging trends and future directions," *Wireless Communications and Mobile Computing*, vol. 2020, no. 7235362, pp. 19, 2020, doi:10.1155/2020/7235362.

[68] L. K. Ramasamy, F. Khan, A. L. Imoize, J. O. Ogbebor, S. Kadry, and S. Rho, "Blockchain-based wireless sensor networks for malicious node detection: a survey," *IEEE Access*, vol. 9, pp. 128765–128785, 2021, doi:10.1109/ACCESS.2021.3111923.

Chapter 14

A deep neural network for the identification of lead molecules in antibiotics discovery

Michael Idowu Oladunjoye[1], *Olumide Olayinka Obe*[1] *and Olufunso Dayo Alowolodu*[2]

Abstract

In this study, we develop a deep neural network (DNN) model, multi-layer perceptron (MLP) to classify the molecules into "active" and "inactive" compounds using a ligand-based virtual screening approach for the lead compounds identification at the early stage of the antibiotic discovery. Lead identification as a major part of virtual screening in the drug discovery process is mostly performed by the quantitative structure–activity relationship (QSAR)-based method. The purpose of applying an artificial intelligence (AI) method is to reduce the time and subsequently the costs that are always associated with the process. The MLP model has several stacks of hidden layers and it used a back-propagation algorithm for the training. The dataset of experimentally known bioactivities of the drug-like compounds and their respective target was obtained from ChEMBL database. A biological target of an antibiotic, *dihydrofolate reductase* (DHFR), was searched from the database to get its inhibitors' chemical properties and the IC_{50} values on which the classification was based. One set of the dataset was preprocessed and split into two for the training and validating sets of 80% and 20% respectively. With this approach, the compounds were successfully classified into the desired categories and an accuracy of 0.74 was achieved.

Keywords: Multilayer perceptron; Binary classification; Lead identification; Drug-likeness properties; Artificial intelligence

14.1 Introduction

Lead identification, as one of the virtual screening techniques, is the starting process for drug discovery programs to evaluate a series of drug candidate molecules.

[1]Department of Computer Science, Faculty of Computing, Federal University of Technology, Nigeria
[2]Department of Cyber Security, Faculty of Computing, Federal University of Technology, Nigeria

The process can be carried out using a computational approach to reduce the usual long time of processing and the inadvertent cost of the screening as a result of the time. There are two categories of virtual screening methods namely, structure-based and ligand-based [1]. The ligand-based virtual screening involves searching for molecules of the same function, which starts with a set of a known compounds with various experimental methods.

Over the past years, there has been a vast amount of available chemical and biomedical data with their associated bioactivities as a result of modern experimental techniques and storage methods from various pharmaceutical processes. The availability of this data is a viable source to curate the needed data for the training of the proposed model. The bioactivities of drug-like properties for the data resources that provided the required information that is essential for the screening to train a multi-layer perceptron (MLP) that can classify each molecule into its biological activity. The model was built using the descriptors as the input values and their bioactivities as their output.

The process adopted a ligand-based method because of the nature of the chemical data representation and measurements. A biological target of an antibiotic, *dihydrofolate reductase* (DHFR), was searched from the ChEMBL database [2] to get its inhibitors' chemical properties and the IC_{50} values. The use of chemical libraries is accepted as a useful for lead compound discovery [3].

The rules for the selection of compound drugs based on various numbers of properties for their potency, solubility, distribution in the body, and toxicity have been established. Such rules were derived from the calculated compounds' characteristics with the defined criteria that essentially consider the effectiveness and safety of use [4]. Lipinski Rule of Five was established when the physicochemical properties of some drugs and drug candidates in clinical trials were examined to know if they share certain important characteristics [5]. The rule is to serve as a guideline in identifying compounds that are likely to have the same physicochemical properties as successful drugs. The lead identification is mainly the screening of molecular libraries to extract the compounds in the range of certain measurements in millimolar according to the Lipinski Rule of five. The rule specifies the molecular weight of less than three hundred, the logP of less than three, the hydrogen bond of less than three, and the rotatable bonds of less than three [6]. It focuses on small molecules and compounds that have reached phase II of the clinical trials assuming that the compounds with poor permeability would not have been included [7]. This assumption is duly followed when collecting the data for the research.

Drug-likeness of a compound based on statistics of its physicochemical properties from the database has been used to determine the drug-likeness of other compounds [8]. That is the known statistics can be used to select compounds from screening libraries such as virtual libraries. The combination of the various rules can also be part of the screening of compounds for new drugs [9]. The screening exercise is generally known to be laborious and time-consuming; therefore, it can be better handled with artificial intelligence (AI) techniques since there is an abundance of drug solutions data to filter the compounds of similar properties.

Also, we need to know that bacteria can cause different types of infections ranging from unnoticed to very severe cases [10] and these infections may demand new antibiotics. It is understood that the virtual screening process is cumbersome, time-consuming, and costly, but the capabilities of AI methods can be explored to solve these problems. The availability of data and the ability of some of the AI methods such as a deep neural network (DNN) that can make sense of such data have led to their application in the identification of lead molecules in the early stage of antibiotic discovery [11]. Besides, so much data is to be screened to get the lead compounds and such will need what will accelerate the speed of the search to as quickly as possible get the target [12]. Therefore, the use of AI techniques cannot be overemphasized considering their applicability as seen in various drug discoveries and designs by various researchers to shorten the drug pipeline and reduce cost. So, this study is expected to validate the capability of a trained DNN model, MLP for the classification of lead molecules in the discovery of antibiotics.

14.1.1 DNN and its architecture

A DNN is simply a class of machine learning (ML) algorithms that use several layers of hidden layers for the learning of data representations [13]. It is simply a type of artificial neural network (ANN) that has more than one hidden layer of connected artificial neurons that mimic the human central neural systems and is generally made up of an input layer, hidden layers, and an output layer. In the early stage of its development, there are three types, namely MLP, convolutional neural networks (CNN), and recurrent neural networks (RNN), and are still popularly used today for different applications [14]. The most basic type is the MLP model which can be used for most problems [1]. Structurally, it is made up of the input, some layers of nonlinear functions of a weighted sum (hidden layers), and the output neurons that are fully connected from the prior one as in Figure 14.1.

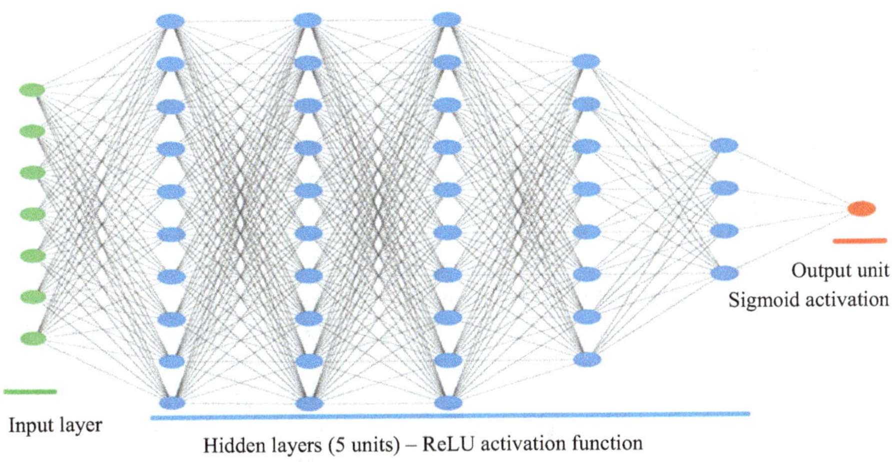

Figure 14.1 The MLP model structure generated during the program runtime with an input layer, five hidden layers, and a binary output structure

14.1.2 Lead identification techniques

A lead is not only a compound that exhibits potential activity against a target but also has other useful properties of drug-likeness. The compounds have a molecular weight of 300 or less in their basic form with other favorable properties. One of the methods that are used traditionally for the identification of lead compounds at the screening stage of drug development is the quantitative structure–activity relationship (QSAR) based model that compares the chemical structures of the compounds by using the database of prior selected active compounds [15]. Mandlik *et al.* [16] describe the model as how the physicochemical properties of a compound and its structure relates to its biological activities. That is the relationship between the structure of the molecule, the relevant physiochemical descriptors, and its biological activity. Ivanciuc [17] considered a drug-likeness prediction based on Lipinski's rule of five as a simple version of QSAR that can classify sets of chemical compounds based on their biological activities as either active or inactive toward a certain biological receptor.

14.2 Literature review

The use of AI methods in the identification of new lead compounds with the desired bioactivities at minimized processing time and cost has been part drug development pipeline [18,19]. ANNs and decision trees are various AI techniques that have been used successfully to improve the speed of the screening exercise with reduced cost [20–23]. Other numerous uses of neural networks include stock market predictions, image classification, and self-driving cars among others [24–26]. These breakthroughs have led to the use of such technologies to increase the efficiency of drug discovery. Their applications have become an important part of many drug discovery programs and this was notable in the drug design that created a novel anticancer drug [27].

Although the use of ANN for the extraction of information for the prediction of molecular properties of drugs from a large dataset was considered poor because of problems of over-fitting and model validation [28] while the application of deep learning to a similar activity was proved better. This assertion was proved when a deep CNN using a structure-based approach successfully predicted new active molecules with no previously known modulators for targets [29]. Stokes *et al.* [30] were inspired by the rapid emergence of antibiotic-resistant bacteria for the use of the CNN model for antibiotic discovery. The study used a deep learning model trained with data from ZINC15 to predict antibiotics using the structure-based method. The process included wet laboratory experimentation for molecular optimization.

The application of deep learning for drug development has been attributed to the availability of a huge amount of chemical and biological data, and the development of high-performance computers [31]. The opportunities in the big data era have also been described as one of the success factors for the application of deep learning in drug design [32]. The big data revolution has contributed to the exploration of the computational tools for the huge numbers of biochemical data in numerous chemical libraries as a result of developments in science and technology in drug discovery [33]. The successes of deep learning methods and their wide applications to almost all domains based on their computational capabilities cannot be overemphasized as motivating factors for their use in drug discovery.

The use of AI techniques with the ligand-based virtual screening option has been demonstrated to facilitate virtual screening where there is little 3D information about the receptor [34]. The applicability of the support vector machine (SVM), decision tree, and KNN algorithms in the process was discussed with positive assumptions.

For a classification problem on experimental medical data, the performance metric of ANN was 0.9713 [35]. Similarly, an ANN model was explored for behavioral science, and a satisfactory prediction success of 90% was recorded against a statistical method [36].

During the COVID-19 pandemics, logistic regression, SVM, and random forest algorithms with QSAR modeling were explored to identify molecules that can be used to block the multiplication of SARS-CoV-2 [37].

ML algorithms, support vector for regression to predict the activity of the compound and random forest for classification with accuracy of 78% for the regression and 92.2% accuracy for the classification indicating the effectiveness methods [38]. Deep learning methods with genomic, phenotypic, and omics databases were considered in study to reduce the risk of failure, cost of production, and time of development of the drug [39].

Recently, Jacobs *et al.* [40] demonstrated the capability of a deep learning model, character Wasserstein autoencoder (cWAE) that trained more than one billion compounds within a few minutes whereas a previous state of the art took a day for about one compound.

A new cascade transfer learning model type of deep learning was developed and trained with a dataset that has similar characteristics to a corona virus database to predict the efficacy of the lead compounds for the drug [41]. Pham *et al.* [42] used a neural network-based method to represent the relationship between the chemical structure and the gene, and also the relationships between genes to predict the difference in their profile. The lead compounds generated when applied the drug for the drug repurposing of COVID-19 were said to be consistent and incongruent with the clinical. Kumari *et al.* [43] proposed a CNN that was considered to be effective and efficient in its approach to virtual screening with an accuracy of 0.86.

A ligand-based approach with a DNN predicted the inhibitory effect of SARS-CoV proteases, potential toxicities, and bioactivity level successfully [44]. Similarly, a trained ANN with SMILES strings accurately classified compounds by varying the number of neurons of hidden layer [45]. Various deep learning-based models were considered for the generation a well-validated data that provides information regarding the biological targets and their interactions with ligands [46]. Bartzatt [47] applied ANN for the prediction of an important molecular property of an anti-EBOLA compound.

A deep learning model was used for the prediction of an antiviral drug that is commercially available for the coronavirus using a drug–target interaction [48]. The model built was pre-trained to identify available drugs that could act on the coronavirus protein. Also, laboratory data samples for the diagnosis of acute kidney disease were used for the training of the multilayer perceptron model and discovered that the process is more effective and faster than other methods in such diagnosis [49]. Table 14.1 summarizes some selected recent work discussed in this section.

Table 14.1 Recent works on the lead compound identification with DNNs

The related studies using DNN for drug discovery				
Year	**Authors**	**Method**	**Achievement**	**Limitations**
2018	Bartzatt [47]	ANN	Prediction of an important molecular property of an anti-EBOLA compound	Molecular weight prediction is not enough to validate the efficacy of drug because it is not universal for all drug-like compounds
2020	Hofmarcher *et al.* [44]	A ligand-based approach with a DNN	The model predicted the inhibitory effect of SARS-CoV proteases, potential toxicities and distance to known actives	The method was limited by the predictive capability of the model as indicated by the value of area under the curve (AUC) in the range of 0.69–0.78
2020	Sharifi *et al.* [49]	Multilayer perceptron-ANN	Effective and faster than other methods in the diagnosis of acute kidney disease	The dataset of 140 observations may be very few for the model training and evaluation
2020	Stokes *et al.* [30]	Deep learning on structure-based methods	An antibiotic of broad spectrum and more molecules with distinct structures was predicted using ZINC15 database	It requires a high-performance processor and a huge amount of money
2021	Hermansyah *et al.* [38]	ML algorithms, support vector for regression to predict the activity of the compound and random forest for classification	The prediction of 0.78 for the regression and 92.2% accuracy for the classification were achieved indicating the effectiveness of the ML methods for the identification of the target inhibitor for the disease	None was stated by the authors but the quality of data may be difficult to obtain

(Continues)

Table 14.1 (*Continued*)

The related studies using DNN for drug discovery				
Year	**Authors**	**Method**	**Achievement**	**Limitations**
2021	Yadav *et al.* [45]	ANNs with SMILES strings	The trained ANN model classified the compounds accurately when it was validated tested with similar compounds that were not previously trained with it	The performance is determined by the number of neurons of hidden layers but some limitations
2021	Zhuang *et al.* [41]	Cascade transfer learning	Prediction of the efficacy of the lead compounds for drug discovery, using COVID-19	The requirements of obtaining a dataset to train the deep learning model
2021	Pham *et al.* [42]	Neural network-based method, DeepCE	Generation of novel lead compounds consistent with clinical evidence	Developed for drug repurposing of COVID-19
2021	Kumari *et al.* [43]	CNN	Virtual screening with an accuracy of 0.86	Structure-based approach
2021	Jacobs *et al.* [40]	cWAE	More than one billion compounds were trained in a few minutes compared to the previous state of the art which took a day for a million compounds	The source of data may be difficult to access and the required systems for implementation
2022	Pan *et al.* [39]	Deep learning methods with genomic, phenotypic, and omics databases	The major contributions are to reduce the risk of failure, cost of production and time of development	The need for large and high quality dataset for the training of the model, and the lack of adequate understanding of the biological values of the prediction
2022	Nag *et al.* [46]	Various machine leaning/deep learning-based models were considered	The development of methods and techniques for the generation a well validated data that provides information regarding the biological targets and their interactions with ligands	The difficulties in proper labeling of the chemical descriptors as a result of to the lack of information regarding their bioactivities

14.3 Materials and methods

The method consists of the collection of datasets of drug-likeness compounds with their relevant features and molecular descriptors, and the development of the model.

The collection of the dataset is a very crucial part of the set of active and inactive compounds. ChEMBL database was the main source of the data. It is a curated chemical database of bioactive molecules that have the properties of drugs to obtain a drug-like compound with their physicochemical properties while their molecular structures were used to generate their descriptors. The DHFR is the potential protein target of which the compounds that bind it are curated. The sample of the bioactivity data downloaded is as in Table 14.2.

The MLP a feed-forward back-propagation network uses the physicochemical properties from which the molecular descriptors were calculated for input values and bioactivity represented as value "0" for inactive and "1" for active compounds as output.

Most drugs are to bind the protein (or enzyme) targets involved in disease. The activities of molecules on their biological targets are measured by varieties of measurements such as IC_{50}, EC_{50}, Ki, and the percent rate of inhibition. These measurements, the physicochemical properties of the molecules, and other relevant data were captured in the ChEMBL database. The sequence of the steps involved in the data collection and pre-processing is illustrated in Figure 14.2.

14.3.1 Dataset preparation and preprocessing

The process will have all the anomalies corrected and get the data standardized. The following methods were used in the pre-processing:

- *Feature clipping*: It seems the dataset contains extreme outliers and applying feature clipping restricts some feature values with a range, especially IC50_value column. For example, the amount needed to inhibit a target cannot be zero and this can be corrected with the min–max method.
- *Data normalization*: This is the adjustment of numbers in each of the columns uniformly in the range of 0–1. It normalizes a vector to have unity

Table 14.2 The raw data of compounds of similar structure and bioactivity

ChEMBL-id	SMILES	IC_{50}-values
CHEMBL25817	CCc1cc(Cc2cnc(N)nc2N)cc(CC)c1O	1.200000e−07
CHEMBL277176	CC(C)(C)c1cc(Cc2cnc(N)nc2N)cc(C(C)(C)C)c1O	6.100000e−08
CHEMBL279455	CCCc1cc(Cc2cnc(N)nc2N)cc(CC)c1OC	4.400000e−08
CHEMBL23338	C/C=C/c1cc(Cc2cnc(N)nc2N)cc(OC)c1OC	9.300000e−08
CHEMBL4635593	Nc1nc(N)c2nc3c4cccnc4c4ncccc4c3nc2n1	5.800000e+00
CHEMBL4635593	Nc1nc(N)c2nc(-c3ccccc3)c(-c3ccccc3)nc2n1	5.800000e+00
CHEMBL443	Cc1cc(NS(=O)(=O)c2ccc(N)cc2)no1	4.700000e+00

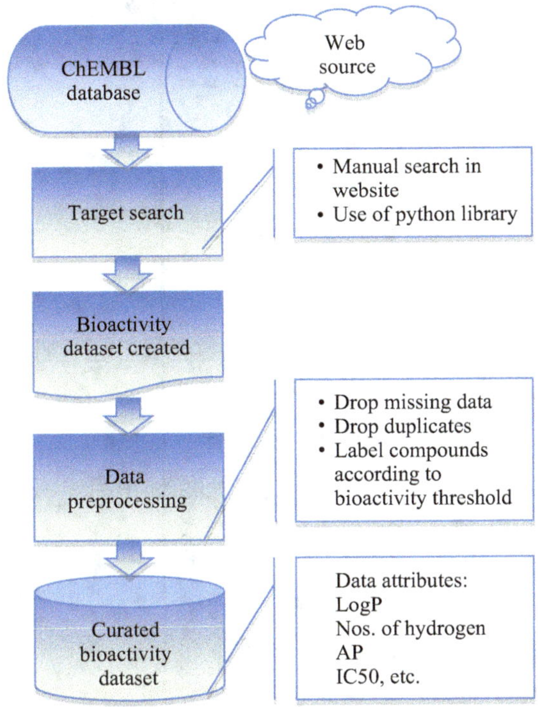

Figure 14.2 The sequence of the steps involved in data collection and pre-processing

variance and zero means. It was applied to convert some of the features into the ranges of 0 and 1 with the min–max method also. The equation for the min–max method:

$$x' = (x - x_{\min})/(x_{\max} - x_{\min}) \tag{14.1}$$

- *Data standardization*: This is achieved with the *Z*-score. The values are centered around the mean with a unit standard deviation. That the distributions have mean = 0 and std = 1. It is used when contemplating a few outliers. The features will be rescaled to reflect a normal distribution with μ=0 and σ=1 with the equation below:

$$z = \frac{x - \mu}{\sigma} \tag{14.2}$$

The preprocessed data with the relevant features for the model training dataset is as in Table 14.4.

The program development environment for the implementation of the work in Python 3.7 and Jupiter notebook is installed through anaconda distribution. The necessary library that allows the handling of the chemical structure such as rdkit

Table 14.3 ChEMBL bioactivity data for DHFR inhibitors

ChEMBL-id	SMILES	IC_{50} values	MW	LogP	NHD	NHA	RB	AP
CHEMBL25817	CCc1cc(Cc2cnc(N)nc2N)cc(CC)c1O	1.200000e−07	272.352	2.06220	3.0	5.0	4.0	0.600000
CHEMBL277176	CC(C)(C)c1cc(Cc2cnc(N)nc2N)cc(C(C)(C)...	6.100000e−08	328.460	3.53240	3.0	5.0	2.0	0.500000
CHEMBL279455	CCCc1cc(Cc2cnc(N)nc2N)cc(CC)c1OC	4.400000e−08	300.406	2.75530	2.0	5.0	6.0	0.545455
CHEMBL23338	C/C=C/c1cc(Cc2cnc(N)nc2N)cc(OC)c1OC	9.300000e−08	300.362	2.28210	2.0	6.0	5.0	0.545455
CHEMBL55911	Nc1nc(O)c(N=O)c(NCCCO)n1	3.500000e+01	213.197	-0.04340	4.0	8.0	5.0	0.400000
CHEMBL56063	Nc1nc(O)c(N=O)c(NCCCS(=O)(=O)...	11.800000e+01	337.361	1.43820	3.0	9.0	7.0	0.521739
CHEMBL443	Cc1cc(NS(=O)(=O)c2ccc(N)cc2)no1	4.700000e+00	253.283	1.36602	2.0	5.0	3.0	0.647059

Table 14.4 The preprocessed data with the relevant features for the model training dataset

MW	LogP	AP	RB	NHA	NHD	IC_{50} values	B-Classes
0.125781	0.153846	1.0	0.137931	0.114286	0.125000	0.000000	0.0
0.169531	0.307692	0.0	0.068966	0.114286	0.125000	0.000000	0.0
0.147656	0.230769	1.0	0.206897	0.114286	0.083333	0.000000	0.0
0.147656	0.153846	1.0	0.172414	0.142857	0.083333	0.000000	0.0
0.256250	0.000000	0.0	0.310345	0.257143	0.208333	0.000002	1.0
0.256250	0.000000	0.0	0.310345	0.257143	0.208333	0.000001	1.0
0.139844	0.076923	1.0	0.172414	0.171429	0.083333	0.000000	0.0

was installed also. The numerical and data handling libraries come with the installation of the anaconda distribution. These helped in the calculation of the molecular properties that are necessary for the development of the ML model.

14.3.2 Model development

This study uses MLP neural network model that contains more than one stack of hidden layers and the back-propagation algorithm to implement the binary classification. It learns a function $f(x) : \mathbb{R}^m \to \mathbb{R}^o$ with a training dataset: m is the number of dimensions for the input and o is the number of dimensions for the output. So, with a set of features $X = x_1, x_2, \cdots, x_m$ and a target y, the model will learn with a non-linear function for either classification or regression.

The MLP network units are a structurally organized sequence of layers where each unit is connected to all the units in the next layer. The stacking of the layers is because of expressive efficiency to express the same function more compactly and efficiently.

Consider the following notations:

L hidden layers, with $H_1, H_2, \ldots, H_L$ hidden units

$W^i_{k,j}$: Weight in layer i, from kth unit in the previous layer to jth unit

Define $W^i_j = \left(W^i_{o,j}, W^i_{1,j}, \cdots, W^i_{H_{i-1},j}\right)^T$

Define $X^D = (x_1, x_2, \cdots, x_D)$

For $i \geq 1$ *define* $h^i = \left(h^i_1,\ h^i_2,\ \cdots,\ h^i_{H_i}\right)$

$$a^i_j = \sum_{k=1}^{L_{i-1}} w^i_{k,j} X^{i-1}_k = W^{iT}_j X^{i-1} \tag{14.3}$$

Applying an activation function gives this:

$$h^i_j = \varnothing_i\left(a^i_j\right) \tag{14.4}$$

For an MLP of four layers, we have:

$$f(x) = w^{4^T} \varnothing_3 \left(w^{3^T} \varnothing_2 \left(w^{2^T} \varnothing_1 \left(w^{1^T} x \right) \right) \right) \tag{14.5}$$

The following equation represents the stacks of fully connected MLP network that receives connections from all the units in the previous layer:

$$h_i^1 = \varnothing_1 \left(\sum_{i=1}^{n} w_1 x_i + b_1 \right) = w_1^{i^T} x$$

$$h_i^2 = \varnothing_2 \left(\sum_{i=1}^{n} w_2 h_i + b_2 \right) = w_2^{i^T} x$$

$$\vdots$$

$$h_i^{n-1} = \varnothing_{n-1} \left(\sum_{i=1}^{n} w_{n-1} h_{n-2} + b_{n-1} \right) = w_{n-1}^{i^{\;T}} x \tag{14.6}$$

Eq. (14.7) represents the final output:

$$y = \varnothing_n \left(\sum_{i=1}^{n} w_n h_{n-1} + b_n \right) \tag{14.7}$$

Each notation refers to various parts of the network, that is ($h_i^1 ... h_i^{n-1}$ are the hidden layers). The activation functions $\varnothing$ to be applied can be Rectified linear Unit (ReLU) but for the output neurons, sigmoid activation is preferred being a binary classification.

14.3.3 Model evaluation

The evaluation of binary classification is often done with a contingency table called a confusion matrix. It visualizes and summarizes the model performance with two dimensions of rows and columns divided into four parts namely True Positive (TP), False Positive (FP), True Negative (TN), and False Negative (FN). The numbers of correctly predicted data are represented by TP and TN while the numbers of data are predicted incorrectly as FP and FN. The overall performance of the model is expressed as the accuracy and is calculated as in Eq. (14.10). Thus:

$$\text{Sensitivity} = \frac{\text{True Positives}}{\text{True Positives} + \text{False Negatives}} \tag{14.8}$$

$$\text{Specificity} = \frac{\text{True Negatives}}{\text{True Negatives} + \text{False Positives}} \tag{14.9}$$

$$\text{Accuracy} = \frac{\text{True Positives} + \text{True Negatives}}{\text{Total Numbers of Samples}} \tag{14.10}$$

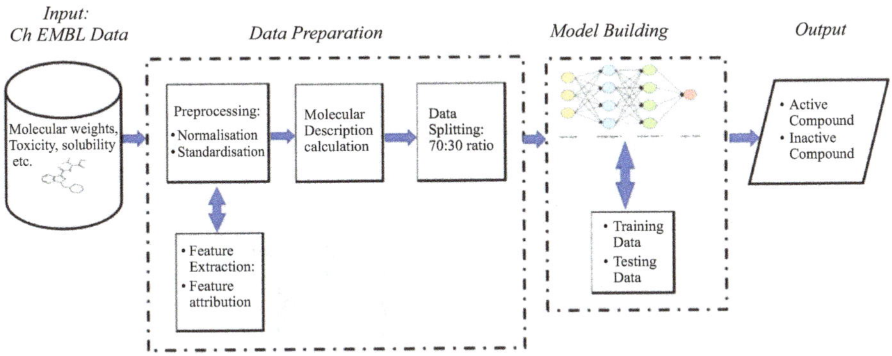

Figure 14.3 The overall MLP model pipeline for the identification of lead molecules

The system architecture for the MLP classification model for lead molecules identification is as in Figure 14.3.

14.4 Results and discussion

We developed an MLP neural network to classify the lead compounds into active and inactive compounds with the physiochemical properties of the molecules for the inhibition of the enzyme, *dihydrofolate reductase* (DHFR). The structure of each molecule obtained from the database is represented with its simplified molecular input line entry system (SMILES) string format that can be read into a Pandas dataframe (Table 14.2). During the program implementation, the SMILES strings were converted into *rdkit* object to enable us calculate the relevant set molecular descriptors as in Table 14.3.

The rdkit was used to compute the molecular weight, LogP and number of rotatable bonds (RB). The aromatic proportion (AP) was calculated with the ratio of the *number of aromatic atoms* to the *total number of heavy atoms* generated with the *rdkit.* Lipinski's rule of five was applied to select the relevant features of eight descriptors for the input into the model and the target feature, bioactivity class was created with the IC_{50} value (inhibition rate) of all "active" entries set to "1" and all the "inactive" set to "0."

Our proposed MLP architecture for the work is as in Figure 14.1 with eight input features followed by four hidden layers and the output layer that receives the active or inactive compound class. All units in hidden layers are activated with the ReLU. The activation function transforms the negative numbers to zero and leaves the others as they were after evaluating the values. The output layer is implemented with the sigmoid activation function being a binary classification method.

There is only one dataset of 2,591 compounds preprocessed and split into two separate datasets for the training and validating sets of 80% and 20%, respectively.

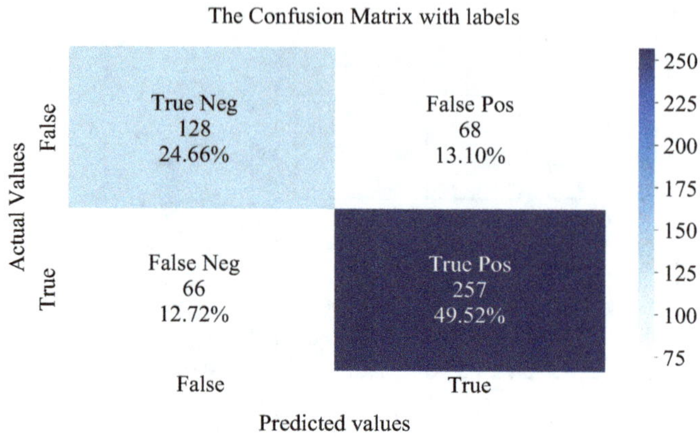

Figure 14.4 The confusion matrix plot

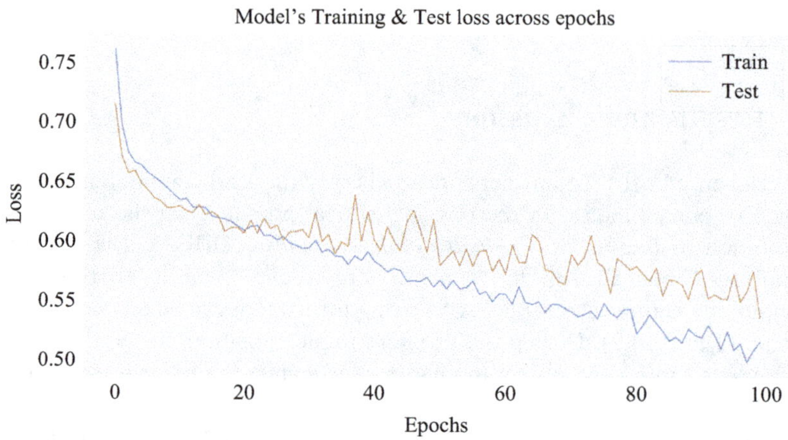

Figure 14.5 The loss of the training and validation datasets over the number of training epochs

The model validated 519 compounds being 20% of the whole dataset and the results of the evaluation representing the TP, FP, TN and FN are as in Figure 14.4.

The MLP classifier model achieved 0.80 sensitivity which means how well the model predicts the positives and 0.66 specificity meaning how well a model predicts negatives.

Figures 14.5 and 14.6 are plotted to review and visualize the performance of the model. The loss on the training dataset decreases across both datasets as represented in Figure 14.5. Figure 14.6 represents consistently increased accuracy with each epoch on both datasets. The small gap between the training and test datasets indicates that there is no over-fitting and the model performs very well on both data. The model

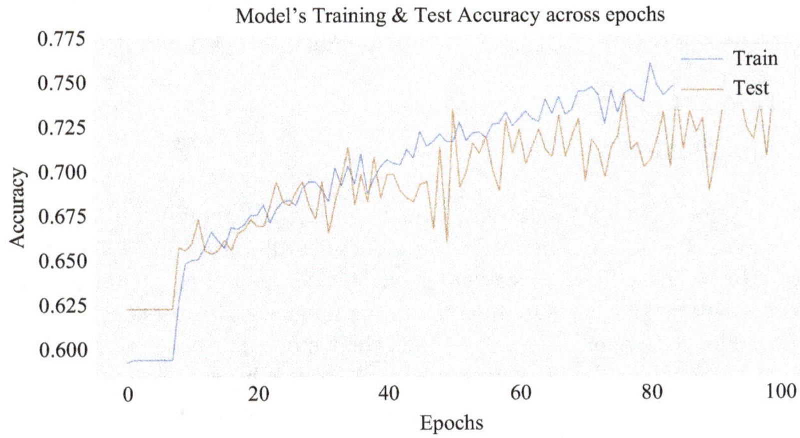

Figure 14.6 The accuracy of the training and validation datasets over the number of training epochs

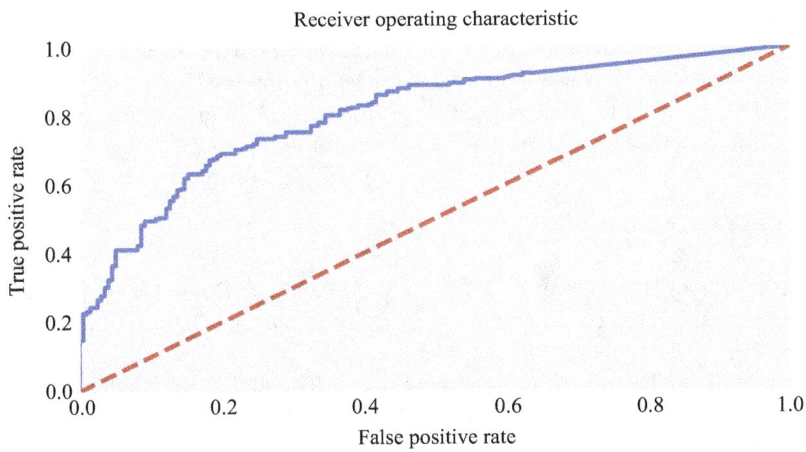

Figure 14.7 The AUC plot

achieved a training accuracy of 0.76 while the validation accuracy is 0.74, which means that the model has good fitting. The AUC score is 0.81 (Figure 14.7), indicating that the model is accurate because the value is close to one.

Lead identification is a product of virtual screening that is usually classified into two types: ligand-based and structure-based screening. The classifier model built, MLP on ligand-based approach where molecules that have same activities are identified, is to shorten the usual long time of the process. The validation accuracy 0.74 has proved that the application of the model for the identification of lead compound at the screening stage of the antibiotic development is desirable. The identified active compounds can be improved upon during their synthesis to avoid the problems that were associated with the prior similar molecules to enhance their potency.

14.5 Conclusion

The MLP neural network model was built on the QSAR approach to identify lead compound using a dataset obtained from a chemical library, ChEMBL. The study adopted a ligand-based virtual screening method because of the nature of the chemical information and measurements. The compounds were obtained with their SMILES codes that were used to calculate their molecular descriptors. The input for the model includes the physicochemical properties based on Lipinski's rule of five and their molecular descriptors desirables for the antibiotics drugs with their bioactivities values. This dataset was preprocessed by cleaning, transforming, and organizing the raw data to the acceptable form suitable for the model.

In conclusion, choosing the DNN model for the lead identification was a result of its known capability for handling the complex patterns and nonlinear relationships of the dataset. This technique has proven effective in the classification of the molecules based on the combined desirable properties of drug-likeness into either "active" or "inactive" to shorten the screening period with an accuracy of 0.74. Therefore, the use of AI-based approach for lead compounds identification in the early stage of antibiotic discovery using biochemical data will be of immense leverage for antibiotic discovery and development.

References

[1] Ramsundar B, Eastman P, Walters P, and Pande V. Deep learning for the life sciences: genomics. In *Microscopy, Drug Discovery*, 1st ed. USA: O'reilly; 2019.

[2] Gaulton A, Hersey A, Nowotka M, *et al.* The ChEMBL database in 2017. *Nucleic Acids Research*. 2016;45(D1):D945–D954. Available from: https://doi.org/10.1093/nar/gkw1074

[3] Mok NY and Brenk R. Mining the ChEMBL database: an efficient chemoinformatics workflow for assembling an ion channel-focused screening library. *Chemical Information and Modeling*. 2011;51:2449–2454. Available from: dx.doi.org/10.1021/ci200260t

[4] Segall M. Rules for drug discovery: can simple property criteria help you to find a drug? In *Drug Discovery World Spring*, 2014. Available from: https://www.ddw-online.com/media/32/rules-for-drug-discovery.pdf

[5] Blass BE. *Basic Principles of Drug Discovery and Development*. USA: Academic Press, Elsevier Inc., 2014.

[6] Lipinski CA. Lead- and drug-like compounds: the rule-of-five revolution. *Drug Discovery Today: Technologies*. 2004;1(4):337–341. Available from: https://doi.org/10.1016/j.ddtec.2004.11.007

[7] Ganesan A. The impact of natural products upon modern drug discovery. *Current Opinion in Chemical Biology*. 2008;12(30):306–317. Available from: https://doi.org/10.1016/j.cbpa.2008.03.016

[8] Kadam RU and Roy N. Recent trends in drug-likeness prediction: a comprehensive review of in silico methods. *Indian Journal of Pharmaceutical Sciences*. 2007;69(5):609–615. Available from: https://doi: 10.4103/0250-474X.38464

[9] Young DC. *Computational Drug Design: A Guide for Computational and Medicinal Chemists*. USA: John Wiley & Sons, Inc., 2009. Available from: https://chemistry-europe.onlinelibrary.wiley.com/doi/epdf/10.1002/cmdc.202100418

[10] Peterson JW. Bacterial pathogenesis. In S. Baron, editor. *Medical Microbiology*. Galveston (TX): University of Texas Medical Branch at Galveston, 1996. Available from: https://www.ncbi.nlm.nih.gov/books/NBK8526/

[11] A Scientific Roadmap for Antibiotic Discovery. 2016 [cited 2022 Jun 26]. Available from: https://www.pewtrusts.org/~/media/assets/2016/05/ a scientific roadmap for antibiotic discovery.pdf

[12] Oladunjoye MI and Obe OO. Deep neural networks in the discovery of novel antibiotics drug molecule: a review. *IJRIAS*. 2020;5(9):147–150. Available from: https://www.rsisinternational.org/journals/ijrias/DigitalLibrary/Vol.5&Issue9/147-150.pdf

[13] Chen H, Engkvist O, Wang Y, Olivecrona M, and Blaschke T. The rise of deep learning in drug discovery. *Drug Discovery Today*. 2018;23(6):1241–1250. Available from: https://doi: 10.1016/j.drudis.2018.01.039

[14] Deep Neural Network: The 3 Popular Types (MLP, CNN and RNN). 2021 [cited 2022 Jun 26]. Available from: https://viso.ai/deep-learning/deep-neural-network-three-popular-types/

[15] Riaz A, Rasul A, Sarfraz I, *et al.* Chemical Biology Toolsets for Drug Discovery and Target Identification. 2020 [cited 2022 Jun 26]. Available from: http://dx.doi.org/10.5772/intechopen.91732

[16] Mandlik V, Bejugam P, and Singh S. Application of artificial neural networks in modern drug discovery. In *Artificial Neural Network for Drug Design, Delivery and Disposition*, Chapter 6.

[17] Ivanciuc O. Drug design with artificial intelligence methods. In Meyers R, editor. *Encyclopedia of Complexity and Systems Science*. New York, NY: Springer; 2009. Available from: https://doi.org/10.1007/978-0-387-30440-3_133

[18] Lake F. Artificial intelligence in drug discovery: what is new, and what is next. *Future Drug Discovery*. 2019;1(2):FDD19. Available from: https://doi.org/10.4155/fdd-2019-0025

[19] Ortega SS, Lopez-Cara LC, and Salvador MK. In silico pharmacology for a multidisciplinary drug discovery process. *Drug Metabolism and Drug Interactions*. 2012;27(4):1–9. Available from: https://doi.org/10.1515/dmdi-2012-0021

[20] Jacob DD and Amaro RE. Machine-learning techniques applied to antibacterial drug discovery. *Chemical Biology and Drug Design*. 2015;85 (1):14–21. Available from: https://doi.org/10.1111/cbdd.12423

[21] Oduntan OE, Adeyanju IA, Falohun AS, and Obe OO. A comparative analysis of Euclidean distance and cosine similarity measure for automated essay-type grading. *Journal of Engineering and Applied Sciences*. 2018;13 (11):4198–4204.

[22] Jarrah YA, Asogbon YA, Samuel OW, *et al.* A comparative analysis on the impact of linear and non-linear filtering techniques on EMG signal quality of transhumeral amputees. In *IEEE International Workshop on Metrology for Industry 4.0 and IoT, MetroInd 4.0 and IoT 2021 – Proceedings*. 2021, pp. 604–608.

[23] Asogbon YA, Oluwarotimi WS, Nsugbe E, *et al.* A deep learning based model for decoding motion intent of traumatic brain injured patients' using HD-sEMG recordings. In *2021 IEEE International Workshop on Metrology for Industry 4.0 & IoT* (*MetroInd4.0&IoT*), 2021, pp. 609–614. Available from: https://doi:10.1109/MetroInd4.0IoT51437.2021.9488440

[24] Oken A. *An Introduction To and Applications of Neural Networks*, 2017 [cited 2022 Jul 5]. Available from: https://www.whitman.edu/Documents/Academics/Mathematics/2017/Oken.pdf

[25] Akinloye FO, Obe OO, and Boyinbode O. Development of an affective-based e-healthcare system for autistic children. *Scientific African*. 2020;9 (1):1–9. Available from: https://doi.org/10.1016/j.sciaf.2020.e00514

[26] Fagbola TM, Adejanju IA, Oloyede A, *et al.* Development of mobile-interfaced machine learning-based predictive models for improving students' performance in programming courses. *Computer Science*. 2019;9 (5):105–115.

[27] Prada-Graciaa D, Huerta-Yépezb S, and Moreno-Vargasb LM. Application of computational methods for anticancer drug discovery, design, and optimization. *Boletin Medico del Hospital Infantil de Mexico*. 2016;73(6):411–423. Available from: http://dx.doi.org/10.1016/j.bmhimx.2016.10.006

[28] Schneider G. Neural networks are useful tools for drug design. *Neural Networks*. 1999;13(1):15–16. Available from: https://doi.org/10.1016/s0893-6080(99)00094-5

[29] Wallach I, Dzamba M, and Heifets A. AtomNet: A Deep Convolutional Neural Network for Bioactivity Prediction in Structure-based Drug Discovery, 2015. Available from: https://doi.org/10.48550/arXiv.1510.02855

[30] Stokes JM, Yang K, Swanson K, *et al.* A deep learning approach to antibiotic discovery. *Cell*. 2020;180(4):688–702.e13. Available from: https://doi.org/10.1016/j.cell.2020.01.021

[31] Lipinski CF, Maltarollo VG, Oliveira PR, da Silva ABF, and Honorio KM. Advances and perspectives in applying deep learning for drug design and discovery. *Frontiers in Robotics and AI*. 2019;6:108. Available from: https://doi.org/10.3389/frobt.2019.00108

[32] Jing Y, Bian Y, Hu Z, Wang L, and Xie XS. Deep learning for drug design: an artificial intelligence paradigm for drug discovery in the big data era. *The AAPS Journal*. 2018;20:58. Available from: https://doi: 10.1208/s12248-018-0210-0

[33] Nantasenamat C and Prachayasittikul V. Maximizing computational tools for successful drug discovery. *Expert Opinion on Drug Discovery*. 2015;10 (4):321–329.

[34] Lavecchia A. Machine-learning approaches in drug discovery: methods and applications. *Drug Discovery Today*. 2015;20(3):318–331. Available from: https://doi.org/10.1016/j.drudis.2014.10.012

[35] Cömert Z and Kocamaz AF. A study of artificial neural network training algorithms for classification of cardiotocography signals. *Journal of Science and Technology*. 2017;7(2):93–103. Available from: https://www.dergipark.ulakbim.gov.tr/beuscitech/

[36] Reby D, Lek S, Dimopoulos J, Joachim J, Lauga J, and Aulagnier S. Artificial neural networks as a classification method in the behavioural sciences. *Behavioural Processes*. 1996;40:35–43.

[37] Jha N, Prashar D, Rashid M, *et al.* Deep learning approach for discovery of in silico drugs for combating COVID-19. *Journal of Healthcare Engineering*. 2021;2021:6668985. Available from: https://doi: 10.1155/2021/6668985

[38] Hermansyah O, Bustamam A, and Yanuar A. Virtual screening of dipeptidyl peptidase-4 inhibitors using quantitative structure–activity relationship-based artificial intelligence and molecular docking of hit compounds. *Computational Biology and Chemistry*. 2021;95:107597. Available from: https://doi.org/10.1016/j.compbiolchem.2021.107597

[39] Pan X, Lin X, Cao D, *et al.* Deep learning for drug repurposing: methods, databases, and applications. *WIREs Computational Molecular Science*. 2022;12:4. Available from: https://doi.org/10.1002/wcms.1597

[40] Jacobs SA, Moon T, McLoughlin K, *et al.* Enabling rapid COVID-19 small molecule drug design through scalable deep learning of generative models. *The International Journal of High Performance Computing Applications*. 2021;35(5):469–482. Available from: https://doi: 10.1177/10943420211010930

[41] Zhuang D and Ibrahim AK. Deep learning for drug discovery: a study of identifying high efficacy drug compounds using a cascade transfer learning approach. *Applied Sciences*. 2021;11:7772. Available from: https://doi.org/10.3390/app11177772

[42] Pham T, Qiu Y, Zeng J, Xie L, and Zhang P. A deep learning framework for high-throughput mechanism-driven phenotype compound screening and its application to COVID-19 drug repurposing. *Nature Machine Intelligence*. 2021;3:247–257. Available from: https://doi.org/10.1038/s42256-020-00285-9

[43] Kumari M and Subbarao N. Deep learning model for virtual screening of novel 3C-like protease enzyme inhibitors against SARS coronavirus

diseases. *Computers in Biology and Medicine*. 2021;132:104317. Available from: https://doi.org/10.1016/j.compbiomed.2021.104317

[44] Hofmarcher M, Mayr A, Rumetshofer E, *et al.* A large-scale ligand-based virtual screening for SARS-CoV-2 inhibitors using deep neural networks. 2020. Available from: https://dx.doi.org/10.2139/ssrn.3561442

[45] Yadav M and Jujjavarapu SE. Neural network methodology for the identification and classification of lipopeptides based on SMILES annotation. *Computers*. 2021;10:74. Available from: https://doi.org/10.3390/computers10060074

[46] Nag S, Baidya ATK, Mandal A, *et al.* Deep learning tools for advancing drug discovery and development. *3 Biotech*. 2022;12:110. Available from: https://doi.org/10.1007/s13205-022-03165-8

[47] Bartzatt R. Prediction of novel anti-ebola virus compounds utilizing artificial neural network (ANN). *World Journal of Pharmaceutical Research*. 2018;7 (13):16–34.

[48] Beck BR, Shin B, Choi Y, Park S, and Deargen KK. Predicting commercially available antiviral drugs that may act on the novel coronavirus (SARS-CoV-2) through a drug-target interaction deep learning model. *Computational and Structural Biotechnology Journal*. 2020;18:784–790. Available from: https://doi: 10.1016/j.csbj.2020.03.025

[49] Sharifi A and Alizadeh K. A novel classification method based on multilayer perceptron-artificial neural network technique for diagnosis of chronic kidney disease. *The Annals of Military and Health Science Research*. 2020; 18(1):e101585. Available from: https://dx.doi.org/10.5812/amh.101585

Chapter 15

Statistical test with differential privacy for medical decision support systems

Yuichi Sei[1], *Akihiko Ohsuga*[1] *and Agbotiname Lucky Imoize*[2,3]

Abstract

Several statistical testing methods have been employed to offer accessible analysis regarding medical data for medical decision support systems (MDSSs), with the Chi-squared test among the most widely used option. Critics have noted, however, that presenting such data risks exposing individual attribute values. This chapter will demonstrate how the findings of statistical analysis can inadvertently reveal individual attribute values. It will then show how advanced differential privacy systems, such as those utilized by companies including Google and Apple, can be used to protect individual attribute values while conducting extremely precise Chi-squared tests.

Keywords: Explainable artificial intelligence; Medical data; Differential privacy; Chi-squared test

15.1 Introduction

In the fields of medicine and pathology, it is essential that data analysis methods provide insights that are accessible to human researchers [1–3]. While statistical methods have occasionally been misapplied, their properties are generally easy to comprehend, making them an essential part of medical data analysis [4].

Genes are analyzed by examining a range of gene groups [5,6]. Single nucleotide polymorphisms (SNPs) are genomic variations that occur at a single-base position in DNA. This can involve at least three groups in certain circumstances. To work out if SNPs can be deemed different to a significant degree, the

[1]Department of Informatics, Graduate School of Informatics and Engineering, The University of Electro-Communications, Japan
[2]Department of Electrical and Electronics Engineering, Faculty of Engineering, University of Lagos, Nigeria
[3]Department of Electrical Engineering and Information Technology, Institute of Digital Communication, Ruhr University, Germany

Chi-squared (χ^2) test is one of the leading forms of statistical analysis employed. Researchers and government agencies consistently collaborate in sharing their results to aid future research.

Risk of disease and factors affecting genetic disorders are among the pieces of sensitive, personal information that can be contained in a genome. There is only a 0.1% difference in each person's genome in terms of their individual attributes, with 99.9% of all human genomes being identical. The difference between people at a single location in a DNA sequence is represented by an SNP. Genome-wide association studies (GWASs) are a technique for determining the statistical link between SNPs and illnesses. This uses the Chi-squared test to identify SNPs corresponding to specific diseases. For instance, Homer *et al.* discovered that if an attacker is aware of the victim's SNPs and the total allele frequency of a particular illness group, he may be able to determine whether the victim belongs to that group [7].

Due to the growing availability of affordable genotyping services, it is realistic to assume that an attacker would know the victim's SNPs, with only a tiny blood sample needed [8,9]. It is also possible to determine the allele frequency of group SNP values using fundamental statistical information including *p*-values and Chi-squared values, as proposed by Wang *et al.* [10]. Therefore, when publishing SNP datasets containing Chi-squared values, anonymization techniques should consistently be implemented [10–12].

Data sharing forms an essential part of genomic research [13]. Techniques to ensure that privacy is protected should therefore be implemented on GWAS results to avoid information leaks. While existing research utilizes noise to a significant degree to ensure privacy protection as it relates to GWAS results, we aim to retain the same level of protection without the same level of reliance. Simply put, we can use privacy-preserving Chi-squared testing that matches the privacy protection of prior research while making GWAS results more functional.

While other approaches have also been applied to recent GWAS results, such as mixed linear model-based methods, Chi-squared testing remains the most important and widely used technique [14–18]. This also applies to recent studies relating to COVID-19, highlighting the importance of further research into Chi-squared testing [19–24].

The Kruskal–Wallis test and the Wilcoxon test have also been used in relation to GWAS [25,26]. For these techniques, differential privacy approaches have been suggested by Couch *et al.* [27], highlighting the importance of addressing such methods in future research.

In privacy circles, ε-differential privacy [28], which has been studied in great depth [29–32], is the dominant privacy metric. Methods by which to share Chi-squared values while remaining within the limits of ε-differential privacy have been proposed by a number of researchers, including [11,33,34]. The major limitation of these approaches, however, is that they are only capable of being used in relation to 2 2 or 2 3 contingency tables, while SNPs based on an I J contingency table are required. Prior research has assessed greater degrees of freedom within contingency tables [12,35], but such approaches have proven inaccurate, especially in cases containing sample sizes limited to several hundred samples of fewer [36–38].

Due to measures like the General Data Protection Regulation (GDPR), gaining access to sensitive information remains extremely challenging and patient biomedical data cannot be distributed unless consent has been provided [39]. In terms of rare diseases, gaining access to sensitive data is even more challenging [40,41], as is accumulating large sample numbers for new diseases, like COVID-19, that require rapid analysis. Individuals may offer their sensitive information in the absence of any form of privatization, but this is not typically the case [42,43]. Several researchers have considered the use of contingency tables larger than 2×3 [44–46], highlighting the importance of the issue regarding private Chi-squared tests for large contingency tables with small samples.

In this chapter, we introduce the proposed method RandChiDist for obtaining the differential private Chi-squared values of $I \cdot J$ contingency table with a low sample size and evaluate the method through experiments on a real data set. The RandChiDist method is based on the contingency table and combines minimized Laplace noise and true Chi-squared values and controls the ratio of Type I errors such as false positives. Both synthetic and real datasets are used in the evaluation process, including two genomic datasets. As well as strictly reducing Type I error ratios, RandChiDist also cuts down on Type II errors, such as false negatives, to a greater degree than existing approaches for controlling Type I error ratios. While a number of alternative approaches are superior to RandChiDist in terms of cutting down the frequency of Type II errors, these methods are incapable of controlling Type I error ratios.

RandChiDist determines the global sensitivity of the Chi-squared value with respect to the Chi-squared test and can add noise based on that value. Based on the Laplace mechanism theorem [28], the resulting required noise is minimized.

The purpose behind this chapter can be summed up thus.* While Chi-squared testing is used across a range of data analysis processes, including SNP identification in relation to specific diseases, the publication of Chi-squared values can compromise privacy. As a result of the challenges regarding the collection of large samples for rare and new diseases, we propose a chi-squared testing algorithm with differential privacy for sample numbers of less than 1,000.

We present the subsequent sections of this chapter: Section 15.2 outlines the chi-squared hypothesis test and differential privacy, as well as describing related work. Section 15.3 introduces our system. Section 15.4 details the simulation results. Section 15.5 examines the experimental results and discusses the need to study large contingency tables constructed from small samples. Section 15.6 sums up the overall findings.

15.2 Related work

15.2.1 Chi-squared hypothesis test

A contingency table consisting of I rows and J columns was considered. $[i,j]$ indicates the ith row and jth column's cell. $\mathcal{V}_{i,j}$ denotes cell $[i,j]$'s value. $\mathcal{W}_{i,j}$ indicates the expected value of cell $[i,j]$.

*An earlier version of the chapter appeared in Ref. 47.

Table 15.1 Case–control analysis

	SNP1 and SNP2 allele type combinations				
	(M,M)	(M,m)	(m,M)	(m,m)	Total
Case	$\mathcal{V}_{1,1}$	$\mathcal{V}_{1,2}$	$\mathcal{V}_{1,3}$	$\mathcal{V}_{1,4}$	m_1
Control	$\mathcal{V}_{2,1}$	$\mathcal{V}_{2,2}$	$\mathcal{V}_{2,3}$	$\mathcal{V}_{2,4}$	m_2
Total	s_1	s_2	s_3	s_4	n

M: major; m: minor.

Let $m_i = \sum_j \mathcal{V}_{i,j}$, $s_j = \sum_i \mathcal{V}_{i,j}$, and $n = \sum_i m_i = \sum_j s_j$. An example of a contingency table is shown in Table 15.1.

The following equation represents the Chi-squared value:

$$\chi^2 = \sum_{i=1}^{I} \sum_{j=1}^{J} U_{i,j} \tag{15.1}$$

$$\text{where } U_{i,j} = \frac{(\mathcal{W}_{i,j} - \mathcal{V}_{i,j})^2}{\mathcal{W}_{i,j}}, \ \mathcal{W}_{i,j} = s_j \cdot \frac{m_i}{n}. \tag{15.2}$$

The level of significance α (the likelihood of a Type I error) and the null hypothesis H_0 were both calculated in advance. Chi-squared was determined according to (15.2), while the *Chi-squared distribution table* was used to determine whether to reject H_0. The probability density function of the Chi-squared distribution is represented by χ_v^2, with v degrees of freedom. The percentage point $P(\chi_v^2 > x) = \alpha$ for multiple combinations of v and α is presented in the Chi-squared distribution table.

15.2.2 Privacy model

The widely accepted standard regarding privacy metrics over recent history has been ε-differential privacy [28, 39–50].

The privacy level is represented by ε, with higher values indicating lower privacy levels. Neighboring databases are deemed to represent two databases, diverging by one record maximum. The definition of ε-differential privacy is as follows:

Definition 1 (ε-differential privacy): *T and T' represent neighboring databases. The randomized mechanism $\mathcal{A}$ meets the ε-differential privacy if the following holds*:

$$P(\mathcal{A}(T) \in Y) \leq e^{\varepsilon} P(\mathcal{A}(T') \in Y), \tag{15.3}$$

for any $Y \subseteq Range(\mathcal{A})$ and any neighboring databases.

Theorem 1 is satisfied by the Laplace mechanism [28], which contributes noise generated using a Laplace distribution. By defining the concept of global sensitivity, this mechanism can be delineated.

Definition 2 (Global sensitivity): *Let f be a function $f : \mathcal{T} \rightarrow \mathbb{R}^d$, where $\mathcal{T}$ is a domain of databases. If f satisfies T and T' for any neighboring databases*

$$\Delta f = \max_{T,T'} ||f(T) - f(T')||_1, \tag{15.4}$$

Δf is the global sensitivity of f.

Theorem 1 (Laplace mechanism [28]): *Let $\mathcal{A}$ be a randomized mechanism that outputs $f(T) + Lap(\Delta f/\varepsilon)$, where $Lap(v)$ returns independent Laplace random variables with the scale parameter v. $\mathcal{A}$ realizes ε-differential privacy.*

Differential privacy can be used not only for MDSS but also for text [51], voice [52], images [53], etc. In addition, this chapter assumes a scenario in which there is an entity holding raw data, and that entity uses differential privacy to provide statistical information about that raw data to a third party. On the other hand, a technology called local differential privacy has been proposed, and there exists a scenario in which each individual processes raw data using local differential privacy technology and aggregates the processed data.

In addition, differential privacy and local differential privacy tend to protect privacy too strongly and reduce data usefulness. Therefore, several concepts that relax differential privacy have been proposed. However, in the case of handling extremely important personal data such as genomic information, strong protection is required, so in this chapter, normal differential privacy is used.

15.2.3 ε-Differentially private Chi-squared test

A contingency table like Table 15.2 applies to chi-squared testing. It can be presented as Table 15.3, with Tables 15.2 and 15.3 being equivalent. Databases like the ones shown in Table 15.3 for privacy-preserving Chi-squared testing were utilized. With the exception of one record, Tables 15.3 and 15.4 contain the same data, making them neighboring databases.

Table 15.2 A contingency table

	Condition C_1	**Condition C_2**
Group G_1	19	31
Group G_2	22	28

Table 15.3 A raw database

ID	Group & Condition
0	G_1 & C_2
1	G_2 & C_1
2	G_2 & C_2
...	...
99	G_1 & C_2

Table 15.4 Example of a neighboring database of Table 15.3

ID	Group & Condition
0	G_1 & C_2
1	G_2 & C_1
[rgb]0.9, 0.9, 0.9 2'	G_1 & C_2
...	...
99	G_1 & C_2

As shown by Yu *et al.* [11], the global sensitivity of the Chi-squared value is

$$\Delta_Y = \frac{n^2}{m_1 m_2}\left(1 - \frac{1}{\max\{m_1, m_2\} + 1}\right), \tag{15.5}$$

when $2 \cdot 3$ contingency is considered and m_1 and m_2 are published.

As shown by [33,34], if $m_1 = m_2$, the Chi-squared value's global sensitivity can be calculated as

$$\Delta_F = \frac{4n}{n+2}. \tag{15.6}$$

Δ_F and Δ_Y are of only use in $2 \cdot 2$ and $2 \cdot 3$ contingency tables.

A unit circle mechanism capable of achieving a high level of accuracy was put forward by Kakizaki *et al.* [54,55], but this approach only applies to $2 \cdot 2$ contingency tables. In addition, the differential private results of the Chi-squared test based on a given significance level α were published, but the differential private Chi-squared values could not be presented. Therefore, data holders who wish to make Chi-squared test results private at multiple α values (e.g., $\alpha = 0.05, 0.001$) would have to independently implement the privatization measure multiple times. In accordance with the composition theorem [56], the privacy level that results from a privacy mechanism outputting K times based on ε-differential privacy is therefore $K\varepsilon$. From a data analysis perspective, the publication of p-values is crucial [57].

The studies detailed above were all conducted on the basis that m_i ($i = 1,\ldots,I$) does not constitute sensitive information and that privatization schemes are not required for the sharing of such values.

A number of approaches regarding arbitrary contingency tables were put forward by Gaboardi *et al.* [12]. A relatively simple method is to avoid adding Laplace noise to the Chi-squared value and instead add it to the contingency table's each cell, making the global sensitivity of 2. This method will henceforth be referred to in this paper as RandCell. It is also known as SNPpval, after its proposition by Jonson and Shmatikov [58]. RandCell yields many false positives because of the size of the Chi-squared value. To counter this, a number of approaches were put forward by Gaboardi *et al.* These include PrivIndep, MCIndep with Gaussian mechanism, and MCIndep with Laplace mechanism. The latter of these demonstrated the best performance and will be outlined in detail in this chapter. For brevity, this method will henceforth be referred to as MCIndep.

MCIndep compares Chi-squared values by randomly generating many contingency tables using the m_i and s_j of that with added Laplace noise. If the Chi-squared value of the contingency table to which RandCell added Laplace noise is greater than the upper $100 \cdot \alpha\%$ of the Chi-squared values of the generated contingency table, we can say that the original contingency table rejected H_0. Relaxing the ε-differential privacy as the privacy metric has also been proposed as an alternate method [59]. This chapter focuses on ε-differential privacy, while the application of our method to (ε, δ)-differential privacy is an area of potential future research.

A number of theorems regarding differentially private Chi-squared testing were put forward by Sei *et al.* [60]. However, this study offered no in-depth proofs regarding their equations, nor any Chi-squared test experiments.

A range of Chi-squared test algorithms for differentially private Chi-squared testing of independence based on local differential privacy were recently proposed by Gaboardi *et al.* [35] (LocalNoiseIND, LocalExpIND, and LocalBitFlipIND). The first of these three was developed by Kifer and Rogers [61]. The best performing of the three across most parameter settings was LocalExpIND. All three approaches can be applied to arbitrary contingency tables, assess local privacy models, and assume the lack of a trusted entity. This chapter assumes that all the raw data belongs to a trusted entity.

The sample complexity bounds regarding ε-differentially private tests for distinguishing between two distributions were determined by Canonne *et al.* [62]. Differentially private change-point detection was also applied in this study. Unlike our approach, which is designed for nonparametric settings, this study was designed for a parametric setting in which the two distributions are perfectly known.

An algorithm designed to privately test the closeness of two distributions was developed by Csail *et al* .[63]. This method can also be used to test two random variables' independence, though no privacy-preserving Chi-squared testing approach is outlined.

ε can alter differentially private hypothesis testing accuracy. This was demonstrated by Liu *et al.* [64], whose approach for calculating an appropriate value for ε is also valuable for calculating our method's ε value. This chapter is not concerned with his goal, however.

A differentially private hypothesis testing approach to Kruskal–Wallis and other important tests was developed by Couch *et al.* [27]. However, this approach is designed for ordinal or interval scale data, not nominal scale data.

Prior approaches regarding arbitrary $I \cdot J$ contingency tables have performed relatively poorly in terms of accuracy. This is especially true in cases involving small population samples. Section 15.4 presents an overview of these approaches.

15.2.4 Adversarial model

In an adversarial model, the server aims to disclose Chi-squared test results to potential hackers. While the attacker follows server protocols and is therefore regarded as a semi-honest entity, they may nonetheless attempt to gain access to Chi-squared test results and identify personal information. Moreover, the attacker may have some knowledge about a person's SNPs and other attribute values. We assume that the attacker is attempting to extract sensitive information about that individual from the published Chi-squared value.

15.2.5 Other privacy models

In this chapter, we focus on differential privacy because it is considered the de facto standard privacy model in this field. However, many other privacy models exist. This section will briefly review them.

Today, many researches employ the privacy model known as k-anonymity [65], which was first proposed for use when releasing medical data. The model assumes that there is an attribute set that the attacker may know (which are called quasi-identifiers) and an attribute set that the attacker does not know, respectively. These unknown attribute values are called sensitive attributes, and the original database is anonymized so that the attacker cannot know the sensitive attribute values of a person. The k-anonymity algorithm can be used to guarantee that an attacker with background knowledge about quasi-identifiers has less than $1/k$ chance of correctly guessing a person's record when looking at the database after anonymization.

k-Anonymity requires there to be k or more records with multiple identical quasi-identifiers. This is because an attacker with knowledge about the quasi-identifier values about a given user is assumed and is intended to protect privacy against such an attacker. However, as the number of quasi-identifiers increases, k-anonymity cannot be satisfied without abstracting each quasi-identifier value to a considerable extent. Therefore, there is the possibility that the data may become entirely useless following anonymization. Therefore, with it being desirable to satisfy k-anonymity in regard to a combination of an arbitrary number of quasi-identifiers m, a privacy model moderating k-anonymity was proposed. This is called k^m-anonymity [66].

Although k-anonymity and k^m-anonymity can prevent identity disclosure, the sensitive attributes of candidate records do not necessarily differ. It is sometimes possible for all candidate records to possess the same sensitive attributes. Therefore, among groups of records where all quasi-identifiers are identical, l-diversity was proposed as a privacy model, showing that more than l types of sensitive attributes are present [67]. More accurately, this is called Distinct l-diversity, but other models such as Entropy l-diversity and (c, l)-diversity also exist.

The privacy models described above are models for discussing whether an attacker, aware that some User A is present in the anonymized database, can learn of User A's sensitive attribute values upon viewing the anonymized database. However, whether User A is present in a database or not may also be private information. For example, when tabulating a database of users suffering from rare diseases, being able to determine an individual's presence in that database can be considered a leak of private information. Moreover, where there is a tabulated database of users with test scores above the average, proving one is not present in that database may also be problematic. Therefore, a privacy model, [68] δ-presence, $(\delta = (\delta_{\min}, \delta_{\max}))$ has been proposed as a model of whether an attacker who does not know whether User A is present in a database when looking at the anonymized database is able to assume the presence of User A with a degree of confidence equal to or greater than $\delta_{\min}$ and less than $\delta_{\max}$. Therefore, it is thought that there are a certain number of users in the world with exactly the same quasi-identifier values.

(ρ_1, ρ_2)-privacy has been proposed [69,70] as a privacy model measuring the extent to which an attacker's prior and posterior knowledge about a user's sensitive attributes changes when that given user's anonymized information is disclosed. In regard to what sensitive attributes an attacker knows about User A, prior knowledge is known only with a confidence level ρ_1; however, by viewing the anonymized data where User A's sensitive attributes can be known with a confidence level greater than ρ_2, (ρ_1, ρ_2)-privacy is considered to be infringed upon.

15.3 Proposed algorithm

15.3.1 Outline

The global sensitivity of the $I \cdot J$ contingency table's Chi-squared value is required for the RandChiDist method. This is because the RandChiDist approach involves adding Laplace noise to the Chi-squared value generated from a contingency table. Section 15.3.2 provides an outline of how to calculate global sensitivity.

Determining whether or not to reject H_0 is done by using the Chi-squared distribution table. We need a modified Chi-squared distribution table, however,

Table 15.5 Notations

n	Number of records in database
I	Number of rows of a contingency table
J	Number of columns of a contingency table
m_i	Total observed value of ith row
s_j	Total observed value of jth column
$[i,j]$	Cell of ith row and jth column
$\mathcal{V}_{i,j}(T)$	Observed value of cell $[i,j]$ in a contingency table T
$\mathcal{W}_{i,j}(T)$	Expected value of cell $[i,j]$ in a contingency table T

because noise is added to the Chi-squared value in the RandChiDist method. Section 15.3.3 outlines how this can be calculated, with the modified table used to accept or reject H_0. Binding Type I errors at most α is considered a constraint that must be satisfied.

Table 15.5 provides an overview of the main notations.

15.3.2 Global sensitivity of Chi-squared value

As in other studies, we assume that the data analyzer is also provided with m_i ($i = 1,\ldots, I$). Contingency tables T_1 and T_2, generated from neighboring databases, are considered. Because these databases are similar, the contingency tables differ only by two cells. T_2's cell $[a, k]$ is 1 greater than T_1's cell $[a, k]$. T_2's cell $[a, l]$ is one less than T_1's cell $[a, l]$ (s.t. $l \neq k$).

The databases collected in Definition 1 only include those that satisfy the values of m_i as the values of $m_i(i = 1,\ldots, I)$ are publicly available. The maximum potential Chi-squared value difference across tables T_1 and T_2 is therefore calculated.

Based on Theorem 1, adding Laplace noise with global sensitivity means that RandChiDist satisfies differential privacy. We therefore propose RandChiDist as follows:

$$\Delta_R = \begin{cases} \dfrac{(m_\zeta + m_\eta)n}{m_\zeta(1 + m_\eta)} & J \geq 3 \\ \dfrac{n^2}{m_\zeta(n - m_\zeta + 1)} & J = 2, \end{cases} \tag{15.7}$$

where

$$\zeta = \operatorname*{argmin}_i m_i \text{ and } \eta = \operatorname*{argmin}_{i \neq \zeta} m_i, \tag{15.8}$$

to the original Chi-squared based on (15.2). The following theorem is obtained.

Theorem 2: *RandChiDist realizes ε-differential privacy.*

T_1

				Total
	0	0		m_1
	m_a-1	1		m_a
	0	0		m_3
	0	m_b		m_b
	0	0		m_5
	0	0		m_6
	0	0		m_I
Total s_1	s_k	s_l	s_J	

T_2

				Total
	0	0		m_1
	m_a	0		m_a
	0	0		m_3
	0	m_b		m_b
	0	0		m_5
	0	0		m_6
	0	0		m_I
Total s_1	s_k+1	s_l−1	s_J	

Figure 15.1 Neighboring databases that maximize the difference between the T_1's Chi-squared value and T_2's Chi-squared value for contingency tables where $J \geq 3$

Proof: Let $\mathcal{V}_{i,j}(T)$ represent cell $[i,j]$'s observed value in contingency table T, and let $\chi^2(T)$ represent the T's original Chi-squared value. We consider neighboring databases T_1 and T_2, which satisfy the following equations:

$$\begin{cases} \mathcal{V}_{a,k}(T_2) = \mathcal{V}_{a,k}(T_1) + 1 \\ \mathcal{V}_{a,l}(T_2) = \mathcal{V}_{a,l}(T_1) - 1, \end{cases} \tag{15.9}$$

where $k, l \in 1,\ldots,$ J$\}$ and $k \neq l$.

Based on Proposition 2, giving the value a, T_1 and T_2 that satisfy the following constraints maximize the difference between the T_1's Chi-squared value and the T_2's Chi-squared value (see Figure 15.1) when J is greater than 2:

$$\mathcal{V}_{a,k}(T_1) + 1 = \mathcal{V}_{a,k}(T_2) = m_a \tag{15.10}$$

$$\mathcal{V}_{a,l}(T_1) - 1 = \mathcal{V}_{a,l}(T_2) = 0 \tag{15.11}$$

$$\mathcal{V}_{b,l}(T_1) = \mathcal{V}_{b,l}(T_2) = m_b, \text{ where } b \neq a \tag{15.12}$$

$$\mathcal{V}_{i,k}(T_1) = 0, \text{ where } i \neq a \tag{15.13}$$

$$\mathcal{V}_{i,l}(T_1) = 0, \text{ where } i \neq a, b \tag{15.14}$$

$$\mathcal{V}_{i,j}(T_1) = \mathcal{V}_{i,j}(T_2) = \text{arbitrary values that satisfy the constraint} \tag{15.15}$$

$$\sum_j \mathcal{V}_{i,j} = m_i, \text{ where } [i,j] \neq [a,k], [a,l], \text{ and } [b,l]. \tag{15.16}$$

Regarding constraint (15.16), the sum of the kth column of T_2 (i.e., s_k of T_2) is understood to be equal to m_a. $U_{i,j}(T)$ represents $U_{i,j}$ in (15.2) for database T. Symbol b denotes any integer from 1 to I and does not denote a. Symbol l denotes any integer from 1 to J and does not denote k. Satisfying the constraint (15.16), the calculation used to work out difference between chi-squared values of tables T_1 and T_2 is as follows:

$$
\begin{aligned}
&\sum_{i}(U_{i,k}(T_2)+U_{i,l}(T_2)-U_{i,k}(T_1)-U_{i,l}(T_1)) \\
&= U_{a,k}(T_2)+\sum_{i\neq a}U_{i,k}(T_2)+U_{b,l}(T_2)+\sum_{i\neq b}U_{i,l}(T_2) \\
&= -U_{a,k}(T_1)-\sum_{i\neq a}U_{i,k}(T_1)-U_{a,l}(T_1)-U_{b,l}(T_1)-\sum_{i\neq a,b}U_{i,l}(T_1) \\
&= \frac{\left(m_a\frac{m_a}{n}-m_a\right)^2}{m_a\frac{m_a}{n}}+\sum_{i\neq a}\frac{\left(m_a\frac{m_i}{n}\right)^2}{m_a\frac{m_i}{n}}+\frac{\left(m_b\frac{m_b}{n}-m_b\right)^2}{m_b\frac{m_b}{n}}+\sum_{i\neq b}\frac{\left(m_b\frac{m_i}{n}\right)^2}{m_b\frac{m_i}{n}} \\
&-\frac{\left((m_a-1)\frac{m_a}{n}-(m_a-1)\right)^2}{(m_a-1)\frac{m_a}{n}}-\sum_{i\neq a}\frac{\left((m_a-1)\frac{m_i}{n}\right)^2}{(m_a-1)\frac{m_i}{n}} \\
&-\frac{\left((m_b+1)\frac{m_a}{n}-1\right)^2}{(m_b+1)\frac{m_a}{n}}-\frac{\left((m_b+1)\frac{m_b}{n}-m_b\right)^2}{(m_b+1)\frac{m_b}{n}}-\sum_{i\neq a,b}\frac{\left((m_b+1)\frac{m_i}{n}\right)^2}{(m_b+1)\frac{m_i}{n}} \\
&= \frac{(m_a+m_b)n}{m_a(1+m_b)}
\end{aligned}
\tag{15.17}
$$

T_1		Total
0	m_1	m_1
m_a-1	1	m_a
0	m_3	m_3
0	m_4	m_b
0	m_5	m_5
0	m_6	m_6
0	m_I	m_I
Total s_k	s_l	

T_2		Total
0	m_1	m_1
m_a	0	m_a
0	m_3	m_3
0	m_4	m_b
0	m_5	m_5
0	m_6	m_6
0	m_I	m_I
Total s_k+1	s_l-1	

Figure 15.2 Neighboring databases that maximize the difference between the T_1's Chi-squared value and T_2's Chi-squared value for contingency tables where $J = 2$

Consequently, global sensitivity is represented by (15.17), based on a and the value of J being greater than 2. When J is greater than 2 and a is not given, global sensitivity is represented by (15.7), based on Proposition 1.

Neighboring databases meeting the constraints detailed below will maximize the difference between the T_1's chi-squared value and T_2's chi-squared value from Proposition 3 when $J=2$ and a is given (see Figure 15.2):

$$\mathcal{V}_{a,k}(T_1) + 1 = \mathcal{V}_{a,k}(T_2) = m_a \tag{15.18}$$

$$\mathcal{V}_{a,l}(T_1) - 1 = \mathcal{V}_{a,l}(T_2) = 0 \tag{15.19}$$

$$\mathcal{V}_{i,k}(T_1) = \mathcal{V}_{i,k}(T_2) = 0 \text{ for all } i \text{ except for } i = a \tag{15.20}$$

$$\mathcal{V}_{i,l}(T_1) = \mathcal{V}_{i,l}(T_2) = m_i \text{ for all } i \text{ except for } i = a \tag{15.21}$$

Satisfying the constraint (15.21), the differences between the Chi-squared values of tables T_1 and T_2 can be calculated as

$$\begin{aligned}
&\sum_i (U_{i,k}(T_2) + U_{i,l}(T_2) - U_{i,k}(T_1) - U_{i,l}(T_1)) \\
&= U_{a,k}(T_2) + \sum_{i \neq a} U_{i,k}(T_2) + U_{a,l}(T_2) + \sum_{i \neq a} U_{i,l}(T_2) \\
&\quad - U_{a,k}(T_1) - \sum_{i \neq a} U_{i,k}(T_1) - U_{a,l}(T_1) - \sum_{i \neq a} U_{i,l}(T_1) \\
&= \frac{\left(m_a \frac{m_a}{n} - m_a\right)^2}{m_a \frac{m_a}{n}} + \sum_{i \neq a} \frac{\left(m_a \frac{m_i}{n}\right)^2}{m_a \frac{m_i}{n}} + \frac{\left((n - m_a)\frac{m_a}{n}\right)^2}{(n - m_a)\frac{m_a}{n}} \\
&\quad + \sum_{i \neq a} \frac{\left((n - m_a)\frac{m_i}{n} - m_i\right)^2}{(n - m_a)\frac{m_i}{n}} - \frac{\left((m_a - 1)\frac{m_a}{n} - (m_a - 1)\right)^2}{(m_a - 1)\frac{m_a}{n}} - \sum_{i \neq a} \frac{\left((m_a - 1)\frac{m_i}{n}\right)^2}{(m_a - 1)\frac{m_i}{n}} \\
&\quad - \frac{\left((n - m_a + 1)\frac{m_a}{n} - 1\right)^2}{(n - m_a + 1)\frac{m_a}{n}} - \sum_{i \neq a} \frac{\left((n - m_a + 1)\frac{m_i}{n} - m_i\right)^2}{(n - m_a + 1)\frac{m_i}{n}} \\
&= \frac{n^2}{m_a(n - m_a + 1)}.
\end{aligned} \tag{15.22}$$

Eq. (15.7) represents the global sensitivity when J is equal to 2 and a is not given, as $n^2/(m_a(n - m_a + 1))$ decreases when m_a decreases.

Δ_R is identical to Δ_Y when using a $2 \cdot 3$ contingency table. Δ_R is identical to Δ_F when using a $2 \cdot 3$ contingency table.

Propositions 1 and 2, which were used in to prove Theorem 2, are described below.

Proposition 1: *Δ_R in (15.7) is maximized when the minima (15.8) are satisfied.*

Proof: By differentiating (15.7) with respect to m_a, we obtain

$$-\frac{m_b n}{m_a^2(1+m_b)}. \tag{15.23}$$

By differentiating (15.7) with respect to m_b, we obtain

$$-\frac{(m_a-1)n}{m_a(1+m_b)^2}. \tag{15.24}$$

To maximize (15.7), m_a and m_b from (15.23) and (15.24) should therefore be minimized.

Let ψ represent the minimum value of m_i, and let $\psi+\phi$ represent the second smallest value of m_i, where $\phi \geq 0$ and $i = 1,\ldots I$. If m_a is ψ and m_b is $\psi+\phi$, (15.7) can therefore be expressed as

$$\frac{n(\psi+\phi)}{\psi(1+\psi+\phi)}. \tag{15.25}$$

If m_a is $\psi+\phi$ and m_b is ψ, (15.7) can then be expressed as

$$\frac{n(\psi+\phi)}{(1+\psi)(\psi+\phi)}. \tag{15.26}$$

As (15.25) is always greater than or equal to (15.26), the value of Δ_R in (15.7) is maximized when (15.8) is satisfied.

Proposition 2: *If $J \geq 3$ and a value of a is given, the difference between the T_1's Chi-squared value and T_2's chi-squared value is maximized by neighboring databases with* (15.16).

Proof: While many neighboring databases satisfy (15.9), we prove that neighboring databases satisfying the constraints (15.16) provide the most significant difference, $\delta(T_1, T_2)$, between $\chi^2(T_1)$ and $\chi^2(T_2)$ when $J \geq 3$. m_i is assumed to be a fixed value for all i values.

$\mathcal{V}_{i,j}(T_1)$ is written as $\mathcal{V}_{i,j}$ for simplicity. Based on Lemma 1, $\mathcal{V}_{a,k}$ should maximize $\delta(T_1, T_2)$. Due to the constraints (15.27), $\mathcal{V}_{a,k}$'s value consequently becomes $m_a - 1$.

Based on (15.9), the constraints are as follows:

$$\mathcal{V}_{a,k}(T_1) \leq m_a - 1 \text{ and } 1 \leq \mathcal{V}_{a,l}. \tag{15.27}$$

Based on Lemma 2, $\delta(T_1, T_2)$ should be maximized for all i values except $i = a$ by making $\mathcal{V}_{i,k}$ zero.

Based on Lemma 3, $\delta(T_1, T_2)$ should be maximized by minimizing$\mathcal{V}_{a,l}$. Due to the constraints, $\mathcal{V}_{a,l}$'s value therefore becomes 1 (15.27).

Based on Lemma 4, $\mathcal{V}_{\mu,l}$ ($\mu \neq a$) should be m_μ and $\mathcal{V}_{i,l}$ for all i, except for $i = a$, and $i = \mu$ should be zero to maximize $\delta(T_1, T_2)$.

By replacing μ in Lemma 4 with b, we can consequently maximize $\delta(T_1, T_2)$ when tables T_1 and T_2 satisfy the constraints (15.16).

Lemma 1: *$\mathcal{V}_{a,k}$ should be maximized to accordingly maximize $\delta(T_1, T_2)$. $\mathcal{V}_{a,r}$ should also be adjusted for all r to satisfy m_a*

Proof: We have

$$\delta(T_1, T_2) = \chi^2(T_2) - \chi^2(T_1) \tag{15.28}$$

$$= U_{a,k}(T_2) - U_{a,k}(T_1) + \sum_{i \neq a} (U_{i,k}(T_2) - U_{i,k}(T_1)) \tag{15.29}$$

$$+ U_{a,l}(T_2) - U_{a,l}(T_1) + \sum_{i \neq a} (U_{i,l}(T_2) - U_{i,l}(T_1)) \tag{15.30}$$

$$= -2 + \frac{m_a}{n} + \frac{n(-\mathcal{V}_{a,k}^2 + s_k + 2s_k\mathcal{V}_{a,k})}{m_a s_k(1 + s_k)} + \sum_{i \neq a} \frac{m_i^2 s_k(1 + s_k) - n^2\mathcal{V}_{i,k}^2}{m_i n s_k(1 + s_k)} \tag{15.31}$$

$$+2 - \frac{m_a}{n} + \frac{n(\mathcal{V}_{a,l}^2 + s_l - 2s_l\mathcal{V}_{a,l})}{m_a(s_l - 1)s_l} + \sum_{i \neq a} \frac{n^2\mathcal{V}_{i,l}^2 - m_i^2(s_l - 1)s_l}{m_i n(s_l - 1)s_l}. \tag{15.32}$$

By differentiating (15.32) with respect to $\mathcal{V}_{a,k}$, we obtain

$$\frac{n(\mathcal{V}_{a,k} - s_k)^2(1 + 2s_k)}{m_a s_k^2(1 + s_k)^2} + \sum_{i \neq a} \frac{n\mathcal{V}_{i,k}^2(1 + 2s_k)}{m_i s_k^2(1 + s_k)^2}, \tag{15.33}$$

because we have

$$\frac{\partial s_k}{\partial \mathcal{V}_{a,k}} = 1. \tag{15.34}$$

As (15.33) is always ≥ 0, (15.32) increases as $\mathcal{V}_{a,k}$ increases.

We can therefore conclude that $\mathcal{V}_{a,k}$ should be increased to maximize $\delta(T_1, T_2)$. Consequently, we have $\mathcal{V}_{a,k} = m_a - 1$.

Lemma 2: *$\mathcal{V}_{i,k}$ should be minimized to maximize $\delta(T_1, T_2)$. $\mathcal{V}_{i,r}$ should also be adjusted to satisfy m_i for all rfor all i values, with the exception of $i = a$.*

Proof: We focus on $\mu \in 1, \ldots, I\}$ such that $\mu \neq a$. By differentiating (15.32) with respect to $\mathcal{V}_{\mu,k}$, we obtain

$$\frac{n(\mathcal{V}_{a,k} - s_k)(\mathcal{V}_{a,k} + s_k + 2s_k\mathcal{V}_{a,k})}{m_a s_k^2(1 + s_k)^2} \tag{15.35}$$

$$+ \frac{n\mathcal{V}_{\mu,k}(\mathcal{V}_{\mu,k} + 2s_k\mathcal{V}_{\mu,k} - 2s_k(1 + s_k))}{m_\mu s_k^2(1 + s_k)^2} + \sum_{i \neq a,\mu} \frac{n\mathcal{V}_{i,k}^2(1 + 2s_k)}{m_i s_k^2(1 + s_k)^2}, \tag{15.36}$$

because we have

$$\frac{\partial s_k}{\partial \mathcal{V}_{\mu,k}} = 1. \tag{15.37}$$

Let $\Theta = \sum_{i \neq a,\mu} \mathcal{V}_{i,k}^2 / m_i$. By solving Eq. (15.36)= 0 for Θ, we obtain

$$\frac{m_\mu(s_k - \mathcal{V}_{a,k})(\mathcal{V}_{a,k} + s_k + 2s_k\mathcal{V}_{a,k}) + m_a\mathcal{V}_{u,k}(2s_k - \mathcal{V}_{u,k} + 2s_k(s_k - \mathcal{V}_{u,k})}{m_a m_\mu(1 + 2s_k)}. \tag{15.38}$$

The value of (15.38) is greater than zero. The value of (15.36) is less than 0 when Θ= 0 in (15.36).

Therefore, (15.36) is less than zero when Θ is less than (15.38). In much the same way, (15.36) is greater than zero when Θ is greater than (15.38). The value of $\mathcal{V}_{\mu,k}$ should therefore either be minimized or maximized to maximize (15.32). Based on this observation, $\mathcal{V}_{i,k}$ should either be reduced to zero or maximized to m_i for all i except for $i = a$ to maximize (15.32).

Based on Lemma 1, $\mathcal{V}_{a,k} = m_a - 1$ is what we have. Consequently, when $\mathcal{V}_{i,k}$= 0 for all i except $i = a$, $s_k = m_a - 1$ is what we have. $\delta(T_1, T_2)$ is therefore

$$-2 + \frac{m_a}{n} + \frac{n}{m_a} + \frac{1}{n}\sum_{i \neq a} m_i + U_{a,l}(T_2) - U_{a,l}(T_1) + \sum_{i \neq a}(U_{i,l}(T_2) - U_{i,l}(T_1)). \tag{15.39}$$

When $\mathcal{V}_{i,k} = m_i$ for all i except $i = a$, however, $s_k = \sum_i m_i - 1 = n - 1$. $\delta(T_1, T_2)$ in this case is

$$-2 + \frac{m_a}{n} + \frac{n}{m_a} + \sum_{i \neq a} \frac{m_i}{n} - \frac{(n - m_a)^2}{m_a(n-1)n} + \sum_{i \neq a} \frac{m_i}{n - n^2}. \tag{15.40}$$

By subtracting (15.40) from (15.39), we obtain

$$\frac{(n - m_a)^2}{m_a(n-1)} + \sum_{i \neq a} \frac{m_i}{n-1}. \tag{15.41}$$

$\mathcal{V}_{i,k}$ for all i except $i = a$ should be zero based on (15.41) always being more than zero.

Lemma 3: *$\mathcal{V}_{a,l}$ should be minimized to maximize $\delta(T_1, T_2)$ accordingly. To satisfy m_a, $\mathcal{V}_{a,r}$ for all r except $r = k, l$ should also be adjusted.*

Proof: By differentiating (15.32) with respect to $\mathcal{V}_{a,l}$, we obtain

$$\frac{n(\mathcal{V}_{a,l} - s_l)^2(1 - 2s_l)}{m_a(s_l - 1)^2 s_l^2} + \sum_{i \neq a} \frac{n\mathcal{V}_{i,l}^2(1 - 2s_l)}{m_i(s_l - 1)^2 s_l^2}, \tag{15.42}$$

because we have

$$\frac{\partial s_l}{\partial \mathcal{V}_{a,l}} = 1. \tag{15.43}$$

Eq. (15.32) increases as $\mathcal{V}_{a,l}$ decreases based on (15.42) always being less than zero.

Lemma 4: $\mathcal{V}_{\mu,l}$ $(\mu \neq a)$ *should be maximized to maximize* $\delta(T_1, T_2)$. *Accordingly, for all r except* $r = k, l$, $\mathcal{V}_{\mu,r}$ *should be adjusted to satisfy* m_μ. *Furthermore,* $\mathcal{V}_{i,l}$ *should be minimized for all i except* $i = a$ *and* $i = \mu$. *Accordingly, for all r except* $r = k, l$, $\mathcal{V}_{i,r}$ *should be adjusted to satisfy* m_i.

Proof: By differentiating (15.32) with respect to $\mathcal{V}_{\mu,l}$ $(\mu \neq a)$, we obtain

$$\frac{n(s_l - \mathcal{V}_{a,l})(2s_l\mathcal{V}_{a,l} - \mathcal{V}_{a,l} - s_l)}{m_a(s_l - 1)^2 s_l^2} \tag{15.44}$$

$$+\frac{n\mathcal{V}_{\mu,l}(\mathcal{V}_{\mu,l} - 2s_l - 2s_l\mathcal{V}_{\mu,l} + 2s_l^2)}{m_\mu(s_l - 1)^2 s_l^2} \tag{15.45}$$

$$+\sum_{i\neq a,\mu}\frac{n\mathcal{V}_{i,l}^2(1 - 2s_l)}{m_i(s_l - 1)^2 s_l^2}, \tag{15.46}$$

because we have

$$\frac{\partial s_l}{\partial \mathcal{V}_{\mu,l}} = 1. \tag{15.47}$$

Let $\Theta = \sum_{i\neq a,\mu}\mathcal{V}_{i,l}^2/m_i$. We have $\mathcal{V}_{a,l} = 1$ from Lemma 3. By solving (15.46) = 0 for Θ, we obtain

$$\frac{m_\mu(s_l - 1)^2 + m_a\mathcal{V}_{u,l}(2s_l(s_l - \mathcal{V}_{\mu,l} - 1) + \mathcal{V}_{\mu,l})}{m_a m_\mu(2s_l - 1)}. \tag{15.48}$$

Eq. (15.48) is always greater than zero. When $\Theta = 0$ and $\mathcal{V}_{a,l} = 1$ in (15.46), (15.46) can be expressed as

$$\frac{n(m_\mu(s_l - 1)^2 + m_a\mathcal{V}_{\mu,l}(2s_l(s_l - \mathcal{V}_{\mu,l} - 1) + \mathcal{V}_{\mu,l}))}{m_a m_\mu(s_l - 1)^2 s_l^2} \geq 0. \tag{15.49}$$

Thus, Eq. (15.46) is greater than zero when Θ is less than or equal to (15.48). Eq. (15.46) is less than zero when Θ is greater than Eq. (15.48), Therefore, the value of $\mathcal{V}_{\mu,l}$ should either be reduced to zero or maximized to m_μ to maximize Eq. (15.32).

The value of $\mathcal{V}_{\mu,l}$ should therefore either be minimized or maximized to maximize $\delta(T_1, T_2)$. For example, $x = \sum_{i\neq\mu}\mathcal{V}_{i,l}$. If $\mathcal{V}_{\mu,l}$ is maximized, such as

$(\mathcal{V}_{\mu,l} = m_\mu)$, then $\delta(T_1, T_2)$ is

$$2 - \frac{m_a + m_\mu}{n} + \frac{m_\mu n}{m_\mu + x - 1} - \frac{n + m_a m_\mu n}{m_a(m_\mu + x)} \tag{15.50}$$

$$+ \sum_{i \neq a,\mu} \left(\frac{\mathcal{V}_{i,l}^2 n}{m_i(m_\mu + x - 1)(m_\mu + x)} - \frac{m_i}{n} \right). \tag{15.51}$$

When $\mathcal{V}_{\mu,l}$ is minimized, however, (i.e., $\mathcal{V}_{\mu,l}$= 0), $\delta(T_1, T_2)$ is

$$2 - \frac{m_a + m_\mu}{n} - \frac{n}{m_a x} + \sum_{i \neq a,\mu} \frac{\mathcal{V}_{i,l}^2 n^2 - m_i^2 (x-1)x}{m_i n(x-1)x}. \tag{15.52}$$

If (15.52) is subtracted from (15.51), we therefore obtain

$$\frac{m_\mu n(-1 + m_\mu + x + m_a x)}{m_a x(-1 + m_\mu + x)(m_\mu + x)} - \sum_{i \neq a,\mu} \frac{n m_\mu \mathcal{V}_{i,l}^2 (-1 + m_\mu + 2x)}{m_i(x-1)x(m_\mu + x - 1)(m_\mu + x)}. \tag{15.53}$$

The second term of (15.53) is zero when I= 2. Eq. (15.53) is therefore always greater than zero and Lemma 4 stays true when I= 2.

Considering situations where $I \geq 3$, we assume that $\mathcal{V}_{i,l}$ is zero for all values of i except $i = a$ and $i = \mu$. The second term of (15.53) is zero in this case, while the first term of (15.53) is greater than zero. We can consequently say that (15.53) is always greater than zero. Therefore, for all values of i except $i = a$ and $i = \mu$, $\mathcal{V}_{\mu,l}$ should be maximized to m_μ when $\mathcal{V}_{i,l}$ is zero.

Focusing on v so that $v \in 1,\ldots, \mathrm{I}\}$ and $v \neq a, \mu$, when $\mathcal{V}_{v,l}$ is maximized to m_v we prove that Eq. (15.53) always represents ≤ 0. Furthermore, when I= 3, the second term of (15.53) is minimized. We therefore obtain

$$(15.53) \leq -\frac{(m_a - 1)m_\mu n}{m_a(1 + m_v)(1 + m_v + m_\mu)} < 0, \tag{15.54}$$

because $x = m_v + 1$.

Thus, with the exception of $i \neq a, \mu$, each $\mathcal{V}_{i,l}$ for all i should be minimized to zero. When $I \geq 3$, Lemma 4 also holds based on this observation.

Proposition 3: *Neighboring databases satisfying the constraints* (15.21) *maximize the difference between the Chi-squared values of tables* T_1 *and* T_2 *when* J *equals* 2 *and* a *is given.*

The proof can be conducted in a way comparable to Lemma 2.

15.3.3 Differentially private hypothesis testing

The anonymized Chi-squared value of an original table can now be determined,

$$\chi^{2*} = \chi^2 + Lap(\Delta_R/\varepsilon), \tag{15.55}$$

where χ^{2*} is the anonymized Chi-squared value.

Based on the definitions of both Laplace distribution and Chi-squared distribution, a Chi-squared value's probability density function possessing v degrees of freedom with Laplace noise added and global sensitivity Δ can be presented thus

$$g_{v,\Delta,\varepsilon}(x) = \int_{\mu=-\infty}^{\infty} \mathcal{L}_{\mu,\beta}(x)\mathcal{Z}_v(\mu)d\mu, \tag{15.56}$$

where

$$\beta = \Delta/\varepsilon, \tag{15.57}$$

$$\mathcal{L}_{\mu,\beta}(x) = \begin{pmatrix} \dfrac{\exp\left(-\dfrac{x-\mu}{\beta}\right)}{2\beta} & x \geq \mu \\ \dfrac{\exp\left(-\dfrac{\mu-x}{\beta}\right)}{2\beta} & \text{otherwise,} \end{pmatrix} \tag{15.58}$$

and

$$L_v(u) = \begin{pmatrix} \dfrac{2^{-v/2}\exp(-u/2)u^{-1+v/2}}{\Gamma(v/2)} & x > 0 \\ 0 & \text{otherwise,} \end{pmatrix} \tag{15.59}$$

where $\Gamma(v/2)$ represents the $v/2$ gamma function, that is,

$$\Gamma(v/2) = \int_0^{\infty} x^{v/2-1}e^{-x}dx. \tag{15.60}$$

RandChiDist rejects H_0 when the significance level is set to α if the Chi-squared value that is calculated using (15.2) and Laplace noise, as well as the scale Δ_R/ε, is equal to or exceeds α. This is calculated based on the following equation:

$$\int_{x=t}^{\infty} g_{v,\Delta,\varepsilon}(x) = \alpha. \tag{15.61}$$

Finally, the χ^{2*} value determined using (15.55) was compared with the t value determined using (15.61). RandChiDist rejects null hypothesis H_0 when χ^{2*} is greater than or equal to t. In all other instances, it fails to reject the null hypothesis.

Algorithm 1 displays the RandChiDist algorithm.

Algorithm 1. Algorithm of RandChiDist

Input: Privacy parameter ε, Original cross table T, Significance level α
Output: Differentially private Chi-squared test result
1: Obtain original Chi-squared value based on (15.2)
2: Obtain global sensitivity ΔR based on (15.7)

3: $\chi^{2*} \Leftarrow \chi^2 + \text{Lap}(\Delta R/\varepsilon)$
4: Obtain value of t from (15.61)
5: **if**$\chi^{2*} \geq t$ **then**
6: Output “reject the null hypothesis H_0”
7: **else**
8: Output “fail to reject the null hypothesis H_0”
9: **end if**

RandChiDist determines and produces an anonymized version of the p value as follows:

$$\int_{x=\chi^{2*}}^{\infty} g_{v,\Delta,\varepsilon}(x). \tag{15.62}$$

By implementing an arbitrary α that compares (15.62) and α, a Chi-squared hypothesis test can therefore be conducted.

15.3.4 Complexity analysis

A complex range of $O(I \cdot J)$ arises from calculating the original Chi-squared. The two largest values of m_i (i= 1,…,I) are required to determine global sensitivity ΔR, making $O(I)$ the computational complexity. An integration, like a Monte Carlo integration, is required to determine (15.61) and (15.62). The cross table does not influence computation complexity, and this approach can be determined at great speed [71].

In terms of our algorithm, the computational complexity is therefore $O(I \cdot J + M)$, with M representing the integration's computational complexity.

15.4 Evaluation

As outlined in Section 15.2, a number of methods, including RandChiDist, RandCell, MCIndep, and LocalExpIND were compared.

LocalExpIND has a wider application than RandChiDist because the former was designed specifically for local privacy. This privacy model is an area of interest to be considered for future research.

Furthermore, we also considered the RandChi method, which uses global sensitivity Δ_R and does not use the private Chi-squared distribution table's value to clarify the impact of obtaining the differentially private Chi-squared distribution table's value (as described in Section 15.3.3).

The source codes for both these approaches can be found as follows: https://uecdisk2.cc.uec.ac.jp/s/qasHCJerBNJW8Sa.

When performing multiple Chi-squared testing, Bonferroni's corrected threshold should be used [72]. While many Chi-squared tests were conducted in this chapter, each is considered to be independent, and so Bonferroni's corrected

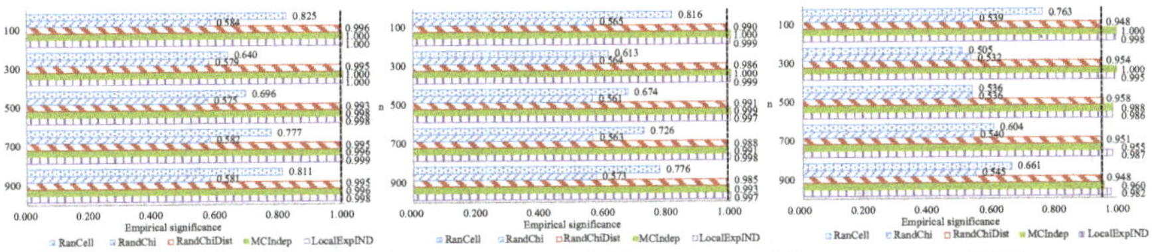

(a) $\alpha = 0.005$ (b) $\alpha = 0.01$ (c) $\alpha = 0.05$

Figure 15.3 Significance results of $2 \cdot 2$ *contingency tables. The dashed lines show* $1 - \alpha$*. The value of* ε *was set to 0.1*

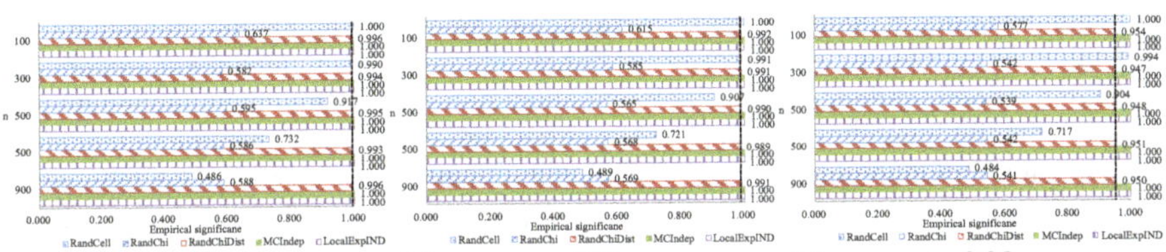

(a) $\alpha = 0.005$ (b) $\alpha = 0.01$ (b) $\alpha = 0.05$

Figure 15.4 Significance results of $4 \cdot 4$ *contingency tables. The dashed lines show* $1 - \alpha$*. The value of* ε *was set to 0.1*

threshold was not applied for comparison purposes. The average results of every independent Chi-squared test are presented in this chapter. It should be noted that a number of prior studies also declined to use Bonferroni's corrected threshold [11,12,33–35,54,55].

The n values varied between 100 and 900. The α values ranged from 0.005 to 0.05 and the ε values were between 0.01 and 10. The parameters of MCIndep were set based on [12].

15.4.1 Significance results

Confirmation that RandChiDist assures a minimum significance of $1 - \alpha$ was established based on significance evaluation. Using a multinomial distribution with probabilities of 0.25, 0.25, 0.25, and 0.25, 1,000 $2 \cdot 2$ contingency tables were generated randomly to assess if each method fails to reject the null hypothesis H_0 as the output. The results using an ε value of 0.1 are presented in Figure 15.3, with the significance determined to be around $1 - \alpha$.

For RandChiDist, MCIndep, and LocalExpIND, significance levels for any n, ε, and α values were maintained at approximately $1 - \alpha$. When ε was less than 1, values much lower than $1 - \alpha$ were recorded for both RandCell and RandChi.

Using a multinomial distribution with probabilities of $1/16,..., 1/16$ on randomly generated $4 \cdot 4$ contingency tables, identical experiments were performed. The results with $\varepsilon = 0.1$ are presented in Figure 15.4. For RandChiDist, MCIndep, and LocalExpIND, the significance values were around $1 - \alpha$, much like the $2 \cdot 2$

contingency tables. The significance values were less than $1 - \alpha$ for both RandCell and RandChi. This was particularly true when ε was lower.

In RandCell, where a Laplace noise is added to every cell, a higher number of cells increases the likelihood of at least one Laplace noise becoming very large. In large contingency tables, this leads to small significance results and a high volume of false positives. Conversely, cell values with noise can be below five or negative if the Laplace noise has a large negative value. Using the rule of thumb, RandCell fails to reject the null hypothesis in this instance. Consequently, when n is large, the $4 \cdot 4$ table results for RandCell are smaller compared to for $2 \cdot 2$ tables.

Table size n does not produce large variations in RandChi's significance results because the global sensitivity, based on equation (15.8), also remains similar regardless of either n or table size.

15.4.2 Power results

Next, the power of each method was calculated. For parameters α, ε, and n, the values were the same as in the significance experiments, though $2 \cdot 2$ contingency tables were generated randomly based on a multinomial distribution with probabilities of $(1/4 + 0.01, 1/4 - 0.01, 1/4 - 0.01, 1/4 + 0.01)$ and $(1/4 + 0.15, 1/4 - 0.15, 1/4 - 0.15, 1/4 + 0.15)$. These contingency tables are referred to as Tables A and B. To calculate if RandChiDist can be successfully applied to unbalanced tables, another probability set $(0.3 + 0.15, 0.3 - 0.15, 0.2 - 0.15, 0.2 + 0.15)$ was also implemented (referred to as Table C). Using on a multinomial distribution with probabilities of $(1/12 + 0.07, 1/12 - 0.07, 1/12, 1/12, 1/12 - 0.07, 1/12 + 0.07, 1/12, \ldots, 1/12)$, $3 \cdot 4$ contingency tables were randomly generated to determine if each approach successfully rejected the null hypothesis H_0 (referred to as Table D). The results for $\varepsilon = 0.1$ are presented in Figure 15.5.

The findings of the multinomial distribution experiment with Table A show that the empirical power of Non-private ranged from 0 to 0.2, which is extremely low. RandChiDist performs marginally better than other privacy-preserving algorithms, though all such algorithms capable of controlling Type I errors fail to achieve high empirical power.

Regarding other multinomial distributions, MCIndep also has low empirical power in relative terms. This method rapidly produced an output failing to reject H_0 after producing a large volume of contingency tables from the original and detecting at least one cell with a value lower than five. MCIndep is therefore likely to produce this output even if all the cells in the target contingency table have values that are close to five.

RandCell and RandChi generated high empirical power, but in both cases, empirical significance was lost in the process. MCIndep's empirical power remained high under certain circumstances; namely, a high volume of samples and uniformly distributed data. Increased empirical power based on fewer samples was achieved by RandChiDist.

Type I errors should be avoided in hypothesis testing that involves chi-squared tests, and the α value (e.g., 0.05) is changed to adjust the probability of errors.

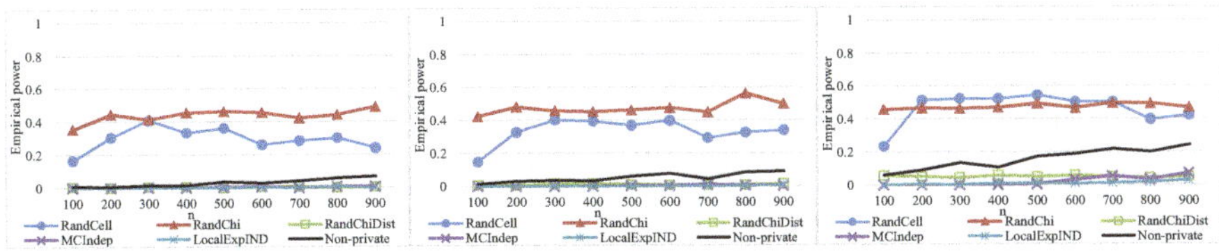

(a) $\alpha = 0.005$ (b) $\alpha = 0.01$ (c) $\alpha = 0.05$

Results of Table A

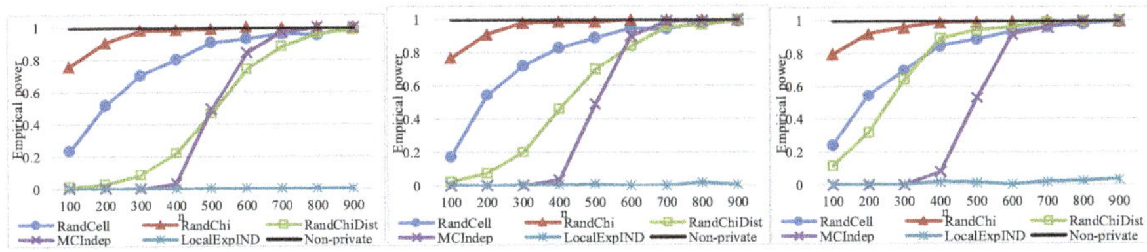

(d) $\alpha = 0.005$ (e) $\alpha = 0.01$ (f) $\alpha = 0.05$

Results of Table B

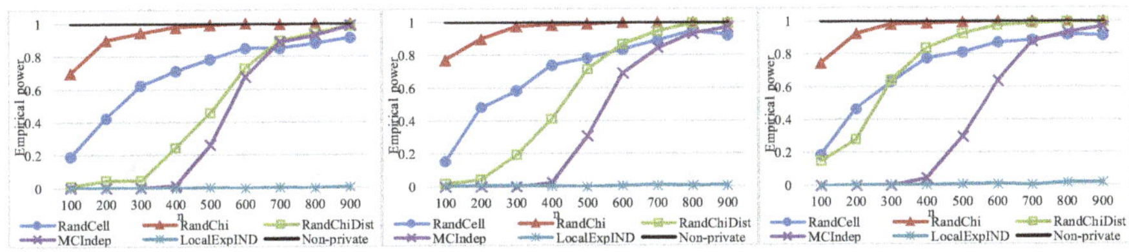

(g) $\alpha = 0.005$ (h) $\alpha = 0.01$ (i) $\alpha = 0.05$

Results of Table C

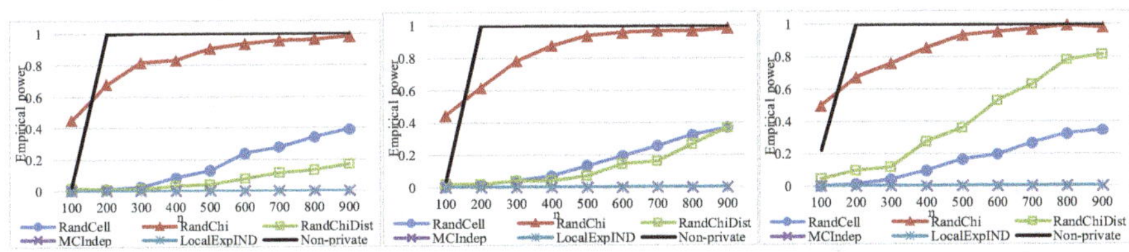

(j) $\alpha = 0.005$ (k) $\alpha = 0.01$ (l) $\alpha = 0.05$

Results of Table D

Figure 15.5 Empirical power results with $\varepsilon = 0.1$

When the empirical significance is lower than $1-\alpha$, the algorithm is useless, even with high empirical power. While for RandCell and RandChi, empirical power was higher than for RandChiDist, the empirical significance values of the first two were both above $1-\alpha$. This means that they often fail to limit Type I errors. RandChiDist can therefore be said to outperform both RandCell and RandChi. Of all the approaches that can control Type I errors (RandChiDist, MCIndep, and LocalExpIND), the highest power is produced by RandChiDist.

15.4.3 Results of real datasets

There were two genomic datasets used.[†] The Human Genome Diversity Project genotype dataset (HGDP) [73] is the first of them. This has 1,244 records left behind after deleting the 2,834 SNPs with uncertain values. The second piece of data is the genotype dataset from the International Haplotype Map Project (HapMap), which was previously used by [74] 1,853 SNPs and 420 full data are included.

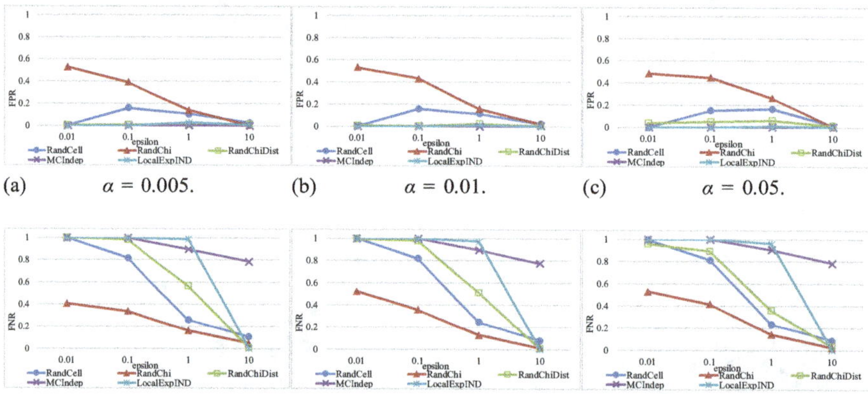

(a) $\alpha = 0.005$. (b) $\alpha = 0.01$. (c) $\alpha = 0.05$.

(d) $\alpha = 0.005$. (e) $\alpha = 0.01$. (f) $\alpha = 0.05$.

Figure 15.6 Results of the HGDP genotype datasets. (a)–(c) represent false positive rate and (d)–(f) represent false negative rate

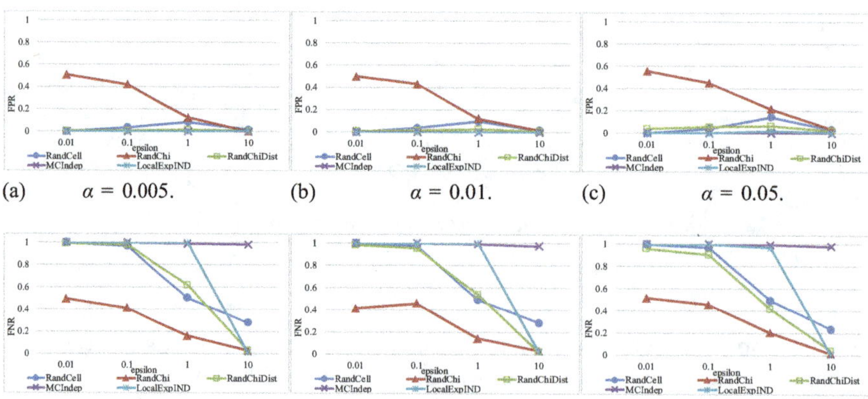

(a) $\alpha = 0.005$. (b) $\alpha = 0.01$. (c) $\alpha = 0.05$.

(d) $\alpha = 0.005$. (e) $\alpha = 0.01$. (f) $\alpha = 0.05$.

Figure 15.7 Results of the HapMap genotype datasets. (a)–(c) represent false positive rate and (d)–(f) represent false negative rate

[†]https://web.stanford.edu/group/rosenberglab/hgdpsnpDownload.html (accessed May 26, 2017).

Contingency tables with four columns and rows were produced randomly for both datasets to be used in the linkage disequilibrium analysis. When values were established that were less than five, another contingency table was made based on the rule of thumb. Then, before using privacy-preserving methods, the initial contingency tables were subjected to standard Chi-squared testing. Before determining the mean rates of false positives and false negatives, Chi-squared testing was carried out 100 times on the contingency tables generated.

The results for the HGDP genotype and HapMap genotype datasets are shown in Figures 15.6 and 15.7, respectively. In most of the parameter settings described in this chapter, RandChiDist performed better than both MCIndep and LocalExpIND.

15.5 Discussion

Both RandCell and RandChi failed to limit the Type I error ratio, based on the results RandChiDist, MCIndep, and LocalExpIND, by contrast, successfully controlled the number of false positives. Of these three methods, the fewest number of Type II errors was recorded by RandChiDist.

Data analyzers will calculate the significance level α in advance when testing a hypothesis. In other words, a true null hypothesis with a probability that is not above α will be rejected. False data interpretations stem from high false-positive rates, which occur when the true null hypothesis is rejected with a probability that is higher than α. RandChiDist therefore presents the best method for avoiding false interpretations.

When it comes to non-private Chi-squared testing, multiple methods can be implemented. Section 15.2.1 presents the most basic approach. The global sensitivity of the Chi-squared value of the simplest Chi-squared testing is determined in both RandChi and RandChiDist, with the minimum amount of noise added based on the Laplace mechanism theorem (Theorem 1).

RandCell, on the other hand, adds noise to each value of every cell based on its global sensitivity. This accumulation of noise can therefore grow to be very large in volume. As outlined in Section 15.2, MCIndep offers an alternate non-private Chi-squared testing method. After removing the parameters of the underlying multinomial distribution generating the samples, this approach produces more than $1/\alpha$ contingency tables. The accuracy of MCIndep is low when working with a small number of samples, however, due to poor accuracy regarding the estimated parameters of the underlying multinomial distribution. A third approach, LocalExpIND, works on the assumption that there is no trusted entity and that each piece of data is individually anonymized. The total volume of noise in this method is high because noise is added to every data point.

According to Sharpe, Chi-squared hypothesis testing should be avoided for contingency tables larger than 2·2 if possible [75]. As Sharpe conceded, however, there are times when this is not possible and, according to the American Psychological Association, around Chi-squared tests for contingency tables larger than 2·2 account for around 30% of such tests. GWAS and a range of other personal database tests are among those to involve Chi-squared hypothesis testing [76–78].

Several works [44–46,79–81], meanwhile, have involved contingency tables with sizes greater than 2·3. Obtaining differentially private Chi-squared values for contingency tables larger than 2·3 therefore represents a topic of great importance.

The approach presented in this paper can be applied to both GWAS and additional small-sample forms of private data analysis. One such example is COVID-19 patients, who have been analyzed through Chi-squared testing. For example, based on an α of 0.05, with $n = 403$, the number of patients who died was 100 while the number of patients who recovered was 303 [82].

COVID-19 patients with and without an acute pulmonary embolism were studied by Poyiadi *et al.* [83]. A total of $n = 328$ patients were studied based on Chi-squared testing, again with α being 0.05. In another case, Jacob *et al.* [84] studied COVID-19's influence on sexual activity based on 868 samples. These works highlight both the demand for small sample size testing and how challenging the accumulation of large sample numbers is in scenarios demanding fast analysis.

It is assumed that data holders will publish both m_i and the differentially private Chi-squared value. Broadly speaking, accurate interpretation of Chi-squared values requires both m_i and a sample size [85]. In instances where every m_i is large, small Chi-squares values can still be generated, even if the differences between two datasets are tiny [86]. This highlights how valuable m_i is for data analysis.

A range of other forms of information can also be delineated just from the publication of m_i. $\mathcal{V}_{i,j}$ for all j are known to be lower than or equal to m_i, for example. That said, even when the values of both m_i and the Chi-squared value are known, we still cannot know each value of $\mathcal{V}_{i,j}$, nor which value of ($\mathcal{V}_{i,j}$ or $\mathcal{V}_{i,j'}$) is greater for any j or j'. As the method proposed in this paper uses differential privacy to protect Chi-squared values, reconstructing the original cross table is therefore not an option. As far as we are aware, the idea that publishing m_i could create privacy issues has not been put forward by any prior researchers.

In this chapter, we assume a scenario in which the entity holding the raw data provides a third party with a Chi-squared value calculated from the raw data with an error based on differential privacy. There are also scenarios in which the entity holding the raw data stores the data on an external cloud, rather than on a server it owns. This also has the effect of preventing information leakage in the event of an attack on its own server. However, when the cloud that manages data is an external third party, there is a risk that information in the cloud may be leaked either intentionally or inadvertently. In the future, it will be necessary to have an information management agent that manages information in consideration of privacy. However, an increase in the time required for data retrieval and other processing on the cloud must be avoided. Therefore, it will be necessary to guarantee data security and to perform operations such as searches at high speed on the cloud managed by a third party. Consider, for example, retrieval. As a simple mechanism, data and keywords for search can be encrypted and registered in the cloud. The user sends the encrypted keywords to the cloud, and the cloud returns a set of data that matches the keywords. However, in this case, the cloud can calculate the frequency of occurrence of each keyword and identify which keywords have more or less keywords in common among the data. As a result, the risk of the cloud administrator being able to guess the contents of the encrypted keywords increases. It is possible to address this issue by utilizing methods that use secret computation and the Bloom Filter [87]. There are several other privacy issues in the use of data for MDSS. It is hoped that these issues will be carefully resolved one by one in the future.

This chapter did not cover deep neural networks. Deep neural networks make it possible to analyze data with high accuracy. Conventionally, models learned by deep neural network techniques are black boxes and lack the explanatory power to explain why such outputs are produced. However, many techniques have now been proposed to enhance the explainability of deep neural networks. On the other hand, there are privacy risks associated with the use of deep neural networks. For example, there are techniques that infer the data used for training based on a learned model. Especially in the field of MDSS, privacy protection for deep neural network models must be strictly considered because learning is performed using sensitive information. The concept of differential privacy used in this chapter is also effective from this perspective [88].

15.6 Conclusion

Statistical analysis differs from machine learning analysis in that it is easier for humans to understand how it works and the results. The Chi-squared test is widely used in data analysis such as GWAS. We proposed the RandChiDist method to obtain differentially private Chi-squared values in contingency tables. It is easy to perform accurate statistical analysis if the number of samples required for data analysis is large; existing privacy-preserving Chi-squared test methods such as MCIndep are suitable when the number of samples n is large, but when n is small, RandChiDist outperforms existing methods.

Assessing additional datasets and applying our approach to further hypothesis testing methods represent future research goals. This chapter also focused on the Chi-squared test. However, the same approach could be used for other statistical tests. It is hoped that more research will be developed to derive statistical test results that satisfy differential privacy from small data samples.

Acknowledgment

The work of Agbotiname Lucky Imoize is supported in part by the Nigerian Petroleum Technology Development Fund (PTDF) and in part by the German Academic Exchange Service (DAAD) through the Nigerian-German Postgraduate Program under Grant 57473408.

References

[1] van der Velden BHM, Kuijf HJ, Gilhuijs KGA, *et al.* Explainable artificial intelligence (XAI) in deep learning-based medical image analysis. *Medical Image Analysis*. 2022;79(102470):1–21.

[2] Yang G, Ye Q, and Xia J. Unbox the black-box for the medical explainable AI via multi-modal and multi-centre data fusion: a mini-review, two showcases and beyond. *Information Fusion*. 2022;77:29–52.

[3] Nyrup R and Robinson D. Explanatory pragmatism: a context-sensitive framework for explainable medical AI. *Ethics and Information Technology*. 2022;24(1):1–15. Available from: https://link.springer.com/article/10.1007/s10676-022-09632-3.

[4] Mata DA and Milner DA. Statistical methods in experimental pathology: a review and primer. *The American Journal of Pathology*. 2021;191(5):784–794.

[5] Wu X, Dong H, Luo L, *et al.* A novel statistic for genome-wide interaction analysis. *PLoS Genetics*. 2010;6(9):e1001131.

[6] Hoh J and Ott J. Mathematical multi-locus approaches to localizing complex human trait genes. *Nature Reviews Genetics*. 2003;4(9):701–709.

[7] Homer N, Szelinger S, Redman M, *et al.* Resolving individuals contributing trace amounts of DNA to highly complex mixtures using high-density SNP genotyping microarrays. *PLoS Genetics*. 2008;4(8):e1000167.

[8] Dorfman R, Mamzer-Bruneel MF, Vogt G, *et al.* Falling prices and unfair competition in consumer genomics. *Nature Biotechnology*. 2013;31 (9):785–786.

[9] Savage N. Privacy: the myth of anonymity. *Nature*. 2016;537(7619):S70–S72.

[10] Wang R, Li YF, Wang X, *et al.* Learning your identity and disease from research papers: information leaks in genome wide association study. In: *Proceedings of the ACM CCS*, 2009. p. 534–544.

[11] Yu F, Fienberg SE, Slavkovi ć AB, *et al.* Scalable privacy-preserving data sharing methodology for genome-wide association studies. *Journal of Biomedical Informatics*. 2014;50:133–141.

[12] Gaboardi M, woo Lim H, Rogers R, *et al.* Differentially private chi-squared hypothesis testing: goodness of fit and independence testing. In: *Proceedings of the ICML*, 2016. .

[13] Pereira S, Gibbs R, and McGuire A. Open access data sharing in genomic research. *Genes*. 2014;5(3):739–747. Available from: http://www.mdpi.com/2073-4425/5/3/739.

[14] Schmidt-Kastner R, Guloksuz S, Kietzmann T, *et al.* Analysis of GWAS-derived schizophrenia genes for links to ischemia-hypoxia response of the brain. *Frontiers in Psychiatry*. 2020;11:393.

[15] Lee KY, Leung KS, Ma SL, *et al.* Genome-wide search for SNP interactions in GWAS data: algorithm, feasibility, replication using schizophrenia datasets. *Frontiers in Genetics*. 2020;11.

[16] Yuan J, Xing H, Lamy AL, *et al.* Leveraging correlations between variants in polygenic risk scores to detect heterogeneity in GWAS cohorts. *PLoS Genetics*. 2020;16(9):e1009015.

[17] Lirakis M, Nolte V, and Schlötterer C. Pool-GWAS on reproductive dormancy in Drosophila simulans suggests a polygenic architecture. *G3: Genes, Genomes, Genetics*. 2022;12(3):1–10. Available from: https://academic.oup.com/g3journal/article/12/3/jkac027/6523974.

[18] Id YJ, Id QW, Id RC, *et al.* Integration of multidimensional splicing data and GWAS summary statistics for risk gene discovery. *PLoS Genetics*. 2022;18 (6):e1009814. Available from: https://journals.plos.org/plosgenetics/article?id=10.1371/journal.pgen.1009814.

[19] Asselta R, Paraboschi EM, Mantovani A, *et al. ACE2* and *TMPRSS2* variants and expression as candidates to sex and country differences in COVID-19 severity in Italy. *SSRN Electronic Journal*. 2020, pp. 1–27. Available from: https://papers.ssrn.com/abstract=3559608.

[20] Galmés S, Serra F, and Palou A. Current state of evidence: influence of nutritional and nutrigenetic factors on immunity in the COVID-19 pandemic framework. *Nutrients*. 2020;12(9):2738. Available from: https://www.mdpi.com/2072-6643/12/9/2738.

[21] Das R and Ghate SD. Investigating the likely association between genetic ancestry and COVID-19 manifestations. medRxiv, 2020.

[22] Shelton JF, Shastri AJ, Ye C, *et al.* Trans-ancestry analysis reveals genetic and nongenetic associations with COVID-19 susceptibility and severity. *Nature Genetics*. 2021;53(6):801–808. Available from: https://www.nature.com/articles/s41588-021-00854-7.

[23] Shelton JF, Shastri AJ, Fletez-Brant K, *et al.* The UGT2A1/UGT2A2 locus is associated with COVID-19-related loss of smell or taste. *Nature Genetics*. 2022;54(2):121–124. Available from: https://www.nature.com/articles/s41588-021-00986-w.

[24] Kotur N, Skakic A, Klaassen K, *et al.* Association of vitamin D, zinc and selenium related genetic variants with COVID-19 disease severity. *Frontiers in Nutrition*. 2021;8(689419):1–10.

[25] Ren WL, Wen YJ, Dunwell JM, *et al.* pKWmEB: integration of Kruskal–Wallis test with empirical Bayes under polygenic background control for multi-locus genome-wide association study. *Heredity*. 2018;120(3):208–218.

[26] Casto AM and Feldman MW. Genome-wide association study SNPs in the human genome diversity project populations: does selection affect unlinked SNPs with shared trait associations? *PLoS Genetics*. 2011;7(1):e1001266.

[27] Couch S, Kazan Z, Shi K, *et al.* Differentially private nonparametric hypothesis testing. In: *Proceedings of the ACM CCS*, 2019, pp. 737–751.

[28] Dwork C, McSherry F, Nissim K, *et al.* Calibrating noise to sensitivity in private data analysis. In: *Proceedings of the Theory of Cryptography (*TCC*)*, 2006, pp. 265–284.

[29] Sei Y, Onesimu JA, Okumura H, *et al.* Privacy-preserving collaborative data collection and analysis with many missing values. *IEEE Transactions on Dependable and Secure Computing*. 2022, pp. 1–17.

[30] Wang J, Han H, Li H, *et al.* Multiple strategies differential privacy on sparse tensor factorization for network traffic analysis in 5G. *IEEE Transactions on Industrial Informatics*. 2022;18(3):1939–1948.

[31] Xin B, Geng Y, Hu T, *et al.* Federated synthetic data generation with differential privacy. *Neurocomputing*. 2022;468:1–10. Available from: https://doi.org/10.1016/j.neucom.2021.10.027.

[32] Cheu A, Smith A, and Ullman J. Manipulation attacks in local differential privacy. In: *Proceedings of the IEEE Symposium on Security and Privacy*, 2021, pp. 883–900.

[33] Fienberg SE, Slavkovic A, and Uhler C. Privacy preserving GWAS data sharing. In: *Proceedings of the IEEE International Conference on Data Mining Workshops*, 2011, pp. 628–635.

[34] Uhlerop C, Slavković A, Fienberg SE, *et al.* Privacy-preserving data sharing for genome-wide association studies. *The Journal of Privacy and Confidentiality*. 2013;5(1):137–166.

[35] Gaboardi M and Rogers R. Local private hypothesis testing: chi-square tests. In: *Proceedings of the ICML*, 2018, pp. 1626–1635.

[36] Kohutek ZA, Wu AJ, Zhang Z, *et al* . FDG-PET maximum standardized uptake value is prognostic for recurrence and survival after stereotactic body radiotherapy for non-small cell lung cancer. *Lung Cancer*. 2015;89(2):115–120.

[37] Sq S, White M, Borsetti H, *et al.* Molecular analyses of circadian gene variants reveal sex-dependent links between depression and clocks. *Translational Psychiatry*. 2017;6(3):e748.

[38] Möckel M, Schindler R, Knorr L, *et al.* Prognostic value of cardiac troponin T and I elevations in renal disease patients without acute coronary syndromes: a 9-month outcome analysis. *Nephrology, Dialysis, Transplantation: Official Publication of the European Dialysis and Transplant Association-European Renal Association*. 1999;14(6):1489–1495.

[39] Kim JW, Jang B, and Yoo H. Privacy-preserving aggregation of personal health data streams. *PLoS One*. 2018;13(11):e0207639. Available from: https://dx.plos.org/10.1371/journal.pone.0207639.

[40] Schieppati A, Henter JI, Daina E, *et al.* Why rare diseases are an important medical and social issue. *The Lancet*. 2008;371(9629):2039–2041. Available from: http://www.eurordis.org.

[41] Nguengang Wakap S, Lambert DM, Olry A, *et al.* Estimating cumulative point prevalence of rare diseases: analysis of the Orphanet database. *European Journal of Human Genetics*. 2020;28(2):165–173. Available from: https://doi.org/10.1038/s41431-019-0508-0.

[42] Capponi A, Fiandrino C, Kantarci B, *et al.* A survey on mobile crowdsensing systems: challenges, solutions, and opportunities. *IEEE Communications Surveys and Tutorials*. 2019;21(3):2419–2465.

[43] Gao H, Xu H, Zhang L, *et al.* A differential game model for data utility and privacy-preserving in mobile crowdsensing. *IEEE Access*. 2019;7:128526–128533.

[44] Bosu A, Carver JC, Bird C, *et al.* Process aspects and social dynamics of contemporary code review: insights from open source development and industrial practice at Microsoft. *IEEE Transactions on Software Engineering*. 2017;43(1):56–75.

[45] Pantforder D, Vogel-Heuser B, Grams D, *et al.* Supporting operators in process control tasks – benefits of interactive 3-D visualization. *IEEE Transactions on Human-Machine Systems*. 2016;46(6):895–907.

[46] Mukherjee P and Jansen BJ. Information sharing by viewers via second screens for in-real-life events. *ACM Transactions on the Web*. 2017;11(1):1–24.

[47] Sei Y and Ohsuga A. Privacy-preserving chi-squared test of independence for small samples. *BioData Mining*. 2021;14(6):1–25.

[48] Ren X, Yu CM, Yu W, *et al.* LoPub: high-dimensional crowdsourced data publication with local differential privacy. *IEEE Transactions on Information Forensics and Security*. 2018;13(9):2151–2166.

[49] Torra V. Random dictatorship for privacy-preserving social choice. *International Journal of Information Security*. 2019;19:1–9. Available from: https://doi.org/10.1007/s10207-019-00474-7.

[50] Grining K, Klonowski M, and Syga P. On practical privacy-preserving fault-tolerant data aggregation. *International Journal of Information Security*. 2019;18(3):285–304. Available from: https://doi.org/10.1007/s10207-018-0413-5.

[51] Weggenmann B, Rublack V, Andrejczuk M, *et al.* DP-VAE: human-readable text anonymization for online reviews with differentially private variational autoencoders. In: *Proceedings of the ACM WWW*, Association for Computing Machinery, Inc., 2022, pp. 721–731. Available from: https://doi.org/10.1145/3485447.3512232.

[52] Nautsch A, Jiménez A, Treiber A, *et al.* Preserving privacy in speaker and speech characterisation. *Computer Speech & Language*. 2019;58: 441–480.

[53] Liu C, Yang J, Zhao W, *et al.* Face image publication based on differential privacy. *Wireless Communications and Mobile Computing*. 2021;2021 (6680701):1–20.

[54] Kakizaki K, Fukuchi K, and Sakuma J. Differential privacy based on geometrical interpretation of chi-squared testing. In: *Computer Security Symposium*, 2016, pp.1199–1206.

[55] Kakizaki K, Fukuchi K, and Sakuma J. Differentially private chi-squared test by unit circle mechanism. In: *Proceedings of the ICML*, 2017, pp. 1761–1770.

[56] McSherry F and Talwar K. Mechanism design via differential privacy. In: *Proceedings of the IEEE FOCS*, 2007, pp. 94–103.

[57] Banerjee A, Chitnis UB, Jadhav SL, Bhawalkar JS, and Chaudhury S. Hypothesis testing, type I and type II errors. *Industrial Psychiatry Journal*. 2009;18(2):127.

[58] Johnson A and Shmatikov V. Privacy-preserving data exploration in genome-wide association studies. In: *Proceedings of the ACM KDD*, 2013, pp. 1079–1087.

[59] Dwork C, Kenthapadi K, McSherry F, *et al.* Our data, ourselves: privacy via distributed noise generation. In: *Proceedings of the Eurocrypt*, vol. 4004, 2006, pp. 486–503.

[60] Sei Y and Ohsuga A. Privacy-preserving chi-squared testing for genome SNP databases. In: *Proceedings of the 39th International Conference of the IEEE Engineering in Medicine and Biology Society* (*IEEE EMBC*), 2017.

[61] Kifer D and Rogers R. A new class of private chi-square tests. In: *Proceedings of the International Conference on Artificial Intelligence and Statistics (AISTATS)*, 2017, pp. 991–1000.
[62] Canonne CL, Kamath G, McMillan A, *et al.* The structure of optimal private tests for simple hypotheses. In: *Proceedings of the ACM STOC*, 2019, pp. 310–321.
[63] Csail MA, Diakonikolas I, Kane D, *et al.* Private testing of distributions via sample permutations. In: *Proceedings of the NIPS*, 2019, pp. 10878–10889.
[64] Liu C, He X, Chanyaswad T, *et al.* Investigating statistical privacy frameworks from the perspective of hypothesis testing. In: *Proceedings of the PET*, 2019, pp. 233–254.
[65] Sweeney L. Achieving k-anonymity privacy protection using generalization and suppression. *International Journal of Uncertainty, Fuzziness and Knowledge-Based Systems*. 2002;10(05):571–588.
[66] Terrovitis M, Mamoulis N, Liagouris J, *et al.* Privacy preservation by disassociation. *Proceedings of the VLDB*. 2012;5(10):944–955.
[67] Machanavajjhala A, Kifer D, Gehrke J, *et al.* l-Diversity: privacy beyond k-anonymity. *ACM TKDD*. 2007;1(1):3.
[68] Nergiz ME, Atzori M, and Clifton C. Hiding the presence of individuals from shared databases. In: *Proceedings of the ACM SIGMOD*, 2007, pp. 665–676.
[69] Evfimievski A, Srikant R, Agrawal R, *et al.* Privacy preserving mining of association rules. *Information Systems.* 2004;29(4):343–364.
[70] Evfimievski A, Gehrke J, and Srikant R. Limiting privacy breaches in privacy preserving data mining. In: *Proceedings of the ACM PODS*, 2003, pp. 211–222.
[71] Atanassov E and Dimov IT. What Monte Carlo models can do and cannot do efficiently? *Applied Mathematical Modelling*. 2008;32(8):1477–1500.
[72] Cabin RJ and Mitchell RJ. To Bonferroni or not to Bonferroni: when and how are the questions. *Bulletin of the Ecological Society of America*. 2000;81(3):246–248.
[73] Conrad DF, Jakobsson M, Coop G, *et al.* A worldwide survey of haplotype variation and linkage disequilibrium in the human genome. *Nature Genetics*. 2006;38(11):1251–1260.
[74] Pemberton TJ, Jakobsson M, Conrad DF, *et al.* Using population mixtures to optimize the utility of genomic databases: linkage disequilibrium and association study design in India. *Annals of Human Genetics*. 2008;72 (4):535–46.
[75] Sharpe D. Your Chi-square test is statistically significant: now what? *Practical Assessment, Research & Evaluation*. 2015;20(8):1–10.
[76] Angelopoulos S, Brown M, McAuley D, *et al.* Stewardship of personal data on social networking sites. *International Journal of Information Management.* 2021;56(102208):1–11.
[77] Rodríguez-Sabiote C, Álvarez-Rodríguez J, Álvarez-Ferrandiz D, *et al.* Using chi-squared automatic interaction detection modelling to identify

student opinion profiles regarding same-sex couples as a family structure. *Heliyon*. 2021;7(3):e06469.

[78] Chen YC, Li TC, Chang YW, *et al.* Exploring the relationships among professional quality of life, personal quality of life and resignation in the nursing profession. *Journal of Advanced Nursing*. 2021;77(6):2689–2699. Available from: https://onlinelibrary.wiley.com/doi/full/10.1111/jan.14770 https://onlinelibrary.wiley.com/doi/abs/10.1111/jan.14770 https://onlinelibrary.wiley.com/doi/10.1111/jan.14770.

[79] Makrinioti H, Hasegawa K, Lakoumentas J, *et al.* The role of respiratory syncytial virus- and rhinovirus-induced bronchiolitis in recurrent wheeze and asthma – a systematic review and meta-analysis. *Pediatric Allergy and Immunology*. 2022;33(3):e13741. Available from: https://onlinelibrary.wiley.com/doi/full/10.1111/pai.13741 https://onlinelibrary.wiley.com/doi/abs/10.1111/pai.13741 https://onlinelibrary.wiley.com/doi/10.1111/pai.13741.

[80] Mizobe F, Takahashi Y, and Kusano K. Risk factors for jockey falls in Japanese thoroughbred flat racing. *Journal of Equine Veterinary Science*. 2021;106:103749.

[81] Ahmad I, Dar MA, Fenta A, *et al.* Spatial configuration of groundwater potential zones using OLS regression method. *Journal of African Earth Sciences*. 2021;177:104147.

[82] Luo X, Xia H, Yang W, *et al.* Characteristics of patients with COVID-19 during epidemic ongoing outbreak in Wuhan, China. *medRxiv*. 2020, pp. 1–17. Available from: https://doi.org/10.1101/2020.03.19.20033175.

[83] Poyiadi N, Cormier P, Patel PY, *et al.* Acute pulmonary embolism and COVID-19. *Radiology*. 2020;201955:1–9. Available from: http://pubs.rsna.org/doi/10.1148/radiol.2020201955.

[84] Jacob L, Smith L, Butler L, *et al.* COVID-19 social distancing and sexual activity in a sample of the British Public. *Journal of Sexual Medicine*. 2020;17(7):1229–1236. Available from: https://doi.org/10.1016/j.jsxm.2020.05.001.

[85] Bearden WO, Sharma S, and Teel JE. Sample size effects on chi square and other statistics used in evaluating causal models. *Journal of Marketing Research*. 1982;19(4):425–430.

[86] Bentler PM and Bonett DG. Significance tests and goodness of fit in the analysis of covariance structures. *Psychological Bulletin*. 1980;88(3):588–606.

[87] Sei Y, Takenouchi T, and Ohsuga A. An efficient algorithm for encrypted text searching in cloud computing. *IPSJ Journal*. 2015;56(10):1977–1987.

[88] Sei Y, Okumura H, and Ohsuga A. Privacy-preserving publication of deep neural networks. In: *Proceedings of the IEEE DSS*, 2017, pp. 1418–1425.

Chapter 16

Automated decision support system for diagnosing sleep diseases using machine intelligence techniques

Santosh Kumar Satapathy[1], Bidita Khandelwal[2,3], Amik Garg[4] and Akash Kumar Bhoi[3,4,5]

Abstract

Sleep is one of the human health's most vital yet often underrated components. Sleep studies are crucial for unearthing various abnormalities associated with sleep, widely prevalent in today's world and bound to increase over the years. An increasingly rapid lifestyle makes short sleeping hours surprisingly too common. Sleep deprivation can heavily impact humans and their quality of life. Diagnosing sleeping issues accurately during the initial stages is one of the significant challenges faced by the medical community. Sleep stage scoring is the primary step in detecting sleep abnormalities or dividing a person's entire sleep duration into different categories according to muscle movements, brain activity, eye movements, etc. Polysomnography is the scientific test that records these human activities during sleep through electrodes connected to patients. A hypnogram that results from this test is the graphical form of the sleep scoring done by technicians. For ages, this process has been carried out manually, which is frequently prone to error, requires ample time, effort, and training, and is susceptible to inter-scorer differences. Therefore, it is essential to devise an automated system for sleep staging. This experimental study involves machine learning techniques to classify the different sleep stages. The reported results proved that the proposed sleep staging model was well performed for the five-class classification task with improved accuracy using the ensemble learning classification model.

[1]Department of Information and Communication Technology, Pandit Deendayal Energy University, India
[2]Department of General Medicine, Sikkim Manipal Institute of Medical Sciences, Sikkim Manipal University, India
[3]Directorate of Research, Sikkim Manipal University, India
[4]KIET Group of Institutions, Delhi-NCR, India
[5]AB-Tech eResearch (ABTeR), India

Keywords: Electroencephalogram (EEG); Sleep staging; Ensemble Learning Algorithm

16.1 Introduction

Sleep is known scientifically as a natural, reversible, and repetitive condition of diminished receptivity to external stimulus in conjunction with complicated and anticipated physiological changes. It is one of the most vital components central to human functioning. However, in keeping up with the rapid pace of today's civilization, a growing number of people have become habitual of sacrificing sleep and falling victim to numerous sleeping conditions that have the potential to lead to far more dangerous consequences if not addressed early on.

An increasing amount of data, primarily from the Western populations, point to a drastic decrease in the average length of sleep and an increased occurrence of sleep irregularities [1]. About 62% of the adult population worldwide have reported continued dissatisfaction with their sleep quality due to various factors (Philips Global Sleep Survey, 2019). Statistics show that insomnia affects about 38–40% of older adults, and about 15–30% of males and 10–30% of females have been diagnosed with Obstructive Sleep Apnea (OSA) [2,3]. Insufficient sleep has become a pervasive problem in modern society. It is now a global public health crisis frequently underestimated and overlooked and has a relatively sizeable economic aftermath. An absence of sufficient sleep over long durations can lead to many physiological and psychological dysfunctions. Moreover, timely detection of sleep-related illnesses can serve as an early signal to various underlying health issues before their aggravation. Reduced sleep has been associated with 7 out of 15 primary causes of death in the United States, including vehicular accidents, hypertension, cerebrovascular disorders, and cardiovascular diseases [4]. Moreover, problems in falling asleep, as well as daytime drowsiness and exhaustion, impact even wider sectors of the community, creating a significant encumbrance when it comes to sickness and death, besides considerable societal costs in industrialized countries by adversely affecting its human capital, from taking toll of its healthy citizens to reducing their capacity of contributing to the nation's social and economic welfare [5]. The correct diagnosis and treatment of sleeping disorders have thus become of supreme importance in a world plagued by the lack of sleep.

Sleep stage classification is often the first and foremost step toward identifying and addressal of potential sleep-related issues. The quality of a patient's sleep can often be evaluated by analyzing the pattern obtained by segmenting his entire sleep duration into stages, each of which possesses characteristics unique to its classification. Known as the current gold standard for measuring sleep objectively, polysomnography is the scientific technique that takes in raw physiological data by utilizing several surface electrodes which take note of a multitude of physiological and behavioral changes during sleep, ranging from brain dynamics which are tracked by electroencephalography (EEG), to eye movements measured by electrooculography (EOG), to the varying heart rate evaluated by electrocardiography

(ECG). Muscle activity is taken care of by electromyography (EMG) [6]. As per the American Academy of Sleep Medicine, sleep is conventionally classified into two predominant categories, i.e., the Rapid Eye Movement (REM) stage and the Non-Rapid Eye Movement (NREM) stage. NREM further comprises N1, N2, and N3. A thorough analysis of the graphical representation of sleep stages plotted against time, known as a hypnogram, helps provide clinicians insight into a person's sleep quality and is used to observe anomalies, serving as the preliminary aid in recognizing sleep-related sleep disorders. The brief diagnosis steps of different sleep-related disorders are presented in Figure 16.1.

This method of categorization to identify sleep dysfunctions has been around for years. However, to date, the task of segmenting sleep duration into stages continues to be performed manually, as was the case in earlier days. This technique of conducting sleep study is often prone to error, owing to the complete reliability of a human practitioner, which is based on unrealistic expectations regarding accuracy and detail. Being solely dependent on the visual pattern detection by a human specialist makes sleep stage classification extremely time-consuming, laborious, and subjective [7]. The process also requires a large staff of highly trained sleep technicians, which is not always available, leading to significant mismatches in the diagnosis results if compared. Several studies have looked into inter-rater dependability and found that scorer correspondence is far from optimal [8].

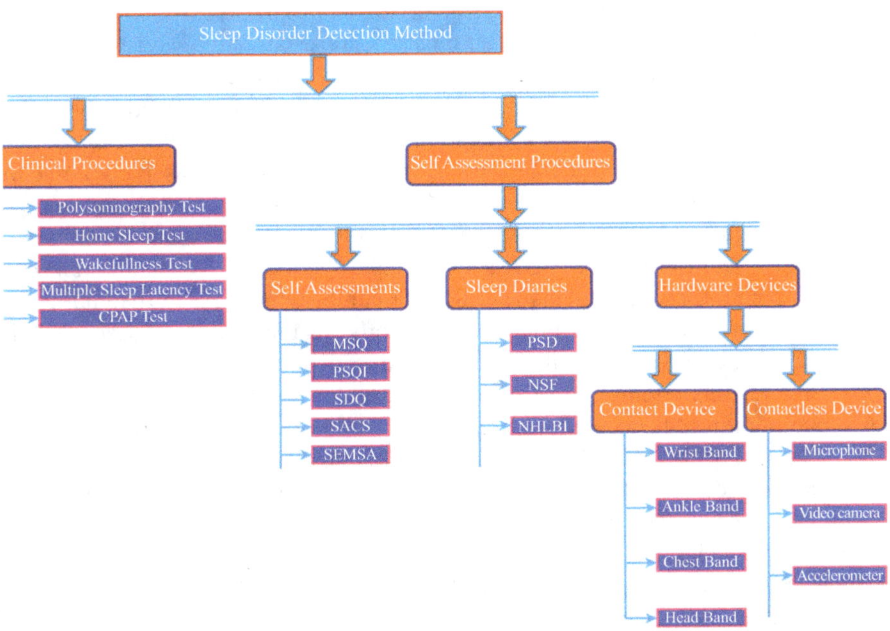

Figure 16.1 Pictorial representation of differential diagnoses of sleep-rel

The advent of machine learning (ML) [1–52] has opened up the possibility of automatizing sleep stage classification with a vision to establish common classification standards and eliminate the human tendency for error through better categorization algorithms and more time-efficient methods. Advances in computerized scoring to the point of its actual feasibility in real-life scenarios can lead to a revolutionary future for the effective and efficient diagnosis and treatment of sleep-related illnesses, mainly reducing the complexity and inaccuracy of the whole procedure [9]. While several ML approaches have been employed for this purpose, and significant study is still being conducted on account of the gravity and scope of this subject in modern times, it remains a challenge to achieve the same results through artificially designed models as those obtained by highly trained clinicians performing the study. Several factors are responsible for this, ranging from the right choice of ML algorithms to the method employed to extract features from the sample data. Handcrafted features and the establishment of mathematical models often make automatization trickier [10]. The abrupt and unprecedented transitions between sleep cycles frequently result in hazy feature extractions, which might result in an incorrect evaluation. Moreover, the pattern of human brain impulses is much more complex than the current human ability to comprehend, leading to information loss and contributing to the complexity of feature extraction through ML techniques [11].

As a counter approach to manual feature extraction, which requires a lot of experimentation and expertise to figure out the optimal features possessing the most excellent weightage, researchers nowadays are resorting to the application of deep learning (DL) algorithms [53–83] involving neural networks to classification models, which make them capable of automatic feature extraction, and able enough to deduce and tune the essential features from the supplied data alone without human intervention. However, as sound as a DL model may theoretically seem, several shortcomings hinder its applicability to practical scenarios. One of the significant drawbacks of using a DL model is that it requires a massive volume of data to perform better or even at par with other technologies. This limits its usage in medicine to a great degree since medical data is often scarce, as most of it is never made public. Additionally, DL models are complicated to train by their complexity and high computing requirements leading to a cumbersome cost to the end-users. Therefore, it is paramount to build a sufficiently accurate model based on ML approaches that can be practically implemented in the current diagnosis.

Therefore, this research aims to demonstrate the findings of an experimental study incorporating ML techniques to resolve the traditional problem of sleep stage classification. For this work, we have carried out several trials involving various algorithms and datasets to decide upon the best possible method and verify the accuracy of the approach across a wide variety of situations. Random Forest (RF) [9], K-nearest neighbors (KNN) [10], decision tree [12], and support vector machine (SVM) [14] are some significant classifiers on which our proposed model has been tested. Sufficient training on a wide variety of sleep data from patients suffering from various disorders and healthy subjects has allowed the model to become experienced enough in detecting a host of sleep abnormalities.

This review-based research thus aims to inspire multiple opportunities for future work involving ML and DL in sleep medicine and to open up avenues for new experimentation by progressing from the results achieved herein. Upcoming sections of the paper entail a comprehensive discussion and analysis of the approach adopted and its observed superiority, the results obtained, and the scope of further improving upon this work.

16.2 Related work

Huy Phan *et al.* [4] proposed a DL approach and followed a method called an end-to-end framework for having a particular segment-to-segment sleep staging. The two networks are used named SeqSleepNet+ and DeepSleepNet+ were used in this study, and both networks are upgraded versions of SeqSleepNet and DeepSleepNet, respectively. Three datasets were used for testing: Sleep-EDF-SC, Sleep-EDF-ST, and Surrey-cEEGrid. The purpose of using three different datasets was to cover all data from different sections to obtain optimized results. The results of such different datasets are knowledge transfer and better accuracy. The accuracy obtained on Sleep-EDF-SC, Sleep-EDF-ST, and Surrey-cEEGrid datasets signifies that SeqSleepNet+ out-performs DeepSleepNet+ by 1.7 %, 6.6 %, and 17.3 %, respectively.

Nico Surantha *et al.* [5] studied sleep stage classification using MIT BIH polysomnographic dataset. The dataset consists of multichannel physiologic signals recorded during the sleep period. The dataset contains recordings of over 80h, each with ECG signals. Three algorithms were used to carry out the study: extreme learning machine, SVM, and integration of ELM with PSO. The preprocessing of the dataset was done in all methods, and feature extraction was carried out. For training purposes, 70% of the dataset was considered, while 30% for the testing. ELM and PSO algorithms surpassed ETM and SVM algorithms, and the accuracy obtained for ELM and PSO algorithms is 82.1%, 76.77%, 71.52%, and 62.66% for two classes, three classes, four classes, and six classes, respectively.

Emadeldeen Eldele *et al.* [6] proposed an architecture named Attnsleep, which uses a DL algorithm and single-channel EEG signal to classify sleep stages. The architecture is mainly divided into two modules: multi-resolution convolutional neural network (MRCNN) and adaptive feature recalibration (AFR). MRCNN is used to extract features based on low to high frequencies, and by modeling the interdependencies between the components, the AFR can increase the quality of the extracted features. The second module, TCE, uses a multi-head focus approach to identify the temporal connections between derived parts. Three different datasets were used to perform the research: Sleep-EDF-20, Sleep-EDF-78, and SHHS, and the accuracy obtained by Attnsleep were 84.4, 81.3, and 84.2, respectively.

G. Naveen Sundar *et al.* [7] performed the study using a DL approach and ML algorithms for sleep stage classification using EEG signals. The dataset used in the study is the PhysioNet Sleep-EDF dataset which contains the data of 197 Polysomnograms. The PSG is sleep recordings performed at 100Hz. The data preprocessing helps extract EEG signals of 30sec from EEG epochs. The technique used in this approach is called CNN-BiLSTM-CRF, and the accuracy obtained is 90.7% which is far better than AlexNet, ResNet, GGNet, and LeNet. The F1 score obtained is 90.6%, while precision and recall are 90.5% and 92.7%, respectively.

Chandra Bhushan Kumar *et al.* [8] proposed a model named as SCL-SSC for sleep stage classification. The method is divided into two parts first is feature learning, and the second one is feature classification. W and N2 sleep stages are challenging to categorize. They used a weighted softmax cross-entropy loss function to overcome the problem, and an oversampling technique was applied to the dataset. The datasets used are Sleep-EDF sleep 2013 and 2018, which are available openly. The model obtained an accuracy of 94.1% with a Kappa score of 0.9197. They also got an F1 score of 92.64%.

Cong Liu *et al.* [9] obtained EEG signal of a single channel driven by data. They used EEMD to decompose EEG epochs and extracted numerous features from signals and decomposed IMFs. After removing the data, the model was trained using the XGBoost algorithm. The testing was done using a five-fold cross-validation technique on the Dreams, Sleep-EDF, and SHHS databases. Their outputs showcased that the four-class classification obtained 86.4%, 93.1%, and 87.5% accuracy for three databases, respectively, and the five-class sort obtained 83.4%, 91.9%, and 85.8% for all three datasets, respectively. They also observed that EEG signals could be recorded by placing dry electrodes on the forehead instead of prefrontal derivations.

Liangsheng Zhang *et al.* [10] considered EEG data into sub-datasets using the bootstrap sampling method and a sample array. L2 norm based for weight optimization of ensemble classifier was integrated with the decision by weighted fusion of ensemble classifier. This method was assessed on public databases, and the comparison was made with the results of general classification techniques. The two datasets obtained 88.80% and 86.53% accuracy, respectively, the highest of all the classification methods. Also, the standard deviation of the ensemble classification method was the lowest.

Ozal Yildirim *et al.* [11] presented a technique of DL utilizing PSG signals, and by using EEG and EOG signals, the one-dimensional CNN model was developed. The model results were tested on two famous public datasets: Sleep-EDF and Sleep-EDFX. The performance was calculated for two to six sleep classes classification problems and accuracy obtained were 98.06%, 94.64%, 92.36%, 91.22%, and 91.00% for Sleep-EDF dataset and 97.62%, 94.34%, 92.33%, 90.98%, and 89.54% for Sleep-EDFX dataset. The results for Sleep-EDFX were the highest, and they claimed that the proposed model was ready for use in clinical conditions and could be evaluated with huge PSG data.

Geetika Kaushik *et al.* [12] showcased a unique approach to analyzing the capability of the Hjorth parameter for detecting seizures utilizing EEG signals. For the decomposition of EEG, TQWT is applied to different subbands at distinct levels. Activity, mobility, and complexity are some Hjorth parameters analyzed over decomposed parameters. The dataset used to validate the methodology was proposed by the University of Bonn in Germany. The outputs showcased that estimation of Hjorth characteristics is efficient and suitable for automated sleep stage classification. Also, evaluating this method on various combinations gave a very high classification accuracy. The state-of-the-art approaches were also compared with the proposed methodology, where this method stood out with the highest efficiency.

Farideh Ebrahimi *et al.* [13] reviewed an automated sleep stage classification technique by cardiorespiratory signals. The proposed review identifies specific essential points that need to be considered to improve sleep staging accuracy. Four central points were highlighted via analysis. First, the 30-sec epoch of the signal does not carry enough information for feature extraction; the minimum length of the call should be 4.5min if 30sec epoch is considered. Second, the delay in time should be regarded as between cardiorespiratory signals and signals from the central nervous system. Third, the ECG signal data could help increase the efficiency of sleep staging, and lastly, CNN and LSTM should be used to structure the model for better accuracy.

Huafeng Wang *et al.* [14] kept the importance of sleep for human beings' neural and physiological development and his team suggested a multiscale dual attention network (MSDN) for better extraction of features automatically using a single direction of CNNs. The network was found to be reliable with complex EEG signals well. Sleep EDF and its expanded version were used to train the model. Out of 197 subjects, 157 were used for training, while 40 were used for validation and testing purposes—two techniques, *k*-fold cross validation where k=20 and hold-out validation, were used in the algorithm. The accuracy obtained was 90.35% and 88.38%, respectively. It was also observed that when the model was trained and tested by the expanded version of the Sleep-EDF database, it showed better accuracy and efficient performance. The result's betterment was "more the data; more the model can learn all the features."

Sarun Paisarnsrisomsuk *et al.* [15] suggested a CNN model with higher accuracy and efficiency than previous works and research. In the suggested model, by observing the responses of the CNN's inner layers, it was observed that the CNN filters are used for extracting the salient features (time and frequency domain) from both the signals. The dataset used two EEG and one EOG signal per training epoch. The overall accuracy attained by the model is 81%. The accuracy of this model was quite more than the model proposed in the previous works of sleep staging using CNN.

Mehdi Abdollahpour *et al.* [2] thinking about the need for faster, more accurate, and automated sleep stage classification, developed a model using high-level convolutional layers. The dataset utilized was Sleep-EDF and Sleep-EDF expanded version. The Sleep-EDF dataset consists of two EEG signals and one EOG signal per subject. The convolutional network was used to extract features from these signals of the dataset. Two sets are created: one consists of parts just from EEG and another consists of features combined of EEG and EOG, which are then converted into horizontal visibility graph (HVG). It was observed that transfer learning has substantially increased the accuracy and made the whole training process faster. The precision attained was 98.77%, and Cohen's kappa coefficient was 0.899. Various classifiers like SVM, logistic regression (LR), and RF were also used. The highest accuracy was TLCNN, and the lowest was SVM, around 97.35%.

Nicola Michielli *et al.* [4] performed the study using single-channel EEG signals available in the Sleep-EDF dataset. The main aim of this study is to make recognizing sleep disorders automated and adversely increase the accuracy and performance of models for all six stages. The proposed methodology suggests an architecture consisting of novel cascaded RNN and LSTM. Around 55 features were invented by the model using RNN. Both time and frequency domain parts were excavated. Two fused RNN and LSTM networks were used. The first structure attained an accuracy of 90.8% for 4-class categorization, and the other structure's performance was 83.6% for 2-class categorization.

16.3 Experimental dataset

For experimental purposes, in this study, we have acquired the sleep recordings from ISRUC-Sleep public repository. The set of sleep experts combined recorded the sleep behavior of the different categories of subjects and the entire recording were done at the Hospital of Coimbra University (CHUC) [25]. The different subgroups of recordings are contained. One hundred subjects' recordings were acquired in subgroup-I, who were suffered little bit sleep problems, 8 subjects' recordings were obtained in subgroup-II, mainly the subjects were affected with different sleep-related disorders. Finally, in subgroup-III, only 10 healthy controlled subjects were collected. For this experimental work, we used only subgroup-I of the ISRUC-Sleep dataset. A brief description of the ISRUC-Sleep dataset is given in Table 16.1. Table 16.2 provides information about the different physiological signals and channels. Table 16.3 presents recorded sleep epochs from the different sleep stages. Details of the study subjects are given in Table 16.4. Figures 16.2 and 16.3 show a sample signal collected from a sleep disorder subject engaged in a single session of sleep data recording; Figure 16.1 presents a sample EEG signal that indicates the sleep behavior of a matter affected by a sleep problem that engaged in two different recording sessions. This study includes three categories of subjects. One is affected by sleep problems and engaged in one sleep data recording session. The subjects in the second category also had sleep disorders but engaged in two data recording sessions. Finally, the third category consists of healthy subjects. We obtained the EEG channel C3-A2 signals of 6 subjects. The stage annotations are also provided in the data repository according to AASM rules. The sleep epochs are labeled W, N1, N2, N3, and R for a wake, N-REM1, N-REM2, N-REM3, and REM, respectively. Unscored ages were not considered for further analysis. Each epoch has a length of 30sec. For the sleep behavior analysis, we presented the samples of all EEG signals extracted from a different sleep stages category of subjects. The sleep EEG signals change according to the sleep stage of the subject.

Every stage of sleep is characterized by unique behavior in a sleeping brain, such as low amplitude, mixed frequency, or sawtooth waveforms, low amplitude muscle movements, and rapid eye movements. The behavior of EEG signals is complex since they are not periodic and also due to the continuous changes of amplitude, frequency, and phase range observed according to the sleep stage. Some characteristics of the different sleep stages are represented in Figures 16.4 and 16.5.

Table 16.1 Summary information with recent research developments

References	Classifier	Signal	No. of classification levels	Results (%)
[36]	Least square support vector machine (LS-SVM)	EEG2 channels	Six	96%
				96.74%
[65]	Ensemble learning algorithm	EEG1 channel	Five	96.67%
[65]	RF	EEGEEG+EOG+EMGEEG +EOG+EMG +ECG8 channels	Five	76.05% 85.30%
				86.24%
[19]	CNN	EEGEOG4 channels	Five–Two	91.00% 91.22% 92.36% 94.64% 98.06% 89.54% 90.98% 92.33% 94.34% 97.62%
[70]	RF+ Hidden Markov Model (HMM)	EEG1 channel	Five–Two	77.01% 79.12% 86.04% 95.47% 72.79% 78.48% 85.77% 92.43%
[71]	One-dimensional convolutional neural network	EEG2 channels	Five	92.66%
[72]	RF	EEGEOGEMG3 channels	Five	92%
[73]	RNN	EOGR–R interval (RR) signals2 channels	Five	84.4%
				74.3%
[74]	SVM	EEG1 channel	Five	90%
[75]	Bagged Trees	EEG1 channel	Five–Two	79.90% 82.08% 88.22 % 96.48% 81.65% 84.68% 90.54% 96.18% 89.54%

(Continues)

Table 16.1 (Continued)

References	Classifier	Signal	No. of classification levels	Results (%)
				90.98%
				92.33%
				94.34%
				97.62%
[76]	CNN	EEG+EOG+ECG+EMG4-channels	Five	93.7%
				82.8%
[77]	RNN	EEGEOGEMG15 channels	Five	89.9%
				88.7%
[78]	RNN	EEG1 channel	Five	83.7%
		EEG+EOG2 channels		83.9%
[79]	Convolutional neural network	EEG1 channel	Five	93.58%
				93.16%
[80]	KNN	EEG1 channel	Six	93.57%
[81]	Temporal Convolutional neural network	EEG2 channels	Five	85.39%
				82.46%
[82]	Bootstrap aggregating (bagging)	EEG1 channel	Five	89.39%
				90.66%
				94.09%
				94.17%
				97.50%
				87.41%
				88.73%
				90.14%
				92.00%
				96.78%

Table 16.2 ISRUC-sleep subgroup-I dataset structure

Recorded signals	Channel information
Electro_Eencephalogram	EEG_{C3-A2}, EEG_{C4-A1}, EEG_{F3-A2}, EEG_{F4-A1}, EEG_{O1-A2}, EEG_{O2-A1}
Electro_Ocoulogram	EOG_{LOC-A2}, EOG_{ROC-A1}
Electro_Myogram	EMG_{Chin}, $EMG_{Left\ leg}$, $EMG_{Right\ Leg}$

Table 16.3 Sleep epochs information

Patient ID/subgroups	WA	N-1	N-2	N-3	RE
Sub-1(SG-I)	164	64	174	232	119
Sub-2 (SG-I)	232	73	227	148	75
Sub-9 (SG-I)	73	144	316	137	85
Sub-16 (SG-I)	129	126	281	121	98

Table 16.4 Percentage of epochs per individual sleep stages

Patient ID	Wake (%)	N1 (%)	N2 (%)	N3 (%)	R (%)
Sub-1	22.00	8.40	23.07	30.80	15.73
Sub-2	30.80	9.60	30.13	19.60	9.87
Sub-9	9.6	19.06	42	18.13	11.2
Sub-16	17.07	16.67	37.33	16.00	12.93

Table 16.5 Medical history of the subjects

Patient ID	Age	Gender	Disease
1	63	Male	Depression
2	51		Restless leg syndrome
9	62		Cheyne–Stokes
16	51		No

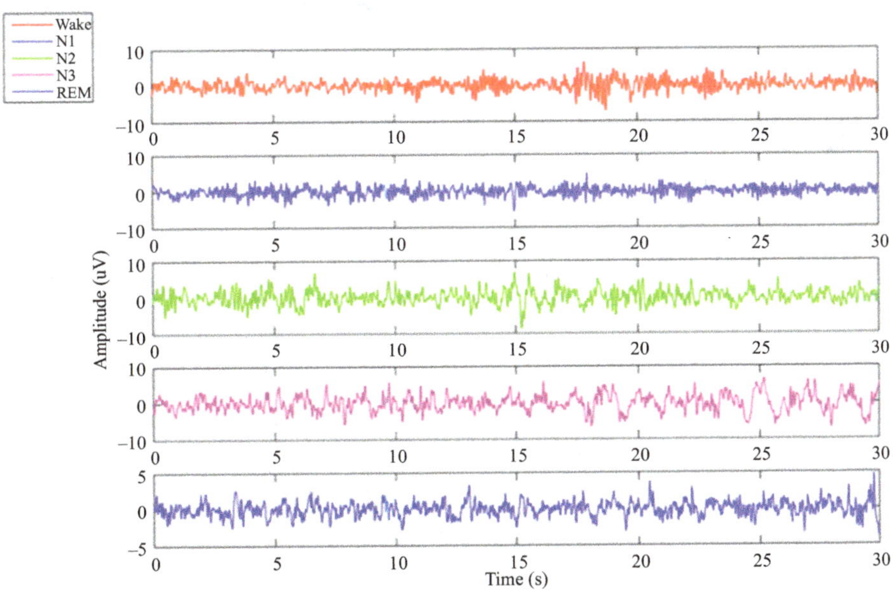

Figure 16.2 Sleep behavior of the subject-16

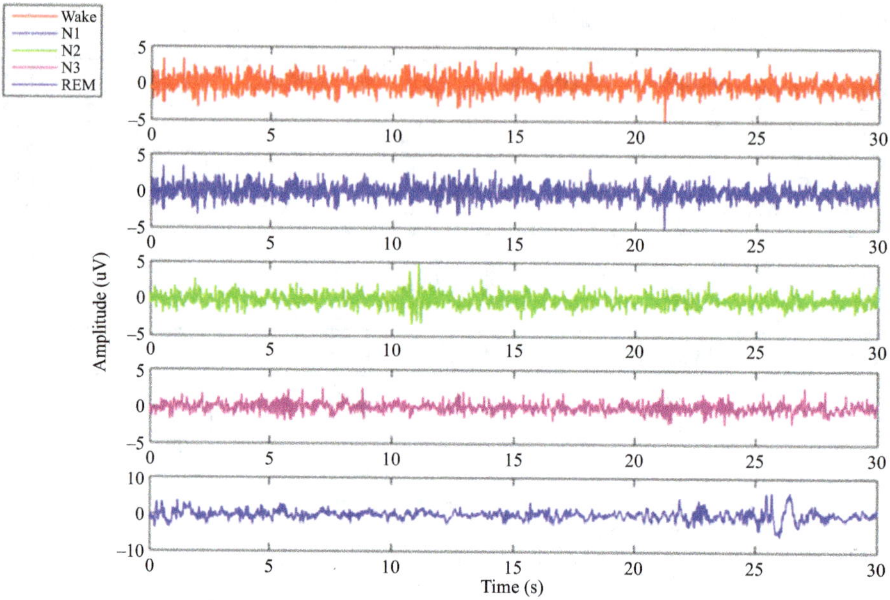

Figure 16.3 Sleep behavior of the subject-1

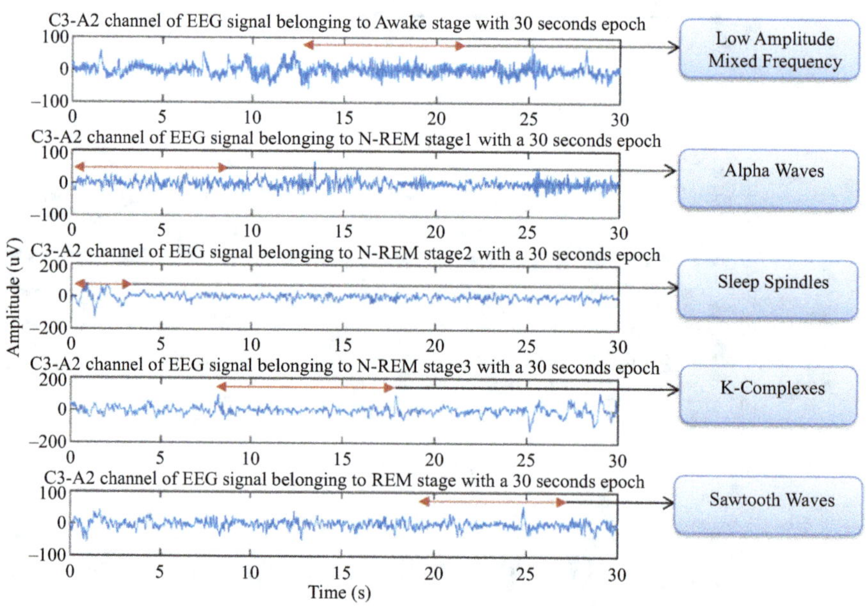

Figure 16.4 C3-A2 channel EEG signal: subject-16 (sleep disordered subject)

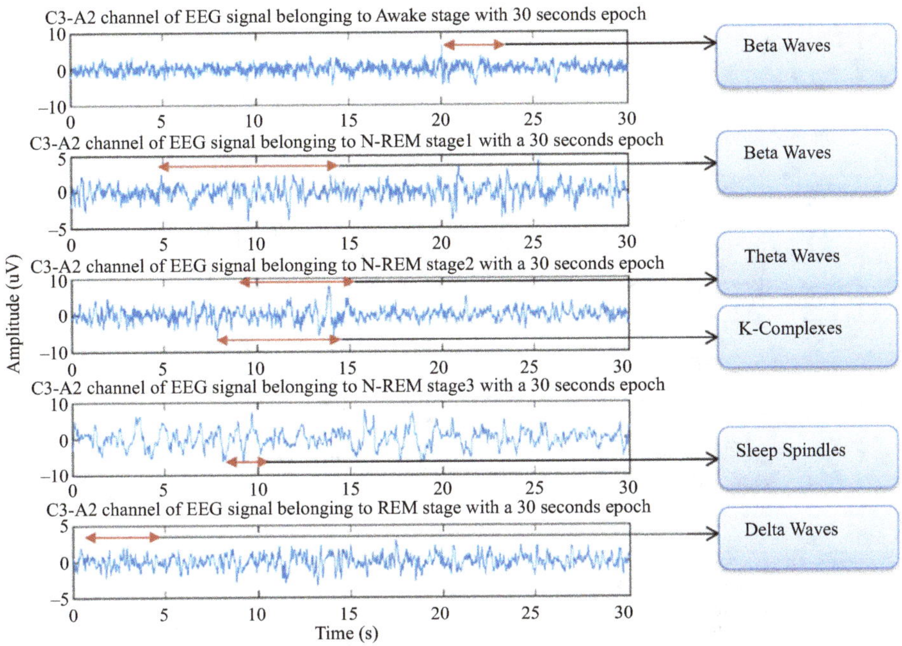

Figure 16.5 C3-A2 channel EEG signal: subject-05 (healthy subject)

16.4 Proposed automatic sleep stage detection method

This chapter proposes an improved and robust sleep staging framework using single-channel EEG signals based on ML techniques. Mainly, in existing studies, there are two significant concerns such as (i) lack of available generalized classification models and (ii) misclassified the N1 and REM sleep stages due to similar characteristics of the sleep stages. This work addressed these issues with improved sleep staging classification accuracy. In this proposed model, we have considered a novel framework by obtaining an ensemble learning algorithm for five-class classification problems. Mainly the whole proposed methodology is carried out into four phases: The block diagram of the proposed sleep staging model is presented in Figure 16.6.

16.5 Classification

16.5.1 SVM

A separating hyperplane can be defined as

$$f(x) = w_i . x_i + b \tag{16.1}$$

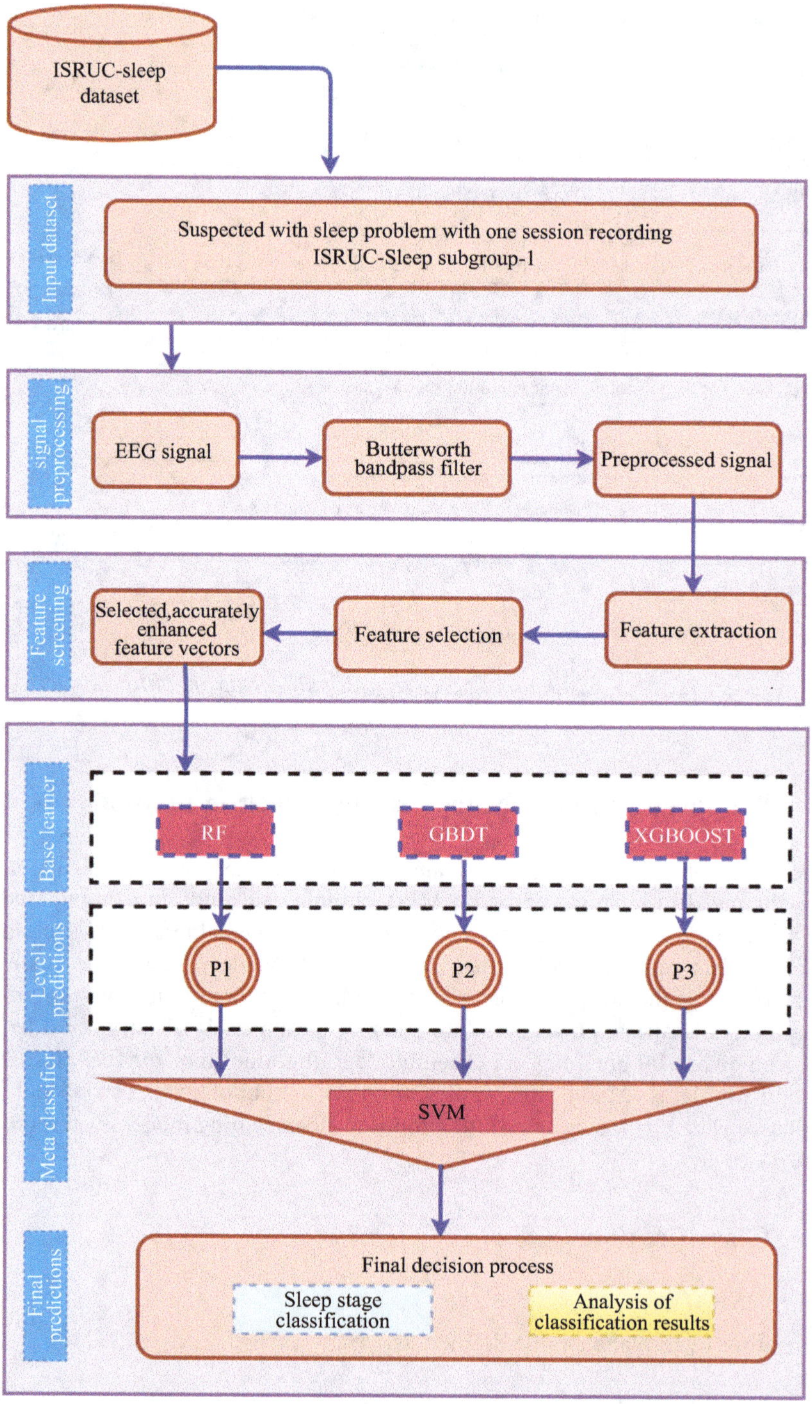

Figure 16.6 Complete structure of proposed sleep staging model

Table 16.6 Retrieved features

Feature type	Feature names	Feature order
Time domain	Mean	Fet_1
	Maxi	Fet_2
	Mini	Fet_3
	Sta_Dev	Fet_4
	Median	Fet_5
	Var	Fet_6
	Zero-Crossing Rate	Fet_7
	75 percentile	Fet_8
	Sig_Skew	Fet_9
	Sig_Kurt	Fet_10
	Sig_Acti	Fet_11
	Sig_Mobi	Fet_12
	Sig_Comp	Fet_13
Frequency domain	Rel_Spect_Pow_δ	Fet_14
	Rel_Spect_Pow_θ	Fet_15
	Rel_Spect_ Pow_ α	Fet_16
	Rel_Spect_ Pow_ β	Fet_17
	Pow_Rat _δ/β	Fet_18
	Pow_Rat_ δ/θ	Fet_19
	Pow_ Rat_ θ/α	Fet_20
	Pow_ Rat_ θ/β	Fet_21
	Pow_Rat_ α/β	Fet_22
	Pow_ Rat_ α/δ	Fet_23
	Pow_Rat_ $(\theta+\alpha)/(\alpha+\beta)$	Fet_24
	Band_pow_δ	Fet_25
	Band _pow_θ	Fet_26
	Band_pow_α	Fet_27
	Band_pow_β	Fet_28

where x_i the set of training is sample features and $w_i \in R_n$ is the weights for one each feature is vector and b is a scalar bias.

The hyperplane H_0 contains information on regions of vectors x_i,which satisfied the equation $f(x) = 0$, H_1 and H_2 are the two hyperplanes, which are placed parallel to H_0, which defined the information on $f(x) = 1$ and $f(x) = -1$, respectively [24].

$$H_1 = w_i.x_i + b = +1 \tag{16.2}$$

$$H_2 = w_i.x_i + b = -1 \tag{16.3}$$

The distance between H_1 and H_2 is called the margin. The margin of a separating the hyperplane is d+ + d−. The pictorial representation of the two-state classification model are presented in Figure 16.7. The separating distance width (d) between H_1 and H_2 is

$$d = (w + x^+ + b - 1) - (w - x^- + b + 1) \tag{16.4}$$

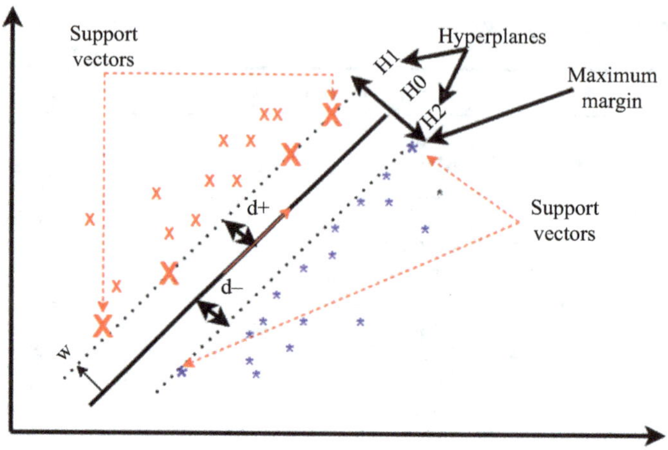

Figure 16.7 Representation of two-state classification

$$d = w.(x^{+} - x^{-}) \tag{16.5}$$

$$d = \frac{2}{\|w\|} \tag{16.6}$$

The optimal hyperplane is the particular hyperplane for which the margin of separation d is maximized.

The optimal separating hyperplane H* that maximizes the margin:

$$(w^{*}, b^{*}) = \text{argmax}_{w,b} \quad \min_{i\in\{1,2,3,\ldots\ldots,N\}} \frac{\langle w,|x_i\rangle + b}{\|w\|^{2}} \tag{16.7}$$

In general, the sleep recordings are affected by some outliers recordings. Sometimes it may cause wrong interpretation. The consideration of outliers may decrease the size of the margin, then the solution may not considerable so well and the sample patterns may not be any longer linear separable. For considering these outliers, we can soften the rules of decision boundaries by considering a slack positive variable $\in_i$ for each training vector. Thus we can modify the equations in the following way:

$$w * x_i + b \geq 1 - \in_i \text{ for } y_i = +1 \tag{16.8}$$

$$w * x_i + b \leq -1 + \in_i \text{ for } y_i = -1 \tag{16.9}$$

To overcome the trivial solution of large $\in_i$, we consider the a penalty cost in the objective function and now the equations become

$$\text{Minimize}\left(\frac{1}{2}\|w\|^2 + c\sum_{i=1}^{L}\in_i\right) \text{ subject to } y_i(w.x_i + b) \geq 1 - \in_i \forall\, i \tag{16.10}$$

where c is denoted as penalization constant, which controls the slack variable $\in_i$, when the value of c increases. Less training errors are allowed, but these changes may occur degradation of performance on the generalization capacity. In this scenario, the classifier is called a soft margin classifier. Similarly, it is treated to be a hard margin classifier with value of $c = \infty$.

To address the optimization problem, we used the concept of Lagrange multipliers α_i, which provides for finding the maxima and minima value of a function subjects to its constraints. The one important things with related to SVM classifier is that, maximum training patterns lie on the outside margin and the optimal values of α_i are zero. Only those training sample points are considered that lie inside the margin with non-zero and each non-zero α_i indicates that corresponding x_i is a support vector. The optimization solution has the form:

$$w = \sum \alpha_i y_i x_i \quad b = y_k - w^T x_k \text{ for any } x_k \text{ such that } \alpha_k \neq 0 \tag{16.11}$$

Then the classification functions in the form of:

$$f(x) = \sum_{i=1}^{k} \alpha_i y_i x_i{}^T x + b \tag{16.12}$$

where k denotes the number of support vectors

$$f(x) = \sum_{i=1}^{k} \alpha_i y_i \varnothing(x) \cdot \varnothing(x_i) + b \tag{16.13}$$

The dot product of input vectors is substituted by a kernel function

$$K(x_i, x_j) = \varnothing(x_i) \cdot \varnothing(x_j) \tag{16.14}$$

The primary role of the kernel function is to measure the similarity between two sample data points.

In this research work, we have used radial basis function (RBF) as the kernel function for our proposed two-state sleep stages classification. Finally the equation of the SVM algorithm for binary classification with wake and sleep labels to each feature vector x_i through the following equation:

$$y(x) = \text{sgn}\left(\sum_{i=1}^{k} \alpha_i y_i K(x, x_i) + b\right) \tag{16.5}$$

16.5.2 Random Forest (RF)

The RF is most effective for classification tasks in different applications. It has been proven to be quite successful and influential accuracy level, achieving an unparalleled accuracy among other classification algorithms [29]. The algorithm votes in each step in the predictive results, and finally selects the most highly voted predictive result as the final prediction result.

16.5.3 Gradient boosting decision tree (GBDT)

GBDT is a boosting technique that combines individual decision trees [30]. It supports an ensemble learning process for managing weak learners and converts those weak learners to strong learners in further iterative rounds. The sample used in each classifier is related to the learning results of the previous layer. During each round of classification, the errors are minimized in the decision tree, and the errors are further minimized through combinations of individual decision trees in a series. Each iteration focuses on minimizing error and finally constructing a highly efficient, accurate model.

16.5.4 eXtreme gradient boosting (XGBoost)

XGBoost is an improved version of gradient boosting decision tree (GBDT) [31]. The main advantages of XGBoost are that it supports linear classifiers and adds a regularization concept to the cost function, which helps to control the complexity of the model. Though it supports parallel processing, the algorithm performs very quickly. This algorithm provides more flexibility to the users, who can set the optimization and evaluation criteria. This algorithm is good at managing missing values and executing pruning operations, which can effectively prevent overfitting.

16.5.5 Stacking ensembling learning

The stacking approach is one of the special architecture in an ensemble learning technique. The main principle of this technique is to build a base classifier and meta-classifier that integrates multiple classification models [32]. The base layer is a collection of multiple classifiers that combined results the average prediction. Then, the meta-classifier model is trained based on the output of the base layers; finally, the meta-classifier predicts the final decision with regard to multiple classifications. The base levels of this algorithm contain different classification algorithms, so it is also referred to as an ensemble learning stacking algorithm. It usually supports heterogeneous learning systems.

16.6 Experimental discussion

In this study, we have used three different feature selection algorithms such as Relief weight(ReF), Fisher Score(FiS), and online streaming feature selection (OSFS) for the purpose of screening the most relevant features. The final selected features using three algorithms are presented in Table 16.7. The selected features were classified through the obtained three base classifiers such as RF, GBDT, and

Table 16.7 Selected features SG-I dataset

Selection algorithm	Screened features with order
FiS	Fet_18,Fet_8,Fet_13,Fet_10,Fet_5,Fet_15,Fet_17,Fet_4,Fet_1, Fet_14,Fet_9,Fet_11,Fet_7,Fet_2,Fet_19,Fet_23,Fet_16,Fet_24, Fet_22,Fet_26,Fet_12,Fet_25,Fet_3,Fet_21,Fet_20,Fet_6, Fet_28,Fet_27
ReF	Fet_8,Fet_18,Fet_5,Fet_14,Fet_10,Fet_9,Fet_1,Fet_4,Fet_13, Fet_15,Fet_7,Fet_6,Fet_3,Fet_11,Fet_2,Fet_22,Fet_16,Fet_24, Fet_17,Fet_26,Fet_27,Fet_25,Fet_21,Fet_28,Fet_23,Fet_19, Fet_12,Fet_20
OSFS	Fet_1,Fet_5,Fet_6,Fet_7,Fet_8,Fet_9,Fet_10,Fet_12,Fet_13, Fet_14,Fet_15,Fet_17,Fet_18,Fet_23,Fet_28,Fet_2,Fet_11, Fet_3,Fet_4,Fet_16,Fet_24,Fet_25,Fet_19,Fet_20,Fet_21, Fet_22,Fet_26

XGBoost algorithms. First, we executed three individual experiments with considering the three base layer classification algorithms. Afterwards, the output of the predicted results from the base-layer classifier is fed into the meta-layer classifier. The final decision was taken with subject to sleep staging from the meta-level layer. Finally, we also made a comparative analysis in between the obtained base classifiers and the stacking ensemble learning results. All the experiments were compiled using MATLAB 2019b version software. In this work, we performed five- to two-classes classification problems. To analysis the effectiveness of the proposed framework, we considered a set of performance indexes like accuracy (Acc) [33], sensitivity (Sen) [34], precision (Pre) [35], and F1score (F1sc) [36].

16.6.1 Feature screening results

16.6.2 Sleep staging performance with ISRUC-Sleep subgroup-I dataset

Tables 16.8, 16.9, and 16.10 present the two-class to five-class classification results based on the three base-layer classifiers such as RF, GBDT, and XGBoost algorithms, respectively. The graphical representation of the confusion matrix for five-class classification problems using the RF classification algorithms and with FiS, ReF, and OSFS selection algorithms are shown in Figures 16.8, 16.9, and 16.10.

The reported confusion matrix for the five-class task using the GBDT classifier is shown in Figures 16.11, 16.12, and 16.13, respectively.

The results of the confusion matrix for five-class are presented in Figures 16.14, 16.15, and 16.16, respectively.

Table 16.8 Classification accuracy results for two-five class classification problems using an RF classifier

Feature selection algorithm	Two class	Three class	Four class	Five class
FiS	99.42%	98.14%	98.46	97.66%
ReF	99.11%	98.92%	98.19%	97.53%
OSFS	99.10%	98.97%	98.37%	97.26%

Table 16.9 Accuracy performance with GBDT classifier

Feature screening algorithm	Two class	Three class	Four class	Five class
FiS	99.42%	98.94%	98.12%	96.21%
ReF	99.22%	98.89%	98.22%	95.76%
OSFS	99.01%	97.69%	96.70%	95.2%

Table 16.10 Accuracy results with XGBOOST classifier

Feature screening algorithm	Two-class	Three-class	Four-class	Five-class
FiS	99.82%	98.77%	98.16%	94.45%
ReF	99.41%	98.82%	98.29%	94.54%
OSFS	99.65%	97.89%	97.45%	94.36%

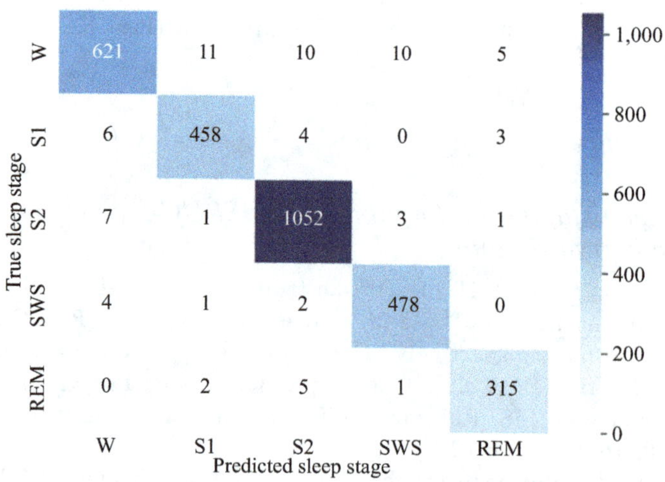

Figure 16.8 Confusion matrix representation: with selected features from FiS algorithm and RF classification model

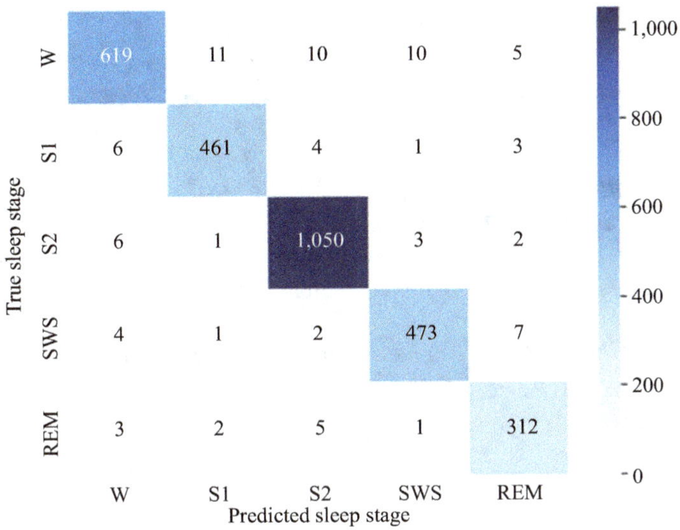

Figure 16.9 Confusion matrix: ReF algorithm and RF classification model

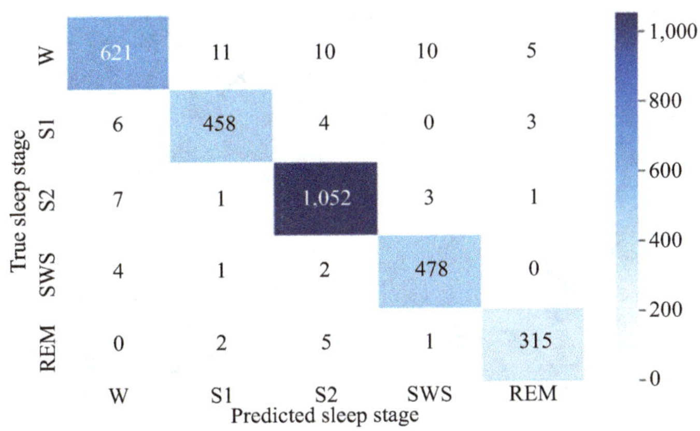

Figure 16.10 Confusion matrix: OSFS algorithm and RF classification model

16.6.3 Automated decision on sleep staging using the ensemble learning stacking algorithm

---

Algorithm Ensemble Learning Stacking Algorithm

Input: *Training Data* $TD = \{(x_1, y_1), \cdots, (x_i, y_i), \cdots, (x_n, y_n)\}$;
Base-level ML based classifier $\mathcal{L}_1, \cdots, \mathcal{L}_3$;
Meta-level classifier $\mathcal{L}$.

***Output: Sleep Staging** Classification result.*

1. ***For each model m=1, ⋯,3 do:***
2. *Use the training dataset TD to train a base-level ML classifier* $H_m = \mathcal{L}_m(TD)$.
3. ***end***
4. *Generate a new dataset* TD'.
5. ***For each individual i=1, ⋯,n do:***
6. ***For each model m=1, ⋯,3 do:***
7. *Use base-level ML learning algorithm* H_m *to classify the training samples* X_i *and return the predicted estimate* $P_{im} = H_m(X_i)$.
8. ***end***
9. *Combine the predicted estimates from 3 base-level ML classification algorithm* $P_i = (p_{i1}, p_{i2}, \cdots, p_{i3})$
   $TD' = TD' \cup \{(P_i, y_i)\}$
10.
11. ***end***
12. *Use the new dataset* TD' *to train the meta layer classifier* $H' = \mathcal{L}(TD')$.
13. *Input a test sample* X_i *and generate sleep staging classification result* $H(x) = H'(h_1(x), h_2(x), \cdots, h_3(x))$.

The reported sleep staging performance using stacking ensemble learning model is presented in Table 16.11. The confusion matrix for the five-state classification problem is presented in Figures 16.17, 16.18, and 16.19 for the RF, GBDT, and XGBoost classifiers, respectively.

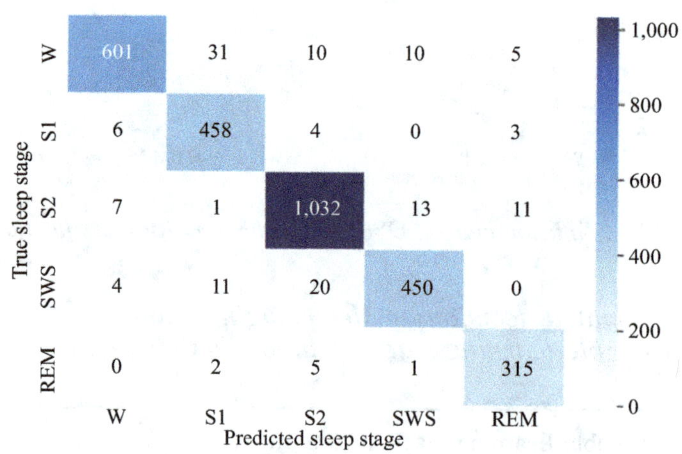

Figure 16.11 Representation of confusion matrix: selected features using FS algorithm and GBDT classifier

From Table 16.12, it has been observed that the accuracy of the proposed model is 98.93% with the FiS algorithm. Similarly, for precision, the model resulted 98.34%, sensitivity as 98.71%, and F1-score as (98.03%). The recall and precision results for the individual sleep stages with FiS, ReF and OSFS screening algorithms using the proposed stacking model are shown in Figures 16.20 and 16.21, respectively. Finally we also made a brief comparative analysis with the

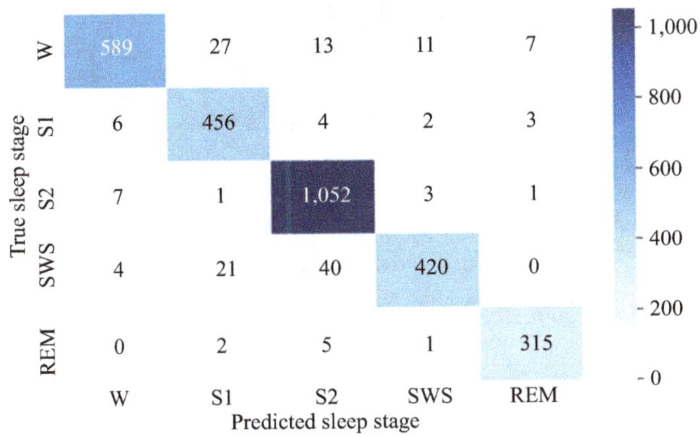

Figure 16.12 Representation of confusion matrix: selected features using ReF-selected features and GBDT classifier

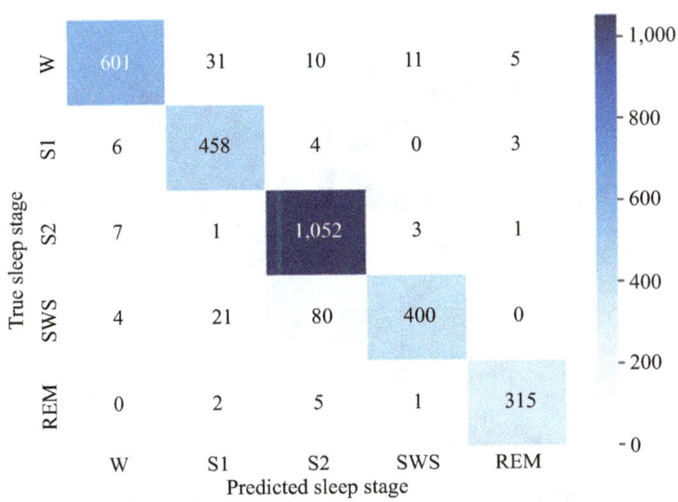

Figure 16.13 Representation of confusion matrix: selected features using OSFS-selected features and GBDT classifier

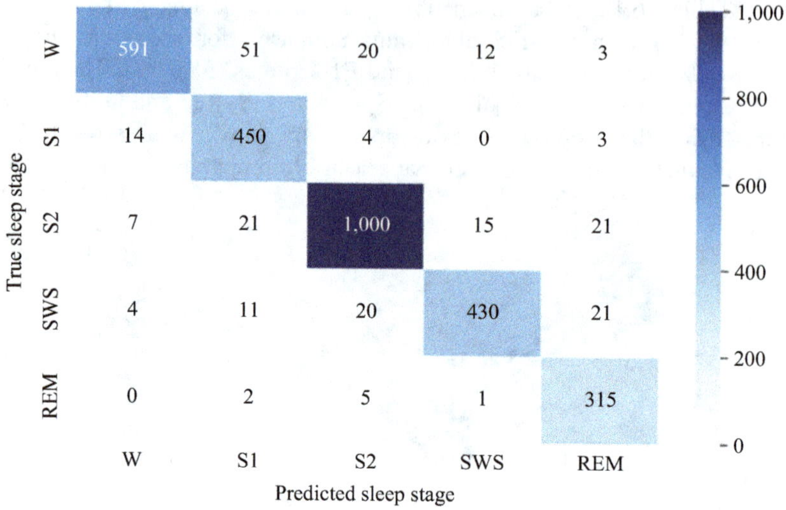

Figure 16.14 Confusion matrix: FiS-selected features and XGBOOST

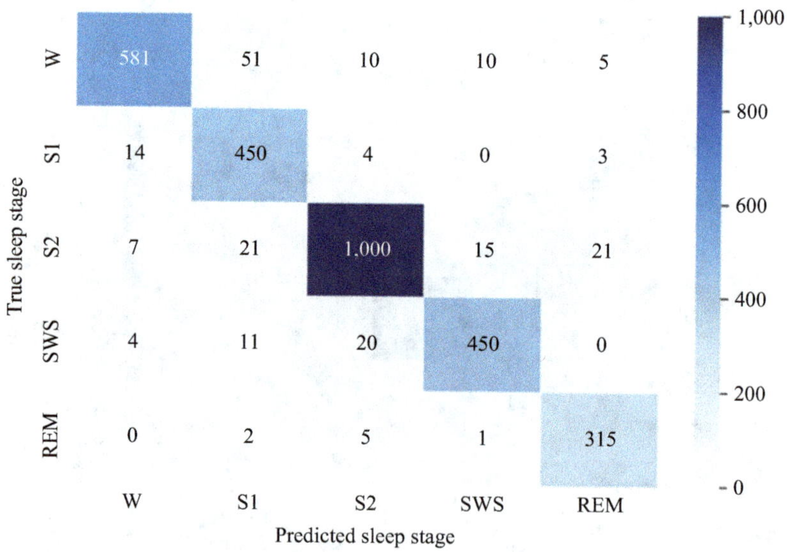

Figure 16.15 Graphical view of confusion matrix: with ReF-selected features and XGBOOST

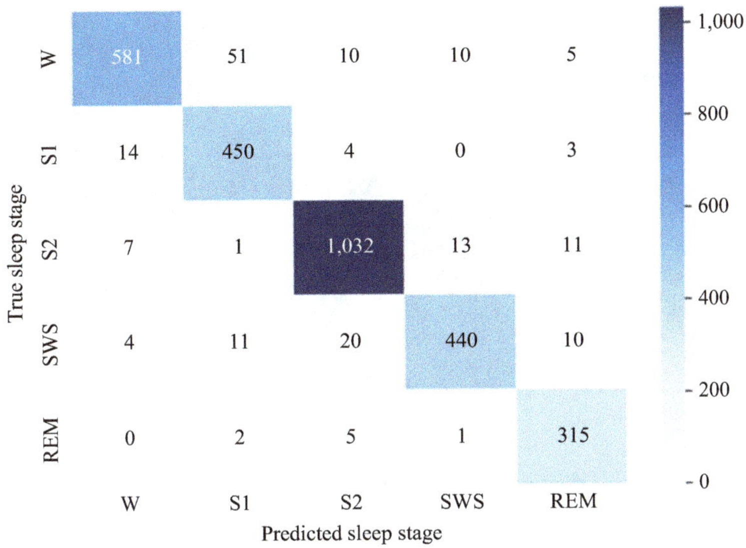

Figure 16.16 Confusion matrix: OSFS-selected features and XGBOOST

Table 16.11 Sleep staging classification performance using stacking model using SG-I subgroup dataset

Feature selection algorithm	Two-class	Three-class	Four-class	Five-class
FiS	99.27%	99.02%	98.98%	97.93%
ReF	99.12%	98.42%	97.68%	97.34%
OSFS	98.97%	98.39%	97.61%	96.86%

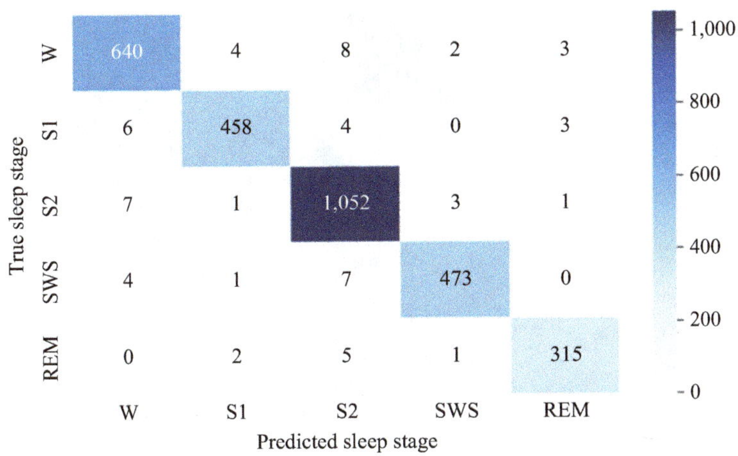

Figure 16.17 Confusion matrix: FS-selected features and stacking ensemble learning algorithm

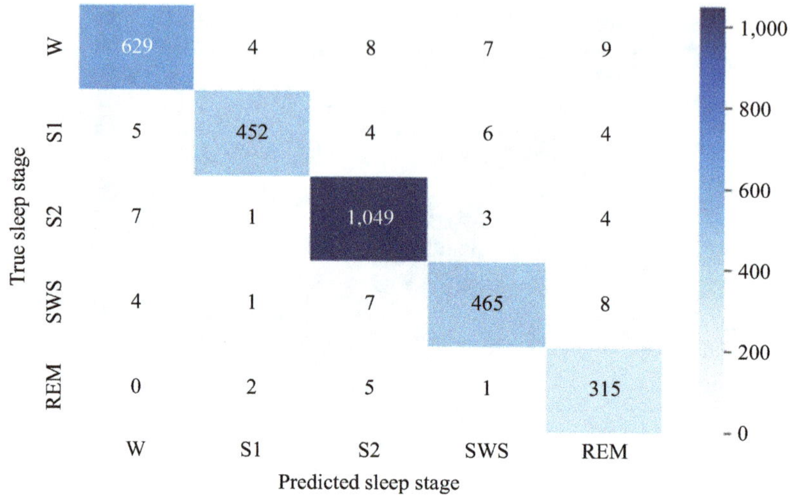

Figure 16.18 Confusion matrix: ReF-selected features and stacking ensemble learning algorithm

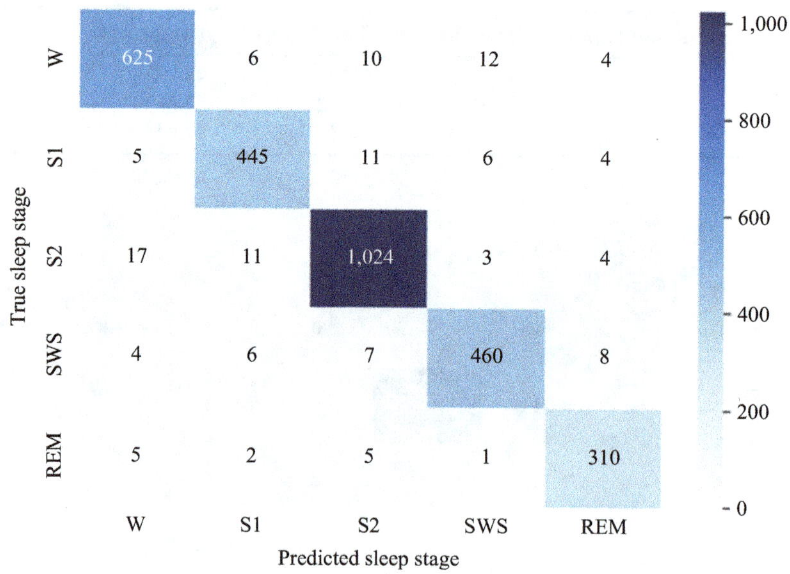

Figure 16.19 Confusion matrix: FiS-selected features and stacking ensemble learning algorithm

Table 16.12 Performance evaluation of sleep staging using stacking ensemble learning model

Performance metrics	**SG-I dataset (five-class)**		
FiS	**ReF**	**OSFS**	
Acc	98.93%	98.11%	96.96%
Pre	98.34%	97.66%	96.45%
Sen	98.71%	97.74%	96.42%
F1-Sc	98.52%	97.70%	96.43%

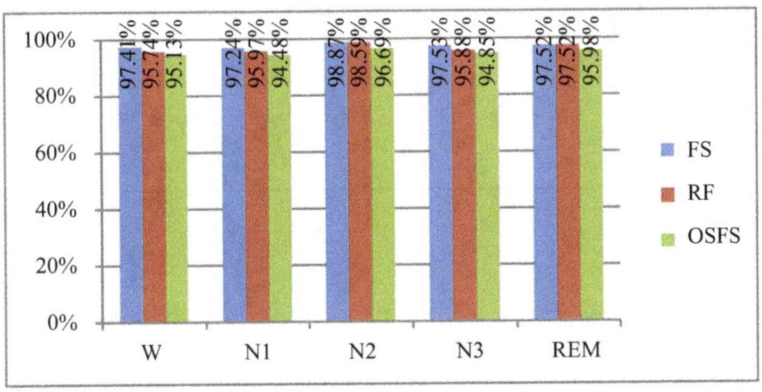

Figure 16.20 Precision results for individual sleep classes using the proposed stacking model

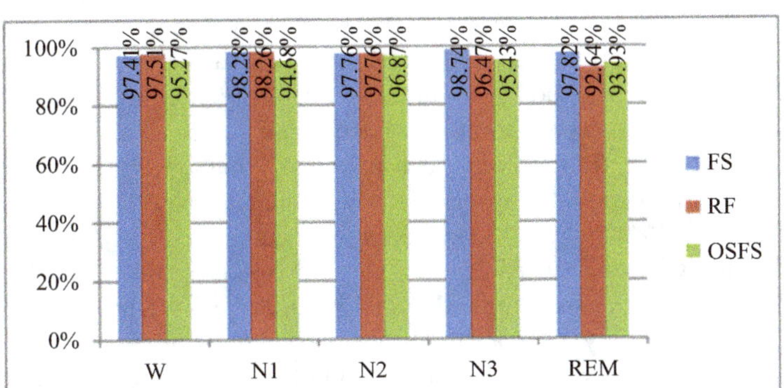

Figure 16.21 Recall results for individual sleep classes using the proposed stacking model

Table 16.13 Comparative analysis with the existing state-of-the-art works

Studies	Classification models	Classification accuracy
[56]	SVM	93.97%
[57]		86.75%
[59]	RF	75.29%
[60]	StackedSparse Auto-Encoders (SSAE)	82.3%
[61]	SVM	83.33%
[63]		92.04%
[64]	Gaussian kernel-based SVM	87.45%
[65]	Stacking Model	96.6
[66]	SVM	91.5%
[19]	RF	95.31%
[68]		97.8%
Proposed work	Ensemble Stacking Model (SG-I)	Fisher Score (FiS) 98.93% ReliefF weight (ReF) 98.11% OSFS 96.96%

other similar published research works based on different traditional ML and DL techniques is presented in Table 16.13.

16.7 Conclusion

Improper sleep causes several disturbances in both physical health and mental conditions. Sleep is essential for our life to maintain activities smoothly. Sometimes it also affects the disorder toward the secretion of hormones. Such continuous sleep deprivation may lead to different types of sleep-related diseases. Currently, it has been found across global that sleep problems increase, and their impact seriously puts in our health. Some researchers came forward and analyzed sleep-related irregularities and their causes. Primary investigations through different surveys by different sleep laboratories have found that significant types of sleep disorders occurred due to improper sleep patterns. To analyze the sleep irregularities, the sleep experts initially observed the sleep behavior by following the edited sleep standards by R&K and AASM through new automated sleep staging. In this chapter, mainly, we have focused how the best way to discriminate and capture the changes in characteristics over the individual sleep stages. We have also executed one experimental work by considering different categories of the subjects affected by different types of sleep-related problems. On the other hand, this chapter also provides basic information about the behavior of the sleep stages, their characteristics, and the challenges of manual inspection. Finally, it focuses on automated sleep staging using ML techniques.

References

[1] Panossian L.A. and Avidan A.Y. Review of sleep disorders. *Medical Clinics of North America*. 2009;93:407–425.

[2] Abdollahpour M., Rezaii T.Y., Farzamnia A., and Saad I. Transfer learning convolutional neural network for sleep stage classification using two-stage data fusion framework. *IEEE Access*. 2020:8:180618–180632.

[3] Satapathy S.K., Loganathan D., Narayanan P., and Sharathkumar S. Convolutional neural network for classification of multiple sleep stages from dual-channel EEG signals. *Soft Computing*. 2020;24:1–16.

[4] Michielli N., Acharya U.R., and Molinari F. Cascaded LSTM recurrent neural network for automated sleep stage classification using single-channel EEG signals. *Computers in Biology and Medicine*. 2019;106:71–81

[5] Surantha N., Lesmana T.F., and Isa S.M. Sleep stage classification using extreme learning machine and particle swarm optimization for healthcare big data. *Journal of Big Data*. 2021;8:14.

[6] Eldele E. An attention-based deep learning approach for sleep stage classification with single-channel EEG. *IEEE Transactions on Neural Systems and Rehabilitation Engineering*. 2021;29:809–818.

[7] Naveen Sundar G., Narmadha D., Amir Anton Jone A., Martin Sagayam K., Dang H., and Pomplun M. Automated sleep stage classification in sleep apnoea using convolutional neural networks. *Informatics in Medicine Unlocked*. 2021;26:100724.

[8] Kumar C.B. SCL-SSC: Supervised Contrastive Learning for Sleep Stage Classification, 2022. *TechRxiv. Preprint.* https://doi.org/10.36227/techrxiv.17711369.v1.

[9] Liu C., Tan B., Fu M., *et al.* Automatic sleep staging with a single-channel EEG based on ensemble empirical mode decomposition. *Physica A: Statistical Mechanics and its Applications.* 2021; 567:125685. ISSN0378-4371, https://doi.org/10.1016/j.physa.2020.125685.

[10] Zheng L., Feng W., Ma Y., *et al.* Ensemble learning method based on temporal, spatial features with multi-scale filter banks for motor imagery EEG classification. *Biomedical Signal Processing and Control.* 2022;76:103634. ISSN 1746-8094. https://doi.org /10.1016/j.bspc.2022.103634.

[11] Yildirim O., Baloglu U.B., and Acharya U.R. A deep learning model for automated sleep stages classification using PSG signals. *International Journal of Environmental Research and Public Health.* 2019;16(4):599. https://doi.org/10.3390/ijerph16040599.

[12] Geetika K., Pramod G., Rishi R.S., and Ram B.P. EEG signal based seizure detection focused on Hjorth parameters from tunable-Q wavelet sub-bands. *Biomedical Signal Processing and Control.* 2022;76:103645. ISSN 1746-8094, https://doi.org/10.1016/j.bspc.2022.103645.

[13] Ebrahimi F. and Alizadeh I. Automatic sleep staging by cardiorespiratory signals: a systematic review. *Sleep Breath.* 2022;26:965–981. https://doi.org/10.1007/s11325-021-02435-814.

[14] Huafeng W., Chonggang L., Qi Z., *et al.* A novel sleep staging network based on multi-scale dual attention. *Biomedical Signal Processing and Control.* 2022;74:103486. ISSN 1746-8094, https://doi.org/10.1016/j.bspc.2022.103486.

[15] Paisarnsrisomsuk S., Sokolovsky M., Guerrero F., Ruiz C., and Alvarez S.A. Deep Sleep: Convolutional Neural Networks for Predictive Modeling of Human Sleep Time-Signals, KDD'18 Deep Learning Day, London, UK, 2018, pp. 1–10.

[16] Diykh M., Li Y., and Wen P. EEG sleep stages classification based on time domain features and structural graph similarity. *IEEE Transactions on Neural Systems and Rehabilitation Engineering.* 2016;24(11):1159–1168.

[17] Gunnarsdottir K.M., Gamaldo C.E., Salas R.M.E., Ewen J.B., Allen R.P., and Sarma S.V. A novel sleep stage scoring system: combining expert-based rules with a decision tree classifier. In: *40th Annual International Conference of the IEEE Engineering in Medicine and Biology Society* (*EMBC*), 2018.

[18] Sriraam N., Padma Shri T.K., and Maheshwari U. Recognition of wake-sleep stage 1 multichannel EEG patterns using spectral entropy features for drowsiness detection. *Australasian Physical & Engineering Sciences in Medicine.* 2018;39(3):797–806.

[19] Memar P. and Faradji F. A novel multi-class EEG-based sleep stage classification system. *IEEE Transactions on Neural Systems and Rehabilitation Engineering.* 2018;26(1):84–95.

[20] Da Silveira T.L.T., Kozakevicius A.J., and Rodrigues C.R. Single-channel EEG sleep stage classification based on a streamlined set of statistical features in wavelet domain. *Medical & Biological Engineering & Computing.* 2016;55(2):343–352.

[21] Wutzl B., Leibnitz K., Rattay F., Kronbichler M., and Murata M. Genetic algorithms for feature selection when classifying severe chronic disorders of consciousness. *PLoS One.* 2019;14(7):e0219683.

[22] Zhu G., Li Y., and Wen P.P. Analysis and classification of sleep stages based on difference visibility graphs from a single-channel EEG signal. *IEEE Journal of Biomedical and Health Informatics.* 2014;18(6):1813–1821.

[23] Braun E.T., Kozakevicius A.D.J., Da Silveira T.L.T., Rodrigues C.R., and Baratto G. Sleep stages classification using spectral based statistical moments as features. *Revista de Informática Teórica e Aplicada.* 2018;25(1):11–22.

[24] Satapathy S.K., Kondaveeti H.K., and Malladi R. Automated sleep staging system based on ensemble learning model using single-channel EEG signal. In: Misra R., Shyamasundar R.K., Chaturvedi A., and Omer R. (eds), *Machine Learning and Big Data Analytics (Proceedings of International Conference on Machine Learning and Big Data Analytics (ICMLBDA) 2021). ICMLBDA 2021. Lecture Notes in Networks and Systems*, Vol. 256. Cham: Springer, https://doi.org/10.1007/978-3-030-82469-3_17.

[25] Khalighi S., Sousa T., Santos J.M., and Nunes U. ISRUC-sleep: a comprehensive public dataset for sleep researchers. *Computer Methods and Programs in Biomedicine.* 2016;124:180–192.

[26] Eskandari S. and Javidi M.M. Online streaming feature selection using rough sets. *International Journal of Approximate Reasoning*. 2016;69:35–57.

[27] İlhan H.O. and Bilgin G. Sleep stage classification via ensemble and conventional machine learning methods using single channel EEG signals. *International Journal of Intelligent Systems and Applications in Engineering*. 2017;5(4):174–184.

[28] Satapathy S.K. and Loganathan D. Automated classification of multi-class sleep stages classification using polysomnography signals: a nine-layer 1D-convolution neural network approach. *Multimedia Tools and Applications*. 2022. https://doi.org/10.1007/ s11042-022-13195-2.

[29] Bajaj V. and Pachori R. Automatic classification of sleep stages based on the time-frequency image of EEG signals. *Computer Methods and Programs in Biomedicine*. 2013;112(3):320–328.

[30] Hsu Y.-L., Yang Y.-T., Wang J.-S., and Hsu C.-Y. Automatic sleep stage recurrent neural classifier using energy features of EEG signals. *Neurocomputing*. 2013;104:105–114.

[31] Zibrandtsen I., Kidmose P., Otto M., Ibsen J., and Kjaer T.W. Case comparison of sleep features from ear-EEG and scalp-EEG. *Sleep Science*. 2016;9(2):69–72.

[32] Berry,R.B., Brooks R., Gamaldo C.E., *et al. The AASM Manual for the Scoring of Sleep and Associated Events: Rules, Terminology and Technical Specifications*. Darien, IL: American Academy of Sleep Medicine 2014.

[33] Satapathy S.K., Sangameswar M.V., and Loganathan D. An automated sleep stages classification using brain EEG signal: a machine learning approaches. In: Chandramohan, S., Venkatesh, B., Sekhar Dash, S., Das, S., Sharmeela, C. (eds), *Artificial Intelligence and Evolutionary Computations in Engineering Systems. Advances in Intelligent Systems and Computing*, Vol. 1361. Singapore: Springer, 2022, https://doi.org/10.1007/978-981-16-2674-6_24.

[34] Satapathy S.K. and Kondaveeti H.K. Prognosis of sleep stage classification using machine learning techniques applied on single-channel of EEG signal of both healthy subjects and mild sleep effected subjects. In: *2021 International Conference on Artificial Intelligence and Machine Vision (AIMV)*, 2021, pp. 1–7, doi:10.1109/AIMV53313.2021.9670967.

[35] Liang S.-F., Kuo C.-E., Hu Y.-H., Pan Y.-H., and Wang Y.-H. Automatic stage scoring of single-channel sleep EEG by using multiscale entropy and autoregressive models. *IEEE Transactions on Instrumentation and Measurement*. 2012;61:1649–1657.

[36] Kim J. A comparative study on classification methods of sleep stages by using EEG. *Journal of Korea Multimedia Society*, 2014;17(2):113–123.

[37] Peker M. A new approach for automatic sleep scoring: combining Taguchi based complex-valued neural network and complex wavelet transform. *Computer Methods and Programs in Biomedicine*. 2016;129: 203–216.

[38] Satapathy S.K. and Loganathan D. Machine learning approaches with heterogeneous ensemble learning stacking model for automated sleep staging.

International Journal of Computing and Digital Systems. University of Bahrain Journals. 2022;11:725–742. http://dx.doi.org/ 10.12785/ijcds/100109.

[39] Satapathy S.K. and Kondaveeti H.K. Automated sleep stage analysis and classification based on different age specified subjects from a single-channel of EEG signal. In: *2021 IEEE Madras Section Conference (MASCON)*, 2021, pp. 1–7, doi:10.1109/MASCON51689.2021.9563485.

[40] Hassan A.R. and Bhuiyan M.I.H. An automated method for sleep staging from EEG signals using normal inverse Gaussian parameters and adaptive boosting. *Neurocomputing*. 2017;219:76–87.

[41] Hassan A.R. and Bhuiyan M.I.H. Automated identification of sleep states from EEG signals by means of ensemble empirical mode decomposition and random under sampling boosting. *Computer Methods and Programs in Biomedicine*, 2017;140:201–210.

[42] Diykh M. and Li Y. Complex networks approach for EEG signal sleep stages classification. *Expert Systems with Applications*. 2016;63:241–248.

[43] Diykh M., Li Y., and Wen P. EEG sleep stages classification based on time domain features and structural graph similarity. *IEEE Transactions on Neural Systems and Rehabilitation Engineering*, 2016;24(11):1159–1168.

[44] Mahvash Mohammadi S., Kouchaki S., Ghavami M., and Sanei S. Improving time–frequency domain sleep EEG classification via singular spectrum analysis. *Journal of Neuroscience Methods*. 2016;273:96–106.

[45] Satapathy S.K., Bhoi A.K., Loganathan D., Khandelwal B., and Barsocchi P. Machine learning with ensemble stacking model for automated sleep staging using dual-channel EEG signal. *Biomedical Signal Processing and Control*. 2021;69:102898, doi:10.1016/j.bspc.2021.102898.

[46] Satapathy S.K. and Loganathan D. A study of human sleep stage classification based on dual channels of EEG signal using machine learning techniques. *SN Computer Science*. 2021;2:157, https://doi.org/10.1007/s42979-021-00528-5.

[47] Obayya M. and Abou-Chadi F.E.Z. Automatic classification of sleep stages using EEG records based on Fuzzy c-means (FCM) algorithm. In: *2014 31st National Radio Science Conference (NRSC)*, 2014, pp. 265–272.

[48] Satapathy S., Loganathan D., Kondaveeti H.K., et al. Performance analysis of machine learning algorithms on automated sleep staging feature sets. *CAAI Transactions on Intelligence Technology*. 2021;6(2):155–174, https://doi.org/10.1049/cit2.12042.

[49] Herrera L.J., Fernandes C.M., Mora A.M., *et al.* Combination of heterogeneous EEG feature extraction methods and stacked sequential learning for sleep stage classification. *International Journal of Neural Systems*, 2013;23(3):1350012.

[50] Radha M., Garcia-Molina G., Poel M., and Tononi G. Comparison of feature and classifier algorithms for online automatic sleep staging based on a single EEG signal. In: *36th Annual International Conference of the IEEE Engineering in Medicine and Biology Society*, 2014, pp. 1876–1880.

[51] Satapathy S.K., Pattnaik S., and Rath R. Automated sleep staging classification system based on convolutional neural network using polysomnography signals. In: *2022 IEEE Delhi Section Conference (DELCON)*, 2022, pp. 1–10, doi:10.1109/DELCON54057.2022.9753132.

[52] Herrera L.J., Mora A.M., and Fernandes C.M. Symbolic representation of the EEG for sleep stage classification. In: *11th International Conference on Intelligent Systems Design and Applications*, 2011, pp. 253–258.

[53] Vanbelle S.A. New interpretation of the weighted kappa coefficients. *Psychometrika*. 2016;81:399–410.

[54] Khalighi S., Sousa T., Oliveira D., Pires G., and Nunes U. Efficient feature selection for sleep staging based on maximal overlap discrete wavelet transform and SVM. In: *Annual International Conference of the IEEE Engineering in Medicine and Biology Society*, 2011.

[55] Simões H., Pires G., Nunes U., and Silva V. Feature extraction and selection for automatic sleep staging using EEG. In: *Proceedings of the 7th International Conference on Informatics in Control, Automation and Robotics*, vol. 3, 2010, pp. 128–133.

[56] Khalighi S., Sousa T., Santos J.M., and Nunes U. ISRUC-sleep: a comprehensive public dataset for sleep researchers. *Computer Methods and Programs in Biomedicine.* 2016;124:180–192.

[57] Sousa T., Cruz A., Khalighi S., Pires G., and Nunes U. A two-step automatic sleep stage classification method with dubious range detection. *Computers in Biology and Medicine.* 2015;59:42–53.

[58] Khalighi S., Sousa T., Pires G., and Nunes U. Automatic sleep staging: a computer assisted approach for optimal combination of features and polysomnographic channels. *Expert Systems with Applications*. 2013;40 (17):7046–7059.

[59] Tzimourta K.D., Tsilimbaris A.K., Tzioukalia A.T., Tzallas M.G., and Tsipouras L.G. EEG-based automatic sleep stage classification. *Biomedical Journal of Scientific & Technical Research.* 2018;7(4):6032–6037.

[60] Najdi S., Gharbali A.A., and Fonseca J.M. Feature transformation based on stacked sparse autoencoders for sleep stage classification. In: *Technological Innovation for Smart Systems*, Hershey, PA: IGI Global, 2017, pp. 191–200.

[61] Kalbkhani H., Ghasemzadeh P., Shayesteh G., and Sleep M. Stages classification from EEG signal based on Stockwell transform. *IET Signal Processing*, 2018;13:242–252.

[62] Satapathy S.K. and Loganathan D. Prognosis of automated sleep staging based on two-layer ensemble learning stacking model using single-channel EEG signal. *Soft Computing*. 2021;25:15445–15462. https://doi.org/10.1007/s00500-021-06218-x

[63] Huang W., Guo B., Shen Y., *et al.* Sleep staging algorithm based on multichannel data adding and multifeature screening. *Computer Methods and Programs in Biomedicine*. 2019;187:105253. doi.org/ 10.1016 /j.cmpb.2019.105253.

[64] Dhok S, Pimpalkhute V, Chandurkar A, Bhurane AA, Sharma M, and Acharya UR. Automated phase classification in cyclic alternating patterns in sleep stages using Wigner-Ville distribution based features. *Computers in Biology and Medicine.* 2020;119:103691. doi:10.1016/j.compbiomed.2020.103691.

[65] Wang Q., Zhao D., Wang Y., and Hou X. Ensemble learning algorithm based on multi-parameters for sleep staging. *Medical & Biological Engineering & Computing.* 2019;57:1693–1707. doi:10.1007/s11517-019-01978-z.

[66] Sharma M., Patel S., Choudhary S., and Acharya U.R. Automated detection of sleep stages using energy-localized orthogonal wavelet filter banks. *Arabian Journal for Science and Engineering.* 2019;45:2531–2544. doi:10.1007/s13369-019-04197-8.

[67] Hassan A.R. and Bhuiyan M.I.H. Computer-aided sleep staging using complete ensemble empirical mode decomposition with adaptive noise and bootstrap aggregating. *Biomedical Signal Processing and Control.* 2016;24:1–10. doi:10.1016/j.bspc.2015.09.002.

[68] Santaji S. and Desai V. Analysis of EEG signal to classify sleep stages using machine learning. *Sleep and Vigilance.* 2020;4:145–152. doi:10.1007/s41782-020-00101-9.

[69] Yan R., Zhang C., Spruyt K., *et al*. Multi-modality of polysomnography 'signals' fusion for automatic sleep scoring. *Biomedical Signal Processing and Control.* 2019;49:14–23. doi:10.1016/j.bspc.2018.10.001.

[70] Ghimatgar H., Kazemi K., Helfroush M.S., and Aarabi A. An automatic single-channel EEG-based sleep stage scoring method based on hidden Markov model. *Journal of Neuroscience Methods.* 2019;324:*108320.* doi:10.1016/j.jneumeth.2019.108320.

[71] Fernandez-Blanco E., Rivero D., and Pazos A. Convolutional neural networks for sleep stage scoring on a two-channel EEG signal. *Soft Computing.* 2020;24:4067–4079. doi:10.1007/s00500-019-04174-1.

[72] Cooray N., Andreotti F., Lo C., *et al* . Detection of REM sleep behaviour disorder by automated polysomnography analysis. *Clinical Neurophysiology.* 2019;130:505–514. doi:10.1016/j.clinph.2019.01.011.

[73] Sun C., Chen C., Fan J., Li W., Zhang Y., and Chen W. A hierarchical sequential neural network with feature fusion for sleep staging based on EOG and RR signals. *Journal of Neural Engineering.* 2019;16(6):066020. doi:10.1088/1741-2552/ab39ca.

[74] Basha A.J., Balaji B.S., Poornima S., Prathilothama, M., and Venkatachalam K. RETRACTED ARTICLE: Support vector machine and simple recurrent network based automatic sleep stage classification of fuzzy kernel. *Journal of Ambient Intelligence and Humanized Computing.* 2021;12:6189–6197. doi:10.1007/s12652-020-02188-4.

[75] Shen H., Ran F., Xu M., Guez A., Li A., and Guo A. An automatic sleep stage classification algorithm using improved model based essence features. *Sensors.* 2020;20(17):4677. doi:10.3390/ s20174677.

[76] Zhu T., Luo W., and Yu F. Convolution- and attention-based neural network for automated sleep stage classification. *International Journal of Environmental Research and Public Health*. 2020;17(11):4152. doi:10.3390/ijerph17114152.

[77] Guillot A., Sauvet F., During E.H., and Thorey V. Dream open datasets: multi-scored sleep datasets to compare human and automated sleep staging. *IEEE Transactions on Neural Systems and Rehabilitation Engineering*. 2020;*1–1*. doi:10.1109/ tnsre.2020.3011181.

[78] Korkalainen H. Accurate deep learning-based sleep staging in a clinical population with suspected obstructive sleep apnea. *IEEE Journal of Biomedical and Health Informatics*. 2020;24(7):2073–2081. doi:10.1109/JBHI.2019.2951346.

[79] Abdollahpour M., Rezaii T.Y., Farzamnia A., and Saad I. Transfer learning convolutional neural network for sleep stage classification using two-stage data fusion framework. *IEEE Access*. 2020;8:180618–180632. doi: 10.1109/ACCESS. 2020.3027289.

[80] Zhang T. Sleep staging using plausibility score: a novel feature selection method based on metric learning. *IEEE Journal of Biomedical and Health Informatics*. 2021;25(2):577–590. doi:10.1109/JBHI.2020.2993644.

[81] Khalili E. and Mohammadzadeh Asl B. Automatic sleep stage classification using temporal convolutional neural network and new data augmentation technique from raw single-channel EEG. *Computer Methods and Programs in Biomedicine*. 2021;204:106063. doi:10.1016/j.cmpb.2021.106063

[82] Huang Z. and Ling B.W.-K. Sleeping stage classification based on joint quaternion valued singular spectrum analysis and ensemble empirical mode decomposition. *Biomedical Signal Processing and Control*. 2022;71:103086. doi:10.1016/j.bspc.2021.103086.

[83] Satapathy S.K. and Loganathan D. Automated classification of sleep stages using single-channel EEG: a machine learning-based method. *International Journal of Information Retrieval Research*. 2022;12(2):1–19. http://doi.org/10.4018/ IJIRR.299941.

Chapter 17

XAI methods for precision medicine in medical decision support systems

Abasiama Godwin Akpan[1], Flavious Bobuin Nkubli[2], Jeremiah Chinonso Mbazor[3], Geofery Luntsi[4] and Offiong Udeme[5]

Abstract

Over the last couple of years, explainable artificial intelligence (XAI) has witnessed tremendous development evidenced by growing research interest in the area. This could be attributed to the increasing role of machine learning, especially deep learning. While these models are highly accurate, they lack explainability and interpretability. There has been limited application of AI systems in vital fields such as precision medicine due to the aforementioned vagueness. The aim of the study is XAI for precision medicine in medical decision support systems (MDSS). The authors outline through an organized examination of literature the application of XAI in MDSS, thus highlighting the several benefits of the use of XAI as reported in the literature such as enhanced decision confidence in precision medicine. The opportunities and challenges of explainable models in MDSS were discussed. Guidelines for the implementation of XAI in MDSS have been recommended in this study while highlighting some of the opportunities and challenges.

Keywords: Artificial intelligence; XAI; Precision medicine; MDSS; Explainability.

17.1 Introduction

The application of artificial intelligence (AI) can lessen the effect of soaring rates of chronic diseases and healthcare costs and increase life expectancy [1]. In

[1]Department of Computer Science & Mathematics, Evangel University, Nigeria
[2]Department of Medical Radiography, University of Maiduguri, Nigeria
[3]Department of Nuclear Power Plant Development, Nigeria Energy Commission Atomic, Abuja and CNERT UNIMAID, Nigeria
[4]Department of Medical Radiography, University of Maiduguri, Nigeria
[5]Department of Psychology, Chukwuemeka Odumegu University, Nigeria

medical applications, AI is an integral part of medical decision support systems (MDSS), helping medical practitioners in the analysis of ailment and healing results. The earliest AI methods were by far understandable; modern societies have experienced an increase in obscure systems, for example, deep neural network (DNN) [2]. Most recent AI models designed with machine learning (ML) and deep learning (DL) are considered by most researchers as "black-box" due to their complex underlying structures, non-linear nature, and lack of interpretability and explainability to medical experts. This vagueness has created the need for explainable artificial intelligence (XAI) architectures [3].

Trust and acceptance of ML models can be enhanced among clinicians by the use of less vogue method. Transparency of methods becomes the best means, which is, accepting how a representation functions as it creates a result [2]. Closely related to explainability is trust since it is based on humans' understanding of the functioning of AI systems. Hence, DNN algorithms should present an explanation for their outcomes, showing the internal workings [4]. By making clear typical features that resulted in the prediction, interpretable models can give details of an exact prediction for a particular patient. In this paper, various jargon for explainability are used [5]. Explainability is interchangeably used to describe interpretability and explainability in studies [6]. In [7], they argued that on the one hand, XAI helps in learning transparency. On the other hand, ML algorithms function by learning models from historical data [8]. Barredo Arrieta [2,9,10] opined that an ML model produces results based on biased.

In this paper, we take the meaning of XAI according to [11], which defines XAI as a means to produce explainable models. This definition is chosen based on the fact that the purpose of explainability is viewed from a user's standpoint. To address challenges related to decision or classification, in different areas, there are several artificial neural network-based designs at different levels. In this discussion, we use the term DNN in place of all these designs or architecture. The practical success of DL models results from a blended proficient learning computing set of rules in addition to vast parametric space. This makes DNN complex models [12]. Conventionally, the possible features of lesions on medical images are reviewed by the doctor manually. This approach takes much of the doctor's time and attention that could have been devoted to patient care for a physician who is required to go through a series of images from just a particular patient examination. However, over the years there has been increased interest in AI-enabled information extraction from images, and medical diagnostics among others [13]. Encouraging outcomes have been recorded in other fields including the medical field too where the performance of reinforcement-based learning techniques and DL techniques that were trained on very large data sets performed better than humans in terms of efficiency [14].

17.1.1 Contributions of the current study

AI has become efficient and is applied in receptive situations with major human insinuations; hence trust is crucial [13]. Medical practitioners should be knowledgeable, reproduce, and control or influence a machine-generated decision-making process on the spot. Hence, there is a growing need to enhance the understanding of decisions originating from ML algorithms that could be reproduced in the context and settings of real-world applications, especially in precision medicine. Therefore, there is a need for systems that promote an easy-to-understand and explain decisions made, with results that can be re-traced on demand. Healthcare professionals work with varied and diverse data sources most of which are unstructured and complex; thus, XAI has the potential to enhance the use of AI and ML in healthcare delivery, thereby promoting trust and transparency within the healthcare system. Therefore, the ability of ML models to justify their decisions will enhance their trust within the healthcare community. Therefore, for ML models to be trusted, their decisions should be justifiable. Rather than the black-box model's decisions which are more intuitive for humans, explanation supports provided by ML models help to make easy decisions made by black-boxes. A major contribution of this study is that it has provided a viewpoint and insight into potential areas of XAI for precision medicine.

17.1.2 Chapter organization

This chapter is organized into the following sections: Section 17.1 focuses on XAI methods for precision medicine in MDSS. In Section 17.2, a literature review of the application of XAI in MDSS is undertaken. In Section 17.3, the opportunities and challenges of explainable models in MDSS were discussed. In Section 17.4, we proposed guidelines and the conclusion.

17.2 Related works

17.2.1 Measurement of XAI in precision medicine

Although the field of XAI can add to dependable AI, it has its restrictions [15]. Some measures that can create trustworthy AI in precision medicine are as follows:

(i) *Reporting data quality*: Ideally, most healthcare data are not tailored toward research and are prone to errors and biases, or may be incomplete in some cases. Therefore, data quality and mode of data collection are equally important as explainability because it allows understanding of the outcome model [16].

(ii) *External validation*: The apprehension of less robust models can be solved through external validation. A paucity of standardized data sets impedes the replication of a prediction model on a data set.

(iii) *Regulation*: Different approaches to regulating AI exist. The initial step is to ensure AI systems meet pre-defined standards or requirements. Even though it is hard to get a complete series of verifiable standards or principles that makes sure the AI system is compliant in terms of law, ethics, and robustness rather than the control of the system, we can take charge of the process of development by ensuring that standard guidelines for development are

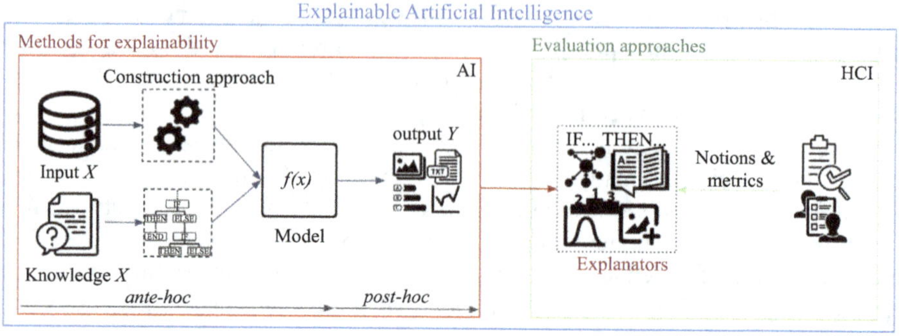

Figure 17.1 Conceptual framework of XAI [20]

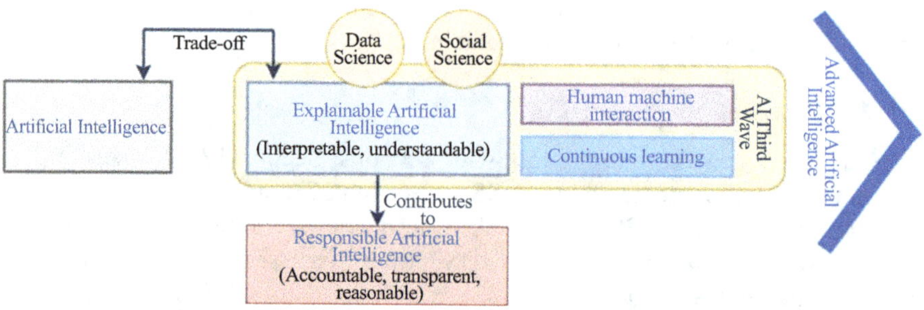

Figure 17.2 Perception of XAI [21]

adhered to. Although it might not be easy to understand the development process as it is also not easy to make a judgment about the quality of a model on the basis of all desired endpoints, this calls for accountability on the part of the professionals [17–19]. On the basis of the system explained the conceptual framework is presented in Figure 17.1.

17.2.2 Concept of explainability and interpretability

Explainability and interpretability are used interchangeably. Doshi-Velez and Kim, as affirmed by [21], define it as a level of human knowledge concerning a particular decision. Based on [22], interpretability is typically associated with perceptions concerning results of a given method.

XAI methods are designed to produce outcomes that are transparent. Research in the area of XAI is exploring different ways in which the characteristics of autonomous AI systems could be understood by humans [21]. In image recognition tasks or pattern recognition tasks, for example, a particular dominant pattern in the image (input) could be a deciding factor for why a system decided that a typical object is made up of an image. Some models do not show the internal logic and the underpinning principles that can be understood by humans. Hence, ML interpretability does not lead to explainability, but is often used synonymously [23,24]. Tonekaboni *et al.* [25] aver that for an AI system to be explainable, the task model will be inherently interpretable. In line with [25], Markus *et al.* [16] differentiated different definitions of interpretability as listed in the following:

(i) Description based on operational aspect – Here, the input object and output object are proven to be correct [26].
(ii) Definitions based on the descriptive assessment of the user – Here, users play a vital role.
(iii) Descriptions based on the latent property – Here, it is a group of factors that are manipulatable, influencing the complexity of a model.

17.2.2.1 Explainability in healthcare

Models developed with AI modalities will interact with the acceptance and adoption of healthcare experts and patients [27]. For a meaningful adoption of AI in the real medical world, practitioners need to be involved in the discussion to support the widespread acceptance and adoption of these solutions [28].

17.2.2.2 Artificial intelligence (black-box) predictions for precision medicine

The major concerns about decisions made by ML models in healthcare are if such decisions should be trusted and in what context are those decisions made by ML and DL? The principles that lead to these are:

(i) *Transparency*: The transparency of a model is based on the understanding [21]. Easy to understand ML models; for example, models built using linear regression are likely to be more transparent, and hence, intrinsically more interpretable as a result of their simple structure.
(ii) *Interpretability:* This is the ability of a model to provide meaning or representation that is easy to understand by humans. Doshi-Velez and Kim stressed, as reported in [21], that models should be domain specified [11,29].
(iii) *Explainability:* This concept is related to the system's dynamics and internal logic of the ML.

17.2.2.3 Explanation models in precision medicine

Markus *et al.* [16] listed three kinds of explanation models in precision medicine as follows:

(i) *Model-based explanations:* This method uses a model to explain the task model. These are categorized under explainable modeling and post hoc explanations. The basis of explainable modeling is to build an inherently interpretable task model for the user. Here, the task model is used as explainable modeling. It is important to note that explainable modeling ensures that the process of decision-making by the model is fully transparent.

(ii) *Attribution-based explanations:* This method quantifies the clarifying ability of the data set and uses the same to clarify the task model. The attribution-based explanation models are also referred to as feature importance, relevance, or influence methods [25,30].

(iii) *Example-based explanations:* By selecting examples from the data sets or by creating new examples in the data sets, for example, by carefully choosing prototypes or models from which other models could be developed; taking note of prominent examples for the model output, or making a counterfactual explanation [31].

17.2.2.4 Explanation of AI (XAI) methods in precision medicine

Explanation AI (XAI) methods in precision medicine are as follows:

(i) *Local interpretable model-agnostic explanations (LIME):* This was anticipated by [32] to aid experts' clarification at all points. The method predicates the behavior of a classifier around the cases to be explained. Initially, LIME uses the perturbation method to produce results from the original data. But in python and R languages, it has different approaches [32,33]. This method can interpret one prediction result. The method is useful and simple, but it has disadvantages. LIME relies on unstable interpretations – where various interpretations can be given for the same prediction. To solve this problem, LIME was improved to DLIME.

(ii) *Shapley additive explanations (SHAP):* This method of explanation tries to improve the understanding of AI outcomes for clinical usage through the calculation of the vital values in individual characters of predictions. It functions for all models but less in giving assumptions on model categories. It generates approximated values than the exact values. Its algorithm is derived from [34] and the objective is to use game concept value.

(iii) *Contextual importance and utility (CIU):* This type of XAI method in precision medicine also uses the utility of characters to elucidate the outcome of the representation [54]. The CIU method is composed of two evaluation approaches; thus

    (a) *Contextual importance (CI):* This evaluation approach estimates the total vital part in the recent context.

    (b) *Contextual utility (CU):* For a given output class, CU makes provision for a judgment or opinion about the favourability or otherwise of the current feature. The fact that the CIU cannot make an intermediate surrogate model is what differentiates it from the LIME and SHAP. Therefore, CI and CU provide explanations and interpretations based on the features contributed by participating data set.

17.2.2.5 Recent works in XAI for precision medicine

S/N	Citations	Purpose	Findings
1.	[35]	The authors argued on an interpretable decision tree	Diagnosis for radiological review
2.	[36]	The authors used a color map overlay with a conventional imaging device	Creation of production for understandable ailment risk
3.	[37]	Authors suggested medically understandable ConvNet architecture	It provided a region of interest
4.	[38]	Authors used visualizations for their explanations to anesthesiologists	Highlights of vital features
5.	[39]	Authors used a self-attention approach	Predicting the acuity score
6.	[40]	Authors employed attention models	Attention scores are high
7.	[41]	Authors deployed the GAN method to prepare a model with mode data	High robustness
8.	[42]	Authors showed an ML-based forecast of metadata	DL and random forest-based metadata were the benchmarks
9.	[43]	Authors attained stability by mounting a blended system	Identification of patients with lofty peril of death

However, XAI studies in MDSS provide avenues for future research as follows:

(i) *Patient-in-the-loop XAI ML:* Caywood *et al.* in [4] explained that in precision medicine patient-in-the-loop and decision-making in concurrent exist. Domain experts use ML decisions to arrive at a final decision on a detailed and exact scenario [35–37,43].

(ii) *Accuracy vs. explainability:* A clear tradeoff exists between prediction accuracy and the lucidity of algorithms. In [11], ML routines such as analytical accuracy and explainability are the least explainable. The balance is by mounting a blended system to optimization [43].

(iii) *Robustness of DL with data augmentation (DA):* Cook in [4] explained that DA provides a means to solve issues of fewer data sets for the training of ML model [51,52].

17.2.2.6 Explainable AI in ML

In [53], a lead is presented for explainability based on DL with the uniqueness of its structure and basic methods for interpretable DL. Choo and Liu [21] provided useful ideas on recent issues in interpretable DL. Their works highlighted the possibilities of DL systems, including generative models driven by the user, progress in the area of visual analytics, reduced use of training data sets, improved robustness of AI, and inclusion of external human intelligence. The studies by Hase and Bansel [55] provide a clear and accurate estimate of the effect of explanations on the simulatability across a wide spectrum of data domains and explanation techniques.

Table 17.1 List of XAI methods in precision medicine [4]

S/N	Authors	ML model	XAI methods
1.	[44]	MAC	Post-hoc
2.	[43]	K-Means	R-group gradients
3.	[4]	SVM	Post-hoc
4.	[37]	CNN	Post-hoc
5.	[40]	RNN	Post-hoc, attention mechanism, visual
6.	[39]	GRU	Post-hoc, attention mechanism, visual
7.	[36]	FCN	Post-hoc, visual
8.	[45]	SRM	Transparent, visual
9.	[46]	Weighted K-nearest neighbor	Transparent, visualization
10.	[42]	Neural network	Post-hoc
11.	[4]	XGBoost	SHAP
12.	[4]	Gaussian process regression	Transparent, global
13.	[4]	Decision tree	Transparent, visual
14.	[16]	Generalized linear regression model	Transparent, rule-based
15.	[47]	Multi-scale CNN	Post-hoc, visual
16.	[30]	Random forests ensemble	SHAP
17.	[38]	DNN	Post-hoc
18.	[48]	Deep CNN (DCNN)	Post-hoc
19.	[4]	DNN	Post-hoc
20.	[4]	SVM	Post-hoc
21.	[41]	Linear SVM	Transparent
22.	[49]	CNN	LIME
23.	[44]	MAC	Post-hoc
24.	[43]	K-means	R-group gradients
25.	[4]	SVM	Visual
26.	[37]	CNN	Visual
27.	[40]	Recurrent neural network-CNN	Post-hoc, attention mechanism, visual
28.	[39]	RNN	Post-hoc, attention mechanism, visual
29.	[36]	FCN	Post-hoc, visual
30.	[50]	Probabilistic/Bayesian framework	Transparent

17.2.2.7 Medical decision support systems

MDSS composed of intelligence agents that avail medical practitioners, patients, and others with vast understanding and filtered clinical data to enhance healthcare [56]. MDSS are prepared with a range of applications, for example, patient analysis, predicting medicinal, recommendation of patient medicinal options, and personalized patient care based on level of risk. It can better the safety of patients, quality of care delivered, and efficiency of healthcare. MDSS are based on existing knowledge and rely on medical or clinical procedures and understanding, but non-knowledge-based MDSS uses ML [56]. MDSS that are ML-based identify patterns in medical data that are historical and use the same to develop predictive models with the ability to predict medical outcomes that are based on new inputs provided.

17.2.2.7.1 Structure of MDSS

The fundamental principles of the design of MDSSs have changed over time. A variety of qualities of MDSSs are related to medical effectiveness, functionality, error prevention, acceptability, portability, and cost-effectiveness. MDSS can be characterized under the following contexts:

(a) **Context axes**
  (i) Inpatient setting
  (ii) Outpatient setting
(b) **Clinical task**
  (i) Diagnostic assistance
  (ii) Therapy assessment and consult
  (iii) Drug prescribed amount or ordering
  (iv) Test selection.
  (v) Alerts and reminders
  (vi) Information retrieval
  (vii) Image recognition and interpretation
  (viii) Prevention
  (ix) Screening
  (x) Professional laboratory equipment
  (xi) Persistent ailment administration
(c) **Understanding axes**
  Understanding aspect composed of basis, superiority, and the set up of the MDSS's awareness and information.
  (i) Scientific knowledge basis
  (ii) Data source
  (iii) Data source intermediary
  (iv) Data coding
  (v) Data customization
  (vi) Update mechanism
(d) **Decision support (DS) axes**
  One of the most important aspects of MDSS is to address a decision-making process that is suitable.
  (i) Reasoning method: Rule-based systems, neural networks, Bayesian network, model-based systems, logical condition, data mining and ML, genetic algorithm
  (ii) Clinical urgency
  (iii) Recommendation explicitness
  (iv) Response requirement
(e) **Information delivery axes**
  (i) Delivery format
  (ii) Delivery mode
  (iii) Action integration
  (iv) Explanation availability
(f) **Workflow axes**

MDSS is a process and an intervention of expertise that will act as a disruption. Synergistic systems with institutionalized process are expected to witness superior practice and confirm to be more successful in optimizing the routine of practitioners.

17.2.2.7.2 Benefits of MDSS

The possible benefits of MDSS in precision medicine are categorized into three extensive categories:

(i) Improved patient safety.
(ii) Improved quality of care.
(iii) Improved efficiency of health.

Modern studies propose that MDSS features are vital to the systems [12]:

(i) MDSS should give DS.
(ii) DS for decision-making should be provided at the time and point of need. This should be simplified in such a way that the clinician's normal pattern of work and patient care is not interrupted and less cumbersome to enhance acceptance.
(iii) The DS ought to be integrated with a bigger system that is previously a component of the experts' specialized regular practice.
(iv) Systems that are useful are better than traditional systems.
(v) Recommendations are supposed to be made by systems rather than just highlighting a patient's assessment.
(vi) A provision should be made for MDSS to demand that clinician note down irregular procedures based on the system's advice or recommendation.
(vii) New data should rather be acquired automatically by systems.
(viii) The system should be user-friendly, for example, with speed.
(ix) Instance and occurrence of prompts are imperative.
(x) Data or information presentation on MDSSs should not be too loaded or too scanty. Researchers are also recommended that blinking icons should be used for important tasks or that interactions should be arranged based on the urgency of need.
(xi) DS outcomes should be made available to both experts and patients.
(xii) Overall, factors from the organization such as the availability of computers at the caring point and the technological perfection of MDSS architecture are critical to development [25]. Markus *et al.* [16] suggest that MDSS's effectiveness remains largely constant when systems proposals are declared more robustly while supporting facts to these are timely and its expansion takes into consideration clinical-specific data. Likewise, when recommendations or suggestions are made more detailed and exact, the effectiveness and functionality remain unaltered. Interestingly, the MDSSs did not achieve the desired outcome as soon as confined experts were integrated into the system cycle [26]. In summary, when MDSSs are developed, certain factors, software, and content should be considered. Key areas of consideration are hardware availability, enough technical know-how, and guidance on how to use the system, the extent of incorporation of the system into the process, and the suitability of the medical information needed.

17.3 Explainable models in MDSS: opportunities and challenges

XAI representation ought to be designed in alliance and contribution from diverse experts from different professions such as sociology, management science, art, psychology, and medical science [32]. Even though XAI can help with recognizing challenges of information, the case posed by formless medical information becomes an impediment to the growth of intelligent systems that are useable. The problem posed by unstructured medical data in the development of useable AI has been discussed with solutions proposed by previous research [33]; some of which include:

(i) Exchange of data between different sources should be enhanced provided that appropriate safeguards for data privacy are ensured.

(ii) Data mining techniques should be considered to elicit vital clinical information that perhaps could have been captured in free text.

(iii) To build up and amass statistics for medical usage, a controlled development process that uses AI should be used. Also, the problem of data availability and heterogenicity might be overcome by using the least available data and information that is simple to begin the design of the structure [57]. Requirements that specific domain should be given consideration as well as a detailed knowledge of the functions of the structure, its routine and understandability of the present systems, extent and the characteristics of the explanations needed [58]. In addition, it has been recommended by [58] that unexplainable methods should be used at the point of need and that priority should also be given to the use of understandable algorithms over multifaceted procedures that requires the function of post-hoc models. In addition, factors such as morals, equality, and safety with knowledgeable skills must be taken into consideration at the stage of choosing the type of explainable model [58]. Further study is required to establish the performance metrics of the model [58]. Recent researches have shown that a greater part of research aimed at prejudiced dimensions, for example, expert contentment, decency, recognition, and confidence in the model [58]. While important insight into a user's experience can be obtained through subjective assessment, the general lack of authenticated and dependable assessment metrics exists. Several quantitative metrics for the evaluation of the properties of explainability for different explanation types have been summarized in a previous study [59]. The authors noted that certain properties such as lucidity, extensive, and wholeness are inadequate in terms of suitable metrics, and so is the category of understanding that is hinged on examples. They also expounded on clinical research for the assessment of ML clarification. Therefore, a conclusion from that study could be paraphrased thus as: "the assessment of ML clarification is a diverse research topic." It is also not possible to define an implementation of evaluation metrics, which can be applied to all explanation methods. The system causability scale was introduced by Holzinger *et al.* [5] as a means to measure the quality of explanations. Another study established inaccurate ML proposals as the main element that influence experts and causes the accuracy of decisions to be low, while interpretation reveals to be lacking in tackling depending on a model that brings out decisions that are untrue [60]. The authors observed that interpretable approaches must be chosen on the foundation of the clinician's previous understanding of the model and with preceding knowledge by professing higher utility from the proposal made by ML.

17.4 Conclusion

The past years have witnessed tremendous development in XAI, evidenced by growing research interest in the area. This could be a result of the increasing role of ML, especially DL. While these models are highly accurate, they lack explainability and interpretability. There has been limited application of AI systems in vital fields such as precision medicine due to the aforementioned vagueness. In this methodical analysis, we based our study on methods and techniques of XAI employed in ML systems applied within the healthcare setting. For ease of comprehension, we further organized and discussed extant literature by proposing and presenting a theoretical structure for the purpose of classification of XAI models. The emphasis in literature based on this study seems to be more on interpretability of the model as the basis for ML methods together with the use of transparency systems, for greater accuracy. Findings from previous studies show greater emphasis on the interpretability of a model as a pre-condition for ML models for better precision. This is following the use of post-hoc and transparent systems. While reports in previous studies try to stress the importance of equilibrium between understandability and correctness, facts abound in study reports of projects' understandability above correctness. This study identifies further study opportunities that could be improved upon paramount among which are aspects of patient-in-the-loop model, the toughness of DL with DA, and the tradeoffs linking correctness and understandability. Also, other study opportunities associated with a variety of detailed scenarios, processes, and practices in the area of XAI remain. Some restrictions of the present study are:

(i) A shortlist of query terms was used.
(ii) The literature review was limited to Internet searches only. While this is consistent by way of reviewing, given that a summary of models in the medical domain, is an all-encompassing review that will require broadening the assessment to other sources of citation such as grey literature among others, other opportunities abound to possibly improve the categorization structure to discuss processes and practices at the point of a construct.

Finally, XAI is an important and growing area of AI study, especially in the medical and healthcare domains. The spread and acceptance of these methods within the medical and healthcare community will not only depend on their efficacy but also on their being able to provide explanations that are meaningful to users within the clinical settings. A viewpoint on the recent development and impending areas of study in XAI is the areas of MDSS has been provided based on the projected categorization structure and the findings from the methodical review of the literatures.

References

[1] Mohseni, S., Zarei, N., and Ragan, E. A survey of evaluation methods and measures for interpretable machine learning [Preprint], April 26, 2020. https://arxiv.org/pdf/1811.11839v4.pdf.

[2] Barredo Arrieta, A., Diaz-Rodriguez, N., Del Ser, J., *et al* . 'Explainable artificial intelligence (XAI): Concepts, taxonomies, opportunities and challenges toward responsible AI'. *Information Fusion*. 2020; (58): 82–115.

[3] Giulia, V. and Luca, L. *Explainable artificial intelligence: A systematic review*. Technological University Dublin, Dublin, Republic of Ireland. 2020; pp. 1–53.

[4] Chakrobartty, S. and El-Gayar, O. 'Explainable artificial intelligence in the medical domain: A systematic review'. In: *Proceedings of AMCIS 2021*.

[5] Holzinger, A., Langs, G., Denk, H., Zatloukal, K., and Müller, H. 'Causability and explanability of artificial intelligence in medicine'. *Wiley Interdisciplinary Reviews*. DMKD. 2019; 9(4): 13–14.

[6] National Academies of Sciences, Engineering and Medicine. *Human AI-Teaming: State-of-the Arts and Research Needs*, Washington, DC: The National Academies Press, 2022. https://doi.org/10.17226/26355.

[7] Schoenborn, J. and Althoff, K. 'Recent trends in XAI: A broad overview on current approaches, methodologies and interactions'. In: *27th International Conference on Case-Based Reasoning*, September 10, 2019.

[8] Suresh, H. and Guttag, J. A framework for understanding unintended consequences of machine learning. *ArXiv: 1901.10002*.

[9] Bhatt, U., Xiang, A., Sharma, S., *et al* .'Explainable machine learning in deployment'. In *Proceedings of the 2020 Conference on Fairness, Accountability, and Transparency, FAT* '20*, New York, NY: ACM, 2020; pp. 648–57.

[10] Gill, N., Hall, P., Montgomery, K., and Schmidt, N. 'A responsible machine learning workflow with focus on interpretable models, post-hoc explanation, and discrimination testing'. *Information*. 2020; 11(3): 137.

[11] Gunning, D. and Aha, D. 'DARPA's explainable artificial intelligence program'. *AI Magazine*. La Canada. 2019; 40(2): 44–58. https://doi.org/10.1609/aimag.v40i2.2850.

[12] London, A.J. 'Artificial intelligence and black-box medical decisions: Accuracy versus explainability'. *Review of Explainable AI in the Medical Domain Twenty-Seventh Americas Conference on Information Systems, Montreal, Hastings Center Report*. 2021; 49(1): 15–21.

[13] Malhi, A., Kampik, T., Pannu, H., Madhikermi, M., and Främling, K. 'Explaining machine learning-based classifications of in-vivo gastral images'. In *International Workshop on Explainable, Transparent Autonomous Agents and Multi-Agent Systems*, New York, NY: IEEE. 2019; pp. 1–7.

[14] Meske, C. and Bunde, E. 'Transparency and trust in human-ai-interaction: The role of model-agnostic explanations in computer vision-based decision support'. In *International Conference on Human-Computer Interaction*, New York, NY: Springer. 2020; pp. 54–69.

[15] Sendak, M., Elish, M., Gao, M., Futoma, J., Ratliff, V., and Nichols, V. '"The human body is a black box" supporting clinical decision-making with deep learning'. *Proceedings of the 2020 Conference on Fairness, Accountability, and Transparency*. 2020; pp. 99–109.

[16] Markus A., Kors, J., and Rijnbeek, P. 'The role of explanability in creating trustworthy artificial intelligence for health care: A comprehensive survey of the terminology, design choices, and evaluation strategies'. *Journal of Biomedical Informatics*. 2021; 113(103655): 1–11.

[17] US Food and Drug Administration. Proposed regulatory framework for modifications to artificial intelligence/machine learning (AI/ML) – based software as a medical device (SAMD). Published: January, 2020. https://www.fda.gov/media/122535/download.

[18] Cortez, N. 'Digital health and regulatory experimentation at the FDA'. *Yale Journal of Law & Technology*. 2019; 21: 4–26.

[19] Malhi, A., Knapic, S., and Främling, K. 'Explainable agents for less bias in human-agent decision making'. In *International Workshop on Explainable, Transparent Autonomous Agents and Multi-Agent Systems*, New York, NY: Springer. 2020; pp. 129–46.

[20] Miller, T. 'Explanation in artificial intelligence: Insights from the social sciences'. *Artificial Intelligence*. 2019; 267: 1–38.

[21] Samanta, K., Avleen, M., Rohit, S., and Kary, F. *Explainable Artificial Intelligence for Human Decision-Support System in Medical Domain*. Helsinki Institute for Information Technology, FCAI, 2021.

[22] Samek, W. and Müller, K. *Towards Explainable Artificial Intelligence*. 2019. http://arxiv.org/abs/1909.12072v1. Accessed January 21, 2022.

[23] Rudin, C. 'Stop explaining black box machine learning models for high stakes decisions and use interpretable models instead'. *National Machine Intelligence*. 2019; 1: 206–15.

[24] Doran, D., Schulz, S., and Besold, T. 'What does explainable AI really mean? A new conceptualization of perspectives'. *ArXiv171000794 Cs*. 2022. https://arxiv.org/abs/1710.00794.

[25] Tonekaboni, S., Joshi, S., McCradden, M., and Goldenberg, A. 'What clinicians want: Contextualizing explainable machine learning for clinical end use'. *Proceedings of Machine Learning and Research*. 2019; 1–21.

[26] Holzinger, A., Carrington, A., and Müller, H. 'Measuring the quality of explanations: The system causability scale (SCS)'. *Comparing Human Machine Explanations*. 2019; ArXiv191209024 Cs.

[27] Topol, E.J. 'High-performance medicine: The convergence of human and artificial intelligence'. *Nature Medicine.* 2019; 25: 44–56.

[28] Marcinkevičs, R. and Vogt, J. Interpretability and explainability: A machine learning zoo mini-tour. 2020. https://arxiv.org/abs/2012.01805v1[cs.LG]. Accessed January 21, 2022.

[29] Linardatos, P., Papastefanopoulos, V., and Kotsiantis, S. 'Explainable AI: A review of machine learning interpretability methods'. *Entropy.* 2021; 23(1): 18.

[30] Elshawi, R., Al-Mallah, M.H., and Sakr, S. 'On the interpretability of machine learning-based model for predicting hypertension'. *BMC Medical Informatics and Decision Making.* 2019; 19(1): 146. https://doi.org/10.1186/s12911-019-0874-0.

[31] Sahiner, B., Pezeshk, A., Hadjiiski, L.M., *et al* . 'Deep learning in medical imaging and radiation therapy'. *Medical Physics.* 2019; 46(1): e1–e36.

[32] Vellido, A. 'The importance of interpretability and visualization in machine learning for applications in medicine and health care'. *Neural Computation Applications.* 2019; 32: 18069–83.

[33] Angehrn, Z., Haldna, L., Zandvliet, A.S., et al. 'Artificial intelligence and machine learning applied at the point of care'. *Frontiers in Pharmacology.* 2020; 11: 759.

[34] Alam, L. and Mueller, S. *The Myth of Diagnosis as Classification: Examining the Effect of Explanation on Patient Satisfaction and Trust in AI Diagnostic Systems*, 2021, Preprint (Version 1). Available at Research Square.

[35] Warman, A., Warman, P., Sharma, A., *et al* . 'Interpretable artificial intelligence for COVID-19 diagnosis from chest CT reveals specificity of ground-glass opacities'. *MedRxiv: The Preprint Server for Health Sciences.* 2020.

[36] Qiu, S., Joshi, P.S., Miller, M.I., *et al* . 'Development and validation of an interpretable deep learning framework for Alzheimer's disease classification'. *Brain.* 2020; 143(6): 1920–33. Doi:10.1093/brain/awaa137.

[37] Liao, W., Zou, B., Zhao, R., Chen, Y., He, Z., and Zhou, M. 'Clinical interpretable deep learning model for glaucoma diagnosis'. *IEEE Journal of Biomedical and Health Informatics.* 2020; 24(5): 1405–12. https://doi.org/10.1109/JBHI.2019.2949075.

[38] Hao, J., Kosaraju, S.C., Tsaku, N.Z., Song, D.H., and Kang, M. 'PAGE-Net: Interpretable and integrative deep learning for survival analysis using histopathological images and genomic data'. *Pacific Symposium on Biocomputing, United States.* 2020; 25: pp. 355–66.

[39] Shickel, B., Loftus, T.J., Adhikari, L., Ozrazgat-Baslanti, T., Bihorac, A., and Rashidi, P. 'DeepSOFA: A continuous acuity score for critically ill patients using clinically interpretable deep learning'. *Scientific Reports.* 2019; 9(1): 1879.

[40] Karimi, M., Wu, D., Wang, Z., and Shen, Y. 'DeepAffinity: Interpretable deep learning of compound-protein affinity through unified recurrent and

convolutional neural networks'. *Bioinformatics*, Oxford, England. 2019; 35(18): 3329–38.

[41] Mirchi, N., Bissonnette, V., Yilmaz, R., Ledwos, N., Winkler-Schwartz, A., and Del Maestro, R. F. 'The virtual operative assistant: An explainable artificial intelligence tool for simulation-based training in surgery and medicine'. *PLoS One*. 2020; 15(2): e0229596.

[42] Fiosina, J., Fiosins, M., and Bonn, S. "Explainable deep learning for augmentation of small RNA expression profiles'. *Journal of Computational Biology*. 2019; 27(2): 234–247.

[43] Kanda, E., Epureanu, B. I., Adachi, T., *et al* . 'Application of explainable ensemble artificial intelligence model to categorization of hemodialysis-patient and treatment using nationwide-real-world data in Japan'. *PLoS One* 2020; 15(5): e0233491.

[44] Kavvas, E. S., Yang, L., Monk, J. M., Heckmann, D., and Palsson, B. O. 'A biochemically-interpretable machine learning classifier for microbial GWAS'. *Nature Communications*. 2020; 11(1): 2580. https://doi.org/10.1038/s41467-020-16310-9.

[45] Anguita-Ruiz, A., Segura-Delgado, A., Alcalá, R., Aguilera, C. M., and Alcalá-Fdez, J. 'EXplainable Artificial Intelligence (XAI) for the identification of biologically relevant gene expression patterns in longitudinal human studies, insights from obesity research'. *PLoS Computational Biology*. 2020; 16(4): e1007792. https://doi.org/10.1371/journal.pcbi.1007792.

[46] Lamy, J.-B., Sekar, B., Guezennec, G., Bouaud, J., and Séroussi, B. 'Explainable artificial intelligence for breast cancer: a visual case-based reasoning approach'. *Artificial Intelligence in Medicine*. 2019; 94: 42–53. https://doi.org/10.1016/j.artmed.2019.01.001.

[47] Caicedo-Torres, W. and Gutierrez, J. 'ISeeU: visually interpretable deep learning for mortality prediction inside the ICU'. *Journal of Biomedical Informatics*. 2019; 98: 103269.

[48] Nagasubramanian, K., Jones, S., Singh, A. K., Sarkar, S., Singh, A., and Ganapathysubramanian, B. 'Plant disease identification using explainable 3D deep learning on hyperspectral images'. *Plant Methods*. 2019; 15: 98. https://doi.org/10.1186/s13007-019-0479-8.

[49] Xiang, A. and Wang, F. 'Towards interpretable skin lesion classification with deep learning models'. In *AMIA Symposium*, 2019; pp. 1246–1255.

[50] Mbazor, J.C. and Marco, T. 'Scalability of MCMC algorithms on different parallel frameworks'. In *The 1st R-CCS International Symposium*, February, 2019, Kobe, Japan.

[51] Shorten, C., and Khoshgoftaar, T. M. 'A survey on image data augmentation for deep learning'. *Journal of Big Data*. 2019; 6(1): 60. https://doi.org/10.1186/s40537-019-0197-0.

[52] Zeng, Y., Qiu, H., Memmi, G., and Qiu, M. 'A data augmentation-based defense method against adversarial attacks in neural networks', 2020 ArXiv:2007.15290 [Cs]. http://arxiv.org/abs/2007.15290.

[53] Xie, N., Ras, G., van Gerven, M., and Doran, D. '*Explainable deep learning: A field guide for the uninitiated*', 2020. *arXiv preprint arXiv*, 14545.

[54] Anjomshoae, S., Kampik, T., and Fr·amling, K. 'Py-ciu: A python library for explaining machine learning predictions using contextual importance and utility'. In: *IJCAI-PRICAI 2020 Workshop on Explainable Artificial Intelligence (XAI)*, 2020.

[55] Hase, P. and Bansal, M. 'Evaluating explainable AI: Which algorithmic explanations help users predict model behavior?', 2020. arXiv preprint arXiv:2005.01831.

[56] Sutton, R.T., Pincock, D., Baumgart, D.C., Sadowski, D.C., Fedorak, R.N., and Kroeker, K.I. 'An overview of clinical decision support systems: Benefits, risks, and strategies for success'. *NPJ Digital Medicine.* 2020; 3: 17.

[57] Antoniadi, M. Yuhan, D., Yasmine, G., *et al* . 'Current challenges and future opportunities for XAI in machine learning-based clinical decision support systems: A systematic review'. *Applied Sciences*. 2021; 11: 5088. https://doi.org/10.3390/ app11115088.

[58] Kenny, E.M., Ford, C., Quinn, M., and Keane, M.T. 'Explaining black-box classifiers using post-hoc explanations-by-example: The effect of explanations and error-rates in XAI user studies'. *Artificial Intelligence.* 2021; 294: 103459.

[59] Jin, D., Jin, Z., Zhou, J.T., and Szolovits, P. 'Is BERT really robust? A strong baseline for natural language attack on text classification and entailment'. In: *Proceedings of the Thirty-Fourth AAAI Conference on Artificial Intelligence, AAAI 2020, the Thirty-Second Innovative Applications of Artificial Intelligence Conference, IAAI 2020, the Tenth AAAI Symposium on Educational Advances in Artificial Intelligence*, EAAI 2020, New York, NY, 7–12 February 2020; AAAI Press: Palo Alto, CA, 2020; pp. 8018–8025.

[60] Jacobs, M., Pradier, M.F., McCoy, T.H., Perlis, R.H., Doshi-Velez, F., and Gajos, K.Z. 'How machine-learning recommendations influence clinician treatment selections: The example of the antidepressant selection'. *Translational Psychiatry*. 2021; 11: 108.

Chapter 18

The psychology of explanation in medical decision support systems

Vitalis Afebuame Iguoba[1] and Agbotiname Lucky Imoize[2,3]

Abstract

Today, an important role is being played by artificial intelligence (AI) in healthcare systems. Many targeted healthcare applications such as medical diagnostics, patient monitoring, and learning healthcare systems are now available with the aid of AI software programs. Clinical decision-making is enabled by AI algorithms and software. The predictive analysis of the AI algorithms is aided by a computerized predictive analysis flowchart that enables it to separate, organize, and check for patterns from complex data and draw a conclusion with some degree of probability, which will enable the healthcare service provider to make a quality decision within a short time. The AI algorithm does not make the final decision going by the existing legal frameworks at the various jurisdictions but rather they are used as supporting tools for diagnosis or a screening tool, instead of doing the usual medical tasks being done by the doctor in a hospital setting. Many studies in the literature are available today on research with patients' electronic health records deployed by AI-assisted data analysis and learning tools. They use an electronic secure computer which does the records keeping instead of the traditional way of paper records. AI applications are being surged by the recent advancement in machine learning (ML), and the improvement of AI applications in health solely depends on the success in designing the AI algorithm, which is called ML. Only a proper and good algorithm design can guarantee the set goals for AI systems. The autonomous system that can perceive, learn, decide, and act on its own will only be possible by continued advances in AI algorithms known as ML. Autonomous machines are simply self-operating machines, which can carry out their assigned task without human intervention. However, the machine's inability to explain their decision and action taken by them to human users has posed a big limitation to

[1]Department of Electrical Engineering, Dangote Cement PLC, Nigeria
[2]Department of Electrical and Electronics Engineering, University of Lagos, Nigeria
[3]Department of Electrical Engineering and Information Technology, Institute of Digital Communication, Ruhr University, Germany

their adoption and effective use. The deployment of more intelligent, autonomous, and symbiotic systems will provide a good solution to the challenges being faced in the healthcare system. Thus, this chapter presents the psychology of explanation in medical decision support systems (MDSS). The psychological perspectives on explanation in healthcare systems with a binocular focus on MDSS are highlighted.

Keywords: Machine learning (ML); Artificial intelligence (AI); Explainable AI (XAI); Psychology of explanation; Clinical decision support system (CDSS); Electronic health records; Medical decision support system (MDSS)

18.1 Introduction

The understanding of trust and effective management in an emerging generation of artificial intelligence (AI) can be broadly defined as the branch of computer science that deals with computational approaches and techniques which allow or enable the machine to perform tasks that usually require some level of human intelligence [1]. There is no doubt that the application of AI tools will aid the doctor in making quality and reliable decisions without error. These AI systems make good medical decisions and sometimes even give a better medical judgment than humans. This medical study is aimed at examining the usefulness of AI in healthcare, data generation and data analysis using AI systems, and diagnosis and treatments that AI systems can offer to the user [2]. Over the years, there has been a remarkable improvement in computing capability and large user data, especially during the COVID-19 pandemic. The growth and development of AI systems applications across various industries will be determined by the optimization of the AI algorithm [3]. The healthcare industries are quite slow in the adoption and application of AI as compared to other industries. Several policies have been proposed by the Chinese government. Such policies include the next-generation AI development plan and the three-year guidance for the Internet plus the AI plan being a targeted plan to promote the development of AI in practice [4]. Due to the nature of healthcare settings which are lifesaving by doctors and their inherent medical knowledge, the physician is allowed to work independently and autonomously. Having this in mind, medical practitioners cannot be persuaded by hospital managers and IT firms, and this poses a big challenge for the adoption and use of these new technologies. The doctors are not attracted to AI systems due to their limited knowledge of the systems. Therefore, the adoption of IT by doctors is an important topic that needs to be influenced by other stakeholders. Pinpointing the factors affecting IT adoption is difficult due to the peculiar characteristics of AI and healthcare settings [5]. The factors that affect the adoption of electronic medical records (EMR) include the physicians' understanding, caregiver identities, and perceived government influence on caregivers. However, the adoption of AI in the healthcare sector is being observed to have limited studies [3].

The stage at which a decision is being made is termed the adoption of AI in healthcare [6]. Healthcare is being revolutionized by AI [7]. To analyze links between prevention or treatment approaches for sand patient outcomes seems to be the main of AI in healthcare. Making healthcare more effective and efficient, cost-saving, and time for diagnosis and disease management is the goal of AI applications in healthcare [8].

The two most rising computer technologies include value-based medical healthcare and digital innovation being driven by blockchain and AI. Differently, both technologies deal with data handling, and secured storage and sharing of data being provided by the blockchain, while the use of insights to generate value from data and the analysis of the data is being done by AI. The physician and caregivers are being supported by an IT tool known as a clinical decision support system (CDSS) which enhances patient care by providing specific medical information about the patient. Specific medical advice is being given to the patient by using CDSS on various items from the patient medical health data [9,10]. The communication mechanism, knowledge base, and inference engine are all components of CDSS [11].

The clinical assistance tools are grouped into two, namely: knowledge and non-knowledge-based. Group of compiled data and rules like if–then statements are used by knowledge-based CDSS. While non-knowledge-based CDSS systems adopt a type of AI instead of using knowledge-based, the patient clinical data are being figured out by the system, allowing it to learn from experience. Data security and ownership should be the focal point of healthcare systems [12]. The intelligibility of the system behavior is a critical factor as this will enable the end user to ascertain the mode of operation of the MDSS. This information will aid the physician in judgment to know whether the system has performed its intended purpose or failed to perform its intended purpose. This is the basis for the physician to determine the reliability of the system.

The purpose of explaining the way intelligent systems reason has been studied. The various works done in this domain have the concept of reasoning by the types of explanations, such as detailed suggestions on why the system made certain suggestions and the confidence that such suggestions are correct. From the previous work we have seen that giving proper explanation can increase users' understanding of the system's functionality [13]. An automated system such as MDSS is frequently misused or disused by the user. According to neural-induced mesenchymal stem cells, in 2020, the global market size of explainable AI was 3.5 billion USD, and its growth is expected to reach 21 billion USD in 2030.

18.1.1 Categories of AI

Basically, the design of intelligent machines can be categorized into four depending on the capacity of the computer system or device.

18.1.2 Artificial narrow intelligence

This is called "Weak AI" or "Narrow AI" – Any AI systems that can perform its intended designed tasks alone and such an approach is termed artificial narrow

Table 18.1 Main definitions of AI terms

AI terms	Definition
AI [33]	Performing an assigned task as a human or even better than a human by a machine, especially computer systems that mimic human intelligence is termed AI. AI is making machines behave like a human. This is achieved by developing an algorithm that studies large data sets to establish a pattern. This data can either be structured or unstructured. The two key components of AI are machine learning (ML) (algorithm) and analytical tool for data analysis.
ML [33,34]	This is the branch of AI whose function is to build systems that can learn or improve the performance of a task based on the existing data. Large data sets are studied to establish a pattern. This data set could be structured or unstructured. ML is simply an algorithm like a flowchart. It studies data and establishes patterns from it. The growth and success of AI solely depend on the design of proper and good algorithms with an accurate data set.
Big data [35]	Data samples that are too large to analyze appropriately by the usual AI technique termed big data. However, another technique like deep neural networks (such as deep learning) is applicable. The data sets are generated from the required area in which the AI tool is to be used. When the correct data are not generated it means the use of such AI tool will be catastrophic.
Neural networks [36]	These are interconnected neurons being arranged hierarchically in layers to learn and perform highly complex tasks from data set using a series of an algorithm. They are arranged in layers based on the required task they are to perform.
Deep learning [37]	A highly complex type of deep neural network with more than three layers being required to estimate the optimal values of parameters for large data is termed deep learning. When the data sample is not too large, decision trees or support vector machines find essential applications.
AI model, AI algorithm, or AI tool	The building block of the AI model for a particular application is called an algorithm, while the output from the ML algorithm is termed the AI model.

intelligence. Repetitive task performance is their hallmark; examples are Siri, Google Translate, and IBM's Watson.

18.1.3 Artificial broad intelligence

This is called "Broad AI" – an AI technique that is responsible for the decision-making when performing the task with two or more narrow AI being combined. An example is self-driving vehicles.

18.1.4 Artificial general intelligence

This is called "Strong AI" or "Deep AI" – the intellectual tasks performed by a human being allowed to be also performed by a machine are termed strong AI. Self-aware and the *theory of mind are available*. They also interact in the same manner as humans. By emotion and with prior knowledge being able to strategize and make a plan.

18.1.5 Artificial super-intelligence

The ability to exceed the human intelligence quotient is aided by hypothetical approaches.

Rules-based AI uses previously validated information (such as clinical guidelines, risk calculators, and published studies) to set up a series of clinically accepted weights or decision steps that lead to a prediction, diagnosis, or recommendation.

Data-based AI, popularly known as machine learning (ML), is trained using sets of labeled input data (called "training data") and uses computer program processes to derive relationships between the inputs and the so-called "labels."

Logic-based AI: The logical statements and the current state goals are represented by the knowledge of agent statements. The specific goals being obtained by logical deduction of these statements and appropriate computational decisions are used as well [14]. The medical decision support systems (MDSS) used a kind of AI ML, to learn from experience to establish clinical data. However, since the process is based on ML the reasons for their conclusions cannot be ascertained or explained.

Knowledge-based AI: The inference engine and knowledge-base components characterized the category of AI, and new decisions are inferred by the inference engine. Declarative, procedural, heuristic structural, or metal knowledge is being used to represent the state of the world of knowledge-based while the inference engine technique consists of rule-based, model-based, and case-based reasoning for generating new knowledge [15]. The key distinction between *machine autonomy* and *intelligent autonomy is being considered in deciding systems' level of intelligence.* In general, models with good features with respect to accuracy and performance do not usually have a good explanation. It covers psychological theories of explanation, and the explanation of AI intelligent operation is given in Figure 18.1.

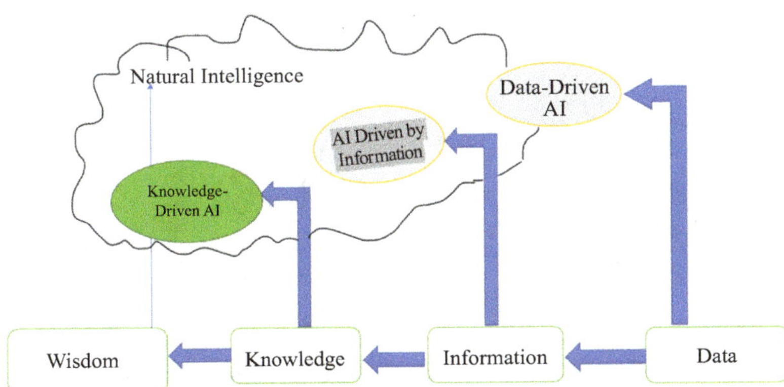

Figure 18.1 Comparison of efforts, precision, and explainability level in AI approaches

18.2 Recent development of XAI in MDSS

Recently, blockchain technology [16] has greatly impacted healthcare, apart from other applications in Bitcoin and Ethereum, supply chain, and Internet of Things and healthcare information technologies (HIT), computer-aided decision support system (CAD), computerized physician order entry system (CPOE), electronic health record (EHR), e-prescription software, and the most recent being AI-enhanced HIT has a more "computational superiority," than the former systems. Improved safety, better quality, and personalization provided by AI-assisted HIT hold great promise for the healthcare industry. The main challenging factor is the inability to implement the outlined HIT results of the extensive studies over the past three decades in real-world medical practice [17].

The limitations of this empirical research are the use of AI in healthcare is still at the infant stage. The medical field is a sensitive area because it deals with human life and health problems. Its adoption and application will be slow until the technology is fully matured. The algorithm is the bedrock of AI, and proper understanding and learning of this algorithm will be an emerging family of technologies that build on ML and computation statistical techniques [3]. The reduction of risks and harmful conditions is termed safety in healthcare. AI-based applications have been key for risk minimization [18]. Figure 18.2 shows the market value of XAI in MDSS.

From 1990 to date, deep learning concepts and foundations have recorded progressive convolutional neural networks, and the accompanying recurrent neural networks, convolutional neural networks, deep reinforcement learning, and adversarial generative networks have recorded a tremendous breakthrough. Explanations of the decisions and actions of the system to human users have been insufficient.

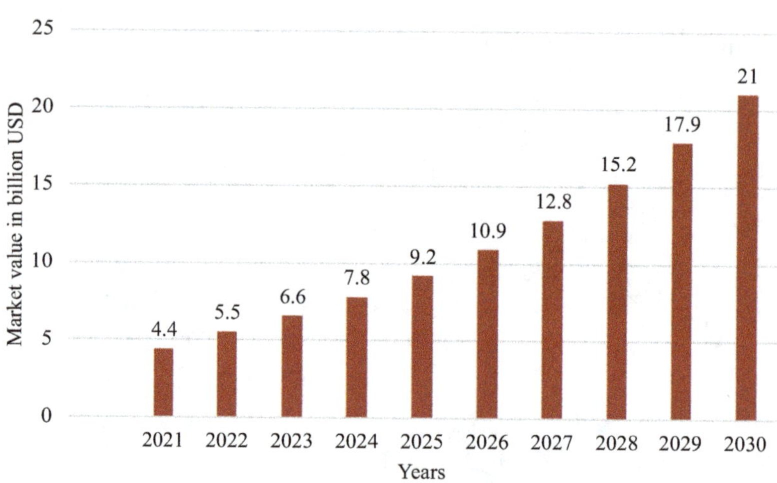

Figure 18.2 The market value of XAI in MDSS

The autonomous and symbiotic systems posed enormous challenges for the US Department of Defense. The systems are becoming smarter on daily basis. The explanation of AI or ML algorithm is necessary to serve as a preview that human-like AI will come shortly [19].

The smart contract and ledger constitute blockchain technology [16, 20]. They both deal with and analyze all the input from the blockchain, and output generate which are usually hash [21]. In this section, the major area of AI applications can be divided into four practices: (1) clinical; (2) research; (3) public health; and (4) administrative. A summary explanation of these four areas of practice is provided in the next section and a summary of the current developments and applications of AI in these four areas is also provided. An automated system like MDSS is either misuse or disused. Because the outcome of a CDSS tool is sometimes wrong, it will be a catastrophe for a user to over-rely on their outcome [13]. Figure 18.3 shows stages of the deployment of AI.

18.2.1 AI in clinical practice

Healthcare data systems have been developed due to the COVID-19 pandemic, and this is related to the way and way data management and control are being done. Several tasks such as generating data, access, and storage of patients' medical information and several automating role functions such as image analysis (like radiology, ophthalmology, dermatology, and pathology) are being played by AI. It also performs signal processing such as electrocardiogram, audiology, and electroencephalography. It can also be used in test and image analysis. In combination with other medical data, AI can produce clinical workflows.

The following sections deal with the possible application of AI in specific areas of medicine that are not regularly reported, e.g., nephrology and personalized

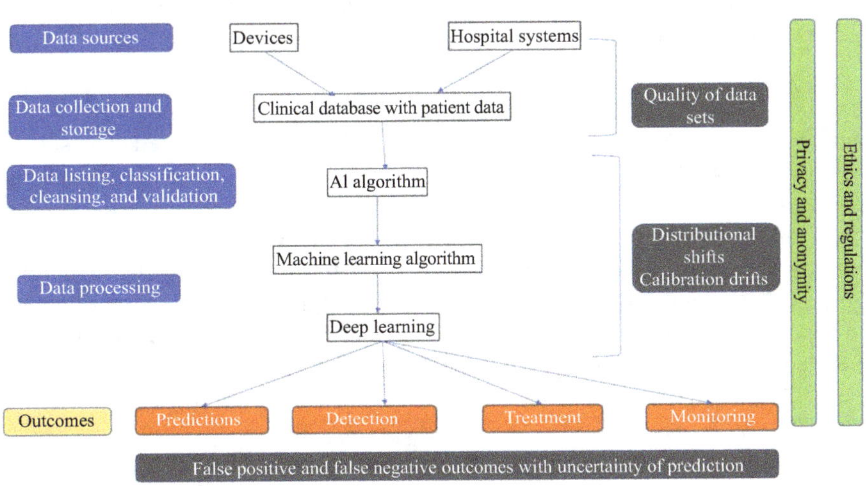

Figure 18.3 Stages of the deployment of AI

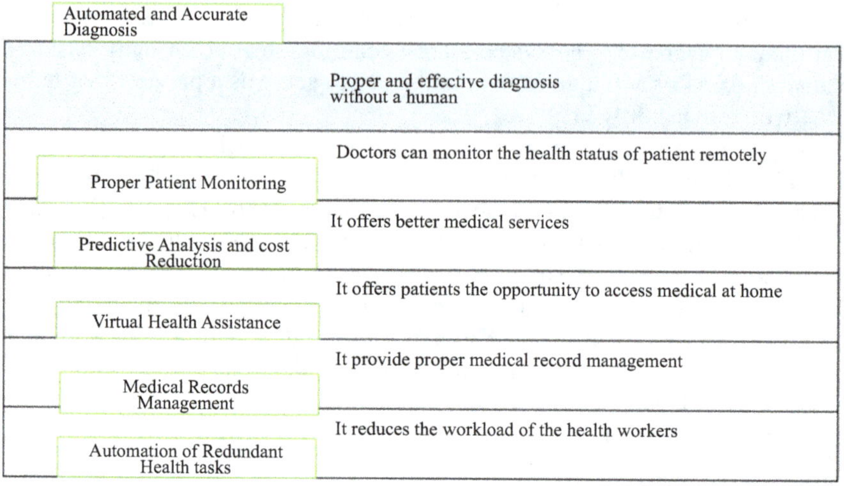

Figure 18.4 The benefits of XAI in MDSS

medicine [9,22]. World Health Organization said cardiovascular diseases result in more than 18 million deaths. The goals of XAI in healthcare include precision medicine, prediction, image analysis, and smart robots [12]. A key summary explanation of XAI in MDSS is given in Figure 18.4.

18.2.2 AI in biomedical research

Biomedical research has derived numerous benefits from AI-derived clinical applications, with recent development making AI applications promising in clinical knowledge retrieval. For instance, a ML algorithm ranks search mainstream medical knowledge resources [23].

18.2.3 AI for public and global health

The importance of AI in public health can never be overemphasized. The science and art of preventing disease, prolonging life, and promoting health through organized efforts are frequently used in that it is of society, organizations, public and private, communities and individuals. Several experiments with relevant AI solutions are ongoing within several public health areas. Prevalence disease or high-risk demographics are being identified by AI applications. AI applications can be used to generate both environmental and occupational using the data generated by sensors and robots, patients' potential contact, and quality medical service [23].

18.2.4 AI in healthcare administration

Blockchain technology is known for transaction cost saving and its deployment in healthcare will provide the following: speed up some healthcare processes and reduce transaction costs following data-handling standards, including privacy, security [24], transmission, exchange, content, and terminology, and due to the peculiarity of the blockchain technology XAI is also used to speed up some healthcare processes.

Many interesting AI applications in healthcare administration are also available. AI application in this domain is less revolutionary compared to patient care. The use of AI is somewhat less potentially revolutionary in this domain as compared to patient care; for instance, an average US nurse spends 25% of work time on regulatory and administrative activities [25].

18.3 Potential benefits of XAI in MDSS

Developing AI systems with interpretability will be beneficial to the users. Investing in explainability will reduce pressures like regulation, accountability, and ethics.

18.3.1 Radiology

Significant AI development has been experienced by radiology. Medical image quantification is used by radiologists in imaging AI technologies. Without much human supervision, deep network models can be deployed to enable automatic localization and delineate the boundaries of anatomical structures or lesions. Priorities and track findings are provided by AI application; this mandate early attention and serves as a guide for radiologists to focus on images with abnormality. An example of such a tool is "cvi42." The Canadian company Circle CVI has been adopted in over 40 countries as a commercialized cardiovascular imaging platform. AI technique has also been useful in image processing techniques called radiomics, although the concept is not readily understood. The aim is to extract quantitative information from diagnostic and treatment planning images. Recently, many studies dealt with the performances of deep learning software and radiologists in the field of imaging-based diagnosis.

Magnetic resonance imaging, computed tomography scan, and X-ray are allowing medical personnel to visualize human internal parts with the aid of medical imaging techniques [12].

18.3.2 Early diagnosis

This deals with the combination of two important ML algorithms, such as principal component analysis and the genetic fuzzy finite-state machine principle being applied. The detection ability is based on movement pattern recognition [26, 27]. The goal of AI is to design systems that can deal with the diagnosis and treatment of diseases. A good example is MYCIN which emerged in 1970 at Stanford for diagnosing blood-borne bacterial infections. Although were not adopted for clinical practice, their performance was not better than human diagnosticians, and as such, they were poorly integrated into the healthcare system medicine, and hepatology is not left out of the steadily progressing AI research in many areas of life. The diagnosis of multiple types of liver disease, most of which are life-threatening, is done by ML models. The accuracy of the recently designed AI neural network for diagnosing is 97.2%.

18.3.3 Emergency medicine

Patient management has benefited immensely from AI emergency medicine at different levels. It provides a quality improvement for patient prioritization during triage and is vast in analysis. The adoption and application of AI in healthcare will enable the doctor to understand which area needs more priority in terms of attention. AI in emergency medicine has a total of 150 studies as reviewed by the recent scoping.

18.3.4 Risk prediction

Risk prediction focuses on assessing the likelihood of individuals experiencing a specific health condition. Analysis of the generated data will enable the AI system to predict using the probability principle the risks ahead for a certain individual given their current health condition. This will enable such individuals to target how to receive specific medical interventions. This system is currently built on regression analysis and the subsets of the available medical data.

18.3.5 Chatbots

Chatbots tend to improve primary healthcare and triage by being powered by AI. Immediate conversational responses and connections are given to patients via chatbots. When implemented chatbots tend to improve the overall patient outcomes and cost savings, in terms of embarking on an unnecessary medical trip.

In Figure 18.5, the application of XAI in MDSS is given. The field of surgery has been revolutionized by AI technology in the capacity of collaborative robots.

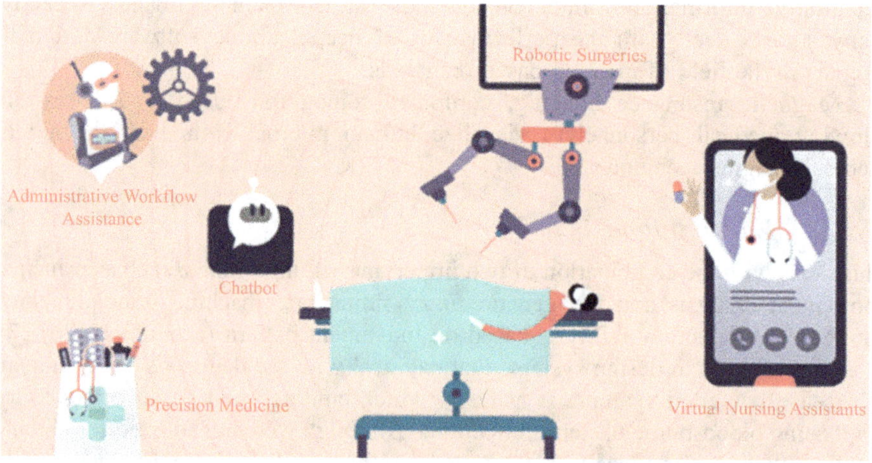

Figure 18.5 Applications of XAI in MDSS

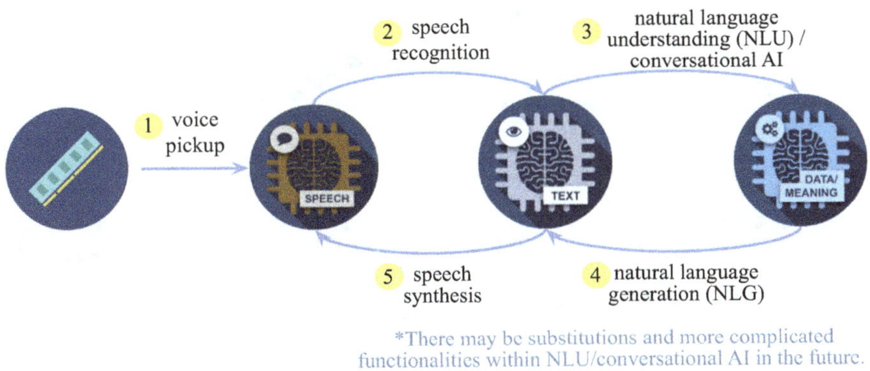

Figure 18.6 The importance of AI in MDSS

18.3.6 Virtual nursing assistance

24/7 virtual nursing assistants are being operated by AI systems. The quick medical response to patients tells the best healthcare setting that will meet their needs. It provides the opportunity for the healthcare workers to monitor their patients, and proper interactions between the doctor and the patient, and it can monitor treatment updates and efficiency. The medical status of the patient can be checked through AI and voice [28]. The importance of AI in MDSS is given in Figure 18.6.

18.3.7 Precision medicine

AI in healthcare has provided precision medicine known to the users. Cheap genome sequencing and the large data being gathered are the building blocks. Deep learning and supercomputing are being used for precision medicine.

18.3.8 Administrative workflow assistance

Automation of AI applications in healthcare administrative workflow is one of the most innovative AI applications in healthcare recently. It enables care providers to decide on urgent tasks, aiding the medical doctor and nurses, and also saving costs.

18.4 Key challenges of XAI in MDSS

CDSS and ethics is being known to have substantial literature. Most of the available literature does not give much information on deep and complex functionalities and issues affecting AI-driven systems [17]. Key challenges of XAI in MDSS are given in Figure 18.7 [9].

18.4.1 Patient harm due to AI errors

Despite the advancement in ML and data analysis, AI-incorporated clinical systems used in healthcare may still fail, their use may be questioned in terms of the end users

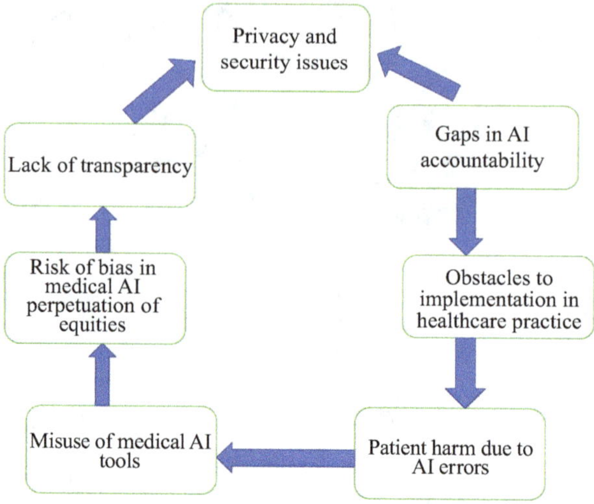

Figure 18.7 Key challenges of XAI in MDSS

safety concern, and the algorithm implemented in AI could lead users to errors in the following ways: (1) false negative appearing as wrong diagnoses, (2) unrequired treatment resulting from the false positive, and (3) making the wrong intervention due imprecise diagnosis, or lack of priority of task in emergency sections.

Assuming the needed large-scale data sets are available to the AI developers with detailed quality training, the following could still be the major source of errors. First, AI tools are seriously affected by noise in the input data during application. For instance, ultrasound scanning – used for imaging modality in clinical practice – is readily prone to scanning errors due to noise. Second, the data set shift resulted from AI misclassifications. Lastly, faulty or erroneous predictions of AI algorithms occur as a result of unexpected change patterns; for example, an AI tool designed to handle a population density of a health facility of 100 persons is now deployed in another hospital setting with a population density of 500.

18.4.2 Misuse of medical AI tools

The risk for human error and human misuse is not being left out in most health technologies, including the application of AI in medical science. Having a good AI design with an accurate and robust algorithm, their usage outcomes solely depend on the way and manner the end user used such an AI system. Proper and correct usage of these AI tools will determine the outcome, and being a medical professional doesn't make one know medical AI tools. The user must understand the mode of operation of the AI tool, to know where and when to use it. Only AI tools' proper usage void of human errors with a suitable application can guarantee the desired result [23].

18.4.3 Risk of bias in medical AI

AI developers want to design AI algorithms that possess fairness and equity. Despite the progressive advancement of AI in medical research and healthcare delivery, inequalities and inequities in medical care exist within most countries of the world. The contributing factors include gender (sex; male or female), age, ethnicity, income, education, and geographical location. An important role is being played by human biases. This is the perception people have generally about certain situations and people. For example, its believed that women are weak vessels because of their nature, when a female is having several medical complaints of pain, after some time the doctor may start relating her weakness to the complaint. These issues are also applicable to AI design. However, future AI applications could embed with this concern not being implemented, enumerated, and controlled [23].

18.4.4 Lack of transparency

Despite the massive improvement of AI applications in the medical field, trust and adaptability become a big challenge as the existing algorithms are seen by individuals and experts as complex and obscure technologies because the technology is difficult to understand.

Much attention is being given to the recent AI algorithm designed by Google for breast cancer screening due to its performance, speed, and robustness required for breast cancer screening. The performance was more than expected. The big question is how the developer arrived at this algorithm; our concern should not only focus on the success of the AI application and its benefits, but we must also consider the steps and procedure that was followed to arrive at it, otherwise the world will be in confusion due to transparency issues in AI applications. With the clarity of the AI, tool transparency issues of explanations are almost completely solved [29].

It is very necessary and important to consider the issues of explanation in AI tool applications in the medical field. The consideration before AI concludes that any approach is very important. Unfortunately, readily available AI tools have big issues with explanations due to their lack of transparency. The absence of transparency will lead to a great lack of trustworthiness, especially in a sensitive area like the medical field which deals with human life. Hence the adoption of AI tools in this area without trust is almost impossible.

Traceability and explainability being used in AI transparency are closely linked. The distinct levels of transparency in terms of the decision taken in the application of AI – explainability while the transparency of usage process and development – traceability. In the real world, there is no availability of traceability for AI tools used in healthcare systems. In providing an "AI passport" for documentation of vital information of an algorithm, a traceability tool is required to monitor the usage of the algorithm when deployed. Finally, there is an urgent to enact some regulatory frameworks regarding this issue [23].

18.4.5 Privacy and security issues

Healthcare is being increasingly widespread with the development of AI solutions. The recent COVID-19 pandemic shows that it is a risk for a lack of data privacy, confidentiality, and protection for patients. This could be a big repercussion because exposure of patient data will violate the rights of such a citizen, especially when the information is used for another purpose instead of medical gains. Because AI uses large data sets, privacy and security associated with such data are imperative. Due to the sensitivity of medical data, there must be detailed consent on it, and the patient must be given the necessary information and informed decision before sharing personal health data. Creating awareness for a patient and the patient's response to this demand is very important in the healthcare industry. The Helsinki Declaration which was formalized and has grown to live with humankind birthed digital technology. This informed consent deals with ethical issues, protection of the patient from harm, and respect for autonomy. The patient may not be able to understand how their data are being used at various levels. The development of opaque AI algorithms seems to have been violated and there is a consent limit on how autonomously and how power is shared in decision-making between patients and clinicians [23].

18.5 The future of XAI in MDSS

The importance of the adoption and implementation of AI systems in healthcare can never be over-emphasized, as this technology will provide a lot of benefits to the end users. AI system application has changed the entire narrative of the healthcare system. Apart from the information management offered by the technology, it has also impacted the administrative system, diagnosis, and treatments of patients in healthcare. At the moment, there is a clear drift of doctors from their usual traditional procedures to the use of AI chatbots that is developing an AI form of doctor, this AI formwork in collaboration with the doctor. The collaboration of patients and AI from the doctor is being studied by a digital healthcare organization in the United Kingdom.

The traditional way of medical procedures is gradually evolving to innovative AI methods of accurately diagnosing and quality treatments, and thus the technology is becoming more popular through developments of ML algorithms and large data analysis. Some of the outcomes have already been tested and implemented in real-time applications. The adoption of ML in XAI will provide a tremendous improvement in MDSS, in terms of error-free medication prescription, eliminating the adverse effect of drug events and other emanating errors medical from the medical process. Continuous clinical trials will give birth to innovative pharmaceutical industries, clinical trials involve failure and success, and the AI tool will help to discover more efficient drugs. Delivery of targeted drugs and vaccines to certain places was actualized during the Coronavirus pandemic, and it can also be used for drug delivery to a geographic area with a prevalence of diseases.

18.6 The research trend of XAI in MDSS

The XAI in MDSS in healthcare has revealed innovations and prospects for quality healthcare delivery. The area of consideration of this research includes pharmaceutical, targeted drug delivery, virtual nursing assistance, automated diagnosis, precision medicine, and healthcare data system optimization. The optimization of ML algorithms and data analysis tools will be the turning point for the AI system. According to Dr Joseph Rieger, "The news that Babylon Health has raised near £50M to build an AI doctor" is a good and innovative development for the healthcare industry; in London, research and trials are on the way to make this Babylon's tech as a replacement for the non-emergency 111 number-CTO of Fujitsu EMEIA. Due to the prevalence of lung disease in China, innovative design by inferring vision is used at Shanghai Changzheng Hospital in China for radiological services The infer vision is an AI innovative design very effective for the diagnosis of lung disease. The machine has interoperability characteristics and high operating speed. The constraints of time and accuracy during diagnosis are eliminated by inferring vision. The existing study on medical reasoning indicated that inexperienced physicians solely depend on their knowledge of pathophysiology in conducting diagnoses [13].

The CDSS interoperability challenges were reviewed. However, the EHRs provided new cloud system architecture and new standards, which have good flexibility of connection with other systems [30]. With the emergence of CPOE and CDSS, the burden on healthcare providers, pharmacists, and nurses to double-check orders has drastically reduced. Initially, the medical personnel does this task, but the task is now better handled by AI tools which produce a better performance than humans. The CDSS created the impression that verifying the accuracy of order is not a task [29].

18.7 The future directions and recommendations

AI tools application will impact healthcare positively, and this progress is connected to ML because ML and large data set analytics tools are the capability that produces precision medicine, diagnosis, disease prevention administrative workflow, and improved quality healthcare delivery [25]. The proper evaluation using the right analytical tools is very important. During the data analysis, it is possible to decide the data that is needed and the considerations in choosing such data. Without a proper data study, the desired pattern cannot be established. In the near when deep learning and AI will reach the same level (to become single), at this level AI will produce machines that will be more intelligent than a human. The two-point issues open new a link for further study.

The CDSS tool has provided many benefits for the medical doctor such as improved decision making, elimination of errors in treatment procedures, the adverse effects of drugs, and false diagnoses. However, for the doctor, persistent use of such an AI tool will develop too much trust in such a CDSS tool as continuous use will make him bring about familiarity. However, the effect of this over-reliance on the CDSS tool by the user is an issue that needs to examine critically by a further study [30–32].

18.8 Conclusions and future scope

An extensive AI legal framework should be put in ML to handle structured data sets (the medical information and images) and NLP that deals with the collection of unstructured data. Suitable algorithms with real data generated from the healthcare system will provide proper illness analysis and recommendations of the best treatment method for the doctors. Despite the AI innovations are essential and considered to transform the pharmaceutical and therapeutical industry, their implementation in the real world has a confrontational barrier. Due partly to the absence of rules and regulations for their operation, these regulations will deal with the issues of safety of their end users. In order to achieve the proper functionality of the system, AI frameworks are needed, starting from the phase of medical examination to management. After the AI is put into use, a continuous data supply is needed to optimize its operation.

Acknowledgment

The work of Agbotiname Lucky Imoize is supported in part by the Nigerian Petroleum Technology Development Fund (PTDF) and in part by the German Academic Exchange Service (DAAD) through the Nigerian-German Postgraduate Program under Grant 57473408.

References

[1] R. L. Kumar, Y. Wang, T. Poongodi, and A. L. Imoize (eds.), *Internet of Things, Artificial Intelligence and Blockchain Technology*, 1st ed. Switzerland AG: Springer Nature, 2021.

[2] Q. Lin, T. Li, P. M. Shakeel, and R. D. J. Samuel, "Advanced artificial intelligence in heart rate and blood pressure monitoring for stress management," *J. Ambient Intell. Humaniz. Comput.*, vol. 12, no. 3, pp. 3329–40, 2021, doi:10.1007/s12652-020-02650-3.

[3] T. Q. Sun, "Adopting artificial intelligence in public healthcare: the effect of social power and learning algorithms," *Int. J. Environ. Res. Public Health*, vol. 18, no. 23, p. 12682, 2021.

[4] K. Young, A. Gupta, and R. Palacios, "Impact of telemedicine in pediatric postoperative care," *Telemed. e-Health*, vol. 25, no. 11, pp. 1083–1089, 2019.

[5] M. A. Adelabu, A. L. Imoize, and K. E. Adesoji, "Enhancement of a camera-based continuous heart rate measurement algorithm," *SN Comput. Sci.*, vol. 3, no. 4, p. 284, 2022, doi:10.1007/s42979-022-01179-w.

[6] J. P. Lomaliza and H. Park, "Improved heart-rate measurement from mobile face videos," *Electronic*, vol. 8, no. 6, 663, 2019, doi:10.3390/electronics8060663.

[7] W. Guo, "Explainable artificial intelligence for 6G: improving trust between human and machine," *IEEE Commun. Mag.*, vol. 58, no. 6, pp. 39–45, 2020, doi:10.1109/MCOM.001.2000050.

[8] C.-C. Lee, C.-W. Hsu, Y.-M. Lai, and A. Vasilakos, "An enhanced mobile-healthcare emergency system based on extended chaotic maps," *J. Med. Syst.*, vol. 37, no. 5, p. 9973, 2013, doi:10.1007/s10916-013-9973-0.

[9] P. K. Anooj, "Clinical decision support system: risk level prediction of heart disease using weighted fuzzy rules," *J. King Saud Univ. Inf. Sci.*, vol. 24, no. 1, pp. 27–40, 2012.

[10] A. Wright, T. T. T. Hickman, D. McEvoy, *et al.*, "Analysis of clinical decision support system malfunctions: a case series and survey," *J. Am. Med. Informatics Assoc.*, vol. 23, no. 6, pp. 1068–1076, 2016.

[11] S. H. El-Sappagh and S. El-Masri, "A distributed clinical decision support system architecture," *J. King Saud Univ. Inf. Sci.*, vol. 26, no. 1, pp. 69–78, 2014.

[12] S. Vijayalakshmi, S. P. Gayathri, and S. Janarthanan, "Blockchain security for artificial intelligence-based clinical decision support tool," in *Internet of Things, Artificial Intelligence and Blockchain Technology*, New York, NY: Springer, 2021, pp. 209–240.

[13] A. Bussone, S. Stumpf, and D. O'Sullivan, "The role of explanations on trust and reliance in clinical decision support systems," in *2015 International Conference on Healthcare Informatics*, 2015, pp. 160–169.

[14] O. A. Osoba and W. Welser, *The Risks of Artificial Intelligence to Security and the Future of Work*, Santa Monica, CA: RAND, 2017.

[15] J. Whittlestone, R. Nyrup, A. Alexandrova, K. Dihal, and S. Cave, "Ethical and societal implications of algorithms, data, and artificial intelligence: a roadmap for research," London: Nuffield Foundation, 2019, pp. 1–59.

[16] L. K. Ramasamy, F. Khan K. P., A. L. Imoize, J. O. Ogbebor, S. Kadry, and S. Rho, "Blockchain-based wireless sensor networks for malicious node detection: a survey," *IEEE Access*, vol. 9, pp. 128765–128785, 2021, doi:10.1109/ACCESS.2021.3111923.

[17] S. Ameen, M.-C. Wong, K.-C. Yee, and P. Turner, "AI and clinical decision making: the limitations and risks of computational reductionism in bowel cancer screening," *Appl. Sci.*, vol. 12, no. 7, p. 3341, 2022.

[18] S. Ellahham, N. Ellahham, and M. C. E. Simsekler, "Application of artificial intelligence in the health care safety context: opportunities and challenges," *Am. J. Med. Qual.*, vol. 35, no. 4, pp. 341–348, 2020.

[19] Y. Duan, J. S. Edwards, and Y. K. Dwivedi, "Artificial intelligence for decision making in the era of Big Data – evolution, challenges and research agenda," *Int. J. Inf. Manage.*, vol. 48, pp. 63–71, 2019.

[20] P. Rana, I. Batra, A. Malik, *et al.*, "Intrusion detection systems in cloud computing paradigm: analysis and overview," *Complexity*, vol. 2022, no. 3999039, p. 14, 2022, doi:10.1155/2022/3999039.

[21] N. Pooranam, G. Ignisha Rajathi, R. Lakshmana Kumar, and T. Vignesh, "Decision support mechanism to improve a secured system for clinical process

using blockchain technique," in *Internet of Things, Artificial Intelligence and Blockchain Technology*, New York, NY: Springer, 2021, pp. 241–258.

[22] R. S. Castillo and A. Kelemen, "Considerations for a successful clinical decision support system," *CIN Comput. Informatics, Nurs.*, vol. 31, no. 7, pp. 319–326, 2013.

[23] T. Evas, "EPRS| European Parliamentary Research Service," *Eur. Added Value Unit. Civ. Liabil. Regime Artif. Intell. Eur. Added Value Assessment. Unione Eur.*, 2020.

[24] V. O. Etta, A. Sari, A. L. Imoize, P. K. Shukla, and M. Alhassan, "Assessment and test-case study of Wi-Fi security through the wardriving technique," *Mob. Inf. Syst.*, vol. 2022, p. 7936236, 2022, doi:10.1155/2022/7936236.

[25] T. Davenport and R. Kalakota, "The potential for artificial intelligence in healthcare," *Futur. Healthc. J.*, vol. 6, no. 2, p. 94, 2019.

[26] A. L. Imoize and O. O. Orodeji, "Development of a low-latency wireless telemetry system for monitoring patients heart rates," *Int. J. Electr. Eng. Appl. Sci.*, vol. 3, no. 2, pp. 61–73, 2020.

[27] A. L. Imoize, T. Oyedare, M. E. Otuokere, and S. Shetty, "Software intrusion detection evaluation system: a cost-based evaluation of intrusion detection capability," *Commun. Netw.*, vol. 10, no. 04, pp. 211–229, 2018, doi:10.4236/cn.2018.104017.

[28] A. L. Imoize and A. E. Babajide, "Development of an infrared-based sensor for finger movement detection," *J. Biomed. Eng. Med. Imaging*, vol. 6, no. 4, pp. 29–44, 2019, doi:10.14738/jbemi.64.7639.

[29] G. Phillips-Wren and L. Jain, "Artificial intelligence for decision making," in *International Conference on Knowledge-Based and Intelligent Information and Engineering Systems*, 2006, pp. 531–536.

[30] X. Schelling and S. Robertson, "A development framework for decision support systems in high-performance sport," *Int. J. Comput. Sci. Sport*, vol. 19, no. 1, pp. 1–23, 2020.

[31] D. S. Char, N. H. Shah, and D. Magnus, "Implementing machine learning in health care—addressing ethical challenges," *N. Engl. J. Med.*, vol. 378, no. 11, p. 981, 2018.

[32] P. Doupe, J. Faghmous, and S. Basu, "Machine learning for health services researchers," *Value Heal.*, vol. 22, no. 7, pp. 808–815, 2019.

[33] T. Panch, P. Szolovits, and R. Atun, "Artificial intelligence, machine learning and health systems," *J. Glob. Health*, vol. 8, no. 2, 020303, 2018.

[34] T. S. Ajani, A. L. Imoize, and A. A. Atayero, "An overview of machine learning within embedded and mobile devices – optimizations and applications," *Sensors*, vol. 21, no. 13, p. 4412, 2021, doi:10.3390/s21134412.

[35] A. L. Beam and I. S. Kohane, "Big data and machine learning in health care," *JAMA*, vol. 319, no. 13, pp. 1317–1318, 2018.

[36] J. A. Anderson, *An Introduction to Neural Networks*, London: MIT Press, 1995.

[37] Y. Lecun, Y. Bengio, and G. Hinton, "Deep learning," *Nature*, vol. 521, no. 7553, pp. 436–444, 2015, doi:10.1038/nature14539.

Index

www.ingramcontent.com/pod-product-compliance
Lightning Source LLC
LaVergne TN
LVHW011547200926
848484LV00001B/294